ELEMENTS OF ELECTRICAL AND ELECTRONIC INSTRUMENTATION

ELEMENTS OF ELECTRICAL AND ELECTRONIC INSTRUMENTATION

An Introductory Textbook

Kurt S. Lion

Professor Emeritus
Senior Lecturer
The Massachusetts Institute of Technology

McGraw-Hill Book Company

New York St. Louis San Francisco Auckland Düsseldorf Johannesburg
Kuala Lumpur London Mexico Montreal New Delhi Panama Paris São Paulo
Singapore Sydney Tokyo Toronto

ELEMENTS OF ELECTRICAL AND ELECTRONIC INSTRUMENTATION
An Introductory Textbook

1 2 3 4 5 6 7 8 9 0 K P K P 7 9 8 7 6 5

Library of Congress Cataloging in Publication Data

Lion, Kurt Siegfried.
Elements of electrical and electronic instrumentation.
1. Electric meters. 2. Electronic instruments.
3. Transducers. I. Title.
TK301.L53 621.3′028 74-16172
ISBN 0-07-037976-9

This book was set in 3A by Bi-Comp, Incorporated. The editors were Robert H. Summersgill and J. W. Maisel; the cover was designed by Pencils Portfolio, Inc.; the production supervisor was Charles Hess. The drawings were done by J & R Services, Inc.
Kingsport Press, Inc., was printer and binder.

To my wife
Elsa née Strauss

Contents

Preface

Instrumentation is the science that deals with the means and methods of obtaining and processing physical and chemical information. Instrumentation methods are applied to physical and chemical problems, to biological and medical problems, and to countless problems in the engineering sciences, as well as in the social sciences, archaeology, geology, astronomy, and many other sciences. Instrumentation is practiced in research institutes, colleges, schools, in hospitals, in industry and in many government and military research and testing laboratories. Probably the largest relative amount of funds and manpower in almost any research project is absorbed by instrumentation requirements. A working familiarity with the principles of modern instrumental methods is, therefore, essential for investigators in many scientific fields. The book is an introduction to the elements of electrical and electronic instrumentation.

The word *instrumentation* means different things to different people. For some it means electric meters and recorders, for others it means electronic circuitry, for still others it means electrochemical apparatus or the application of physical instruments to biological or medical problems (bioengineering or bioinstrumentation).

It has been my intention in this book to cover more or less the entire field of electrical and electronic instrumentation. I believe that such a book is more interesting and useful than a book limited to special topics. It should make it clear to the student that usually there is not one but a variety of solutions for a given problem. This fact should contribute to the reader's innovation and creativity. The book has as its goal a broad coverage of the methodology in the field of electrical instrumentation. It is primarily a book in instrumentation physics rather than in instrumentation technology. It is written for those who are familiar with the concepts of elementary physics, but it provides material to clarify the basic concepts. This should help to fill in voids for those who were never taught correctly or for those who have forgotten these concepts. The mathematical level is limited to that of a simple differential equation. Mathematics is used sparingly since mathematical manipulations sometimes can obscure the physics behind the equation.

The presentation of the subject matter is essentially descriptive, for I believe that this is desirable for a study of instrumentation. The teaching of instrumentation at an introductory level should be directed primarily to the formation of concepts, i.e., of "mental images of a thing formed by generalization from particulars." The book is written primarily for a scientist who uses and occasionally designs instruments. The reader may be a chemist, a biologist, a medical research worker, a mechanical engineer, a textile engineer, a metallurgist, or any investigator in an experimental science—also occasionally a physicist, an electrical engineer, or an electronics technician. The book should also be useful for junior or senior students in an introductory instrumentation course.

Most of the material is presented in the form of *instrumentation elements*. An instrumentation element is a functional unit, or a module, or a single block in a block diagram, as described in further detail in Chapter 1. The concept of technical modular units was used first by C. von Bach at the end of the nineteenth century. Faced with the need for an adequate logical system to cover a great variety of machines, he created the very successful concept of *machine elements*. In a similar vein, I have used the instrumentation elements as a means to reduce the large amount of material to be covered and to describe the basic solutions for the construction of instruments.

Some topics require no detailed descriptions; one can measure time with a stopwatch without knowing the constructional details of the watch. Instrumentation elements are described in detail when the element is based on a principle that is new within the sequence of the text and when such a principle appears to have possible applications in a variety of problems, or when the detailed knowledge is needed for the correct use or operation of the element. This book differs in scope and in content from my book "Instrumentation in Scientific Research, Input Transducers" (McGraw-Hill, New York, 1959). The latter deals with input transducers only and treats them on a handbook level. The present book is an introductory treatise of the most important electrical and electronic instrumentation elements, i.e., transducers with electrical input or electrical output as well as modifiers (information processing elements).

Some teachers have claimed that knowledge in the field of instrumentation is proliferating so fast that a course given one year is outdated the next year. In my opinion an introductory course that is outdated in one year is not worthy of being taught in the first place. A course or a textbook of instrumentation should be a collection of topics that are timeless. It is just as wrong to neglect the classical methods and techniques as it is to omit those modern methods that promise to be of lasting value.

My main goal has been to simplify the subject matter and to convey to the reader basic and convincing concepts that can be used in a creative manner. It is not my intention to teach the student how to build an amplifier or how to repair a cathode-ray oscilloscope. In most cases it is more economical to buy a commercially available piece of equipment than to build it; and a research worker can hardly compete with the service man of the manufacturer.

I gratefully acknowledge the help and the numerous contributions of my long-time collaborator, Dr. Tibor L. Földvári, now Vice President of Harvard Apparatus Company and Adjunct Professor of Biology at Boston University. I should also like to express my gratitude to my former collaborator, now at Arizona State University, Professor Peter K. Stein, for his numerous suggestions and improvements. I am also grateful to my colleague at the Massachusetts Institute of Technology, Professor David N. Hume, for his help, particularly with the chapter on electrolytic systems, and to Dr. James Ross for his contribution on oxygen electrodes. I am most grateful to Mrs. E. Duncan and N. L. Duncan, who have been of invaluable help with the styling and editing of the manuscript.

I am particularly grateful to the National Science Foundation for providing me with the financial means to develop a student laboratory to teach instrumentation. The laboratory exercises described in the Appendix of this book stem from this development and are reproduced here with the kind permission of the National Science Foundation.

Kurt S. Lion

1 Definitions and Organization

The purpose of instrumentation is the procurement of *information* or the control of an object or a process in accordance with such information.

INFORMATION AND SIGNALS

Information is carried or conveyed by signals. In a broad sense, a signal is a physical or chemical quantity (e.g., a voltage or a wavelength) or a combination of physical quantities (pattern) which can exist in a variety of meaningful states, at least in two states (on or off). An unchangeable or only randomly varying voltage does not represent a signal. Information can be expressed and carried in *analog* form, i.e., in a continuously variable form, such as the position of a meniscus relative to a scale. Alternatively, information can be expressed or coded in *digital* form, i.e., in discontinuous steps, such as by a group of figures or characters.

The desired information may be a physical or chemical quantity (e.g., the strain in a turbine blade or the concentration of hydrogen ions in solution) or it may have the general character of a message. If the information has the character of a message, the relationship between signal and information is usually laid down in a code (e.g., Morse code or pattern code: "one if by land and two if by sea"). If the information is of a physical or chemical nature, the relationship between the information and the signal is usually that of a physical law expressed by a mathematical equation, or by an empirical calibration characteristic, or by a table. An ill-defined relationship between signal and information wanted may cause the information derived from an instrumentation system to be unsatisfactory. For instance, the psychogalvanic reflex meter, an instrument which measures the electric resistance of the skin (or the electrical polarization of the skin), may furnish a correct output signal reflecting the skin resistance or its variations. Used as a lie detector, it suffers from the ambiguous relationship between the skin resistance and the emotional response of the subject.

In accordance with the usual classification of physical phenomena, six different forms of physical signals can be distinguished: mechanical, thermal, magnetic, electric, optical, and molecular (chemical) signals.

INSTRUMENTATION ELEMENTS

All instruments are composed of basic units called *instrumentation elements** or *operating components.*† An instrumentation element is a modular unit which performs a specific task and is represented by a building block in a function diagram. For example, a photoelectric cell with the necessary supply source and load resistance converts light into a corresponding electric current. Another example is an amplifier, which converts a small voltage into a large one. There are two groups of instrumentation elements: *transducers* and *modifiers.*

A transducer converts a signal of *one* physical form into a corresponding signal of *another* physical form. For example, a piezoelectric transducer converts a *mechanical signal* (force) into a corresponding *electric signal* (voltage).

A modifier, or signal-processing element, converts a signal of *one* physical form into a modified signal of the *same* form. For instance, an electrical integrator produces at its output terminals an *electric* signal which is proportional to the time integral of the *electric* signal applied to its input terminals.

Since there are six different forms of input signals and six different forms of output signals, all instrumentation elements can be classified in a logical system as shown in Table 1-1.

An example of an electrical instrument composed of three elements is the vibration meter shown in Fig. 1-1. The vibration is converted into an electric voltage or current signal by a mechanical-electrical transducer. The output signal from the transducer is led to an amplifier; the amplifier output is connected to a recorder which plots the displacement-time diagram. The arrows signify the

* K. S. Lion, "Instrumentation in Scientific Research," McGraw-Hill, New York, 1959, distinguished among three groups of instrumentation elements: input transducers, output transducers, and modifiers. The distinction between input and output transducers seems unnecessary and has since been abandoned. The present book is based on only two groups of instrumentation elements.

† C. S. Draper, W. McKay, S. Lees, "Instrument Engineering," vol. 1, p. 8, McGraw-Hill, 1952.

Table 1-1 Synopsis of transducers and modifiers
Examples of transducers (*T*), groups of transducers or transducer systems as well as modifiers (*M*) and their logical grouping for different input and output signals.

Out \ *In*	*1 Mechanic*	*2 Thermal*	*3 Magnetic*	*4 Electric*	*5 Optic*	*6 Molecular*
1 Mechanic	(*M*) Gravity balance Ballistic pendulum Bellows	Thermometers Bimetal strip	Magnetometer Magnetostriction transducer	Electrometers Electrostrictive transducer	Crooke's radiometer Radiation pressure systems	Hydrometer Electrodeposition cell
2 Thermal	Adiabatic systems Friction calorimeter	Mixing calorimeter	Eddy-current transducer	Thermal converter	Bolometer	Combustion calorimeter
3 Magnetic	Magnetoelastic transducer	Paramagnetic transducer devices	Magnetic recorder	Electromagnetic transducer	Curie-effect radiation meter	Magnetic resonance system
4 Electric	Resistive, Inductive, Capacitive, Piezoelectric transducer	Thermistor Thermoelement	Magnetoresistance, Hall-effect transducer	(*M*) Transformers (*T*) Langmuir probe Charge collectors	Photoelectric cell Photoresistive cell Photogalvanic cell	Transducers based on potentiometry, conductimetry, polarography
5 Optic	Interferometer Photoelastic transducer	Thermal radiation transducer	Faraday cell	Kerr cell Electroluminescent transducer	(*M*) Geometric-optical systems	Transducers based on emission, absorption spectroscopy
6 Molecular	Ultrasonic chemical transducer	Thermal dye indicators	Paramagnetic Oxygen Analyzer	Electrolytic integrator	Photographic emulsion	Chemical systems modifiers

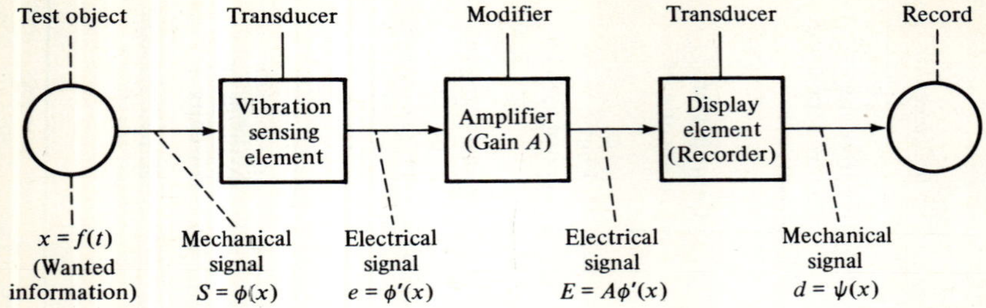

Fig. 1-1

flow of information. If only the amplitude of the vibration is to be measured, a peak voltmeter may be used instead of the recorder. If the frequency spectrum of the vibration is to be analyzed, a tuned filter may be inserted between the amplifier and the output meter. If the physical properties of the instrumentation elements are known, an instrument can be designed, its function can be predicted, and an optimum solution for a given instrumentation problem can usually be found.

It is sometimes advantageous to use several elements in sequence for the conversion of signals from one form into a more suitable form. The example in Fig. 1-2 illustrates elements to be used for the sequential conversion of pressure signals into electric signals. These elements are to be used if single-step transducers of suitable range or sensitivity are not available. Any one of the five elements in the upper horizontal row of Fig. 1-2 can be used for the conversion of a pressure signal p into a displacement d of a membrane, a spiral, or other member. Any one of the elements in the lower horizontal row can be used for the conversion of the displacement d into an electric output signal E_o. The

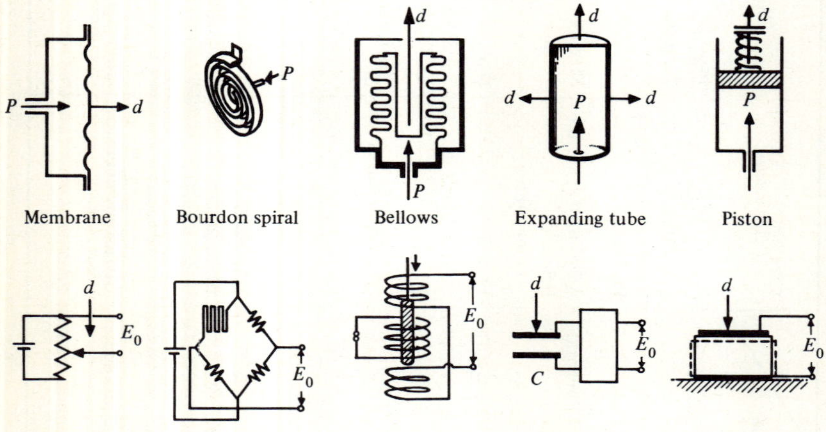

Fig. 1-2

ten basic elements in Fig. 1-2 can be used for the construction of 25 different pressure-to-electrical signal transducers.

GENERAL CRITERIA OF INSTRUMENTATION ELEMENTS

All instrumentation elements are characterized by the following criteria.

Type of input This refers to the physical form of the signal that is applied to the input of the element and that carries the desired information. In general, the applied signal has the dimension of energy or power, i.e., it contains two components, such as force and displacement or voltage and current. Frequently, however, one of the components of the signal can be neglected. For instance, the displacement of a stiff spring under the action of an applied mechanical force and displacement signal may be very small so that the spring may be considered responsive to force only. A capacitive transducer, which requires very little force to produce a measurable change of capacitance, may be called a *displacement responsive transducer*, or *displacement transducer*.

Input impedance This determines the amount of force, energy, or power required from the preceding stage or from the object under investigation. If the physical mechanism involved in the operation of an element is such that the input quantity *actuates* or *drives* the element (e.g., a galvanometer), then maximum power is transferred from the preceding stage to the element if the input impedance of the element matches the output impedance of the preceding stage (source or object). However, maximum power transfer is not always important. Many applications require an input impedance which is large compared with the output impedance of the preceding stage (see Chap. 2).

Type of output If the output signal is of an electrical nature, the information may be contained in:

1. The signal level (voltage or current, e.g., the output from a photoelectric cell)
2. The impedance (resistance, inductance, capacitance, e.g., the output signal from a resistance wire strain gauge)
3. The time function of a voltage or a current (e.g., frequency, phase or time derivative, e.g., the output of a frequency modulator)

Output impedance The output impedance determines the amount of power that can be transferred from the element to the succeeding element or stage at a given output level.

Transfer function The relationship between the output signal Q_o and the input signal Q_i,

$$Q_o = f(Q_i) \tag{1-1}$$

A linear transfer function is frequently desirable; however, linearity of the overall transfer function of a complete system may be obtained even though nonlinear elements are used in some of its stages. This is done by using compensating circuits and feedback or servosystems.

The derivative of the transfer function $dQ_o/dQ_i = S$ is the *sensitivity* of the instrumentation element.

Instrument error In general, an instrument will not follow Eq. (1-1) but will have an output

$$Q_o' = f(Q_i) + F \tag{1-2}$$

where F is the (absolute) error of the result or of the output quantity, i.e., the deviation of the observed output quantity Q_o' from the correct value Q_o

$$F = Q_o' - Q_o$$

The error can be expressed either in terms of the output or input quantity. The fractional error is $F_f = F/Q_o$, which is approximately equal to F/Q_o'; the percentage error is $F_{\%} = F/Q_o \times 100$.

The error is usually complex. It is practical to distinguish the following components:

1. *The scale error.* (*a*) The observed output may deviate from the correct output by an amount which is constant throughout the entire range of the instrument (additive constant, zero displacement). (*b*) The observed output may deviate from the correct value by a constant factor. (*c*) The experimentally observed transfer function may deviate from that postulated by theory (*nonconformity*). In particular, if a linear relationship between input and output is postulated but not experimentally realized, the error is called *nonlinearity or nonlinear distortion.* (*d*) The output may not only depend on the applied input but also on the past history of the element, i.e., on the input formerly applied to the element (*hysteresis error*).
2. *The dynamic error.* The output does not correctly follow the variations with time of the input, or it depends on the time function, such as a time derivative or the frequency of the input quantity.
3. *Noise and drift.* A signal originating in the element and varying with time appears at the output terminals or is superimposed on the output signal. The magnitude of this noise or drift output is, in principle, independent of the magnitude of the signal applied to the input. If information is available about the statistical nature of the noise output, it is possible sometimes to distinguish between the desired output signal and the undesirable noise.

If a human observer reads the output, he is a part of the instrument system, and his reaction must be included in the consideration of errors. For instance, a scale error can be caused by parallax in scale reading, or a dynamic error can be caused by the observer's reaction time or psychological anticipation of an expected result (the "personal equation" of an observer).

Response to environmental influences The performance of an instrumentation element is fully described by the transfer function and the errors, as mentioned above, provided the instrument is in a constant environment and not subjected to external disturbances. If instruments are subjected to environmental influences (such as changes of temperature, pressure, acceleration, magnetic or electric fields, or to changes of the supply voltages), variations of the transfer function and of the errors may result. Although such variations may lead to deviations in the instrument output, they are described separately, for practical reasons, from the instrument errors described in the paragraph on instrument error above

Useful range or level In general, the lower limit of the useful range is imposed by: the unavoidable random variation of the output signal (noise) or other unwanted variations contained in the output signal or originating in the element and appearing in the output, and the tolerable error of the wanted signal. If n is the mean value of the noise level and if $F_{f,\max}$ is the maximum tolerable fractional error of the output signal Q_o, then the lower limit for which the element can be used is, in terms of the output signal,

$$Q_{o,\min} \geq \frac{n}{F_{f,\max}} \tag{1-3}$$

and in terms of the input signal Q_i

$$Q_{i,\min} \geq \frac{n}{SF_{f,\max}}$$

The upper limit of a useful input level is reached, in general, when th input signal tends to damage the element or when excessive distortion of th signal sets in.

In Table 1-1 instrumentation elements are grouped according to the function they serve. In the following chapters they are grouped according to the physical principles involved in their operation.

REFERENCES

Lion, K. S.: Mechanical Input Transducers, in "Instrumentation in Scientific Research," McGraw-Hill, New York, 1959.

Stein, Peter K.: "Measurement Engineering," vol. I: Basic Principles, 6th ed., Stein Engineering Services, Inc., Phoenix, Ariz., 1970.

Neubert, H. P. K.: "Instrument Transducers," Oxford, Clarendon Press, 1963.

Norton, Harry N.: "Handbook of Transducers for Electronic Measuring Systems," Prentice-Hall, Englewood Cliffs, N.J., 1969.

Instrument Society of America: "Basic Instrumentation Lecture Notes and Study Guide," 313 Sixth Ave., Pittsburgh, Pa. 15222.

Rohrbach, Christof: "Handbuch für elektrisches Messen Mechanischer Grössen," VDI Verlag, Düsseldorf, 1968.

2
Basic Electrical Concepts

VOLTAGE AND CURRENT SOURCES

An ideal voltage source or constant-voltage source furnishes at its output terminals a voltage that is independent of the load connected to the source or of the current drawn from the source. The voltage may or may not vary in time; in fact, it must vary in time if it represents a signal.

If a constant-voltage source producing an output voltage of E volts is applied to a resistance of R ohms, as shown in Fig. 2-1*a*, it will cause in the circuit a current of I amperes:

$$I = \frac{E}{R} \tag{2-1}$$

(Ohm's law, Fig. 2-1*a*.)

A constant-voltage source in a strict sense does not exist; the output voltage will always vary with the current drawn from the source. A practical embodiment of a constant-voltage source is a source with an internal resistance which is negligibly small compared to the load resistance R to which the source is connected. Practical constant-voltage sources can also be constructed with the help of electronic systems.

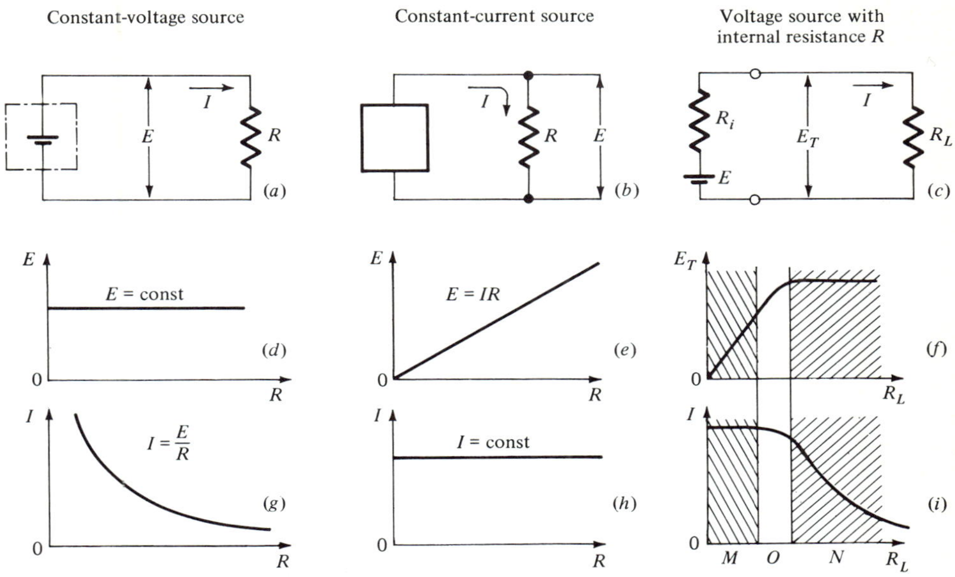

Fig. 2-1 First column (*a*), (*d*), (*g*) constant-voltage source. Second column (*b*), (*e*), (*h*) constant-current source. Voltage source with internal resistance (*c*), (*f*), (*h*).

A constant-voltage source must be distinguished from a regulated- or stabilized-voltage source. A regulated-voltage source furnishes an output voltage which should be constant in time, but may or may not vary with the current drawn from the source.

An ideal current source or constant-current source furnishes a current that is independent of the magnitude of the load resistance to which it is connected, or that remains constant even if the load resistance varies. The current produced by the constant-current source may or may not vary with time.

If a constant current I passes through a resistor R, as shown in Fig. 2-1*b*, a potential difference E (*voltage drop*) occurs between the terminals of the resistor.

$$E = IR \tag{2-2}$$

(Ohm's law, Fig. 2-1*b*.)

In a strict sense, a constant-current source does not exist. A practical constant-current source contains a voltage source E' which causes the current. Furthermore, it contains an internal resistance R_i' which is large compared to the load resistance R. The current in such a system then is determined primarily by the magnitude of E' and R_i'. Constant-current sources can also be constructed with the help of electronic circuits.

Ordinary sources, like batteries or generators, generally consist of a voltage source E in series with an internal resistance R_i connected to a load resistance R_L, as shown in Fig. 2-1*c*. The current in such a circuit is determined by the voltage

E and the sum of the resistances R_i and R_L

$$I = \frac{E}{R_i + R_L}$$

If the internal resistance R_i is large compared to the external or load resistance R_L (for instance, if $R_i = 10^3\ \Omega$ and $R_L = 1\ \Omega$), then R_L can be neglected and the current is, in a first approximation,

$$I = \frac{E}{R_i}$$

i.e., the current is practically independent of R_L and the source acts as a constant-current source (see region M of Fig. 2-1*c*).

If the internal resistance R_i is small compared to the load resistance R_L (for instance, if $R_i = 10^3\ \Omega$ and $R_L = 10^6\ \Omega$), then R_i can be neglected and the current is, in a first approximation,

$$I = \frac{E}{R_L}$$

i.e., the current is independent of the internal resistance R_i and the source acts as a constant-voltage source (see region N of Fig. 2-1*c*). Between regions M and N there is a region O where the circuit acts neither as a constant-current nor as a constant-voltage source.

The concept of a constant-voltage or a constant-current source is not limited to power sources, such as batteries, power supplies, or electronic stimulators, but applies also to signal sources, such as the output from an amplifier, transducer, cardiac potential or membrane potential.

The resistance R in the circuits of Fig. 2-1 is assumed to be independent of the applied voltage or of the current passing through it. The current in such a resistor is strictly proportional to the applied voltage (see Fig. 2-2); the voltage-current characteristic of the resistor is a straight line. Elements with such a characteristic are called *linear circuit elements*. The magnitude of the resistance is expressed by the slope of the voltage-current characteristic (inversely proportional to tan α), specifically,

$$R = E/I \tag{2-3}$$

(Ohm's law, Fig. 2-2.)

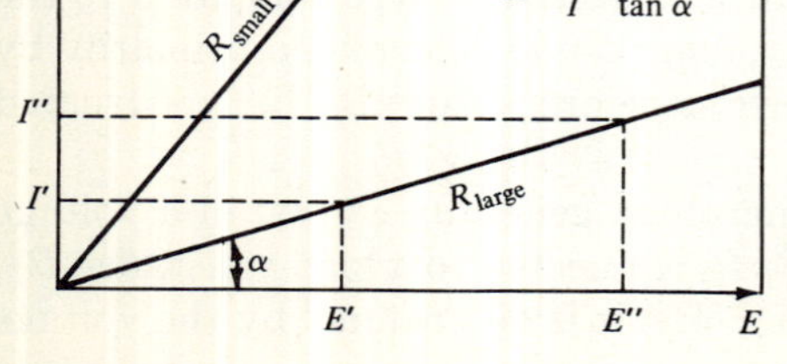

Fig. 2-2 Voltage-current characteristics of two resistances R_{small} and R_{large} (linear circuit element).

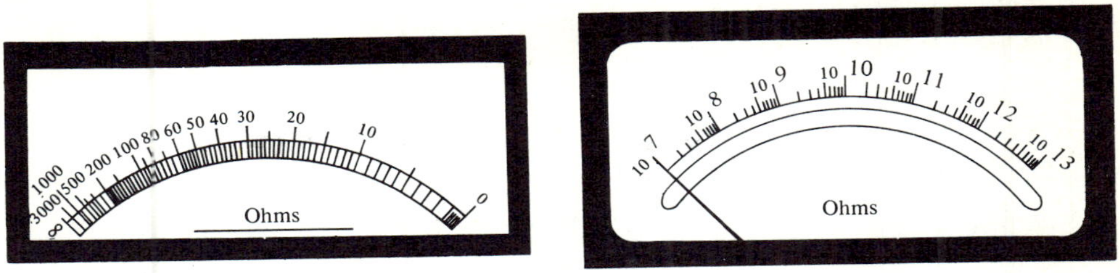

(a) (b)

Fig. 2-3 Scale character of ohmmeters. (a) Linear-current meter characteristic, hyperbolic resistance scale; (b) logarithmic resistance scale.

Equation (2-3) furnishes a definition of resistance; it follows that a resistance is not defined experimentally unless a current passes through it. Equation (2-3) also provides the basis for the common method of resistance determination through voltage and current measurement (Fig. 2-4). Under simplified conditions, i.e., by neglecting the influence of the voltmeter resistance and the ammeter resistance, the unknown resistance R can be found from Eq. (2-3). This method permits the measurement of a power resistance under actual operating conditions.

In the commercial ohmmeter, the voltage E is kept constant and the current I is measured; the scale of the current meter can then be calibrated directly in ohms, taking into account the internal resistance of the voltage source and the meter resistance. The scale is hyperbolic as shown in Fig. 2-3a. The use of special current meters or of electronic circuits permits the construction of ohmmeters with linear or logarithmic scales (Fig. 2-3b).

Equation (2-3) is derived for dc voltages and currents; however, it is applicable also if E and I are time-variable quantities (ac values). The magnitude of E/I at ac is not only dependent on the resistance but also on the inductances and capacitances in the circuit. Under these circumstances, and assuming that there

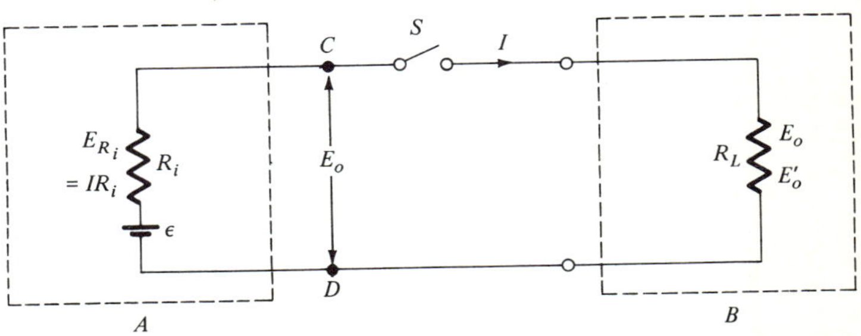

Fig. 2-4 Source with internal resistance R_i connected to load resistance R_L; see text for explanation.

are only linear elements in the circuit, Eq. (2-3) can be written

$$\frac{E}{I} = Z$$

where Z is the impedance of the circuit.

Table 2-1 shows the orders of magnitude of some voltages, currents, and resistances.

Table 2-1

Voltages:	
10^{-6} V	Microvolt region (μV) radio signals, brain waves
10^{-3} V	Millivolt region (mV); thermo emf, cardiac potentials
1 V	1.5-V single dry cell; 100- to 250-V line voltages*
10^{3} V	Kilovolt region (kV); power lines, x-ray machines
10^{6} V	Megavolt region (MV); lightning
Currents:	
1.6×10^{-19} A	1 electron per second
10^{-12} A	Picoampere (pA), measurable with sensitive galvanometer
10^{-6} A	Microampere (μA); sensitive meters
10^{-3} A	Milliampere region (mA); collector current in a transistor; perceptible as electric shock†
1 A	100-W light bulb at 100 V
10^{3} A	Kiloampere (kA), welding
Resistances:	
10^{-4} Ω	Contact resistance (good contact)
0.02 Ω	Copper wire, 1-mm diameter, 1 m long
1 Ω	Standard of resistance (National Bureau of Standards)
10^{3} Ω	1 kilohm (1 kΩ light bulb, 10 W, 100 V)
10^{6} Ω	1 megohm (1 MΩ)
10^{12}–10^{18}	1–10^{6} teraohms (TΩ), insulation resistance

* Electric equipment such as switches, fuses, outlets, or resistors sometimes bear misleading labels (e.g., 250 V, 10 A). These labels do not denote the actually prevailing operating conditions but rather the maximum rating for which these pieces of equipment can be used.

† Contrary to general belief, it is the electric current passing through the body which causes an electric shock, not the applied voltage. However, in order to cause a sufficiently large current through the body, two conditions must be fulfilled: (1) the voltage must be sufficiently high and (2) the resistance in the pathway of the current through the body must be sufficiently low. The resistance of the human body varies widely, from as high as several megohms to as low as 300 Ω—the highest resistance being located in a thin layer on the skin surface. Skin lesions or the presence of perspiration or water (in particular, salt water) reduce the resistance to a low value. The severity of the shock depends: (1) on the current level, (2) on the path of the current, and (3) on the duration of the exposure. At a power frequency of 60 H_3, 1 mA is perceptible; 10 mA may prevent voluntary control of muscles so that the subject is unable to free himself from contact; 100 mA can be fatal if the exposure lasts for more than 1 s and if the current passes the region of the heart. Resuscitation through artificial respiration is frequently successful.

INTERNAL RESISTANCE

Every source has an internal resistance. The resistance may be located, for instance, in the electrolyte between the electrodes of a battery or in the windings of a generator or a transformer. A source can be considered a combination of a constant-voltage source or electromotive force (emf) in series with a resistor R_i (system A in Fig. 2-4, Thevenin's theorem). This equivalent circuit behaves physically like a more complex circuit which may be composed of several potential differences and resistances. In fact, if only the output terminals C and D of the source are accessible and if nothing is known about the detailed construction of the source (the source is in a "blackbox"), measurements made on the output terminals would furnish information of the magnitude of a lumped voltage or emf ε and of a lumped resistance R_i only. The equivalent resistance R_i is the output resistance (or, in the more general case, the output impedance) of the source.

If a load is connected to the output terminals of the source (system B in Fig. 2-4), a current I will flow in the circuit and will cause a voltage drop IR_i across the internal resistance R_i. The terminal or output voltage E_o is then smaller than the emf ε.

$$E_o = \varepsilon - IR_i \tag{2-4}$$

where IR_i is the voltage drop across the internal resistance R_i.

The emf ε and the internal resistance R_i can be measured. First, the output voltage is measured while no current is drawn from the source. Since IR_i is zero, it follows from Eq. (2-4) that

$$E_o = \varepsilon$$

The terminals of the source are then connected (short-circuited). The current under short-circuit conditions can be measured with an ammeter and is $I_s = \varepsilon/R_i$. The internal resistance can be found from

$$R_i = \frac{\varepsilon}{I_s} = \frac{E_o}{I_s}$$

The short-circuit experiment frequently is not applicable; a short circuit may damage the source. A more protective method of measuring the output impedance is the following. First, the output voltage is measured under no-load condition, i.e., with a good voltmeter which draws only a negligible current from the source. This output voltage is E_o. Second, a load resistance R_L is connected to the terminals C and D (by closing switch S) so that a current I flows. The output voltage under this (load) condition is measured and is

$$E_o' = E_o - IR_i$$

The load current I is measured with an ammeter. The internal resistance can be found from

$$R_i = \frac{E_o - E'_o}{I}$$

When testing batteries, their internal resistances have to be taken into consideration. If measured with a high-resistance voltmeter, which draws a very small current, a used battery with a dried-out electrolyte will show almost the same output voltage as a fresh battery. Yet, such a used battery will show almost no output voltage when measured under current drain (under a load). The correct way to measure a battery is to apply a load, such as a bulb, while measuring the output voltage with a voltmeter.

The internal resistances (orders of magnitude) of some common sources are the following:

Lead cell (automotive) batteries	$10^{-3}\ \Omega$
Batteries (normal type D cells)	0.1 to 10 Ω
Rectifier power supplies	10 to 1,000 Ω
Electronically regulated supplies (effective resistance)	10^{-3} to 10 Ω
Standard cells	$10^4\ \Omega$
Glass electrode cells	10^7 to $10^9\ \Omega$

RESISTIVE NETWORKS

If several resistors $R_1, R_2, \ldots, R_n$ are connected in series, their effect in any circuit is equal to that of an equivalent resistance R_s, where

$$R_s = R_1 + R_2 + \cdots + R_n$$

If two resistors R_1 and R_2 are connected in parallel, their equivalent resistance is

$$R'_p = \frac{R'_1 R'_2}{R'_1 + R'_2}$$

If more than two resistors are connected in parallel, their equivalent resistance can be found by a step-by-step procedure

$$R'_p = \frac{R''_1 R''_2}{R''_1 + R''_2} \qquad R''_p = \frac{R'_p R''_3}{R'_p + R''_3}$$

Every complex network consists of nodes and loops. In a node the currents are

$$\Sigma(I_n) = 0 \qquad I_3 = I_1 + I_2$$

The voltages in a loop are

$$\Sigma(E_n + I_n R_n) = 0 \qquad E = E_1 + E_2$$

(Kirchhoff's laws.)

POWER

The power supplied from the system A in Fig. 2-4, or that transferred from system A to system B, or the power dissipated in the resistive load R_L of the system B is

$$P_{\text{watt}} = E_o I = \frac{E_o^2}{R_L} = I^2 R_L$$

The current is

$$I = \frac{\varepsilon}{R_L + R_i}$$

Therefore, the power transferred from A to B is

$$P = I^2 R_L = \varepsilon^2 \frac{R_L}{(R_L + R_i)^2}$$

If the voltage ε and the internal resistance R_i are given, the power depends only on the magnitude of the load resistance R_L. The power P approaches zero for either very small or very large values of the load resistance R_L and is a maximum for $R_L = R_i$. In other words, a maximum of power can be transferred from a source to a load if the internal resistance of the source or the output resistance of the preceding stage is equal to the load resistance or input resistance of the succeeding stage (principle of resistance matching or impedance matching).

The condition for impedance matching is not very critical. Figure 2-5 shows the relative amount of power transferred from one system to another for

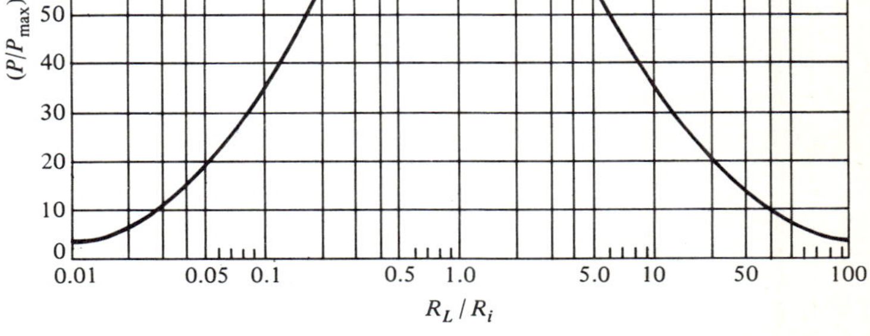

Fig. 2-5 Impedance-matching characteristic. Power transfer versus impedance mismated.

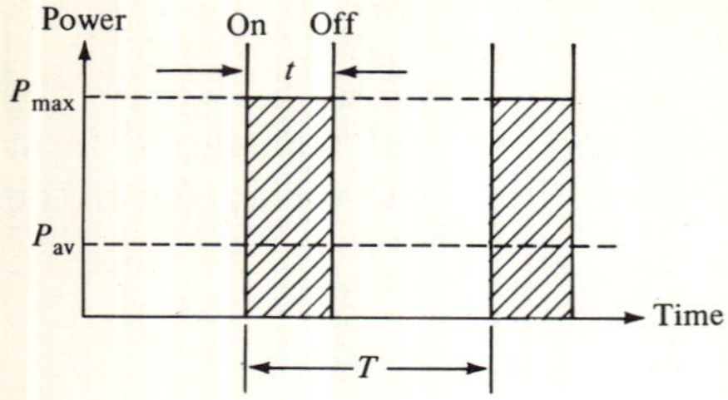

Fig. 2-6 Schematic diagram of pulse modulation.

different ratios R_L/R_i. For a deviation of ± 10 percent from the correct value of impedance matching ($R_L/R_i = 1.1$ or 0.9), the power transfer is still practically 100 percent; for a 20 percent change, the power transfer diminishes to 99 percent. Even for a 100 percent change (ratio of $R_L/R_i = 2$ or 0.5), about 89 percent of the maximum available power is transferred.

It is frequently desirable to control the amount of power transferred from one system to another. This can be accomplished by changing the voltage level applied to the load or the current level in the circuit. A variation of the voltage level is technically difficult at dc and generally requires the use of electronic equipment. A control (a reduction) of the current is relatively simple and is accomplished by inserting a series resistor R_s in the circuit. A disadvantage of the method is the considerable loss of power and the low efficiency. With the values of $R_i = 1\ \Omega$, $R_s = 98\ \Omega$, and $R_L = 1\ \Omega$, the useful power absorbed in the load is 1 W, while the power lost and converted into heat in R_i plus R_s is 99 W. The total power furnished from the source is 100 W. The power efficiency of the system (useful power/total power) is $1{:}100 = 1$ percent. A very efficient way to control the average power transferred from one system to another is by pulse modulation of the power. In the simplest case this can be accomplished by the periodic opening and closing of a switch located between both systems. Using the notation in Fig. 2-6, the average power transfer is

$$P_{av} = \frac{P_{max}t}{T}$$

By variation of t between zero and T, the power can be controlled between zero and the maximum value; the power efficiency is a maximum for any value of t. The application of the method is restricted to loads which are not affected by the periodic interruption of the current.

Power efficiency is generally desirable but is not always the most important factor in instrumentation. Other considerations may make it necessary to sacrifice power for other physical advantages, such as linearity, efficiency, or avoidance of loading of the preceding stage.

ADDING AND SUBTRACTING OF VOLTAGES OR CURRENTS

Figure 2-7*a* shows two voltage sources connected in series. The potential difference between the terminals of one source is E_1, that of the second source is E_2.

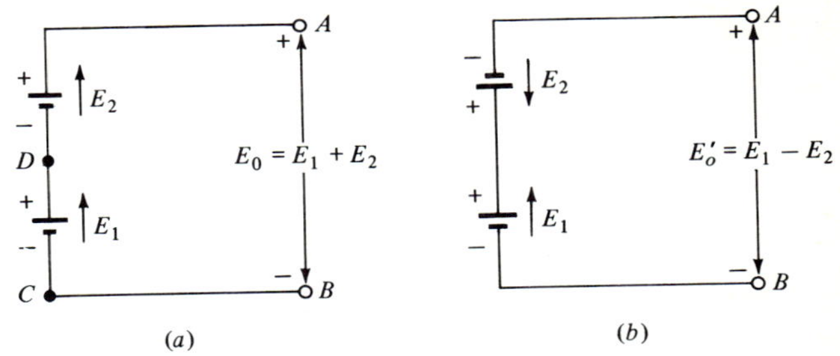

Fig. 2-7 (*a*) Addition of voltage signals; (*b*) subtraction of voltage signals.

The total potential difference (output voltage) between terminals A and B of Fig. 2-7*a* is $E_o = E_1 + E_2$. If the polarity of the source E_2 is reversed (Fig. 2-7*b*), the output voltage is $E_o' = E_1 - E_2$. This simple circuit solves a frequently occurring problem in the analog computer—it acts as an adding or subtracting network for two or more input signals E_1 and E_2, which may vary with time (e.g., temperature difference determination by means of two thermoelements).

The method has the disadvantage that the sources $E_1, E_2, \ldots, E_n$ must be independent of each other. Frequently, the sources $E_1, E_2, \ldots, E_n$ represent the output signals from individual amplifiers, and all have one terminal grounded. Obviously, if terminals C and D in Fig. 2-7*a* are both grounded, the source E_1 is short-circuited, and the circuit is inoperative. To make the circuit work, all sources except one must be isolated from ground (must be floating).

The circuit of Fig. 2-7*a* is redrawn in Fig. 2-8*a*. This redrawn circuit contains the internal resistances R_1 and R_2 of the sources E_1 and E_2 as well as a load resistance R_L (which may be the input resistance of the succeeding stage). Furthermore, the circuit is modified in the equivalent circuit diagram Fig. 2-8*b*.

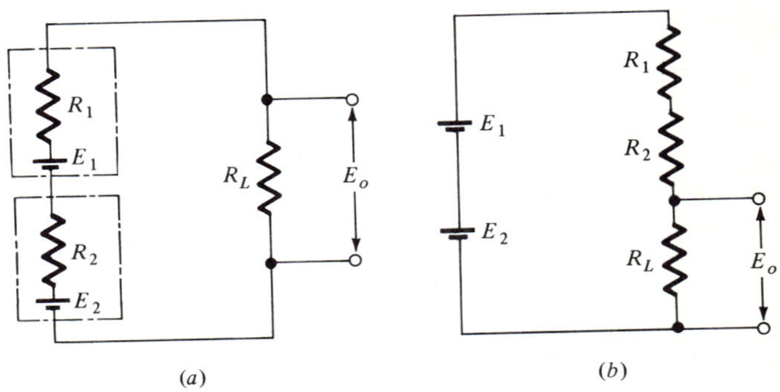

Fig. 2-8 (*a*) Voltage-sum forming network; (*b*) equivalent circuit.

The output voltage is

$$E_o = (E_1 + E_2) \frac{R_L}{R_1 + R_2 + R_L}$$

The input impedance for the sources E_1 or E_2 is the sum of the resistances in the circuit, e.g.,

$$R_{i_1} = R_1 + R_2 + R_L$$

The mutual coupling between two input sources, i.e., the current increment produced at one input by the voltage increment at another input is

$$\frac{\Delta I_2}{\Delta E_1} = \frac{1}{R_1 + R_2 + R_L}$$

and becomes zero for $R_L = \infty$. Electronic methods to render $R_L = \infty$ are described in Chaps. 11 and 12.

Current addition and subtraction can be achieved with parallel connection of the sources (see Fig. 2-9). Resistances R_1, R_2, and R_3 are connected in series with the voltages E_1, E_2, and E_3. Therefore, each parallel branch E_nR_n acts as a constant-current source when R_1, R_2, and R_3 are much larger than R_L. The total current passes through the load resistor R_L and causes a voltage drop, the output voltage E_o. If all series resistances $R_1, R_2, R_3, \ldots, R_n$ are equal and have a value of R, then

$$E_o = \frac{R_L}{R + R_L}(E_1 + E_2 + E_3 + \cdots + E_n)$$

Frequently the problem arises of measuring or recording the small variations of a large voltage, e.g., a voltage which is on the average 100 V but varies between the values 99.5 and 100.5 V. If this voltage were applied to a voltmeter having a total range of 110 V, the variations would hardly be noticeable. It is practical in such cases to use a circuit like that shown in Fig. 2-7*b* and to connect a fixed voltage E_2 (*bucking voltage*) of 100 V with reversed polarity in series with the variable voltage E_1. The output voltage E_o is then proportional

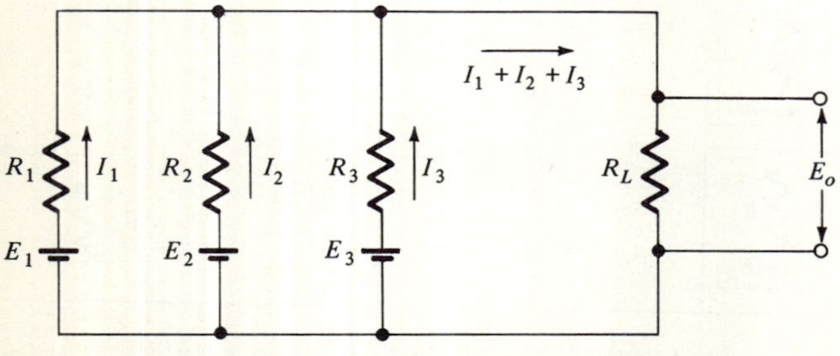

Fig. 2-9 Current-sum forming network; the individual input source connected in parallel.

to the difference between the fixed and the variable voltage, i.e., it varies from +0.5 to −0.5 V. The output voltage can be applied to a voltmeter having a total range of 1 V (zero center, i.e., −0.5 to +0.5 V) and produces a deflection over the full scale of the meter.

POTENTIOMETERS

Figure 2-10 shows a constant-voltage source E_i connected to a resistance R (potentiometer). The current I in the circuit is E_i/R, and the entire potential difference E_i is applied between points A and B. If point B is connected to ground, then its absolute potential is zero (by convention). The potential of point A is then $+E_i$ volts. The potential increases gradually from B to A. For instance, the center of R (point C) is at the potential $+E_i/2$, i.e., the potential difference between C and B is $E_i/2$ volts, point C being positive with respect to B. A potential difference of $E_i/2$ volts also exists between points A and C, point C being negative with respect to A. By moving the contact C from B to A, the potential difference CB increases from zero to the maximum applied voltage of E_i volts.

If r is the resistance between C and B, the voltage between C and B is

$$E_o = Ir = E_i \frac{r}{R} \tag{2-5}$$

The output voltage from the potentiometer depends only on the input voltage and the contact setting, i.e., on the ratio r/R not on the absolute magnitude of r or R. The circuit is commonly used as an attenuator or as a multiplying element with mechanical setting from 0 to 1 in analog computer circuits.

Equation (2-5) is correct only if no load is applied to the output terminals B and C, i.e., if the resistance of any load connected to these points is large compared with r. If such is not the case, the load resistance R_L has to be considered in parallel to r.

Variable resistors are commercially available in the form of linear or circular potentiometers. The resistance is usually made from a resistance wire wound around a mandrel or from carbon. By appropriate shaping of the mandrel the

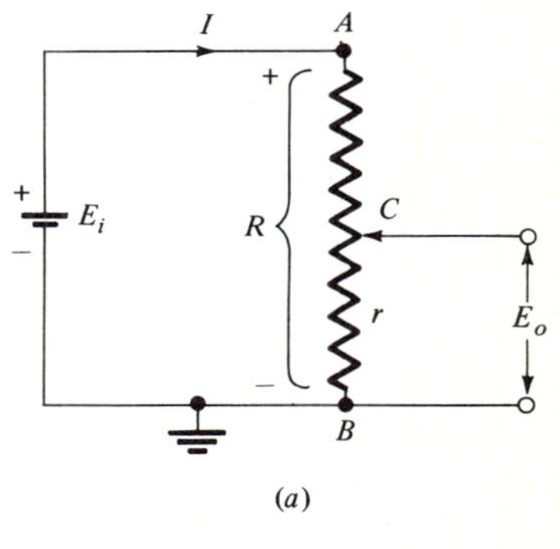

Fig. 2-10 Basic potentiometer circuit.

resistance can be made to increase linearly with the rotation of the potentiometer or in any other function (square, sinusoidal, or logarithmic). The accuracy of the resistance setting, or the resolution of the potentiometer, is somewhat limited by the stepwise movement of the contact from one winding of the wire to the next rather than in a continuous fashion; but the discontinuity is negligible for many applications. Frequently the resistance material is arranged in a helical fashion. This construction permits a fine adjustment of the resistance and the use of a long resistance wire.

Potentiometers frequently are used as transducing elements for converting displacements or forces into electric signals. Such potentiometers are commercially available with resistance values ranging from several ohms to several megohms. The resistance varies linearly with the displacement; the deviation from linearity in precision potentiometers is sometimes not more than 0.05 percent. The torque requirements of potentiometers is, in general, between 10^4 and 5×10^5 dyn·cm. Special potentiometers are built with a torque requirement of not more than 200 dyn·cm. Other special potentiometers for fast response are also available with very small moments of inertia (about 0.04 g·cm^2).

Figure 2-11*a* shows a potentiometer. A voltage source E_i is applied to the input terminals. The internal resistance of the voltage source E_i is assumed to be zero. The output voltage E_o appears between terminals B and C; it is zero when r is zero, i.e., when the variable contact C is at B. The output voltage increases linearly with r and reaches a maximum when the variable contact C is at A. The output resistance R_o, i.e., the resistance an observer would "see" by looking into the output terminals B and C, can be found from the equivalent circuit diagram Fig. 2-11*b*. The partial resistances r and $R - r$ appear to be connected in parallel; their equivalent resistance is

$$R_o = \frac{(R - r)r}{R - r + r} = \frac{(R - r)r}{R}$$

If r is zero, i.e., if the variable contact C is at B, the output resistance R_o is zero. If $r = R$, i.e., if the variable contact C is at A, the output resistance is again

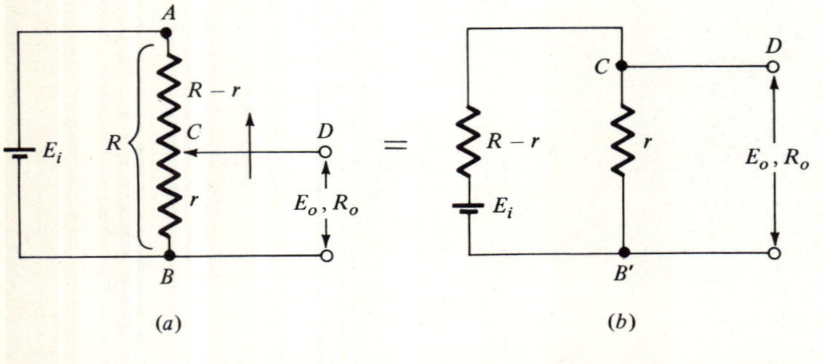

Fig. 2-11 (*a*) Potentiometer voltage divider; (*b*) potentiometer under load.

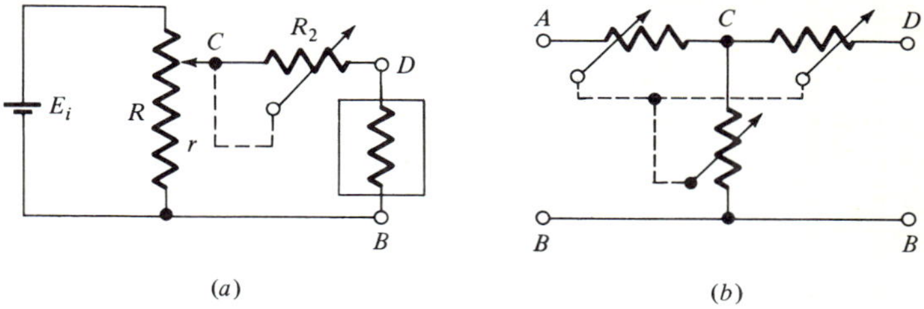

Fig. 2-12 (*a*) L-pad potentiometer; (*b*) T-pad voltage divider.

zero. With increased movement of the variable contact C in the direction of the arrow, the output resistance rises first, reaches a maximum value of $R/4$ at $r = R/2$, and decreases to zero again at $r = R$. For some applications this effect of variable output resistance is undesirable; it can be avoided by the use of an L-pad attenuator, i.e., a system of two mechanically coupled resistors as shown in Fig. 2-12*a*. The mechanical coupling (dotted line) is such that the resistance R_2 decreases as the resistance r increases so that the output resistance between terminals D and B remains constant when the setting of contact C is varied.

Figure 2-12*b* shows another attenuator, the T pad, in which not only the resistance between the output terminals D and B but also that between input terminals A and B remains constant while the ratio of output voltage to input voltage is varied.

The potentiometric circuit is commonly used for the comparison and measurement of voltages and electromotive forces without drawing current from the source to be measured. For this purpose the circuit is set up as shown in Fig. 2-13. E_a is an auxiliary source; its voltage must be larger than the voltage to be measured. The source E_a sends a current I through the resistor R. R is a slide-wire potentiometer with uniform cross section so that the resistance between the fixed contact B and the variable contact C increases linearly with a displacement of C. The voltage to be measured is applied between points B and C. A

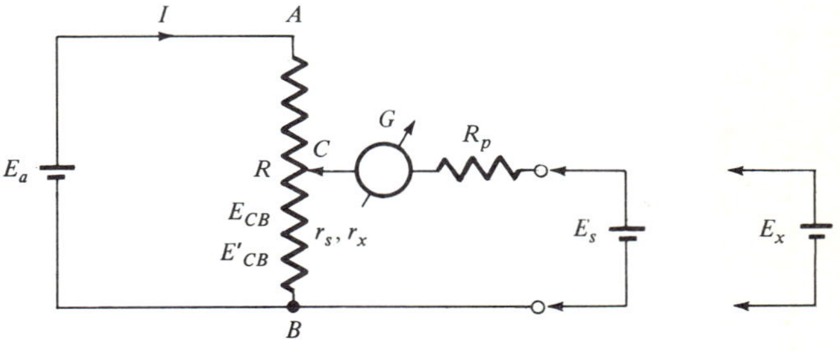

Fig. 2-13 Potentiometer for the measurement of emf.

current detector (galvanometer G) and a protective resistor R_p are connected as shown.

The measurement is carried out in two steps. First, a voltage of known magnitude (standard cell E_s) is applied. The contact C is set until the potential difference between C and B is equal to that between the terminals of E_s, so that no current passes the galvanometer. Then

$$E_s = E'_{CB} = Ir_s \tag{2-6}$$

Second, the standard cell is replaced by a source E_x of unknown voltage. The contact C is again varied until the zero detector indicates no current. Then

$$E_x = E_{CB} = Ir_x \tag{2-7}$$

From Eqs. (2-6) and (2-7) it follows that

$$E_x = E_s \frac{r_x}{r_s}$$

The unknown voltage appears as the product of the voltage of the standard cell and the ratio of two resistances, independent of their absolute magnitude and also independent of any meter calibration. The accuracy of this method is limited only by the accuracy of the standard cell voltage and that of the calibration of r_x and r_s. In addition, practically no current is drawn from the test object at balance. With some care in adjustment, the current drawn from the test object during the balance operation can be negligible as well.

A great number of variations of the potentiometer method are known. For instance, circuits that permit direct reading of the unknown voltage, others that permit the use of decade resistances with digital read-out, and still others with automatic setting of the contact C by means of a servo system: the galvanometer controls the operation of a motor which moves the variable contact until the balance point is reached. Potentiometer methods are primarily applicable where very accurate measurements are to be made (better than four significant figures) or where the current drawn from the test object must be very small.

WHEATSTONE BRIDGES

Figure 2-14*a* shows two potentiometers AB and $A'B'$ connected to the same source E_i. The potential of the variable contacts C and C' can be adjusted from zero (when the contacts are near points B or B'') to a maximum of E_i (when the contacts are near points A or A'). To each potential of point C, a potential of C' can be found which is equal to that of C. If such is the case, a meter M connected between C and C' will show no current.

An arrangement of this type is called a *Wheatstone bridge;* it can be used for the measurement of resistances. The circuit is redrawn as in Fig. 2-14*b*. A current I_1 passes from point A via resistors R_1 and R_2, and a current I_2 passes via resistors R_3 and R_4. The current I_g in the meter diagonal is assumed to be

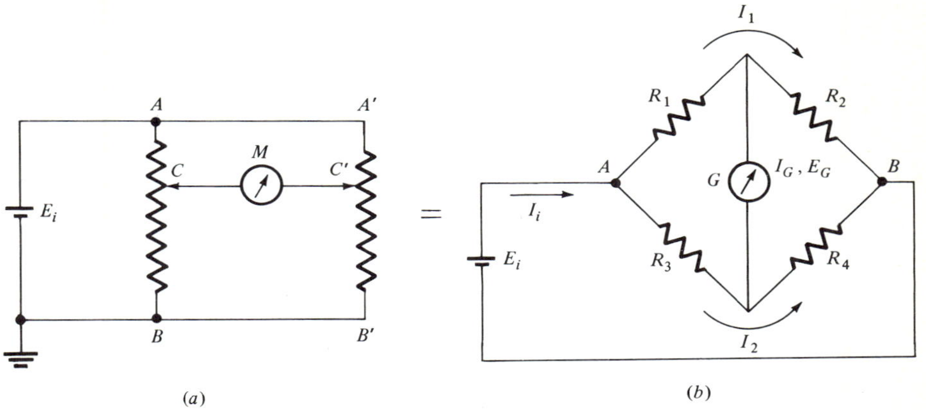

Fig. 2-14 Wheatstone bridge. (*a*) Schematic diagram; (*b*) circuit diagram.

zero. It is then

$$E_1 = E_3 \qquad \text{and} \qquad E_2 = E_4$$
$$I_1R_1 = I_2R_3$$
$$I_1R_2 = I_2R_4$$
$$\frac{R_1}{R_2} = \frac{R_3}{R_4} \qquad \text{or} \qquad R_1 = R_2\frac{R_3}{R_4} \tag{2-8}$$

R_1 is the unknown resistor which can be found from reading the resistance values R_2, R_3, and R_4 at balance. The equilibrium condition (2-8) remains unchanged when the bridge supply source E_i and the galvanometer G are interchanged.

The resistor R_2 (standard resistor) is frequently fixed, and the resistance ratio R_3/R_4 can be made continuously adjustable on a slide wire. Alternatively, the resistance ratio R_3/R_4 is varied in steps of 0.1, 1, 10, . . . while the resistor R_2 is made continuously variable.

Bridges may be used in two ways: in the balanced-bridge method or in the deflection method. The balanced method is the one just described. In the deflection method the unknown resistor R_1 is inserted in arm 1, and the bridge is first balanced by varying R_2, R_3, and R_4 so that the galvanometer current is zero. Any change of R_1 by the amount ΔR_1 causes a bridge unbalance so that a detector current I_g flows, and a deflection of the galvanometer G can be noted. The calibration characteristic $I_g = f(\Delta R_1)$ is experimentally established, and the value of the unknown resistor R_1 is found from the galvanometer deflection and the calibration characteristic. The deflection method is frequently used in percentage bridges, i.e., bridges calibrated to indicate in percent the variation $\Delta R_1/R_1 \times 100$.

Establishing the optimal dimensions of the resistances in a bridge is in principle an elementary matter but in practice a complicated problem. A change of

Table 2-2 Optimum dimensions of bridge resistances*

Primary condition	*Secondary condition*	*Maximum occurrence for*	*Approximate maximum*
Maximum current sensitivity $\frac{\partial I_g}{\partial R_1}$	I_i = const.	$m = \infty$ $n = \sqrt{q+1}$	$R_2 \gg R_1$
	E_i = const.	$m = \sqrt{q/(1+q)}$ $n = 0$	$R_3 \ll R_1$
	I_1 = const.	$m = \infty$ $n = 0$	$R_3 \ll R_1 \ll R_2$
Maximum voltage sensitivity $\partial E_g/\partial R_1$	I_i = const.	$m = \infty$ $n = \infty$	$R_3 \gg R_1$ $R_2 \gg R_1$
	E_i = const.	$m = 1$	$R_1 = R_2$
	I_1 = const.	$m = \infty$	$R_2 \gg R_1$

$m = R_2/R_1$; $n = R_3/R_1$; $p = R_4/R_1$; $q = R_g/R_1$. At balance m also $= R_4/R_3$; $n = R_4/R_2$; $p = mn$.

* From J. Fischer, *Elektrot. u. Maschinenb.*, **48**:1060 (1930) cit. by J. Krönert, *Arch. f. Techn. Mess.* (*ATM*) *J.*, 910–912, 1933.

the unknown resistor R_1 causes a change of the bridge output, for instance, of the output voltage E_g in the detector diagonal or of the output current I_g. The sensitivity of the bridge is the differential quotient $\partial E_g/\partial R_x$ or $\partial I_g/\partial R_x$. The maximum value of this figure depends not only on the resistance values in the bridge but also on secondary conditions, such as whether the input (supply) voltage E_i applied to the bridge is kept constant or the input current I_i. Table 2-2 gives the optimum bridge resistances for a number of secondary conditions.

If the meter G in the bridge diagonal is of the low-impedance (galvanometer) type, the bridge is primarily current-sensitive, and the optimum condition for the bridge will be found in the upper three horizontal columns of Table 2-2. If the meter is of the high-impedance type, the lower three horizontal columns apply.

The sensitivity of the bridge increases with the applied voltage or the applied current. A limit is set by the heating effect of the unknown resistor R_1 and by the bridge current I_1. If the current I_1 is high, the bridge sensitivity is high, but a manipulation of the bridge circuit may lead to an (undesirable) heating effect of R_1. It is then advisable to operate the bridge with the condition I_1 = const., as indicated in the third and sixth horizontal columns of Table 2-2.

DECIBEL NOTATION

The power gain or the power loss in an electrical network can be obtained from measurements on the two input and the two output terminals. If the network is attenuating, the output power P_o will be smaller than the input power P_i. If the network contains amplifiers, the output power may be higher than the input power. The power gain or the power loss in an element or network can be expressed by the ratio P_o/P_i. However, it is sometimes practical to express the gain or loss of power by a logarithmic unit, the decibel (dB), where

$$\text{Number of dB} = 10 \log \frac{P_o}{P_i}$$

Since the power is proportional to the square of the voltage or the current, the gain or loss can also be expressed by the voltage or current ratio

$$\text{Number of dB} = 20 \log \frac{E_o}{E_i} \quad \text{or} \quad 20 \log \frac{I_o}{I_i}$$

Table 2-3 shows corresponding values of dB and voltage or current gain, voltage or current loss, power gain, and power loss.

Table 2-3 Decibel notations

	Voltage or current		*Power*	
dB	*Gain*	*Loss*	*Gain*	*Loss*
0.1	1.01	0.989	1.02	0.977
0.3	1.03	0.966	1.07	0.933
1.0	1.12	0.891	1.26	0.794
3.0	1.41	0.708	1.99	0.501
10.0	3.16	0.316	10.0	0.100
30.0	31.6	0.032	10^3	10^{-3}
100.0	10^5	10^{-5}	10^{10}	10^{-10}

3

Electric Meters, Recorders, and Display Systems

MECHANICAL ANALOG SYSTEMS

Almost all meters contain a system which moves under the influence of an electric input signal against the restoring force of a spring. The meter movement comes to a stop if the mechanical force created by the input signal is equal to the restoring force generated in the instrument.

DYNAMIC BEHAVIOR

The speed with which such a system moves and, therefore, the highest frequency at which the system can follow an applied sinusoidal, square wave, or otherwise varying input signal, depends (in a first approximation) only on the ratio of the restoring force to the inertia of the system. For a rotational movement, the speed is proportional to $\sqrt{D/J}$, where D is the restoring moment and J is the moment of inertia. The movement also depends on the damping of the system.

Such systems exhibit a dynamic error when a transient input signal, such as a sudden change in level, is applied to the input, or when an ac signal of constant amplitude but variable frequency is applied.

Figure 3-1a shows the response of several meter systems to a step function (curve S). An ideal meter would follow such a function without distortion or

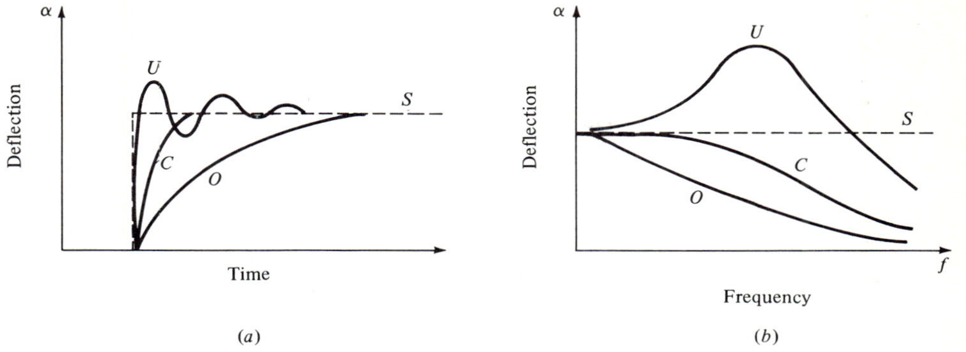

Fig. 3-1 (*a*) Response of a meter system to a step function *S*. *U*, underdamped; *O*, overdamped; *C*, critically damped system. (*b*) Response of meter systems to an alternating voltage of constant amplitude but variable frequency *S*.

time lag. Practical meter systems can follow three different patterns: If the system has little damping (is underdamped), it will move from the initial to the new equilibrium position and will oscillate around the new position as illustrated by curve *U*. If the system has an intermediate damping, it will move to the new position without oscillation and will come to rest in the shortest possible time as illustrated in curve *C* (critical damping, aperiodic movement). If the system is excessively damped (overdamped, curve *O*), the movement to the new equilibrium position will be slow.

The response of the same meter systems to a sinusoidal input signal of constant amplitude but variable frequency is shown in Fig. 3-1*b*. An ideal meter would follow a straight line *S*, i.e., its deflection for a given input signal level would be independent of the frequency of this signal. An underdamped system shows resonance (curve *U*) with the height and width of the resonance peak depending on the damping. The response of an overdamped system regresses continuously as the frequency rises (curve *O*). The deflection of a critically damped meter system (curve *C*) remains independent of frequency up to about 60 percent of the resonance frequency (of the undamped system).

METER CIRCUITS

Voltages are always measured between two points of different potentials. For instance, in the circuit of Fig. 3-2 the voltmeter V_1 measures the source voltage, i.e., the potential difference between terminals *A* and *B* of the source. The voltmeter V_2 measures the potential difference between terminals *C* and *D* of the load.

A voltmeter may draw a current i_1 or i_2. As long as this current is small compared to the current *I* in the circuit drawn in heavy lines, application of the voltmeter will not appreciably change the current and voltage distribution in the entire circuit. The voltage measurement is "correct" insofar as it reflects

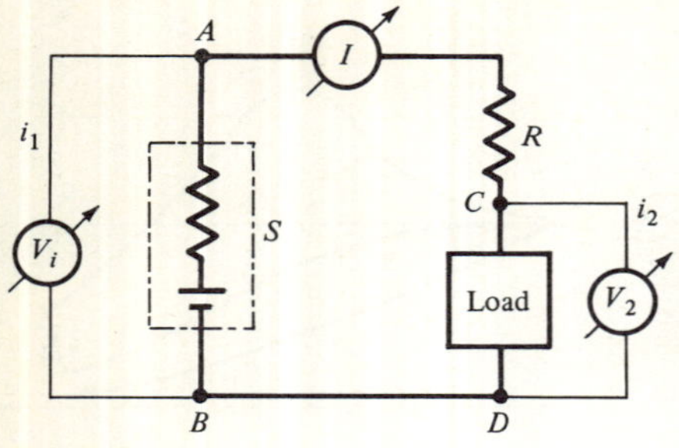

Fig. 3-2 Simple circuit (heavy lines) consisting of source S, ammeter I, resistor R, and load L. The voltmeter V_1 measures the voltage across the source; V_2, the voltage across the load.

the same values of voltages that are present when the voltmeter is not applied. This condition requires that the resistance of the voltmeter must be high compared to the resistance of the circuit to which it is connected. Good mechanical voltmeters have resistances of at least 1,000 Ω/V. Electronic voltmeters have resistances of the order of several to more than a hundred megohms; electronic "electrometers" may have resistances up to the order of 10^{16} Ω. If a voltmeter has a relatively low resistance, it may alter the potential distribution in the circuit under investigation. Even in this case it will measure correctly the voltage between its terminals E_{AB}; however, this voltage is not the same as the voltage between A and B in the absence of the voltmeter.

The range of a voltmeter can be increased by connecting a resistance in series with the meter (Fig. 3-3a). If the range of the voltmeter V (resistance R_v) is E volts, the meter can be used for a range $n \times E$ (for $n > 1$) volts by connecting a resistance R_s in series with it. The magnitude of this resistance should be

$$R_s = (n - 1)R_v$$

(The method is not applicable for electrostatic meters.)

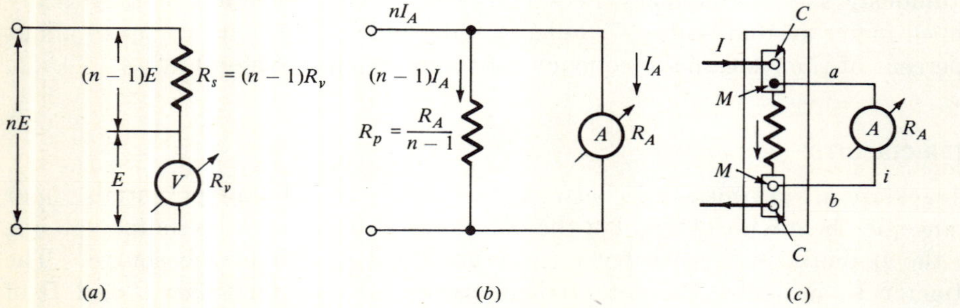

Fig. 3-3 (a) The range E of the voltmeter V is increased by the series resistance R_s to measure the voltage $n \times E$. R_v is the resistance of the voltmeter. (b) The range I_A of the ammeter A is increased by the parallel resistance R_p to measure the current $n \times I_A$. R_A is the ammeter resistance. (c) Shunt resistor with separate current terminals C and meter terminals M to avoid the influence of variable contact resistance.

Similarly, insertion of a current meter in a circuit should not alter the voltage and current distribution in a circuit, i.e., its resistance should be small compared to the other resistances in the same circuit. If the resistance of an ammeter is not small compared to the other resistances in the same circuit, the insertion of the meter will have the effect of reducing the current in the circuit. The current indicated by the ammeter will, nonetheless, be the correct value under this modified condition, i.e., the ammeter indicates correctly the current passing through it.

The range of an ammeter can be increased by connecting a resistance in parallel with the meter (Fig. 3-3*b*). If the range of the ammeter system A is I amperes and if the resistance of the meter is R_A, the meter may be used for a range $n \times I$ amperes (for $n > 1$) by connecting the resistance R_p (*shunt*) in parallel to the meter. The magnitude of the shunt resistance should be

$$R_p = \frac{R_A}{(n - 1)}$$

(The method is not applicable for certain inductive meters at ac.)

Shunt resistances are frequently small and of the same order of magnitude as contact resistances. In order to avoid errors resulting from variations of contact resistances at the terminals, shunt resistors are generally built as shown in Fig. 3-3*c*. Two heavy "current terminals" C are provided with two smaller "meter terminals" M. A variation of the contact resistance at C will cause a change of current I in the entire circuit as well as a proportional change of the meter current i. Therefore, the meter will read I correctly. A variation of resistance of the meter terminals M causes a change of the meter current i, but such change is usually negligible because the contact resistance and its variations are generally of the order of $10^{-3}\ \Omega$, while the meter resistance R_A is of the order of 1 to $10^3\ \Omega$. Therefore, the current i is practically determined by the resistance R_A and does not vary significantly with a change of contact resistance. The instrument should be used with the same connecting cables a and b with which it has been calibrated.

INSTRUMENT TYPES

Moving-coil systems The moving-coil system is the most frequently used of all meter types. A schematic diagram of the system is shown in Fig. 3-4*a*. A coil is suspended in the magnetic field between the poles of a permanent magnet. Electrical connections with the coil are made through the suspension wires, as illustrated, or through two helical springs, or through highly flexible bronze ribbons. A current passing through the coil produces a magnetic field in the coil; the coil acts like a magnet; and the magnetic field of the permanent magnet causes a torque of the coil. The coil rotates until the magnetic torque is equal to the restoring moment exercised by the suspension wires.

If the coil moves in a homogeneous field, the deflection of the coil or of a

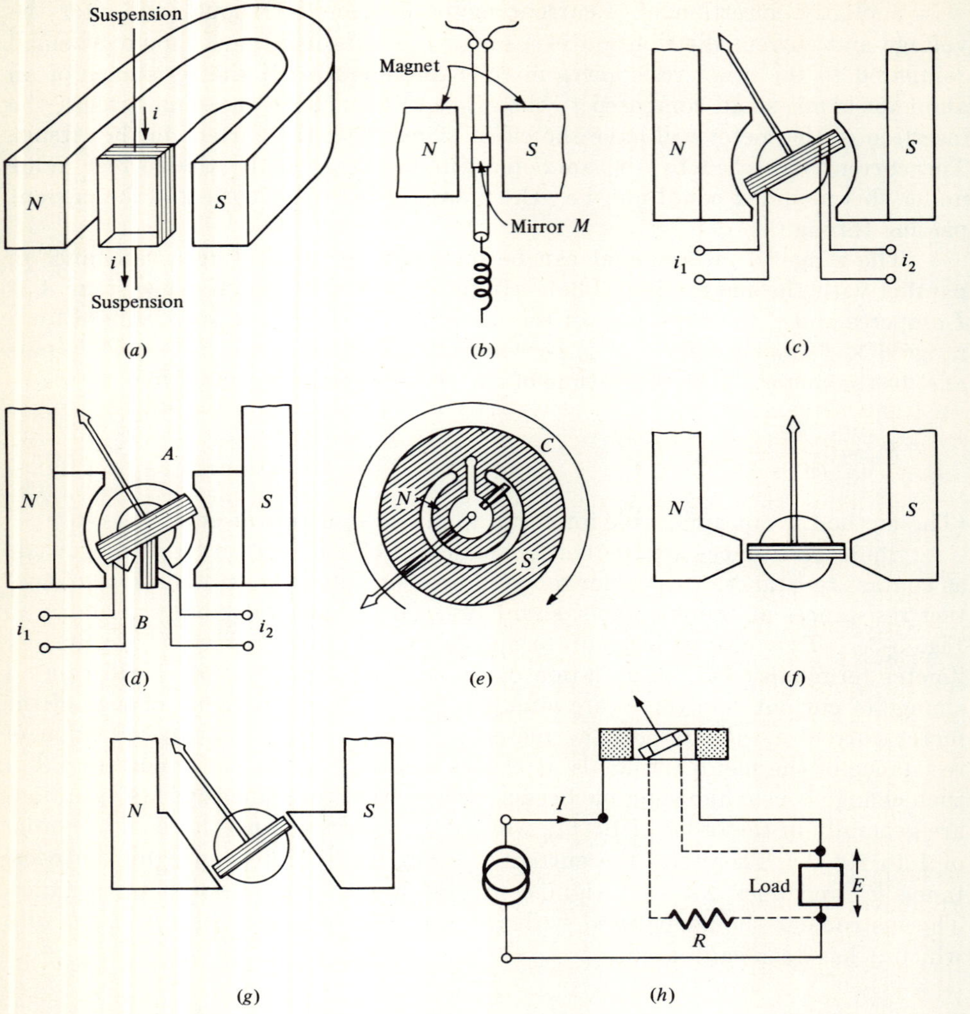

Fig. 3-4 Schematic diagrams of different moving-coil systems. (*a*) Basic moving-coil system. (*b*) Low-inertia moving-coil system consisting of a single-wire loop with a mirror *M*. (*c*) Moving-coil system for the measurement of the sum or the difference of the input signals i_1 and i_2 (differential galvanometer). (*d*) Ratio meter; the deflection is proportional to the ratio $i_1 : i_2$. (*e*) Moving-coil system for large angle of deflection (about 270°). (*f*) Moving-coil system with high sensitivity in the vicinity of the zero position for bridge detector. (*g*) Moving-coil system with approximately logarithmic characteristic. (*h*) Electrodynamometer; for the measurement of electric power (wattmeter).

pointer or mirror rigidly connected to the coil is proportional to the current passing through the coil. The scale characteristic is linear. The moving system is damped by air resistance, by eddy currents arising during motion in the metallic frame on which the coil is wound (this can be avoided by using an open frame, i.e., a frame with a nonconductive gap), and by current induced during motion in the windings of the coil (*electrodynamic damping*). None of these damping systems is very effective for meters which have a high restoring force (as frequently used for mechanical recording). The sensitivity of the system can be reduced by series and parallel resistances, as described above, or by a magnetic shunt which is provided in better instruments.

Voltmeters and ammeters Meters with mechanical read-out vary considerably in construction details. For instance, the magnets of magnetoceramic materials are frequently inside the coil; the coils may be suspended in jewel bearings. The coils may be produced by photographic etching techniques. Meters are built either as portable laboratory instruments or as panel instruments. The lowest level of input for a full-scale deflection is usually of the order of magnitude of 10 mV or 10 μA. The power requirements of a moving-coil meter (without series or parallel resistor) can be as low as 0.1 μW but are, for practical reasons, usually of the order of 1 mW. The power requirements rise, of course, if series or parallel resistors are used. The resistance of the coil is between one and several thousand ohms. A voltmeter is more sensitive the lower its resistance is since, for the same applied voltage, the current of a low-resistance meter system is larger than for a high-resistance system. However, a low-resistance voltmeter is likely to cause a measurement error as explained above. A current meter is more sensitive the higher its resistance is.

Parallactic reading errors can be avoided by using pointers with flat ends (*knife pointers*) and a mirror beneath the scale. The accuracy of laboratory instruments can be fairly high. Laboratory standard meters exhibit an error of about 0.1 percent of the scale-end value; portable laboratory-type meters, an error of 0.2 to 0.5 percent. The error of panel instruments larger than 4 inches in size is of the order of 1 percent, and of smaller meters (3 in and less) of the order of 2 to 5 percent. In order to keep the fractional error small, the meter range should be such that the meter is operated in the upper part of the scale.

The heavy magnet surrounding the system acts as a shield for external magnetic stray fields. However, the shielding is never complete, and errors are likely to arise if several moving-coil meters are arranged in close proximity. The friction error of jewel-bearing instruments varies with the orientation of the moving-coil axis. The error is small for a vertical orientation of the axis, intermediate for horizontal orientation, and large for an orientation under an angle.

Moving-coil systems with optical read-out (*mirror galvanometers*) are used primarily for recording on light-sensitive papers at high speed or high frequencies. The formerly often used low-frequency, high-sensitivity mirror galvanometer has essentially been replaced by the combination of a high-gain amplifier with a meter or recorder. However, sensitive mirror galvanometers can be superior to amplifier-meter systems as far as noise reduction and long-term stability are concerned.

Modern mirror galvanometers are built in a housing containing a light source, a scale, and, frequently, a set of mirrors for multiple reflection of the light beam which is equivalent to a long optical distance between the galvanometer mirror and the scale (*optical magnification*).

The damping of a galvanometer (or any similar electromechanical transducers, such as a speaker system) is predominantly electromagnetic. During the movement of the coil, a voltage is induced in the windings which causes a current through the coil and the external circuit. The smaller the resistance in the circuit, the higher is the damping. Critical damping, which is most desirable, is obtained by adjusting the resistance in the external circuit to a value which depends on the construction of the transducer (*critical external damping resistance*, CXDR). In some cases it is convenient to work with galvanometers that are slightly underdamped. Difficulties are likely to be encountered when the range of a galvanometer is to be changed by series and parallel resistance, since the insertion of such resistances changes the CXDR. In such cases an L-pad or T-pad attenuator, or an Ayrton shunt, as illustrated in Fig. 3-5, can be used for range adjustment without incurring a change of the damping.

High-inertia (integrating) moving-coil systems If a signal pulse of any arbitrary waveshape is applied to a moving-coil system and if this pulse is short compared to the period of the moving-coil system, the system will respond with a deflection. The amplitude of this deflection is proportional to the time integral of the current pulse (*ballistic response*). Ballistic meters are usually mirror galvanometers with a period of about 10 s or more.

Another meter with integrating response is the *fluxmeter*. It contains a moving-coil system that is greatly overdamped and has a negligible restoring force. If a dc signal is applied to the fluxmeter, a current i will flow through the coil, and the system moves with a velocity ω which is proportional to the current. If the current ceases, the movement stops and the pointer indicates a deflection α which is proportional to the time integral of the velocity and also proportional to the current-time integral:

$$\int \omega \, dt = \int \frac{d\alpha}{dt} \, dt = \alpha = k \int i \, dt$$

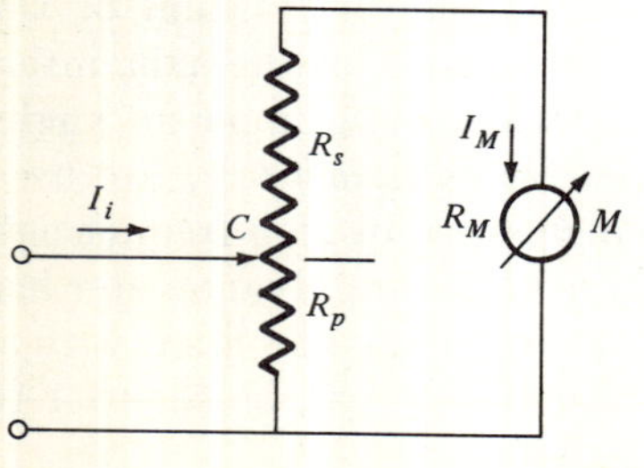

Fig. 3-5 Ayrton shunt for the sensitivity of a meter M without change of the critical external damping resistance. If the output resistance of the preceding stage is much larger than R_M, then the XDR is $R_s + R_p$ regardless of the position of contact C.

where k is the proportionality constant. The instrument has been used for the measurement of magnetic field strengths with an inductive flux coil (hence, the name fluxmeter). This instrument is useful for any current integration over a time span from seconds to about an hour. The system is returned to the zero position by the application of a small current in a direction opposite to that of the signal.

Low-inertia (oscillographic) systems Since the upper frequency limit to which a meter can follow the variations of an applied input signal depends primarily on the restoring force and the inertia of the system, it is possible to extend the frequency range by increasing the restoring force or by reducing the moment of inertia. A system for an extended frequency range is schematically illustrated in Fig. 3-4*b*. The coil is reduced to a narrow loop of fine wire under the tension of a spring. The loop is mounted in a strong magnetic field. A mirror cemented to the wires permits light-beam recording of the system movement. To assure proper damping, the system is sealed in a container filled with a viscous fluid. Usually several systems are combined in a common housing, along with a light source and the necessary optical and photographic equipment. Oscillographs of this type can reproduce input signals of a frequency up to 5,000 Hz with a dynamic distortion of less than 1 percent.

Differential moving-coil system The instrument is basically a moving-coil system with a mechanical restoring force (spring) but contains two coils wound on the same frame as shown in Fig. 3-4*c*. If two input signals i_1 and i_2 are applied to both coils, the total torque will be proportional to the sum $i_1 + i_2$, or if one of the currents is reversed, it will be proportional to the difference $i_1 - i_2$. It is an advantage of the instrument that the sum or difference of two currents can be formed in two electrically separated circuits. A disadvantage is that for difference measurements the torque in the vicinity of the zero position is proportional to the small difference of two currents, while the coils and springs are to be designed for the full current. As a galvanometer with optical read-out, the system is well suited for very accurate resistance measurements.

Ratio meters The instrument consists of two moving-coil systems A and B on a common axis as shown in Fig. 3-4*d*. None of the systems has a mechanical restoring force. One system (B) moves in an inhomogeneous magnetic field in such a way that the restoring torque exercised by this system increases with increased deflection (or increased angle of rotation). This system acts as an "electric spring"; its torque is proportional to the current passing through it and to the angle of rotation. The other coil moves in a homogeneous magnetic field; its torque, in any position, is proportional to the current passing through it. The entire system comes to an equilibrium when the torques are equal. The final deflection is proportional to the ratio of both currents, independent of their absolute magnitude.

Wide-angle moving-coil system For some applications it is desirable to use meters with particularly long scales or with deflections of the pointer over a range of almost 360°. A moving-coil system for such purposes is shown in Fig. 3-4*e*. A hollow cylindrical core N forms one pole of a permanent magnet. It is surrounded by a cylindrical pole piece S which forms the other magnetic pole. The magnetic field in the cylindrical air gap between N and S is radially directed and homogeneous. A rectangular coil C is mounted with one side in the magnetic field; its other side in the (practically field free) center hole. The coil can turn around the central pivot point. Two spiral springs (not shown) furnish the restoring moment and act as connecting leads to the coil. The torque of such an instrument is, of course, only one-half the torque of an ordinary moving-coil system.

Moving-coil systems with nonlinear scale Figure 3-4*f* shows schematically a system with a nonhomogeneous magnetic field. The field is highest in the vicinity of the center (zero) position of the coil, so that in this position the meter has a high sensitivity, while the sensitivity decreases to both sides with increased meter deflection. Meters of this type are useful primarily as zero indicators, e.g., in Wheatstone bridges or potentiometers.

A meter with an approximated logarithmic characteristic is shown in Fig. 3-4*g*. Due to the inhomogeneity of the magnetic field, the sensitivity of this meter is highest in the left-end position of the coil and diminishes gradually as the meter moves into the right-end position. The range of such a logarithmic meter is usually about 1 decade (from 0.1 to 1 V).

Ac-meter systems The meters described above show a deflection which changes direction when the polarity of the applied voltage or current changes. If an alternating signal is applied to these meters and if the inertia of the system is small (or if the natural frequency of the meter system is high compared to the frequency of the applied signal), then the meter will follow the applied signal and will register the instantaneous values of the applied signal. However, if the inertia of the system is large, the meter will indicate the average value of the signal. Since the average value of a symmetric ac signal is zero, these meters cannot be used for ac measurements. In this case either the alternating current has to be rectified, i.e., converted to a direct current which can be measured with these meters, or other types of meters have to be used.

Electrodynamometer A moving-coil meter in which the permanent magnet is replaced by an electromagnet is called an *electrodynamometer*. Figure 3-4*h* shows such a meter. It can be built with and without iron. Electrodynamometers without iron are less sensitive than those containing iron but are among the most accurate systems available for ac measurements. Systems with iron exhibit errors resulting from the magnetization of the iron. These errors generally increase with higher frequency.

The torque produced in electrodynamometers is proportional to the currents passing through each coil. Thus, if the same current passes both the stationary and the movable coil, the deflection, at any time, is proportional to the square of the current and, therefore, independent of the polarity of the current.

Figure 3-4*h* shows a circuit in which an electrodynamometer is used as a wattmeter, i.e., for the measurement of power. The current from the source to the load is led through the stationary coil; the magnetic field produced in this coil is $H_1 = k_1 I$. The voltage across the load is connected through a resistor R to a movable coil; the magnetic field produced in this coil is $H_2 = k_2 E$. The deflection is proportional to the product of both fields and is

$$\alpha = kH_1H_2 = kEI$$

where $k = k_1k_2$, i.e., proportional to the power, in watts, dissipated in the load.

The damping in electrodynamometers is generally accomplished with a pair of aluminum wings rotating with little clearance within a closed chamber (*air damping*) or with an aluminum disk rotating between the poles of a magnet (*eddy-current damping*). External magnetic stray fields are likely to cause errors in these meters and therefore magnetic shielding is often applied.

Moving iron-vane systems There are a great many meter types in which a soft iron element becomes magnetic under the influence of an electric current passing through a coil, and produces a mechanical force and a deflection of a pointer. One form, the radial-vane meter, is shown in Fig. 3-6. Two pieces of soft iron, a fixed and a movable piece, are arranged in a coil. Current passing through the coil magnetizes the iron pieces in the same sense so that they repel each other

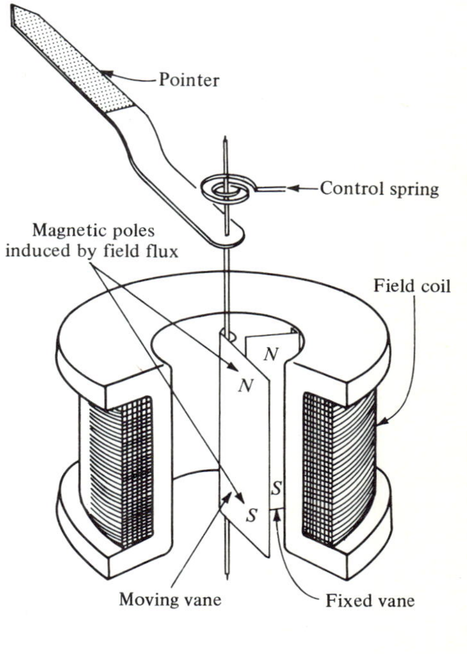

Fig. 3-6 Moving-iron radial-vane meter system for the measurement of ac voltages or currents.

and a deflection occurs. The calibration characteristic is nonlinear, at least in the beginning of the scale.

Like all ac meter types, the moving iron-vane system can be used for direct as well as alternating current; however, hysteresis errors are likely to arise in dc operation through the effect of remanence and coercive force in the iron. The calibration for increasing and decreasing current may be different by as much as 3 percent, but can be kept at about 1 percent in precision instruments. The power consumption is high, about 1 W. For this reason, change of the meter range by series and parallel resistors is not recommended.

Air damping of the moving system is generally applied. The influence of magnetic stray fields can be considerable. Errors through temperature influence are likely to arise, particularly in voltmeters. The error resulting from a change of frequency of the input signal can be kept within 1 percent in precision meters for a change of frequency from 60 to 1,000 Hz.

The meter types described solve certain problems that can also be solved by other electrical or electronic means. For instance, the sum or difference of signals can be measured with the differential moving-coil meter illustrated in Fig. 3-4*c* or with the circuits shown in Figs. 2-11 to 2-13 in combination with a standard moving-coil meter. A meter with a logarithmic characteristic can be obtained by the nonlinear moving-coil system shown in Fig. 3-4*g* or with the combination of a diode (or any other logarithmic converter) with a standard meter as shown in Fig. 11-5. The preferred method depends, in general, on technique considerations (range, accuracy, etc.), on the availability of instrumentation elements, on the personal preference of an experimenter, or sometimes on a general technological trend or "fashion" which may vary from time to time.

RECORDING INSTRUMENTS

Direct recorders Any one of the described meter systems can be used for recording by replacing the pointer and the scale with one of the writing mechanisms and a recording medium as described below. The driving system must have sufficient force to overcome the friction imposed by the writing mechanism and keep the friction error small. The error is generally of the order of 1 percent of the scale-end value.

The frequency range is generally below 100 Hz. Some instruments have a critical frequency of 200 to 300 Hz, and some can be operated even beyond the critical frequency with the help of electronic systems which amplify or reduce the excursions of the moving system in a particular frequency range by compensating mechanisms.

The angular excursion of the moving system is usually a linear function of the applied signal level. However, the trace on the recording paper is a circle as is evident from Fig. 3-7*a*. For some applications and for small angles of excursion, this nonlinearity can be taken in stride. For others, a linearization system must be applied to convert curvilinear into rectilinear deflection. Two examples of linearizers are shown in Fig. 3-7*b* and *c*. In Fig. 3-7*b*, the recording medium is moved over a ridge. The pointer or stylus (such as a heated ribbon,

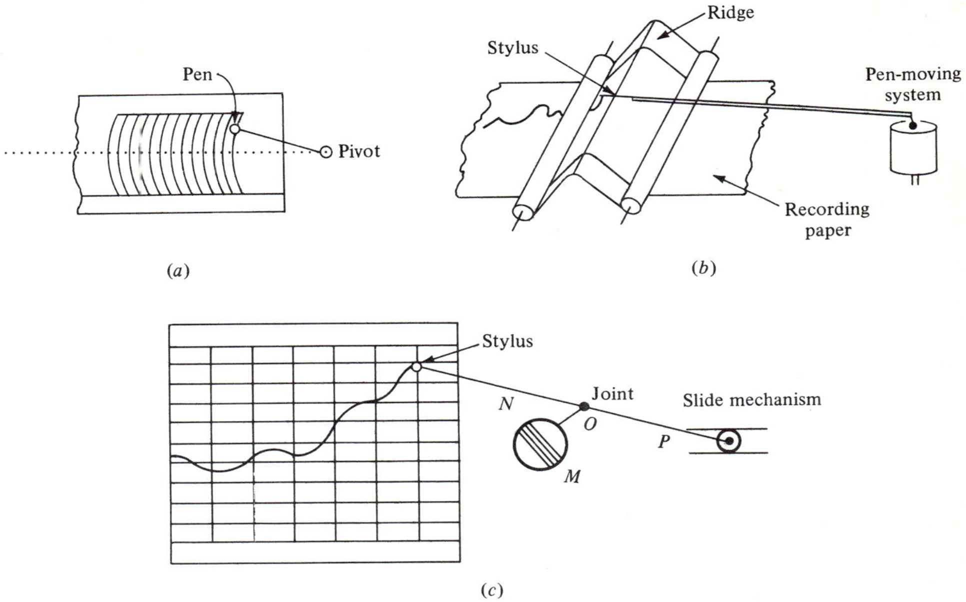

Fig. 3-7 (*a*) Curvilinear recording produced by rotary movement of a pen-moving system. (*b*) Linearizing system; the recording paper moves over a ridge where a stylus makes contact with the paper and causes a rectilinear marking. (*c*) Kinematic linearizing system *N-O-P*. *M*, meter-moving system, such as a moving-coil system.

see below) makes contact with the recording medium in a straight line perpendicular to the direction of the recording movement. The remaining tangent error is usually tolerable. In Fig. 3-7*c*, the moving coil *M* acts on a kinematic system *N-O-P* on the writing stylus and causes a fairly good approximation of a rectilinear trace.

Servosystem recorders The general scheme of a servosystem recorder is shown in Fig. 3-8. The instrument contains a potentiometer consisting of a reference voltage source E_r applied to a resistor R. A motor M can move a sliding contact C along this resistor. Depending on the position of C, a potential difference of a value between zero and E_r will arise between points C and G.

The input signal voltage E_i to be recorded is applied to terminals B and G. A sensing device S, such as a galvanometer or a sensitive relay, responds to any difference of potential between points B and C. For instance, if the potential of B is higher than that of C, the sensing device will furnish an output signal in one direction into amplifier A. The output from this amplifier is fed into motor M which moves by means of a belt drive D, the contact C to a value of higher potential. When the potentials of B and C are alike, the output signal from the sensing device ceases and the motor stops. A writing pen P is connected to motor M by a second belt drive F, and continuously records the position of C

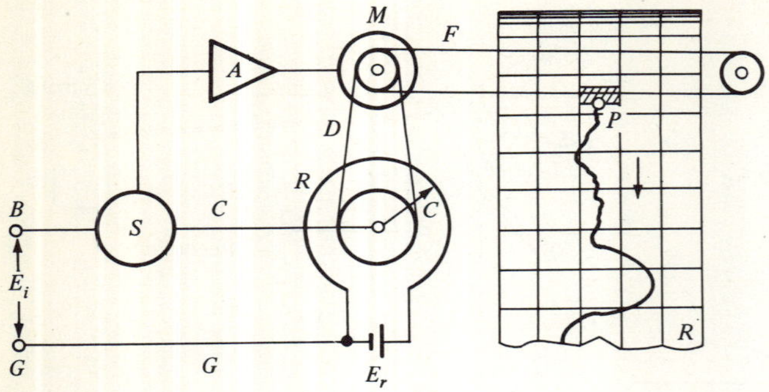

Fig. 3-8 Servosystem recorder. E_i, input signal; S, comparator; A, amplifier; and M, motor. The motor drives a sliding contact C by means of the belt-driving system D and a stylus with pen P by means of the belt-driving system F.

on a paper strip R which moves with uniform velocity. Servosystem recorders vary considerably in their complexity and their speed of response. They usually contain mechanical, electrical and electronic subsystems, such as amplifiers, damping devices, power supplies, and calibrators. They have been built for input signals between 1 μV and 100 V; their calibration characteristic may be linear or logarithmic, encompassing a range up to 4 decades (1 to 10,000). Generally, the error of such recording systems is of the order of a fraction of 1 percent and is imposed frequently by the width of the recording trace. The speed of response is of the order of a fraction of a second for full-scale movement.

***x-y* recorders** These are recording systems that permit the presentation of any function of the form $e_y = f(e_x)$ in an orthogonal coordinate system. Direct writing *x-y* recorders formerly were built in which a light beam was deflected in one plane by a first galvanometer mirror and in a plane perpendicular to the first by a second (larger) galvanometer mirror. The light beam was then displayed on a screen or photographic layer.

Indirect-acting (servosystem controlled) *x-y* recorders consist of two separate servosystems which control the movement of a writing stylus in two orthogonal directions by means of two chain drives. A schematic diagram of an *x-y* recorder is shown in Fig. 3-9. The speed of an *x-y* recorder is about 1 to 10 s for a full-scale movement.

WRITING MECHANISMS

Mechanical methods A simple recording system consists of a meter with a pointer which swings in front of a recording paper without touching it. The pointer is equipped with a stylus; the paper is pressure-sensitive. In regular intervals, a clockwork or a motor presses the stylus for a fraction of a second against the

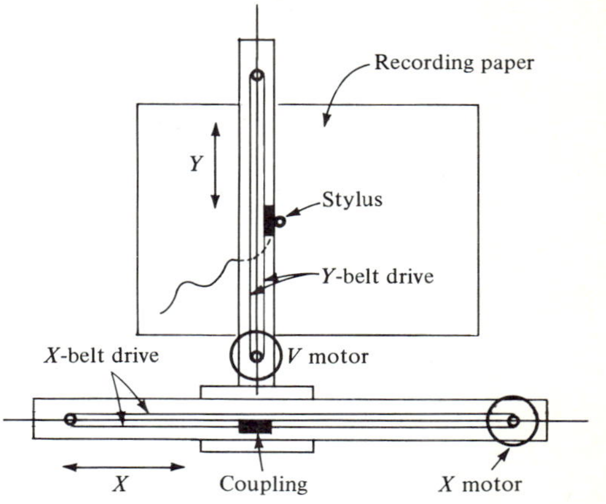

Fig. 3-9 Two rectangular servosystems forming an x-y recorder.

paper, thus causing it to leave a mark on the paper. The recording is a sequence of dots. Friction with the paper is avoided except during the short printing periods. The time resolution is, of course, very modest. This method is used primarily for inexpensive recording systems.

Pressure-sensitive papers for continuous recording are commercially available. They consist of a layer of microscopic wax beads on a dark-colored paper backing. When these beads are pressed, they become transparent exposing the dark base of the backing. A disadvantage of this recording material is primarily the relatively high-pressure requirement with some papers and, thus, the high friction error. Also, the paper is easily marked by humidity or by handling. Methods to reduce the influence of friction have been recommended. Either a small mechanical vibrator is mounted on the recorder (for instance, a buzzer), or a small alternating voltage is superimposed on the signal.

The oldest and still the most widely used method of recording is the pen and ink system. An ink reservoir is mounted on the stylus or in the vicinity of the instrument axis. A duct extends from the reservoir to the pen. The pen consists of a capillary tube or a tube with a reduced lumen that acts as a capillary. The pen is first "primed" by the application of a suction cup so that the duct and the pen are filled with ink. The pen is then brought in contact with the paper. In some systems, a counterweight pulls the pen up, away from the paper. The pen is held in contact with the paper by the cohesion forces within a small ink bridge. The friction is only that which occurs in the small viscous ink bridge, rather than that between pen and paper.

The writing speed of a standard tip can vary from 250 mm/h to 10 m/s with the same pen tip. Finer tips can be used for speeds as low as 1 cm/h. Different colors can be used for the distinction of multiple traces. Ink writer recorders have been built with pressurized pen systems, with ballpoint pens and with fiber pens. They work satisfactorily but may require some maintenance to avoid ink flooding, ink drying, vapor lock, or freezing.

Thermal methods The writing systems have an electrically heated stylus, ribbon, or wire. The recording medium consists of a dark or colored paper backing and a layer of wax or any other easily melting top layer. The heated stylus melts the top layer at the point of contact so that a trace of the colored layer becomes visible. High-speed recording requires higher stylus temperature than low-speed recording. Some recorder systems compensate for this effect. This type of recorder is used primarily in medical instruments such as electrocardiographs.

Electrical methods One of the oldest electrical methods of recording uses a carbon-impregnated, hence, slightly conductive, paper which is coated with a thin white insulating layer. The application of a voltage between the stylus and the back of the conductive layer causes a dielectric breakdown of the insulating layer and an eruption of the carbon-impregnated-layer surface. The resistance of the marked area is reduced by the dielectric breakdown so that a widening of the mark (*image spreading*) can occur at low writing speeds.

A metallized recording paper offers some advantages over the carbon-impregnated paper. The recording material consists of a paper backing which is lacquered, and a thin metal-film layer (Zn, Cd, Al, or Ni, about 0.1 μm thick) deposited on the lacquer layer. Contact with the metal layer is made through a metallic roller. Because of the large surface of the roller electrode the current density in the area of contact is small and no marking occurs. The writing electrode is a tungsten wire pressing against the recording surface with a force varying from 0.3 to 10 g. The current density is high enough to evaporate the metal film at the point of contact. The width of the trace is about 0.3 to 0.8 mm wide and can be controlled by varying the applied voltage.

A different physical mechanism is used in the electrolytic writing method. The paper is impregnated with a salt solution and is kept moderately moist. Contact with the paper is made with a large (inactive) electrode and a small stylus. The stylus reacts with the electrolyte in the paper, some ions go into solution and leave a colored trace on the paper. Alternatively, the electrolyte may contain a neutral salt and a color indicator for acidity which may develop underneath the recording electrode. The system is fast. It has been used for facsimile transmission; it responds to voltages from a fraction of a volt to about 100 V. A disadvantage is the need for a humid paper. For laboratory experiments, a piece of filter paper soaked in a solution of phenolphthalein and sodium chloride and mounted on a metallic phonograph turntable furnishes satisfactory traces for time measurements. A bright red trace appears beneath the negative stylus.

Excellent writing characteristics can be obtained with a semiconducting layer. The recording material is built up from three layers, a paper backing, a metal layer (such as aluminum), and a layer of zinc oxide in an organic binder. The zinc oxide is a semiconductor which has a whitish appearance. A tungsten carbide ball serves as stylus. If a current from the stylus passes the semiconducting layer, the zinc oxide is reduced to free zinc, which has a dark appearance. The mark is in the coating. Writing speeds extend from 10 mm/h to 5 m/s.

Optical methods Photographic recording of a light beam reflected from a galvanometer is several orders of magnitude faster than any one of the previously described writing methods. Standard photographic processes have been used but are rather inconvenient because of the time delay between recording and observation imposed by the chemical processing of the emulsion. Polaroid techniques can reduce this time delay to about 15 s. There are methods of physical (dry) processing which furnish a visible image in a matter of seconds after the latent image is formed. The image is projected to the photographic layer from a high-intensity ultraviolet light source and is developed by subjecting the emulsion to a low-energy light source. The image appears fast and is moderately stable for a few hours at room light. It can be made completely stable by fixing it. Photographic recording can be quite expensive if records are taken over a long time. Photographic methods can also be used for cathode-ray display and for digital recordings.

Cathode-ray oscilloscope Because of the inertia-free operation over a large frequency range, the cathode-ray oscilloscope (CRO) has become the most frequently used instrument for the observation of electric signals, particularly rapidly varying signals. A schematic diagram of a cathode-ray tube (CRT) is shown in Fig. 3-10. Electrons emitted from a heated cathode are accelerated and focused by an electron optical system onto a phosphor-coated screen where they cause a point of visible fluorescence. The electron beam can be deflected by electric and magnetic fields and follows the variation of such fields without noticeable inertia up to frequencies of the order of 10^9 Hz. Also, the intensity of the electron beam and, hence, the brightness of the fluorescent image can be controlled by suitably applied signals.

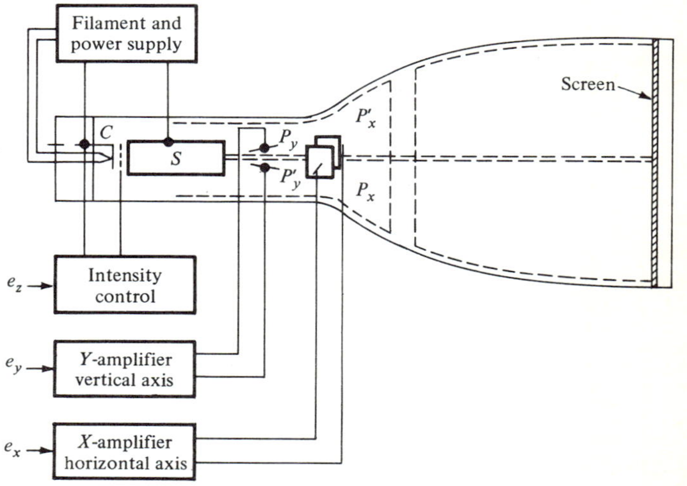

Fig. 3-10 Basic diagram of cathode-ray oscilloscope.

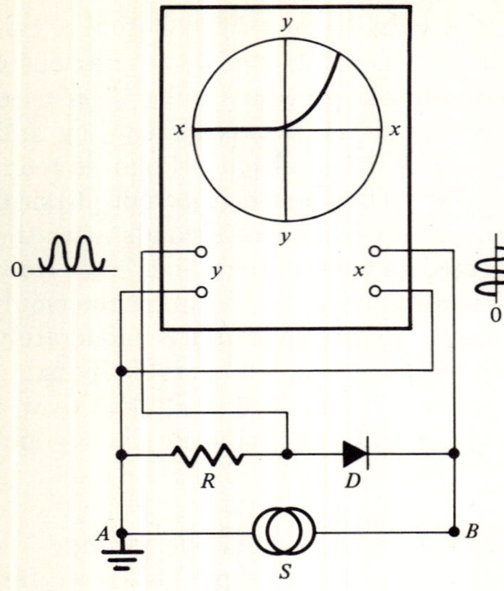

Fig. 3-11 Cathode-ray oscilloscope for two input signals, x and y, serving as an x-y recorder.

By applying a first signal to the X-axis terminals and a second functionally related signal to the Y-axis terminals, the cathode-ray oscilloscope acts as an x-y recorder. An example of such an operation is the use of a cathode-ray oscilloscope for the display of a current-voltage characteristic of a diode as shown in Fig. 3-11. The diode in series with resistor R is connected to an ac source S (S may be the output from a 5-V transformer at 60 Hz). The resistance of R is small so that the current i is directed primarily by diode D.

The voltage source S is connected to the X terminals of the CRO; this has the effect of producing a horizontal trace on the screen of the CRO. The voltage drop across R is proportional to the current I, i.e., practically proportional to the diode current. This voltage is applied to the Y terminals and causes a vertical displacement. Both traces superimposed are displayed 60 times per second in both directions on the screen of the CRO; the EI characteristic appears as a standing figure.

The CRO is used primarily for the display of signals that vary as a function of time. For such applications, a voltage must be generated that rises linearly with time. This voltage is applied to the X-axis deflection plates of the CRT and causes the fluorescent spot on the screen to move with constant velocity from left to right. The signal to be displayed is applied to the vertical, Y axis, deflection plates. Time-base generators are usually built into the CRO electronic systems.

Figure 3-12* is a block diagram of a typical oscilloscope (omitting power supplies). The waveform A to be observed is fed into the vertical amplifier input. The calibrated VOLTS/DIV. control sets the gain of this amplifier. Two output signals, one positive to ground, B, and another one negative to

* Figure 3-12 and related description from: A primer of Wave Forms and their Oscilloscope Displays, Tektronics, Inc., by permission.

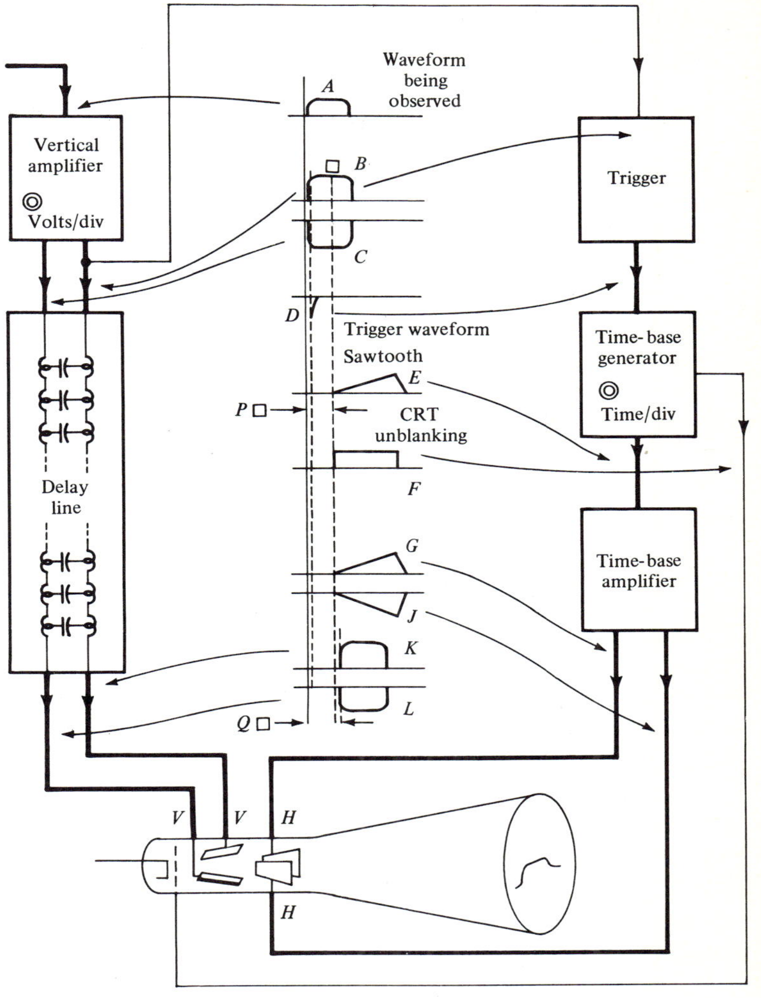

Fig. 3-12 Block diagram for a cathode-ray oscilloscope for the observation of $E = f(t)$ display.

ground, C, are generated in the vertical amplifier system and are fed through a delay line to the vertical-deflection plates of the CRO. The purpose of the delay line is to retard the arrival of the amplified input signal at the vertical-deflection plates, V, by a fraction of a microsecond.

The time-base generator or "sweep generator" develops a sawtooth wave E that is used as a horizontal-deflection voltage. The rising or positive-going part of this sawtooth, called the *run-up* portion of the wave, is linear. This rate of rise is set by the calibrated TIME/DIV. control. The sawtooth voltage is fed to the time-base amplifier. This amplifier supplies two output sawtooth waveforms G and J simultaneously, one of them positive-going like the input and the other negative-going. The positive-going sawtooth is applied to the right-hand horizontal-deflection plate of the cathode-ray tube, and the negative-going sawtooth is applied to the left-hand deflection plate. As a result, the cathode-ray beam is swept horizontally from left to right at a constant velocity.

In order to maintain a stable display on the cathode-ray-tube screen, each horizontal sweep must start at the same point on the waveform being displayed. To maintain a stable display, a sample of the displayed waveform is fed to a *trigger circuit* that gives a negative output voltage spike D at some selected point on the displayed waveform. This triggering spike is used to start the run-up portion of the time-base sawtooth. As far as the display is concerned, then, "triggering" can be taken as synonymous with the starting of the horizontal sweep of the trace at the left-hand side of the screen.

A rectangular *unblanking* wave F derived from the time-base generator is applied to the grid of the cathode-ray tube. The duration of the positive part of this rectangular wave corresponds to the duration of the positive-going or run-up part of the time-base output, so that the beam is switched on during its left-to-right travel and is switched off during its right-to-left retrace.

The leading edge of the waveform being displayed is used to actuate the trigger circuit. The triggering and unblanking operations require a finite time interval of about 0.15 μs. To permit the observation of the leading edge of an applied signal waveform, the trigger pulse D must arrive at the time-base generator about 0.25 μs before the leading edge of the input signal arrives at the vertical-deflection plates or, which is equivalent, the arrival of the signal waveform at the vertical-deflection plate must be delayed by a time of 0.25 μs after the trigger wave D arrives at the time-base generator.

Cathode-ray oscilloscopes vary considerably in their construction and operation. For instance, the acceleration of the electron beam is applied before the beam enters the deflection plates, but a part of the field can be applied between the deflection plates and the screen. CRO's also vary in their deflection sensitivity, electron optical qualities, and other parameters. Most CRO's display single traces. There also are oscilloscopes with more than one CRO system in one envelope which simultaneously display multiple traces. Multiple traces can also be produced by a single-gun system with the help of electronic switches which alternately apply one of two signals to the input of a CRO.

Limitations of the bandwidth, rise time, or response to transients are imposed by both the CRO system and the amplifiers. The amplifiers also may introduce drift and instability. The input impedance of the amplifiers is, in general, that of a resistance of 100 kΩ to 10 MΩ in parallel with a capacitance of the order of 50 pF. Some CRO's have a circuit which permits magnification of the horizontal sweep which can be adjusted so that only a portion of the sweep appears on the screen (appears magnified) while the rest of the sweep is off the screen.

CRO's are commercially available with different screen materials (phosphors). Screens differ in their color, light output, spectral distribution, persistence, and resolution.

Some screens exhibit a blue trace of short persistence during excitation by an incident electron beam (*fluorescence*) and a long persistence yellow afterglow (*phosphorescence*) following the beam exposure. The two colors can be separated by light filters. Photographic recording of the screen trace generally requires a high-intensity light output from the CRT screen of short persistence with a blue color of relatively high photographic efficiency. A fine fluorescent trace is required, not surrounded by a halo of stray light. With such arrangements, photographic recordings can be obtained with as little as 1 percent error. Of course, the photographic efficiency depends also on the physical characteristics of the photographic emulsion.

DIGITAL METERS

The input is an analog signal, usually a dc voltage. If used for other input quantities, such as for ac voltage, resistance, or displacement measurements, such input quantities are generally converted into analogous dc voltages prior to application to the digital voltmeter.

The operation of a digital voltmeter may be explained (in a very simplified manner) as follows. The applied voltage signal is first amplified or attenuated so that the signal applied to the input of the digital voltmeter corresponds to the proper range of the meter. The applied signal is "sampled," i.e., in regular intervals and for short periods of time the signal is used to charge a capacitor so that the capacitor voltage is proportional to the average signal level during each sampling period. Then a voltage that rises linearly with time (*ramp function*) is generated within the electronic part of the digital meter and, simultaneously, a pulse generator emits pulses in regular intervals into a counter. If the ramp voltage has reached a value which is exactly as high as the applied (sampled) voltage, an electronic switch interrupts the flow of pulses into the counter. The counter stops and the signal is displayed by a set of display tubes in the proper decimal notation and (in some instruments) with the proper polarity indication. Details of the counter, converters, and display tubes will be found in later chapters.

Digital meters are particularly useful for their high resolution. Even unskilled human observers can read them fast and with relatively few errors. A disadvantage for some applications may be the sequential (discontinuous) appearance of the display.

REFERENCES

Verhagen, C. J. D. M., and J. F. Tol: in A. E. Pannenborg (ed.), Electrical Instruments, vol. 23 of S. Fluegge (ed.), "Encyclopedia of Physics," Springer-Verlag, Berlin, Heidelberg, New York, 1967.

Davis, D. R., and C. K. Michener: Graphic Recorder Writing Systems, *Hewlett-Packard Journal*, **20**(2):2–6 (1968).

4

Instrumentation Elements Based on Resistance Variation

INTRODUCTION

The instrumentation elements described in this chapter are based on metallic conduction except in a few cases where semiconductors are described for comparison with metallic conductors. Carriers of electricity in all metallic conductors and in a great number of nonmetallic substances are electrons. If a voltage is applied to a conductor as shown in Fig. 4-1, the electrons will be forced to move. If n electrons per cubic centimeter are available for conduction and if e is the charge of the electron, a the total cross-sectional area of the conductor, and v the velocity of the electrons, then the total current is

$$I = neav \tag{4-1}$$

If E is the applied voltage in a conductor of length l, the voltage causes a field of the strength $F = E/l$. The velocity of the electrons in the conductor is proportional to the product Fe and inversely proportional to a constant α (a value which depends on the material of the conductor) so that

$$v = \frac{Fe}{\alpha} = \frac{Ee}{l\alpha} \tag{4-2}$$

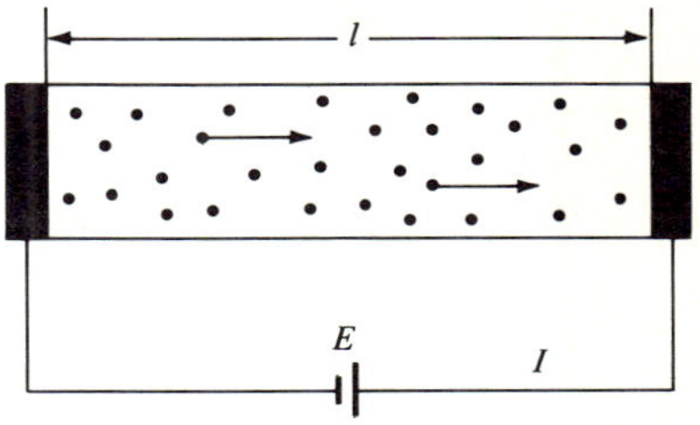

Fig. 4-1 Electron movement in an electronic conductor.

Substituting for v in Eq. (4-1) gives

$$I = nea \frac{Ee}{l\alpha} \tag{4-3}$$

By rearranging the terms on the right side of Eq. (4-3), one obtains

$$I = E \frac{ne^2}{\alpha} \frac{\alpha}{l} = \frac{E}{R} \qquad \text{(Ohm's law)} \tag{4-4}$$

where

$$R = \frac{\alpha}{ne^2} \frac{l}{\alpha} = \rho \frac{l}{\alpha} \tag{4-5}$$

The resistance is the product of two quantities, the resistivity ρ and a geometrical quantity l/α. The dimension of the resistivity is in ohm-centimeters; the reciprocal of the resistivity is the conductivity. Values of resistivities for different materials are given in Table 4-1.

RESISTANCE STRAIN GAUGE (PIEZORESISTIVE GAUGE)

The law expressed in Eq. (4-5) is the basis of a frequently used force or displacement transducer, the resistance strain gauge. The strain gauge consists essen-

Table 4-1 Resistivity and resistance-temperature coefficient of different materials

Material	*Resistivity,* Ω*·cm*	α*, Resistance-temperature coefficient/C*
Ag	0.015×10^{-4}	4.1×10^{-3}
Cu	0.0155×10^{-4}	4.3×10^{-3}
Ni	0.068×10^{-4}	6.8×10^{-3}
Fe	0.087×10^{-4}	6.5×10^{-3}
Pt	0.098×10^{-4}	3.92×10^{-3}
Pb	0.193×10^{-4}	4.2×10^{-3}
Manganin (Cu 86, Mn 12, Ni 2)	0.43×10^{-4}	$\pm 1.01 \times 10^{-3}$
Bi par axis	1.1×10^{-4}	4.4×10^{-3}
H_2SO_4, 40 percent solution	1.42	
Tap water	10^4 to 10^6	
Mica	10^{16}	
Metal oxides (thermistors)		-40×10^{-3} to -70×10^{-3}

tially of a fine wire, usually about 0.001 in in diameter or of a foil strip of about 0.0001 in in thickness. When lengthened within the elastic limit, two physical mechanisms will cause a change in its resistance. First, its geometrical form will vary, i.e., the wire will show an increase of length and a decrease of cross-sectional area. Second, a change of electric resistivity will occur caused by a distortion of the crystal lattice in the polycrystalline metal. This distortion causes a variation of the number of free electrons and their mobility. The result can be an increase or a decrease of the resistance, depending on the material of the wire. The fractional resistance variation of a strain gauge can be expressed by

$$\frac{\Delta R}{R} = S \frac{\Delta l}{l}$$

where l = the length
R = resistance of the unstrained wire
Δl = change of length
ΔR = change of resistance caused by external strain

The quantity S (sensitivity factor of the gauge or gauge factor) can be positive or negative and varies between -12 and $+3.6$ depending on the metal used. Table 4-2 shows some typical values of S.

The effect of resistance variation with strain is small. The resistance of most wire strain gauges varies by not more than 1 percent for a practically useful range of applied strain. Two different forms of wire strain gauges are employed: unbonded and bonded strain gauge.

An example of an unbonded strain gauge is shown schematically in Fig. 4-2*a*. Four sets of strain-sensitive wires A to D are electrically connected to form a Wheatstone bridge. An angular displacement of bar F around a fulcrum causes an increase of stress of wires B and C and a decrease of wires A and D. The changes of resistance will cause an imbalance of the bridge and give rise to an output signal. The range of displacements for practical unbonded strain gauges is of the order of a thousandth of an inch. The force required for this elongation is of the order of 5×10^3 dyn (5 g). The resistance is of the order of

Table 4-2 Strain-gauge factor of different materials

Material	*Composition*	*Strain-gauge factor S*	*Resistance-temperature coefficient/C*
Manganin	Cu 84, Mn 12, Ni 4	0.3 to 0.47	$\pm 0.01 \times 10^{-3}$
Constantan	Cu 60, Ni 40	2.0 to 2.1	$\pm 0.03 \times 10^{-3}$
Nichrome*	Ni 80, Cr 20	2.1 to 2.3	1×10^{-3}
Nickel	Pure	-12.1	6.7×10^{-3}
Alloy 479 (S. Cohn Corp.)	Pt 92, Wo 8	4.0 to 6.0	0.24×10^{-3}
Silicon (doped, *n* or *p*)		-100 to $+200$	

* Trademark of Driver-Harris Corp.

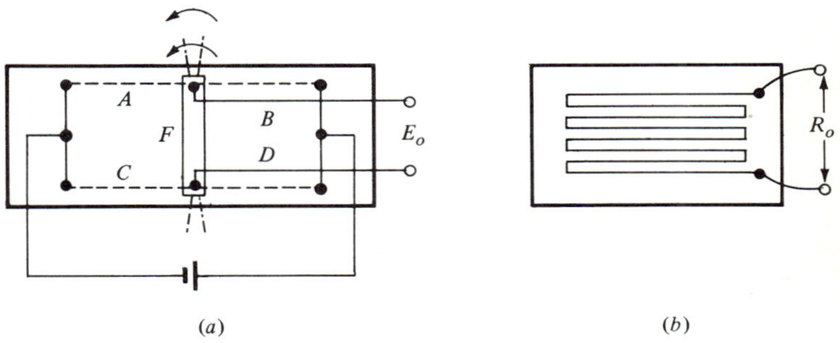

Fig. 4-2 (*a*) Unbonded strain gauge; (*b*) bonded strain gauge.

several hundred ohms. The output voltage from a strain-gauge bridge depends, of course, on the input voltage, but the input voltage cannot be raised too high since excessive heating of the wire would set in. For an input voltage near the maximum safe limit, one can expect an output of the order of 10 to 100 mV.

A bonded strain gauge is shown schematically in Fig. 4-2*b*. The gauge consists of a grid of fine filaments attached to a support. The filaments are held in position by a cement layer; two connecting leads are soldered or welded to the grid. For application, the entire strain gauge is attached to the structure to be investigated and a resistance variation is measured. The forces required to produce a useful output in transducers using bonded strain gauges are about a hundred times larger than in those using unbonded strain gauges. These forces are large compared to those in capacitive transducers. The practically obtained resistance variations are again of the order of 1 percent for a maximum range of displacement.

Strain gauges can also be made by evaporating metal on insulating surfaces or by using foils cut similar to those shown in Fig. 4-3. Very thin strain gauges can have sensitivities that considerably exceed those made from wire in bulk form, but such strain gauges may exhibit more hysteresis, drift, and nonlinearity.

The piezoresistive effect in semiconductors, primarily of silicon and germanium, considerably exceeds that of metals or alloys. Very pure single-crystal silicon with a controlled addition of small amounts of selected impurities (*doped silicon*, see Chap. 11) can be used for strain gauges with positive or negative gauge factors and of a magnitude varying from 50 to 200. However, many such semiconductors exhibit large temperature coefficients which require careful control measures for practical applications. Silicon gauges can be used at temperatures exceeding 500°F; silicon will not show plastic deformation below 1000°F.

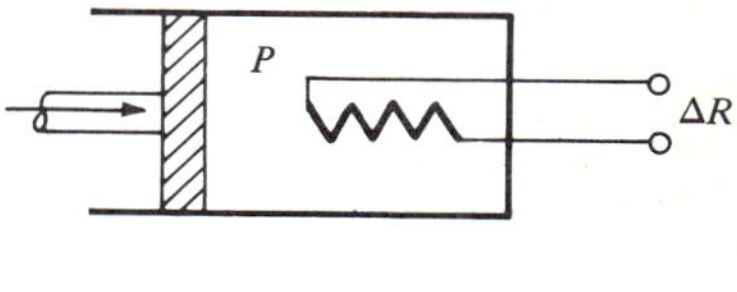

Fig. 4-3 Bridgman-type pressure transducer.

The advantages of resistance strain gauges are their flexibility, small physical size, high stability, and accuracy. Disadvantages are the relatively large force required to elongate the gauge and the small output level. Also, the strain gauge is very sensitive to temperature variations.

There are several methods available for the reduction or elimination of the temperature effect in resistance strain gauges. Some of these methods are generally applicable to the thermal stabilization of other instrumentation elements. The most frequently used methods are the following.

Dummy gauge system A strain gauge is inserted in one arm of a Wheatstone bridge and is exposed to a stressed sample and to temperature change. A similar gauge (*dummy gauge*) is inserted in an adjacent arm of the bridge and exposed to an unstressed sample but is at the same temperature as the first gauge. If both strain gauges are well matched, the influence of the temperature on the bridge equilibrium is reduced over a wide but limited range of temperature. A variation of this method is also applicable to semiconductor strain gauges: a p-type silicon gauge (positive gauge factor) is applied in one bridge arm of a Wheatstone bridge and an n-type gauge (negative gauge factor) in the other arm. Both gauges are exposed to the same test sample and both gauges have positive temperature coefficients, so that the stress sensitivity is increased while the influence of temperature variation is nearly canceled.

Pulse operation If the temperature variation in a strain gauge is caused by the current passing through the strain gauge, the temperature of the gauge and, thus, the temperature error will increase with the square of the average current passing through the gauge. However, the higher the current which can be sent through the gauge, the higher will be the output signal from the gauge. A high output signal with a minimum of heating can be obtained by pulse operation, similar to that described in connection with Fig. 2-14. Strain gauges normally operated with potentials between 3.5 and 14 V dc have been operated satisfactorily with pulses of different durations, repetition rates, and amplitudes. Output signals up to 50 times greater than with dc operation have been obtained.

Thermal self-compensation Some transducers exhibit a thermal characteristic $Q_o = f(t)$, which shows a maximum, a minimum, or an inflection. If operated in such regions, the temperature effect is obviously low or even zero. By combining different materials having appropriate thermal coefficients and geometry, it is sometimes possible to arrive at a construction of a transducer which permits operation in a region where the temperature error approaches zero.

RESISTIVE PRESSURE TRANSDUCER

If a wire is subjected to an increase of external pressure, exerted from all directions as shown in Fig. 4-3, its electric resistance will change. The resistance will usually decrease with increased pressure, although in some metals it increases and in others it decreases first and increases at a higher pressure. The effect has been extensively studied by P. W. Bridgman and forms the basis of a pressure transducer.

For many metals, and over a limited range of pressure variations, the resistance R_o at a pressure of 1 atm changes to the value R when the resistor is exposed to a pressure variation Δp, according to

$$R = R_o(1 + b\Delta p)$$

where b is the pressure coefficient of resistance for the resistor material. Values of b for different materials are given in Table 4-3.

Pressure transducers of the Bridgman type are suitable for very large pressures up to 20,000 kg/cm^2 (350,000 lb/in^2). The change of resistance, for instance, for manganin is 2.3 percent for a change of pressure of 10,000 kg/cm^2. The transfer function $\Delta R/R = f(p)$ is nearly linear for many metals, but can be nonlinear for others, for instance, for bismuth, cesium, rubidium, and sodium. For the construction of a practical transducer, the resistance wire is wound on an insulating core and mounted in the pressure chamber. The Bridgman-type transducer is the only direct-acting pressure transducer in existence. Its advantages over any mechanical gauge are simplicity and ruggedness.

RESISTANCE AND TEMPERATURE

The resistance of most conductors changes with temperature. In general, the resistivity of metals increases with increased temperature (positive resistance-temperature coefficient). The resistivity of electrolytes, semiconductors, and insulators decreases, generally, with increased temperature (negative coefficient). Within narrow ranges of temperature where the resistance-temperature coefficient may be considered constant, the resistance of a conductor at a temperature T is

$$R_T = R_o[1 + \alpha(T - T_o)] = R_o(1 + \alpha\Delta T)$$

where R_o = resistance of the conductor at the temperature T_o

α = temperature coefficient.

Table 4-1 shows values of the coefficient α for different materials at room temperature.

Table 4-3 Resistance-pressure coefficient of different materials

Material	*Pressure coefficient b (cm²/kg)*	*Deviation from linearity*
Aluminum	$-3.8 \times 4.2 \times 10^{-6}$	−0.001
Bismuth	21.4×10^{-6}	(Nonlinear)
Copper	-1.8×10^{-6}	−0.0004
Iron	-2.3×10^{-6}	−0.0005
Lithium	7.72×10^{-6}	
Rubidium	-155×10^{-6}	(Nonlinear)
Manganin	2.3×10^{-6}	

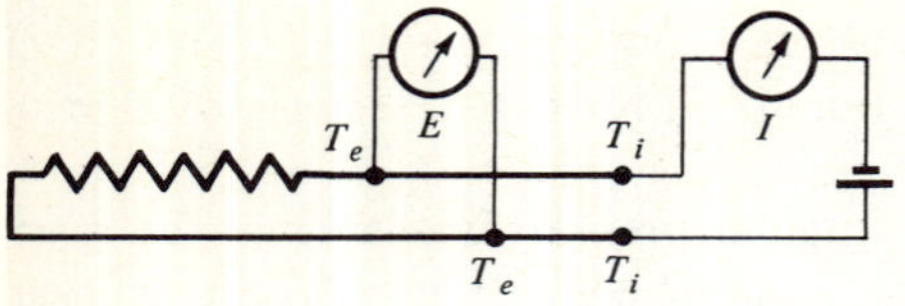

Fig. 4-4 Resistance thermometer.

The described effect is the basis of all electric resistance thermometers. The resistance thermometer consists of a resistance wire on an insulating carrier exposed to the temperature to be measured (see Fig. 4-5). Nickel has a relatively high temperature-resistance coefficient, but is only applicable at temperatures below 300°C. Pure metals have relatively high temperature coefficients. The coefficients of alloys are usually small and can even be negative in certain ranges of temperature. Changing the physical character of the metal (annealing, recrystallization) can cause a change in the temperature coefficient which in some cases (for instance, at the Curie point) can be discontinuous. For very low temperature, phosphor bronze alloys can be used. These materials exhibit a large change of resistivity below 7K.

Of particular interest is the platinum resistance thermometer, which forms the basis of the international temperature scale from -190 to $+660$°C, but which can be used for measurement up to 1000°C and down to -264°C (9K), when corrections are applied.

The resistance is measured either with the Wheatstone bridge or by voltage and current measurements. In order to avoid the influence of the contact resistance between the connecting line and the resistance-temperature element, the resistance-thermometer element is usually equipped with separate voltage and current terminals as shown in Fig. 4-4 (see also Chap. 3, Fig. 3-3*c*). Variation of the contact resistance between the current terminals T_i and the connecting leads causes a variation of the current I and a proportional variation of the measured voltage E and, therefore, does not affect the result $R = E/I$. Variation of the contact resistance between the voltage terminals T_e and the leads connecting the voltmeter is usually small compared to the voltmeter resistance and causes only a negligible error.

Resistance thermometers, particularly those made of platinum, are the most accurate of all temperature-sensing devices. In the vicinity of 0°C or at room temperature, measurements can be made with an accuracy of 10^{-4} °C. Ordinarily the error is several thousandths of a degree. At 450°C the accuracy does not exceed several hundredths of a degree, and around 1000°C the accuracy is not higher than about one-tenth of a degree.

If a Wheatstone bridge is set up for a temperature measurement with a resistive-sensing element, the bridge current may unduly heat the sensing element and may introduce a serious error. Whether or not such an error exists can be tested by balancing the bridge and then varying the bridge-supply voltage or the current. If the bridge remains balanced regardless of the variation of the bridge supply, the self-heating effect is negligible. However, if the meter in the bridge

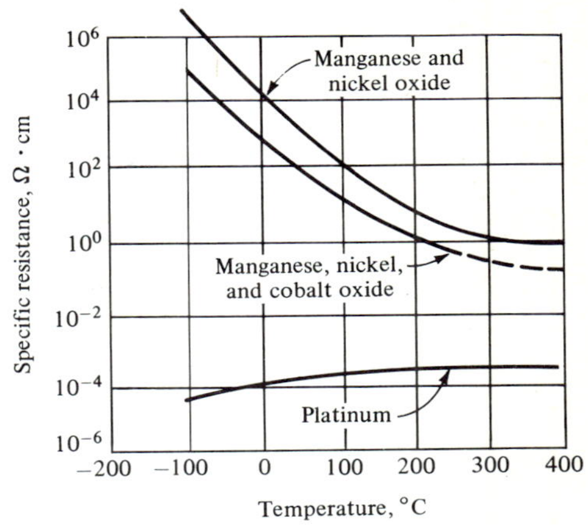

Fig. 4-5 Characteristics of thermistors; platinum for comparison.

diagonal varies in response to the variation of the bridge supply, such an effect indicates that the bridge current heats the temperature-sensing element and causes a variation of the resistance of the sensor which, in turn, causes a bridge imbalance.

The resistance-temperature effect of certain conductors (such as oxides, sulfides, or silicates of Ni, Co, Cu, U, Zn, Ti, Al, or Mg) or that of silicon is many times higher than that of metals. Sensing elements made from such materials are called *thermistors*. They are commercially available in the form of beads, rods, or disks. Depending on the composition, the resistivity of thermistor materials can have values ranging from 10^{-1} to 10^9 Ω · cm. The resistance of commercial thermistors at room temperature is usually in the range between 100 Ω and 1 MΩ. Small admixtures of impurities or physical alterations caused by heat treatment can result in a permanent change of the resistivity by a large factor.

The temperature coefficient of metals is, generally, negative; the resistivity decreases with increased temperature. Figure 4-5 shows the variation of two different thermistor materials and also of platinum for comparison. Between 0 and 300°C the resistance of the thermistor decreases by a factor of about 1,000, while the resistance of platinum increases in the same range by a factor of 2. The temperature coefficient of the resistance α is not constant but usually decreases with increased temperature.

The time constant of a thermistor (i.e., the time in which the thermistor reaches 63 percent of its steady-state resistance value after being exposed to a new temperature) varies for commercial thermistors between 1 and 200 s, although occasionally figures in the millisecond range are found.

Thermistors are commonly used in a range from −100 to +300°C, although for stability requirements an upper limit of 100 to 150°C is frequently recommended. Thermistors made of sintered aluminum oxide have been used at temperatures between 800 and 1100°C.

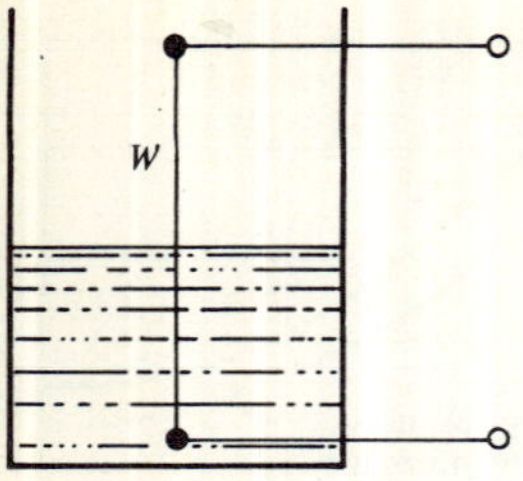

Fig. 4-6 Liquid-level transducer based on thermal effect.

The advantages of thermistors over metal-resistance thermometers are their high-temperature coefficient, relatively small size which makes localized temperature measurements possible, physical ruggedness, and large resistance value which, for many practical purposes, eliminates the influence of the lead wire and contact resistance variation in the measurements.

The variation of resistance with temperature forms the basis of a great number of transducers and modifiers. An example of a liquid-level sensing element is given in Fig. 4-6. A fine wire W extends vertically through a container partly filled with a liquid. The wire is heated by a constant current. The heat transfer from the wire to the liquid is greater than that from the wire to the vapor. Thus, that part of the wire immersed in the liquid will be cooler and its resistance lower than the part extending in the vapor atmosphere. As the level of the liquid recedes, the average temperature of the wire and, thus, its resistance will increase.

The instrument has been used primarily for determining the level of liquefied gases but is applicable to many other liquids as well. In liquid nitrogen the instrument permits the determination of the liquid level to within 0.1 in. The sensitivity increases with decreased wire diameter; platinum wire of 10-μm diameter or platinum ribbons are frequently used. The calibration is nearly linear if the vapor is at the same temperature as the liquid.

Figure 4-7a shows another application of the same principle for the measurement of low pressure in vacuum systems. This instrument, the Pirani gauge,

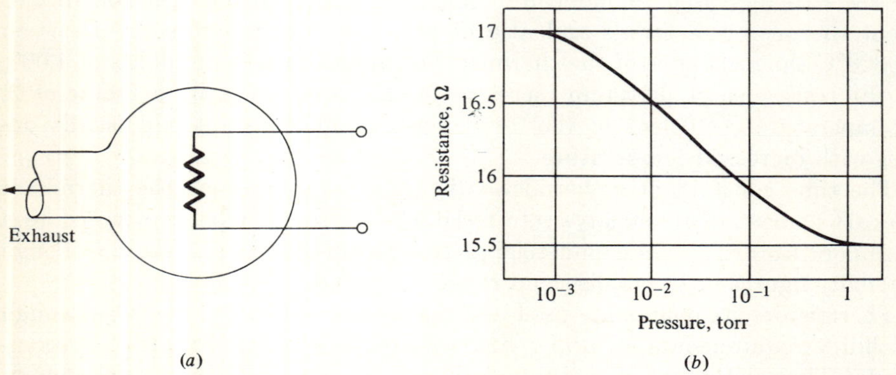

Fig. 4-7 Pirani gauge. (a) Vacuum meter; (b) calibration curve.

usually consists of a fine wire of about 0.001-in diameter mounted in a glass or a metal tube. The tube is joined to the vacuum system; the wire is heated to a temperature between 100 and 400°C; cooling takes place by thermoconduction through the gas surrounding the wire.

At atmospheric pressures the heat conductivity of gases is independent of their pressure. As the pressure is reduced, a point is reached at about 10 torr, at which the heat conductivity begins to decrease with decreasing pressure. The gas molecules striking the hot wire carry off energy to the walls of the vessel. The actual temperature of the wire is then controlled by the rate at which gas molecules collide with it. As the pressure is lowered, the wire will lose heat less rapidly because there are fewer molecules available for heat transfer, consequently, its temperature will increase and so will its resistance.

The Pirani gauge is commonly connected in a resistance bridge. The voltage or the current applied to the bridge is kept constant, and at a given pressure in the gauge the bridge is balanced. A variation of the pressure causes a bridge imbalance, i.e., the appearance of an output voltage. A typical calibration curve is shown in Fig. 4-7*b*. Since different gases have different heat conductivities, the calibration curve for each gas will be different. For some applications, notably in the absence of hydrogen and helium, this difference may be neglected. The Pirani gauge indicates the total pressure, i.e., the sum of all partial pressures of any gases or vapors present. The range of the gauge is generally from about 1 torr down to about 10^{-3} torr. The variations of resistance are in the order of about 10 percent for the full range. A practical gauge may vary from about 15.5 to 17 Ω. The response is in the order of several seconds to minutes. The gauge is, of course, not applicable in gas atmospheres where gases may react with the hot filament.

Since a system like that shown in Fig. 4-7*a* furnishes different calibrations depending on the type of gas present in the vessel, it can be used for gas analysis. For this purpose, the pressure is held constant (atmospheric pressure) and a gas mixture, usually a binary mixture, is admitted to the chamber. The rate of heat transfer from the hot filament to the walls is determined primarily by the thermal conductivity of the gas. Numerical values (relative values, air = 100) of the thermal conductivity of different gases at 100°C are shown in Table 4-4.

The thermal conductivity of gas mixtures does not follow a simple mixing rule. Therefore, empirical calibration of the system is required. The method can be made exceedingly sensitive so that not only small amounts of gases (less than one hundredth of 1 cc) can be measured but also small variations of the gas concentration. The method is primarily useful in gas chromatography as a detector for varying gas concentrations. A system for the comparison of a sample gas and a reference gas whereby four gas-sensing elements are arranged in a Wheatstone bridge arrangement is shown in Fig. 4-8.

Since the cooling of the wire depends on the flow velocity v of the surrounding material, among other parameters, the thermal system can also be used as a gas-flow velocity meter or as an anemometer. Figure 4-9 shows an example of a thermal gas-flow velocity transducer. It consists of a thin wire (such as nickel,

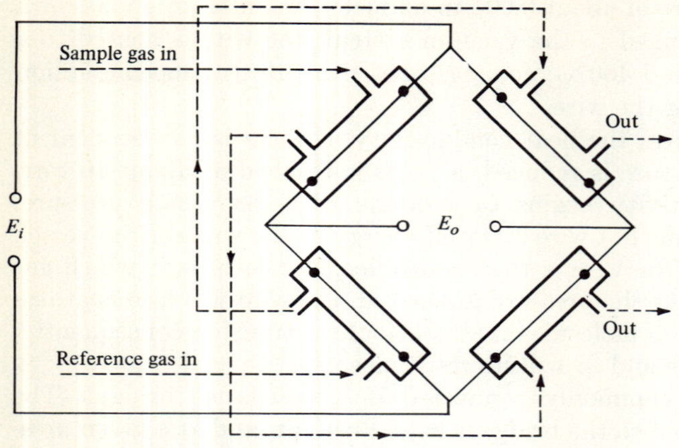

Fig. 4-8 Thermal conductivity cell for gas analysis.

platinum, or tungsten) held between two electrodes and exposed to the flowing gas. The rate of heat loss and, therefore, the resistance variation is very nearly proportional to $\sqrt{v}$. The heat loss also depends on the physical properties of the medium, the geometry, and the surface structure of the wire. The response of an element like that shown in Fig. 4-9 varies with the flow direction relative to the position of the wire; it is maximal if the wire is perpendicular to the flow direction. The directional sensitivity can be enhanced by the use of multiple-wire probes.

The range of the thermal flowmeter transducer extends from 5 cm/s to the supersonic range. The system can measure instantaneous flow rates and fast

Table 4-4 Relative thermal conductivity of gases (air = 100)

Material	*Thermal conductivity*	*Relative sensitivity* for 1% gas admixture*
Air	100.0	
Hydrogen	710.0	−7.19
Helium	553.0	−4.6
Nitrogen	99.6	0.007
Oxygen	101.4	−0.043
Carbon monoxide	96.2	0.1
Carbon dioxide	70.0	1.0
Water	77.5	0.48
Benzene	58.3	1.36
Acetone	55.7	1.35

* Output, in millivolts, of an instrument which furnishes an output of 1.0 for 1 percent admixture of CO_2.

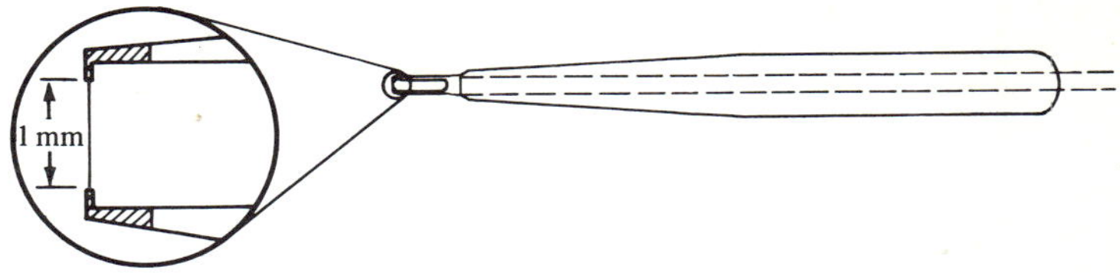

Fig. 4-9 Gas-flow velocity transducer (anemometer).

velocity fluctuations. Transducers of this type have been built with frequency responses exceeding 10,000 Hz.

An entirely different application of the resistance-temperature effect is shown in Fig. 4-10. Here the system is used as a bolometer for the measurement of radiation, in particular, infrared radiation. The bolometer consists essentially of a thin metal strip which absorbs the radiation to be measured and is heated thereby. The temperature increase causes a variation of the electric resistance of the strip. The transfer characteristics $\Delta R/R$ *versus* incident radiation of such bolometers are essentially linear. The sensitivity is proportional to the current passing through the bolometer, at least at low currents. The sensitivity can be increased by reduction of the heat capacity of the strip. Very thin strips such as those made of Wollaston foil (platinum) or evaporated on a nonconducting carrier are frequently used. The reduction of the thickness cannot be carried too far since the resistance-temperature coefficient of very thin metal layers is considerably less than that of metal in bulk form.

The resistivity of electronic conduction for many metals becomes zero at very low temperature (superconduction). In the transition zone from normal conduction to superconduction, the resistance-temperature coefficient attains very high values (order of magnitude several thousand percent per degree) so that bolometers operating in the transition range are very sensitive. Operation at low temperature offers other advantages as well, such as the reduction of noise or thermal reradiation. Superconducting bolometers have been built with tantalum wires operated at a temperature between 3.22 and 3.23K or with columbium nitride operated at 15K. The technical difficulties in operating such transducers are considerable.

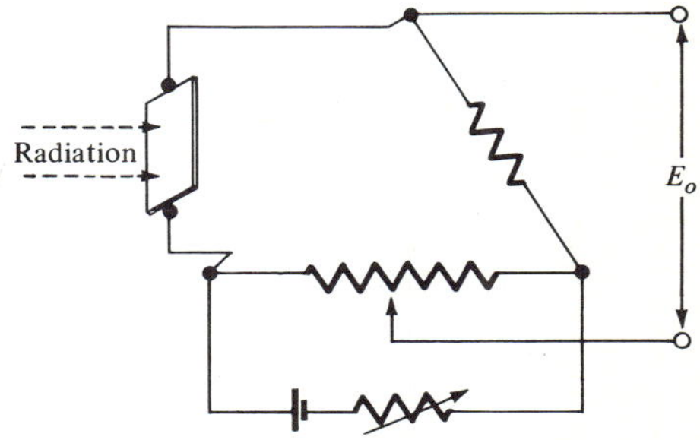

Fig. 4-10 Bolometer for radiation measurement.

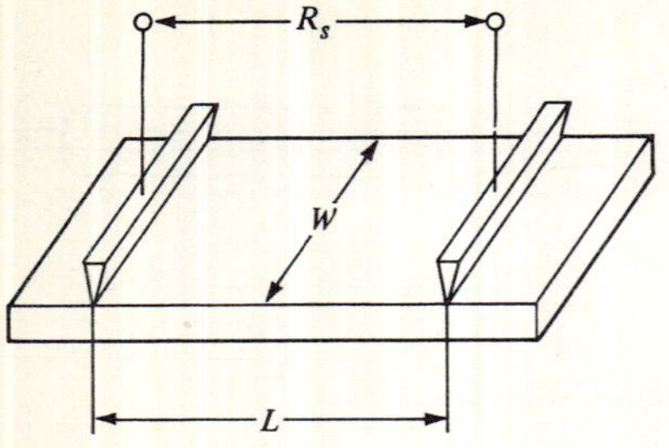

Fig. 4-11 Setup for the measurement of surface resistance.

INSULATION RESISTANCE: VOLUME RESISTANCE AND SURFACE RESISTANCE

The resistivity of a thin surface layer of dielectric materials is different from the volume resistivity, not only because of an adherent humidity layer but also because of contamination, chemical alterations, absorption of gases, or structural modification. The resistance R_t between two electrodes embedded in or attached to a dielectric material is composed of volume resistance R_v and surface resistance R_s:

$$\frac{1}{R_t} = \frac{1}{R_v} + \frac{1}{R_s}$$

The volume resistance R_v is determined by the volume resistivity and the length and cross-sectional area of the test sample. The surface resistance, in the case of a rectangular test sample with two electrodes as shown in Fig. 4-11, is

$$R_s = \sigma \frac{L}{w}$$

where σ is the surface resistivity, in ohms (sometimes, though incorrectly, called ohms per square).

The volume conduction can be measured separately from the surface conduction by means of the guard-ring arrangement shown in Fig. 4-12. The arrangement consists of a circular electrode A (guarded electrode, radius r_1) on one side of the sample, concentrically surrounded by an annular electrode B (guard electrode, inner radius r_2, outer r_3). On the other side of the sample is a circular electrode C (unguarded electrode, radius r_3). The guard ring prevents the surface

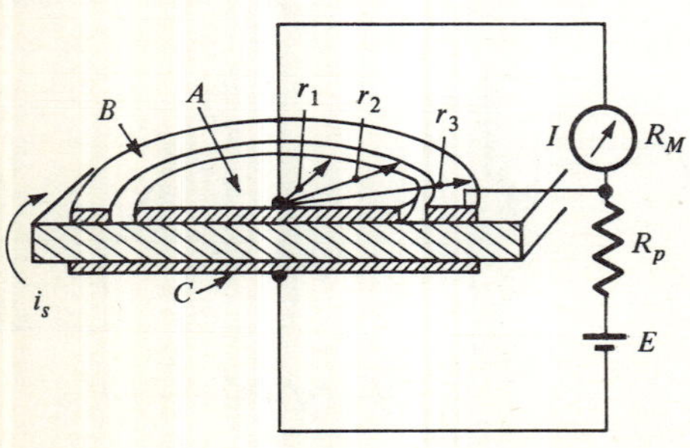

Fig. 4-12 Guard-ring arrangement.

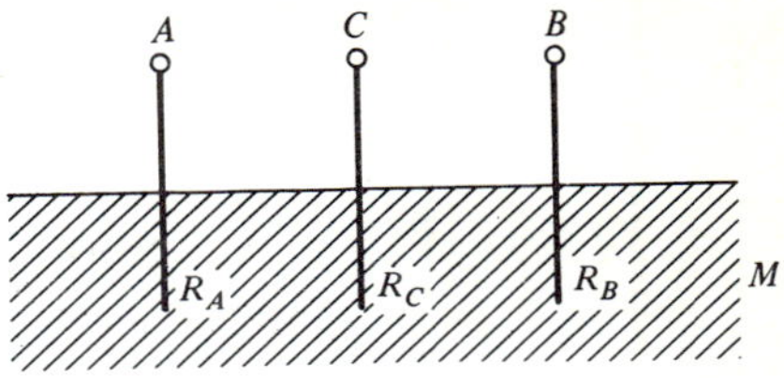

Fig. 4-13 Setup for the measurements of interface (contact) resistance.

current i_s from flowing through the meter. The arrangement is used not only for resistance measurement on dielectric samples but also for the elimination of the effect of surface conduction on photoelectric cells, ionization chambers, and similar elements.

CONTACT RESISTANCE

Figure 4-13 shows an electrode A inserted in a medium M; it may be an electrode driven into ground or applied to a body. Sometimes it is necessary to determine the contact resistance arising at the interface between the electrode and the medium. This may be done by measuring the resistance between A and a second electrode B, which gives the sum of both interface resistances R_A and R_B. If R_A is equal to R_B, then R_A can be found by dividing the measured resistance by 2. If R_A is not equal to R_B, the unknown resistor R_A can be found by the use of a third auxiliary electrode C, and by making the following measurements:

R_1 measured between A and $B = R_A + R_B$
R_2 measured between A and $C = R_A + R_C$
R_3 measured between B and $C = R_B + R_C$

Then

$$\frac{R_1 + R_2 - R_3}{2} = \frac{R_A + R_B + R_A + R_C - R_B - R_C}{2} = R_A$$

COMMERCIAL RESISTORS

A commercial resistor is characterized by two quantities: resistance value (in ohms) and maximum power rating (in watts). Precision resistors, which are sometimes combined in decade units, usually are designed for a maximum load of 0.5 W, preferably 0.1 W, per resistance step. The accuracy of the calibration depends on the resistance value and is, for resistances of 10 Ω and up, of the order of 0.1 percent and better. The power rating of radio resistors is between 0.5 and 2 W (0.1 W for subminiature types). The resistance value and the tolerance are expressed by a color code.

Every resistor exhibits capacitance and inductance which can be of disturbing influence if the resistor is used at alternating currents of higher frequencies. This effect is described in Chap. 7.

5

Electrostatic and Capacitive Elements

BASIC CONCEPTS

If two plates are d centimeters apart and are connected to a voltage source E, an electric field will exist in the space between the plates. The field strength in the central part between the plates is

$$F = \frac{E}{d} \tag{5-1}$$

It is characteristic of the electric field that it exercises a force on an electrically charged body or particle brought in the space between the plates. The magnitude of this force is

$$f = qF \tag{5-2}$$

where q is the electric charge of the body. Under the influence of the force f, the body will move along the force lines. The orthogonals perpendicular to the force lines are the equipotential lines or equipotential surfaces.

In Fig. 5-1, the two plates are connected to a current source S; a current I flows from the source to the plates. If the switch is closed for a time t, the

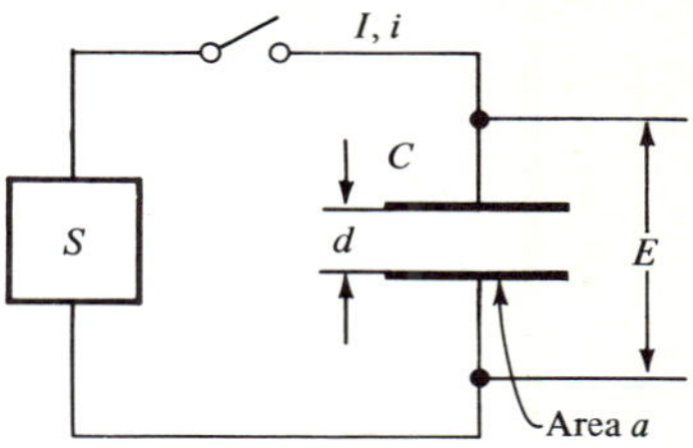

Fig. 5-1 Capacitor C charged from a constant-current source S. I, i, total and instantaneous currents; E, voltage across capacitor.

total amount of electricity, or charge, passing from the source to the plates is $Q = It$ ampere seconds or coulombs. In the more general case of a current i varying with time,

$$Q = \int_0^t i\,dt \tag{5-3}$$

The accumulation of charges at the plates A and B causes a potential difference or voltage E between A and B; the voltage E is proportional to the charge Q:

$$EC = Q \tag{5-4}$$

where C is the capacitance of the plate arrangement. The value of C depends on the geometry of the capacitor and the physical properties (dielectric constant) of the medium between the plates.

In the case of a capacitor with two plane-parallel plates (Fig. 5-1) of an area a, spaced by a distance d, the capacitance is

$$C = k\varepsilon \frac{a}{d}$$

or, numerically,

$$C_{\text{pF}} = 0.0885 \times \varepsilon \frac{a}{d} \tag{5-5}$$

(linear dimensions in centimeters).

The value of the [specific] dielectric constant ε can be found by making two measurements of capacitance: C_1, the capacitance between two plates when the space between the plates is filled with a dielectric material, and C_0, where the space between the plates is empty or filled with air (without changing the position of the plates with respect to each other). The dielectric constant is

$$\varepsilon = \frac{C_1}{C_0}$$

Values for the dielectric constants of different materials are found in Table 5-1.

Equation (5-5) is correct only for very small distances between the plates (d is less than any linear dimension of the plates). If d is not small, the capacitance is larger than that computed from Eq. (5-5) by the stray capacitance arising at the edges of the plates. The effect is illustrated in Fig. 5-2a to c.

Table 5-1 Dielectric constant of various materials

	ε
Argon (0°C, 1 atm)	1.000545
Hydrogen (0°C, 1 atm)	1.000264
Air (19°C, 1 atm)	1.0006
Air (19°C, 100 atm)	1.0548
Benzene (20°C)	2.284
Methanol (20°C)	33.62
H_2O (0°C)	88.00
H_2O (100°C)	55.33
Glass	4.0–6.0
Mica	6.9–9.2
Nylon	3.5
Porcelain	6.0–8.0
Titanium dioxide	14–110

A different form of an electric field is shown in Fig. 5-2*d*. The oscilloscope (or other instrument) S is assumed to be grounded as is the observer O. A wire (*aerial*) is connected to the instrument; its potential is different from ground. There is an electric field between aerial A and the grounded instrument S as well as between aerial A and observer O. Electric field lines begin and terminate at conducting surfaces and in close proximity to the conducting surfaces, they are perpendicular to these surfaces. This is in contrast to magnetic field lines which form closed loops.

If two capacitances C_1 and C_2 are connected in parallel (Fig. 5-3*a*), the equivalent capacitance C_p is equal to the sum of both capacitances

$$C_p = C_1 + C_2$$

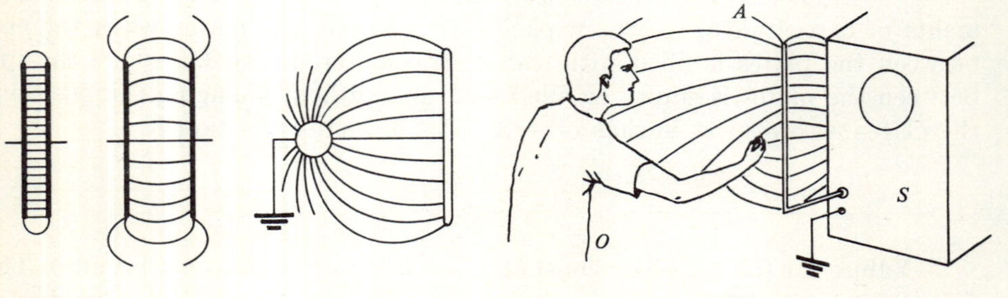

(*a*) (*b*) (*c*) (*d*)

Fig. 5-2 Illustrations of electric stray fields. (*a*) Two-plate capacitor, narrow plate distance; (*b*) plates moved apart; (*c*) electric field between a sphere and a plate; (*d*) electric field between a wire connected to the ungrounded terminal of a cathode-ray oscilloscope, the grounded scope, and an observer.

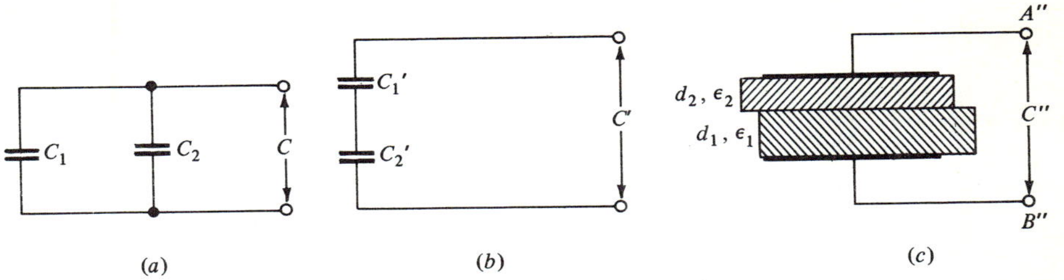

Fig. 5-3 (a) Capacitors C_1 and C_2 in parallel; (b) capacitor C_1' and C_2' in series; (c) capacitor with two dielectric layers of the thicknesses d_1 and d_2 and the dielectric constants ϵ_1 and ϵ_2.

If two capacitances C_1' and C_2' are connected in series (Fig. 5-3b), their equivalent capacitance C_s can be found from

$$\frac{1}{C_s} = \frac{1}{C_1'} + \frac{1}{C_2'}$$

$$C_s = \frac{C_1'C_2'}{C_1' + C_2'}$$

If a capacitor is made up of several layers of dielectric materials (see Fig. 5-3c), it is treated like a series combination of partial capacitors, each partial capacitor having a plate distance d equal to the thickness of one dielectric layer and a dielectric constant of that same layer. For example, the capacitance between terminals A'' and B'' of Fig. 5-3c can be found from

$$C'' = 0.0885 \frac{a}{d_1/\varepsilon_1 + d_2/\varepsilon_2}$$

If a dc voltage is applied to a series combination of two capacitors (see Fig. 5-3b), the voltage distribution is determined by the insulation resistance of each partial capacitor. It is dangerous to apply to such a combination a dc voltage which exceeds the safe operating voltage of any one of these partial capacitors. If the insulation resistance of the two partial capacitors is unequal, the larger partial voltage will be across the partial capacitor with the higher insulation resistance and the capacitor may break down. Thereupon, the entire voltage will be across the other capacitor which will then break down too.

CAPACITIVE FORCE OR DISPLACEMENT TRANSDUCERS

Equation (5-5) forms the basis of the capacitive force or displacement transducer. The capacitance can be varied in response to the physical displacement in three ways: (1) by varying d, i.e., by changing the plate separation; (2) by varying a, i.e., moving the plate in the direction y; and (3) by varying ε, e.g., by moving a body with a dielectric constant higher than air into the gap between the plates.

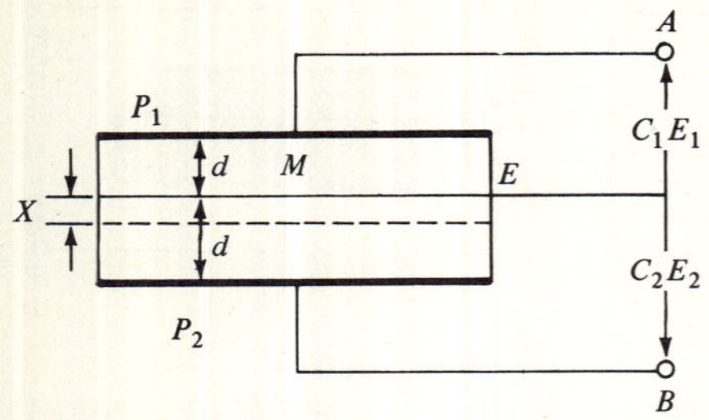

Fig. 5-4 Differential capacitor. P_1, P_2 fixed plates; M, movable plate. The distance between P_1 and P_2 is $2d$.

Change-of-distance systems, two-plate capacitor This is the most frequently used type of capacitive transducer. The response characteristic $C = f(d)$ is hyperbolic, and only approximately linear over a small range of displacement. The sensitivity with variable separation d is

$$S = \frac{\Delta C}{\Delta d} = -\frac{\varepsilon a}{d^2} \tag{5-6}$$

Theoretically, the sensitivity of the transducer can be increased to any desired degree by making the distance d increasingly smaller, but a practical limit is reached when the electric field strength in the air gap exceeds the breakdown limit. The breakdown limit in air at atmospheric pressure is about 30 kV/cm.

Frequently the capacitor is connected to a system that measures not the capacitance variation ΔC, but the fractional change of capacitance $\Delta C/C$. In this case, the sensitivity is

$$S' = \frac{\Delta C}{C\,\Delta d} = -\frac{1}{d} \tag{5-7}$$

i.e., the sensitivity S' increases in inverse proportion to the plate separation d, and is independent of the other dimensions of the capacitor. The response characteristic is linear.

Differential capacitor The capacitor shown in Fig. 5-4 has three plates: two fixed plates P_1 and P_2 and a movable plate M which together form the two capacitors C_1 and C_2. In the center position, the distance between M and either plate is d; the capacitors C_1 and C_2 are equal. If plate M is moved by a small amount x in response to an applied mechanical signal, one capacitor will be increased and the other one decreased. The subsequent stage (which converts a change of capacitance into an output voltage or current) measures either the difference between both capacitors or their ratio.

Change-of-area systems Different capacitors of this type for longitudinal and for rotational movements are shown in Fig. 5-5a and b. Displacement of one plate in the direction of the arrow y causes a variation of the overlapping area a, and thus of the capacitance. The capacitance of the arrangements shown in these figures changes linearly with the displacement, but the characteristic can

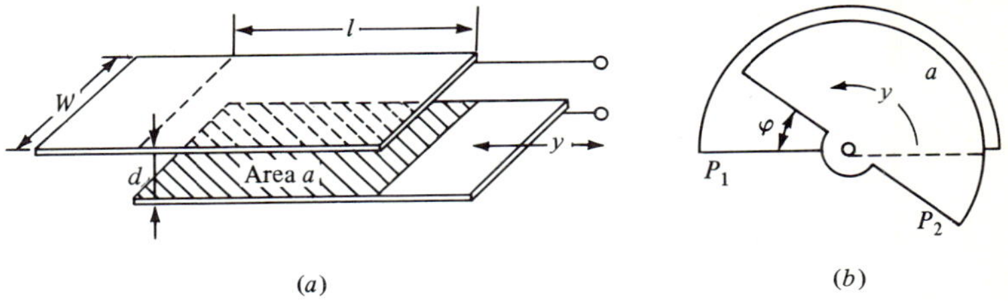

Fig. 5-5 (*a*) Schematic diagram of a capacitive displacement transducer, (*b*) variable capacitor with circular plates as a transducer for rotary movement.

be modified by appropriate shaping of one of the two plates. The change of capacitance is proportional to the width of the plates in Fig. 5-5*a* or to the radius of the plates in Fig. 5-5*b*. The sensitivities $\Delta C/\Delta l$ in Fig. 5-5*a* or $\Delta C/\Delta\varphi$ in Fig. 5-5*b* can be increased by multiple-plate arrangements, but not the fractional sensitivities $\Delta C/C\Delta l$ or $\Delta C/C\Delta\varphi$.

An interesting form of a change-of-area transducer for linear displacement is shown in Fig. 5-6. The serrated transducer has a minimum of capacitance if a tooth of plate P_1 is opposite a hole in plate P_2. Shifting the relative positions of both plates by one tooth width causes a variation (approximately linear) to a maximum of capacitance. A sensitivity of 1 pF/0.0001 in has been obtained with this transducer.

Capacitive transducers have been used for the measurement of extremely small displacements down to the order of molecular dimensions (10^{-9} cm, 4×10^{-10} in). On the other hand, they have been proposed for the measurement of large distances up to about 100 ft in airplane altimeters. The change-of-distance method is generally preferable for the small and the very large ranges; the change-of-area method is used for displacement of the order of 1 to 10 cm.

The force requirements of capacitive transducers can be extremely small. The force between two capacitor plates having an area of 2 cm² which are 1 mm apart and are connected to a potential difference of 100 V is of the order of 1 dyn. The transducer has the principal advantage that the physical mechanism involved in its action does not depend on the physical property of any plate material. Therefore, a high degree of stability can be obtained. A disadvantage in the capacitive transducer is its relatively large output impedance which requires careful shielding and short connections to subsequent stages.

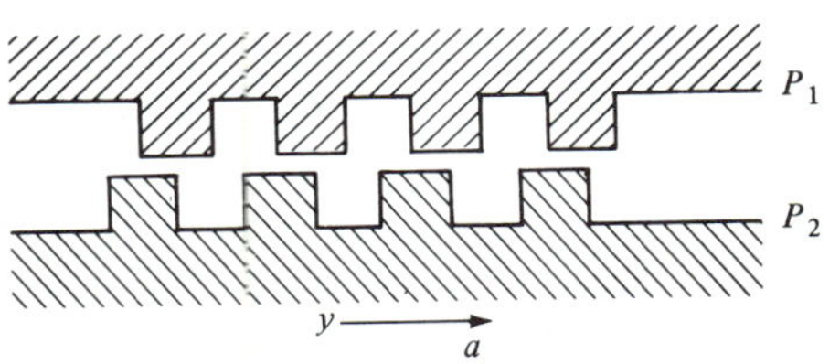

Fig. 5-6 Serrated capacitor; see text for explanation.

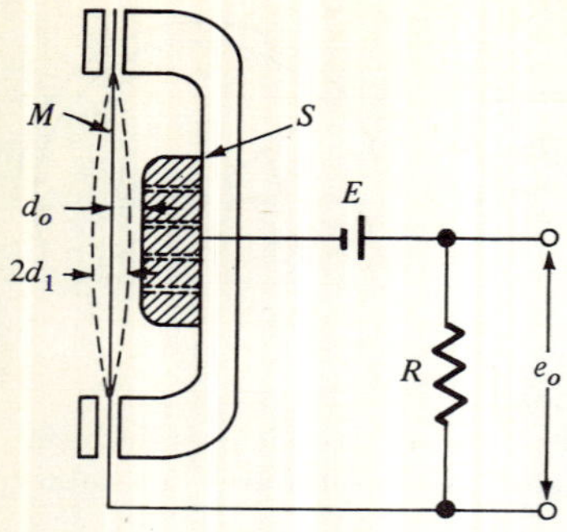

Fig. 5-7 Schematic diagram of capacitance microphone. M, membrane; S, fixed electrode; d_0, distance between M and S if membrane is not moving; $2d$, peak-to-peak movement of membrane.

An example of a capacitive transducer, the condenser microphone, is shown schematically in Fig. 5-7. One capacitor plate has the form of a thin metallic or metallized membrane M and is driven back and forth by an acoustic pressure wave; the other is a solid plate S arranged at a short distance from the membrane. The air space in the capsule and in holes bored in the plate S provides suitable damping of the membrane.

While the membrane is at rest, its distance from the plate is d_0, and the capacitance between M and S is C_0. The membrane oscillates with the frequency f (angular velocity $\omega = 2\pi f$) and with an amplitude d_1, which is small ($d_1 < d_0$), and causes a periodic change of capacitance. The instantaneous value of the capacitance is

$$C = C_0\left(1 - \frac{d_1}{d_0}\sin\omega t\right) \tag{5-8}$$

The condenser microphone is connected in series with a dc source E of about 100 V and a large resistor R of the order of $10^7\ \Omega$. The charge of the capacitor is

$$Q = e_cC = (E - IR)C_0\left(1 - \frac{d_1}{d_0}\sin\omega t\right) \tag{5-9}$$

Differentiation with respect to time of Eq. (5-9) furnishes

$$I = \frac{dQ}{dt} = (E - IR)C_0\left(-\frac{d_1}{d_0}\omega\cos\omega t\right) + C_0\left(1 - \frac{d_1}{d_0}\sin\omega t\right)\left(-R\frac{dI}{dt}\right)$$

The solution of this differential equation is, in good approximation,

$$e_o = ER\omega C_0\frac{d_1}{d_0}\sin(\omega t)$$

where e_o is the instantaneous output voltage appearing between the resistor terminals.

Change of dielectric constant An example of a capacitive transducer based on the variation of a dielectric constant is shown in Fig. 5-8. A cylindrical capacitor

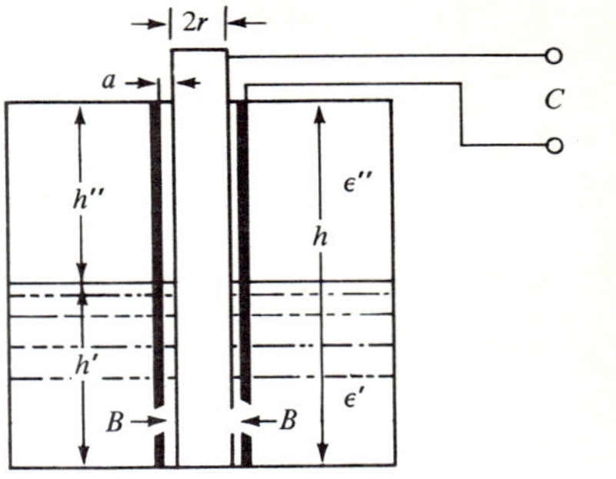

Fig. 5-8 Capacitive liquid-level transducer; see text for explanation.

is immersed in a nonconducting liquid (oil, gasoline); variation of the liquid level causes a change of capacitance. The capacitance for $h > r > a$ is, in good approximation,

$$C = \frac{(h'\varepsilon') + (h''\varepsilon'')}{2\ln\left(1 - \frac{a}{r}\right)}$$

where h' = height of the liquid level
h'' = length of the cylinder above the liquid surface
ε' = dielectric constant of the liquid
ε'' = dielectric constant of the gas above the liquid (usually $\sim$1)
r = outside radius of the inner cylinder
a = distance between inner and outer cylinder

Capacitive methods can also be used for chemical analysis of binary mixtures of nonconducting fluids where both components differ in their dielectric constant. The dielectric constant of the mixture is measured, and it varies with the relative concentration of both components.

Capacitive systems can also be used as pressure transducers in all those cases where the dielectric constant of a medium changes with pressure. For instance, the dielectric constant of benzene changes by 0.5 percent in the range from 1 to 100 atm, that of air at 19°C changes in the same pressure range from 1.0006 to 1.0548. Variation with pressure of the dielectric constant also occurs in solids, primarily in crystals such as MgO, LiF, NaCl, KCl, and KBr.

The dielectric constant of gases changes with a variation of the humidity. The variation is small but measurable. For instance, the dielectric constant of dry air at 45°C is 1.000247, that of air at the same temperature saturated with water is 1.000593. One of the most accurate methods for the determination of moisture in gases is based on this effect; the measurements are carried out by a microwave technique at a frequency of about 10^{10} Hz.

Capacitive transducers are frequently combined with mechanical modifiers for the measurement of volume, density, liquid level, weight, etc. Figure 5-9 shows a differential pressure transducer for measurement in gases. The capsule is symmetrically built, each chamber contains a fixed plate, while the flexible

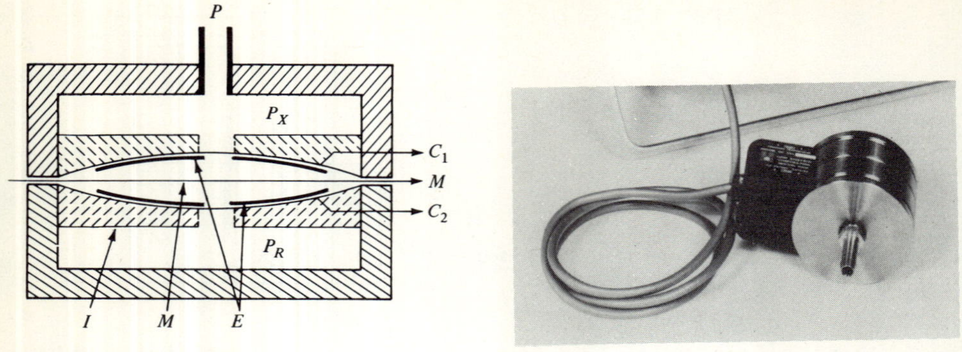

(a) (b)

Fig. 5-9 Capacitive pressure transducer. (*a*) Cross section (schematic diagram showing the membrane M and the two electrodes E which form a differential capacitor with partial capacitances C_1 and C_2. I, insulators; P, pressure port; P_X pressure chamber (for unknown pressure); P_R, reference pressure. (*b*) Practical embodiment of a pressure transducer. (*Lion Precision Corporation, Newton, Mass.*)

membrane M acts as the third, movable plate of a differential capacitor. The system can be used for single-ended input or for differential pressure by connecting the two pressure parts of the transducer to the two pressure systems to be compared. Of course, sequential pressure transducers can be built with other modifiers, such as bellows or Bourdon tubes and with other displacement transducers, such as resistive transducers, strain gauges, or inductive transducers (see Fig. 1-2).

CHARGE AND DISCHARGE OF CAPACITANCES

In Fig. 5-10*a*, a capacitor C in series with a resistance R is connected to a constant-voltage source E. The capacitor may be uncharged initially. If switch S is closed, a current i will flow and will charge the capacitor. The voltage across resistor R is $e_r = Ri$, that across the capacitor is $e_c = 1/C\int i\,dt$. The sum of these voltages must be equal to the applied voltage E

$$iR + \frac{1}{C}\int i\,dt = E \tag{5-10}$$

Since $i = C(de_c/dt)$, Eq. (5-10) can be written

$$RC\frac{de_c}{dt} + e_c = E \tag{5-11}$$

The solution of this differential equation furnishes for the voltage across the capacitor the value

$$e_c = E(1 - \varepsilon^{-t/RC}) \tag{5-12}$$

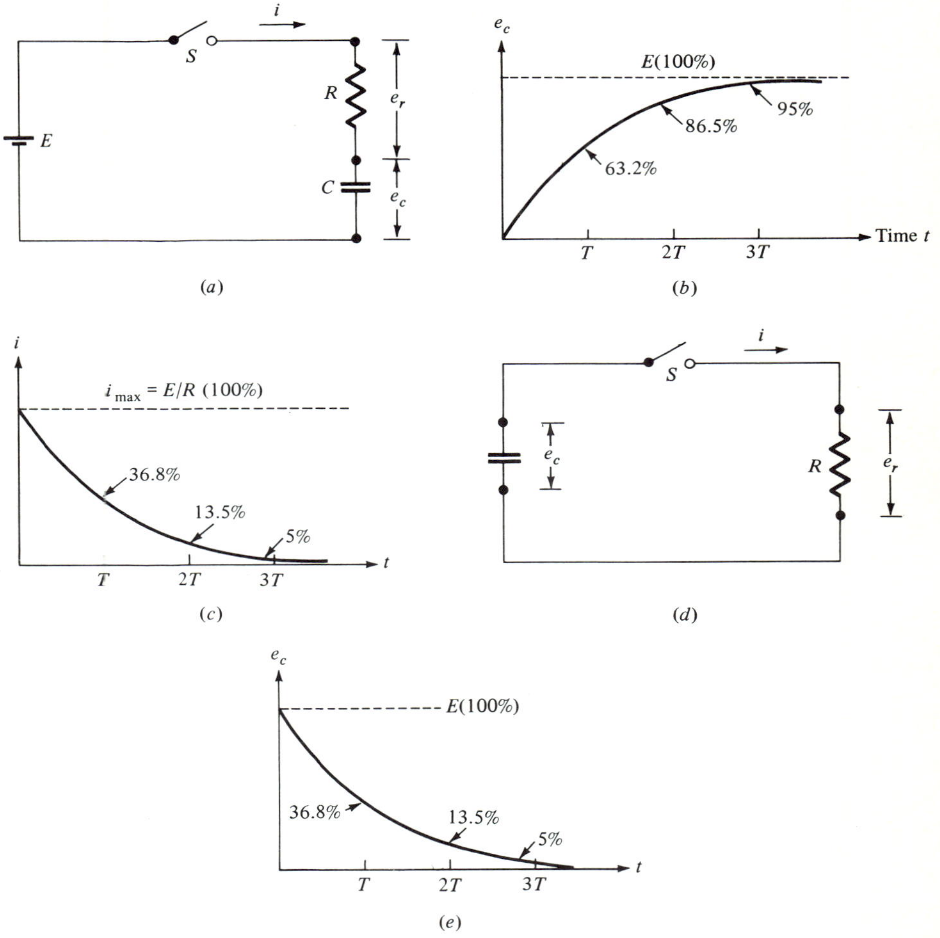

Fig. 5-10 Charge and discharge characteristics of an RC circuit; after switch S is closed the voltage e_c varies with time as shown in (b) and (e); the current i varies as shown in (c).

while the current in the circuit is

$$i = \frac{E}{R} \varepsilon^{-1/RC}$$

i.e., the voltage across the capacitor rises exponentially with time and reaches asymptotically the value of E. The current at $t = 0$ has the value $I_o = E/R$ and decays exponentially with time.

The rate at which the voltage rises or the current decays depends on the product $RC = T$, the time constant of the circuit. T is expressed in seconds (ohms $\times$ farad or megohms $\times$ microfarad) and signifies that time in which the

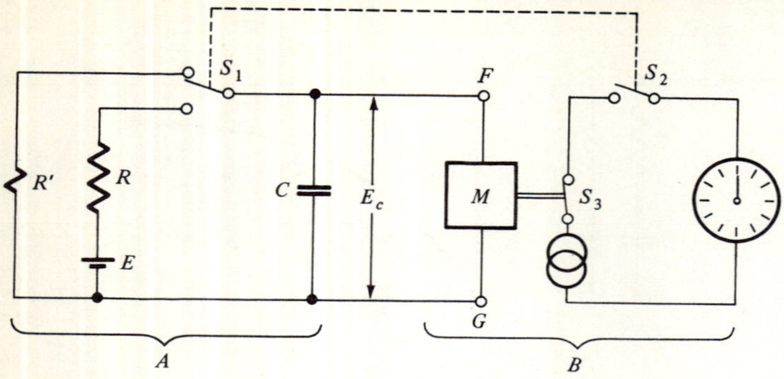

Fig. 5-11 Simple time switch, circuit diagram; see text for explanation.

voltage reaches the value $E(1 - 1/e)$ or 63.2 percent of E, or the time in which the current decays to $I_o \times 1/e$ or 36.8 percent of I_o.

The discharge process is illustrated in Fig. 5-10*d*. A capacitor C initially charged to a voltage E is connected to a resistance R. If the switch is closed, a current will flow in the circuit and the capacitor will get discharged. The situation can be expressed, analogous to Eq. (5-11), by

$$RC \frac{de_c}{dt} + e_c = 0 \tag{5-13}$$

The solution of this differential equation furnishes for the capacitor voltage the value

$$e_c = Ee^{-t/RC} \tag{5-14}$$

and for the current

$$i = \frac{E}{R} e^{-t/RC} \tag{5-15}$$

Both voltage and current decay exponentially with time as illustrated in Fig. 5-10*c* and *e*.

The fact that voltages and currents in resistance-capacitance networks vary with time in a predictable manner makes these circuits suitable for a number of applications such as time switches, time- or frequency-measuring systems, oscillators, and others.

An example of a time switch is shown in Fig. 5-11. It consists of the time-determining network A containing elements R and C, and a voltage-controlled system B, such as an electronic relay system which starts operating as soon as the voltage applied to its terminals reaches or exceeds a predetermined level. The timing event is initiated by closing simultaneously switches S_1 and S_2. The relay contacts S_3 are normally closed (NC). The voltage across the capacitor C rises according to Eq. (5-12) or Fig. 5-10*b*, provided the resistance

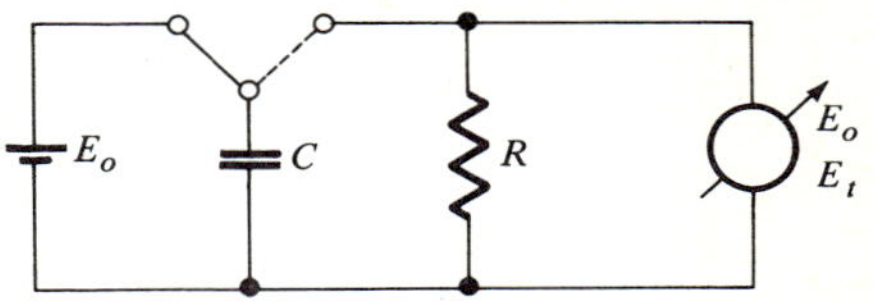

Fig. 5-12 *RC* circuit for the measurement of large resistors or capacitors or very small currents.

between terminals F and G is sufficiently high so that the voltage E_c across the capacitor C is determined only by the circuit elements in the network A. As soon as the voltage at the relay M reaches or exceeds the set operating value, the relay-operated switch will open the contacts S_3. The switching process is then terminated and the timer comes to a stop. Opening switches S_1 and S_2 causes a (fast) discharge of C over the small resistor R' and makes the timer ready for a new operating cycle.

The range of time over which such a timing switch can be operated is limited primarily by the inertia of the relay. With fast electronic systems, switching times in the nanosecond region can be obtained. On the other end, switching times in the range of one to several minutes can be effected.

The time constant is the product RC. It is unimportant whether R is large and C small or vice versa. The only practical limitation for the selection of R and C is that capacitors generally exhibit a leakage resistance of the order of 10^7 to 10^{12} Ω between their terminals. The series resistance R should be small compared to the leakage resistance or to any other resistance in parallel to the capacitor.

The charge or discharge characteristic of capacitors can be used for the measurement of very large resistances or very small currents. For this purpose, a capacitor C of known value is first charged to a voltage E_o as shown in Fig. 5-12. The capacitor is then disconnected from the voltage source, and connected to the unknown resistor R. After a measured time interval t, the voltage on the capacitor has diminished to a value E_t. This voltage is measured electrometrically, i.e., without drawing current from the capacitor. The resistance can then be found from

$$R = \frac{t}{C \ln (E_o/E_t)}$$

If t^* is the time in which the voltage E_t has fallen to one-half of E_o, the resistance is

$$R = 1.422 \times \frac{t^*}{C}$$

The average current during the time t is

$$I_{\text{av}} = \frac{Q}{t} = \frac{C(E_o - E_t)}{t}$$

The method permits, by extending the measurement over a sufficiently long time, measurement of currents of the order of 10^{-19} A, i.e., about 1 electron/s.

DIFFERENTIATING *RC* NETWORKS

If a variable voltage e_i is applied to a capacitor C, as illustrated in Fig. 5-13*a*, the current in the circuit is

$$i = C\frac{de_i}{dt} \tag{5-16}$$

If the circuit is modified according to Fig. 5-13*b* by the insertion of a resistor R which is so small that the current in the circuit is essentially still that expressed by Eq. (5-16), the voltage drop across the resistor will be

$$e_o = iR = RC\frac{de_i}{dt} \tag{5-17}$$

i.e., the output voltage e_o is proportional to the time derivative of the input voltage e_i.

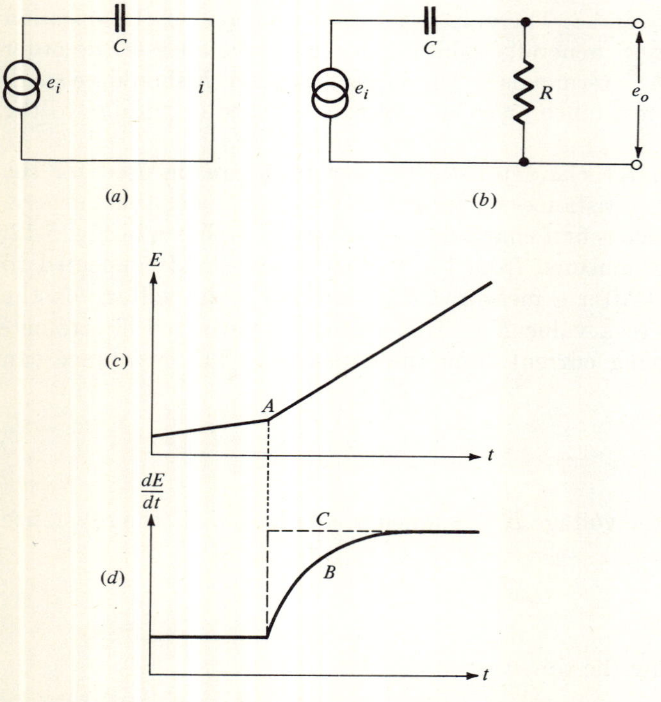

Fig. 5-13 Differentiating circuits. (*a*) The current i. (*b*) The voltage e_o is proportional to the time derivative of the input signal e_i. (*c*) Voltage E varying with time; the slope changes abruptly at A. (*d*) Differentiated signal; B, output from an imperfect differentiator; C, correct output.

Equation (5-17) is only correct if R is small; the voltage across R is then small compared to the input voltage e_i or compared to that across the capacitor e_c, i.e., the circuit acts correctly as a differentiating element only if

$$e_o \ll e_i \qquad \text{or} \qquad e_o \ll e_c \tag{5-18}$$

If the input voltage e_i is a sinusoidal signal of the angular frequency $\omega = 2\pi f$, then the conditions (5-18) are fulfilled if

$$R \ll \frac{1}{\omega C} \qquad \text{or} \qquad R\omega C \ll 1$$

If the above conditions are not fulfilled, an error will arise in the differentiator output which is, in the first approximation,

$$\varepsilon = RC \frac{de_o}{dt}$$

or approximately

$$R^2C^2 \frac{d^2e_i}{dt^2} \tag{5-19}$$

The error is particularly large for transient signals when the last term of Eq. (5-19), the second derivative of the input voltage, assumes large values. This situation is illustrated in Fig. 5-13*c*. The slope of the voltage-time function changes abruptly at point A from the value de_1/dt to de_2/dt. A perfect differentiator would furnish an output like that indicated by the dotted line in Fig. 5-13*d*. The practical RC differentiator furnishes an output as indicated by the solid line, i.e., it approximates the theoretical curve in an exponential function with the time constant RC. The smaller RC is, the smaller the error is, but the smaller the output from the differentiator is too.

For correct differentiation, a capacitor of negligible leakage and negligible dielectric aftereffect (see below) should be used. Capacitors with polystyrene dielectrics are recommended.

INTEGRATING *RC* NETWORKS

If a capacitor C is connected to a current source as shown in Fig. 5-14*a*, a charge Q will accumulate in the capacitor, and the voltage e_o between its terminals will rise as noted in the introduction of this chapter.

$$Q = e_oC = \int i\,dt$$

$$e_o = \frac{1}{C}\int i\,dt$$

If a large resistance R is inserted in the circuit (Fig. 5-14*b*), the current i in the circuit will essentially be e_i/R, and the output voltage will be

$$e_o = \frac{1}{RC}\int e_i\,dt \tag{5-20}$$

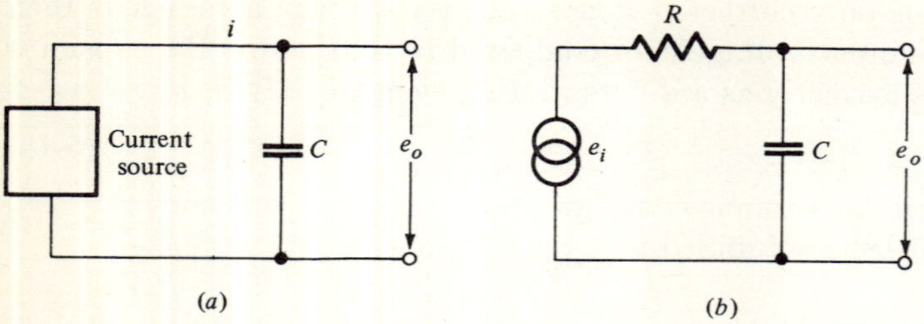

Fig. 5-14 Integrating circuit. (*a*) Capacitive integrator driven by a constant-current source; (*b*) *RC* integrator driven by a constant-voltage source e_i.

i.e., the output voltage will be proportional to the time integral of the input voltage.

Equation (5-20) is correct only if the voltage e_o across the capacitor is small compared to the applied voltage e_i or compared to the voltage e_r across the resistance, i.e.,

$$e_o \ll e_i \qquad \text{or} \qquad e_o \ll e_r$$

or for a sinusoidal input signal of the angular frequency ω if

$$\frac{1}{\omega C} \ll R \qquad \text{or} \qquad R\omega C \gg 1$$

If the above conditions are not fulfilled, an error will arise in the output which is

$$\varepsilon = \frac{1}{RC}\int_0^t e_o\,dt\,\frac{1}{R^2C^2}\int_0^t\int_0^t e_i\,dt\,dt$$

The error increases with the duration of the integration. The larger the time constant *RC* of the integrating network, the smaller is the error, but the smaller also is the output from the integrator. Where integration has to be carried out over long periods of time, it is advisable to use the *RC* integrator for short periods, discharge the capacitor, and charge again repeatedly. The total integral is computed from the number of charge cycles. Correct integrators, like differentiators, require capacitors that are free from leakage and dielectric aftereffect.

COMMERCIAL CAPACITORS

A commercial capacitor is characterized by three quantities: (1) its capacitance value in farads; (2) by the voltage rating, i.e., the maximum voltage that can safely be applied to the capacitor terminals; and (3) by the amount of power absorption or dissipation in the capacitor, a quantity usually expressed by the ratio of resistance to capacitance (*dissipation factor*) or the reciprocal value (*Q value or storage factor*), or by a function (cos, tan) of these quantities. The capacitance, the voltage rating, and the loss of a capacitor depend on the dimensions and the type of the dielectric.

Commercial capacitors are built either as variable capacitors, i.e., with adjustable capacitance, or as fixed capacitors. A wide variety of dielectric materials is being used, such as gases, inorganic materials, impregnated paper, plastic, oxide films, etc.

Vacuum and gas-filled capacitors are designed for high voltage and high frequency. They are unaffected by variations of temperature, moisture, and humidity; they are usually rugged and hold calibrations accurately. Their power dissipation is low; their dielectric is self-healing, i.e., voltage breakdown does not damage them permanently. Vacuum capacitors are manufactured with capacitance values from 50 to about 500 pF and for maximum working voltages of the order of 10 kV.

Capacitors filled with gases at high pressure (about 150 lb/in^2 of air or nitrogen) are used mainly for high voltages of the order of 100 kV; their capacitance is ordinarily of the order of 100 pF.

Capacitors with mica dielectric have very low dielectric losses and are thermally very stable (temperature coefficient as low as 20 ppm/C). They are commercially available in capacitance values from 1 pF to 0.1 μF with voltage ratings of 100 to 5,000 V, and they can be used for both direct and alternating current and for high frequency. Being stable, they are suitable for use as standards.

The most frequently used paper capacitor is made either from metal foil with a dielectric of paper that is impregnated with oil, wax, or synthetic compounds or from metallized paper. Long strips of the materials are rolled together, a procedure which causes a certain amount of inductance, or they are folded (inductance-free mounting). Capacitors of this type have been built with capacitances from 0.001 μF to several microfarads and with voltage ratings from one hundred to several thousand volts. The voltage rating of these capacitors is smaller for ac than for dc application, and again smaller for higher frequencies than for line frequency. It also varies with the duration of the applied voltage and with the operating temperature. Paper capacitors have, in general, a limited lifetime. They are used primarily for dc applications in power supplies and filters, and the smaller capacitance values for audio frequencies.

Capacitors with plastic dielectrics (e.g., polystyrene) have high dc resistance. Their dielectric losses are low even at frequencies as high as 1 Mc. Their temperature coefficient is between -100 and -200 ppm/C; their maximum operating temperature is 85°C. The polystyrene capacitor is particularly suitable for differentiating and integrating circuits because of the absence of dielectric aftereffect.

Dielectric aftereffect (also soaking or residue effect) arises if a capacitor is charged with direct current for some time. Under the influence of the electric field applied to the dielectric layer, charges within this layer will move in both directions, forming in the dielectric a positive space charge at the negative electrode and a negative space charge at the positive electrode. When the electric field is turned off and the capacitor is discharged, these charges move back and cause appearance of a recharge of the capacitor. The effect can be dangerous at high voltages. Capacitors used at high-voltage direct current should be permanently shorted when not in use.

Capacitors with different ceramic dielectrics (usually titanium dioxide) have physical properties suitable for special applications. One type, with a dielectric constant between 6 and 15, has low dielectric losses even at frequencies as high as 10^{10} Hz. Another type, with a dielectric constant of about 90, can be made with a wide range of temperature coefficients, from $+200$ to -800 ppm/C. A third type, with a dielectric constant between 1,000 and 5,000, can be used for capacitors of small size. Their physical properties vary widely with the operating conditions.

Electrolytic capacitors are formed from two foil electrodes, usually aluminum or tantalum that are etched to increase the surface area. Between the electrodes is a highly viscous electrolyte: a thin insulating layer is formed on one electrode by electrolytic action. Because of the thinness of this layer, very high capacitance values can be obtained; commercial electrolytic capacitors are made with capacitances up to several millifarads.

Electrolytic capacitors are used for the application of direct current and are polarized. Application of even a small voltage in the wrong direction is likely to damage the capacitor. The leakage current is considerable, about 10^{-4} A/μF, and increases considerably at voltages near the maximum voltage rating.

REFERENCE

Foldvari, Tibor L., and Kurt S. Lion: Capacitive Transducers, *Instrument and Control Systems*, **37**(11):77–85 (1964).

6
Magnetic and Inductive Elements

BASIC CONCEPTS

The space between the two poles of a magnet is pervaded by a magnetic field; the field exercises on a (hypothetical) single pole of strength **m** force **f** = **mH**. The quantity **H** is the *magnetic field strength*. It is a vector measured in ampere-turn/meter or in oersted whereby 1 A/m = 1.26×10^{-2} Oe. Magnetic fields have been measured from 10^{-6} Oe to the highest produced magnetic fields of the order of 10^{6} Oe. The magnetic field of the earth is of the order of 0.5 Oe; the magnetic south pole is at the terrestrial north pole. Gravitational or electric fields may be superimposed on a magnetic field.

The effect of a magnetic field on a magnetic dipole can be found from the summation of the effects on both poles. On a dipole of moment **M** = **ml** (where **l** is the distance between both poles), the field **H** exercises a torque

$$\mathbf{T} = \mathbf{MH} \sin (\mathbf{M},\mathbf{H})$$

where (**M**,**H**) is the angle between the dipole direction and the field vector **H**. A translating force is never exercised on a dipole in a homogeneous field, only in an inhomogeneous field. In such a field the translating force is proportional to the field strength and to the field gradient, i.e.,

$$\mathbf{f}' = \mathbf{MH}\,\frac{d\mathbf{H}}{\mathbf{dx}}$$

Magnetic fields can be produced by electric currents. A current I through a conductor element $d\mathbf{l}$ furnishes, at a point in a distance r, to the magnetic field strength the contribution

$$d\mathbf{H} = \mathbf{I}\,d\mathbf{l}\,\sin\,(I,r)/r^2$$

in a direction perpendicular to the plane formed by r and dl. The combined effect of any extended current path on the magnetic field strength at a point P is found by the integration over the effects of each path element. For instance, the field in the center of a circular loop of radius r, produced by a current I, is

$$H = 5I/r$$

In a coil with n turns having a length l (with $l > r$), the field strength is

$$\mathbf{H} = nI/l$$

The field is quite homogeneous in the central part of such a coil, so that such coils can be used for the calibration of magnetic transducers. A magnetic field strength **H** of n amperes per meter causes in a vacuum a magnetic induction or field density B of $n\mu_0$ volt-seconds per square meter. The universal constant (*induction constant*) μ_0 has the value 1.256×10^{-6} V·s/A·m. Under static conditions, magnetic field strengths can be measured from their mechanical effect on magnetic dipoles (*magnetometer*), provided the dimensions of the dipoles are small compared to those of the field. Either the (small) deflection of the magnetometer is used as a measure of the field, or the deflection caused by the field is compensated, e.g., by torsional resetting of the original magnetometer position and measuring the torque applied.

Magnetic field strengths can also be measured by the force exercised by the magnetic field on moving charges in an evacuated tube containing an electron gun energized by a power supply. The gun emits an electron beam which is focused so that equal streams of electrons arrive at the two semicircular electrodes. Both electron currents are amplified and their difference formed by a differential amplifier is measured by a meter. Under symmetrical conditions the output is zero. A magnetic field perpendicular to the drawing surface causes a deflection of the electron beam of magnitude

$$y = \frac{K'Bl^2}{\sqrt{E}}$$

where K' = an apparatus constant (order of magnitude 0.1)
B = unknown field density
l = length of the beam (about 1 m)
E = voltage accelerating the electron beam (about 1 kV)

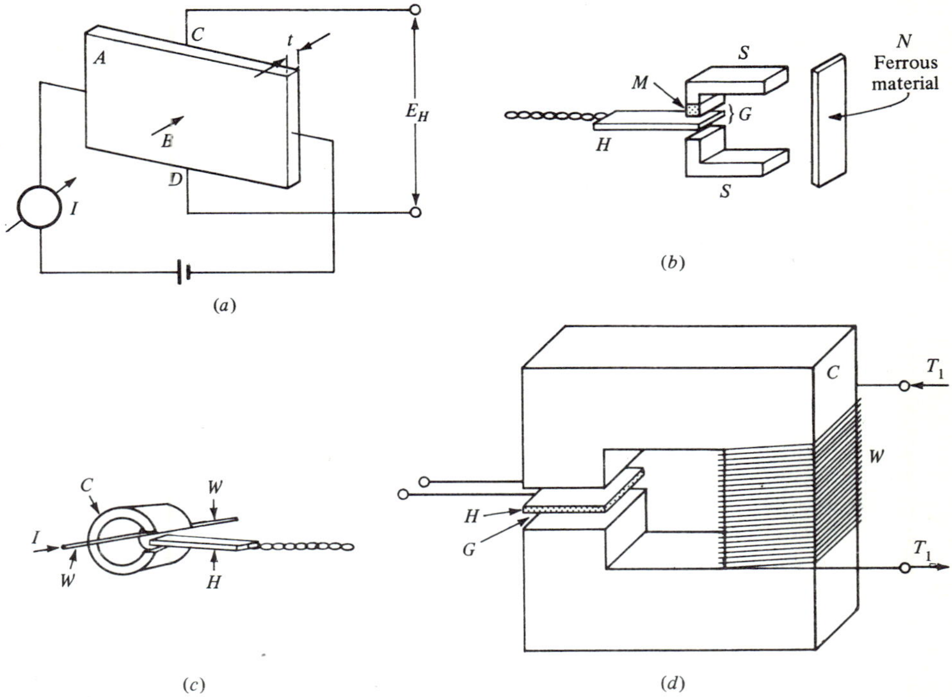

Fig. 6-1 (*a*) Hall-effect transducer; used for the measurement of magnetic fields. (*b*) Application of a Hall-effect transducer *H* for the measurement of the position or the displacement of a ferromagnetic probe *N*. *S*, magnetic structure; *M*, permanent magnet. (*c*) Hall-effect transducer *H* used for the measurement of current *I* in a conductor *W*. *C*, ferromagnetic concentrator; see text for description. (*d*) Hall-effect transducer used as an analog multiplying element.

The sensitivity of this system is extraordinarily high, probably the highest of all magnetic-field-measuring methods. Magnetic field strengths of less than 10^{-5} Oe have been measured with this system, with a signal-to-noise ratio of about 10:1. The experimental difficulties are great.

Also based on the effect of magnetic fields on moving charges is the Hall-effect transducer illustrated in Fig. 6-1*a*. A current *I* passes lengthwise through a plate *A*, made from a semiconducting material such as indium arsenide, indium antimonide, or *n*-type germanium. Two conductors are connected at contacts *C* and *D*. The contacts are adjusted so that there is no voltage between them in the absence of a magnetic field. If a magnetic field *B* traverses plate *A* in a direction perpendicular to the plate, it will tend to deflect the current carriers so that a voltage E_H, the Hall voltage, will arise between *C* and *D*.

$$E_H = K_H \frac{IB}{t} \tag{6-1}$$

where t is the thickness of the semiconductor plate and K_H is the Hall coefficient, a quantity which depends on the material of the semiconductors used (such as germanium, indium arsenide, etc.). The Hall-effect transducer has found a number of applications in the field of instrumentation.

As a magnetic-to-electric transducer, the semiconductor plate is inserted into the magnetic field to be measured; the magnetic force lines are perpendicular to the semiconducting plate. The transducer furnishes an output which is proportional to the field-line density B. The system has the advantage of requiring little space in the field direction, so that it can be inserted in narrow gaps, and it furnishes a continuous electric signal in direct response to the magnetic field strength. Its disadvantages are a high sensitivity to temperature variation and the fact that the Hall coefficient may vary from plate to plate so that each plate needs individual calibration. Also, the Hall coefficient is not strictly constant but varies with the magnetic field strength. The transducer is best suited for monitoring field strengths of the order of 10 to 10^4 Oe.

The Hall-effect transducer can also be used for the measurement of the location or displacement of a structural element, i.e., it can serve as an indirect-acting position displacement or proximity transducer in all cases where a change of the geometry of a magnetic structure causes a change of magnetic field strength. An example is shown in Fig. 6-1*b*. The transducer consists of a ferromagnetic structure S incorporating a permanent magnet M. A Hall-effect transducer H is located within a gap G, adjacent to the magnet M, and furnishes an output which is proportional to the magnetic field strength in G. The field strength, in turn, can be varied by the position of a ferromagnetic body N. The method permits measurement of displacements down to about 0.001 in.

An interesting application of the Hall effect is illustrated in Fig. 6-1*c*. The device serves to measure currents in a conductor without the need for cutting or interrupting the conductor and without making electrical connection between the conductor circuit and the meter. A current I (dc or ac) passes through a conductor W and causes a magnetic field surrounding the conductor. This magnetic field causes in a slotted ferromagnetic tube (*magnetic concentrator*) a magnetic field which is measured with a Hall-effect device H. The system can be used for the measurement of currents from less than a milliampere up to thousands of amperes. At high current levels, the magnetic concentrator can be omitted since the magnetic field in the vicinity of the wire is high enough to be measured directly with a Hall-effect transducer.

Figure 6-1*d* shows another application of the Hall effect, an *analog multiplier*. The system consists of a ferromagnetic core C which has a gap G into which a Hall-effect transducer H is inserted. A coil W is provided. If a current I_1 passes the coil, it will cause a magnetic flux density B in the gap

$$B = \text{const.} \times I_1$$

Substituting B, the output voltage from the Hall-effect transducer becomes

$$E_H = \frac{K_H}{t} \text{const.} \times II_1$$

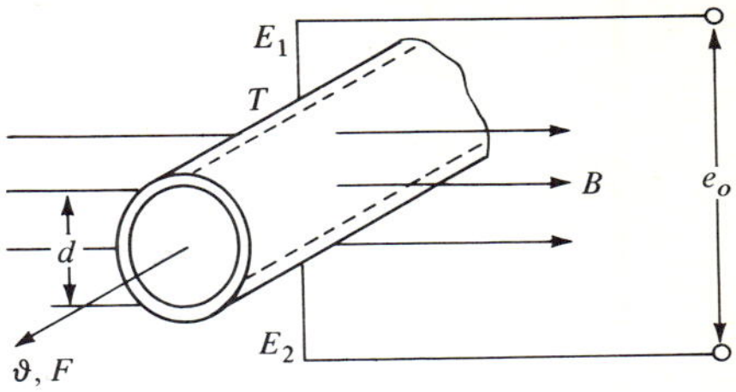

Fig. 6-2 Schematic diagram of an electromagnetic flowmeter.

i.e., the output voltage is proportional to the product $I \times I_1$. Obviously, the output voltage is positive since I and I_1 are both positive or negative. E_H is negative for $+I$ and $-I_1$ or $-I$ and $+I_1$, that is, the multiplier operates correctly in all four quadrants. The multiplier can be used for either dc or ac operation. The current I in practical or commercial multipliers is typically between 50 and 500 mA and is generally limited by the heating of the semiconductor plate. The magnetic field is in the range from 500 to 1,500 G. The output voltage for indium arsenide is about 10 to 100 mV, and the output impedance of the order of 5 Ω. Primary sources of errors in the magnetic circuit are saturation, hysteresis, phase shift, and stray fields.

Figure 6-2 shows a further instrumentation element based on the effect of a magnetic field on a moving charge, the *magnetic-flow velocity transducer*. A liquid flows through a nonconducting tube T (e.g., a blood vessel) that is located within a homogeneous magnetic field. Two electrodes E_1 and E_2 are inserted in this tube, their surfaces flush with the inside wall. If the liquid (which must be at least moderately conductive) flows through the tube with the average linear velocity $\bar{v}$ in centimeters per second (averaged over the cross section of the tube), a voltage arises between the two electrodes of

$$e_o = Bd\bar{v} = \frac{BF}{d}\frac{4}{\pi} \times 10^{-5}\text{ V}$$

where F is the average volume flow rate (in milliliters per second), B the magnetic induction or flux density, and d the internal diameter of the tube, i.e., the electrode distance. The linear-flow velocity range of the transducer extends from 10^{-3} mm/s to the highest velocities; the result is, to a certain extent, independent of the velocity profile, i.e., independent of any turbulence. The output voltage is small, usually in the microvolt range. There are a number of effects which may influence the accuracy of the transducer. To overcome some of them, some investigators have used magnetic fields supplied from generators with sinusoidal or square waves, a process which, of course, causes a sinusoidal or square-wave output voltage.

Induction systems If a coil with n turns and an average area a is placed in a magnetic field (Fig. 6-3a), a voltage e_o will be produced between both terminals

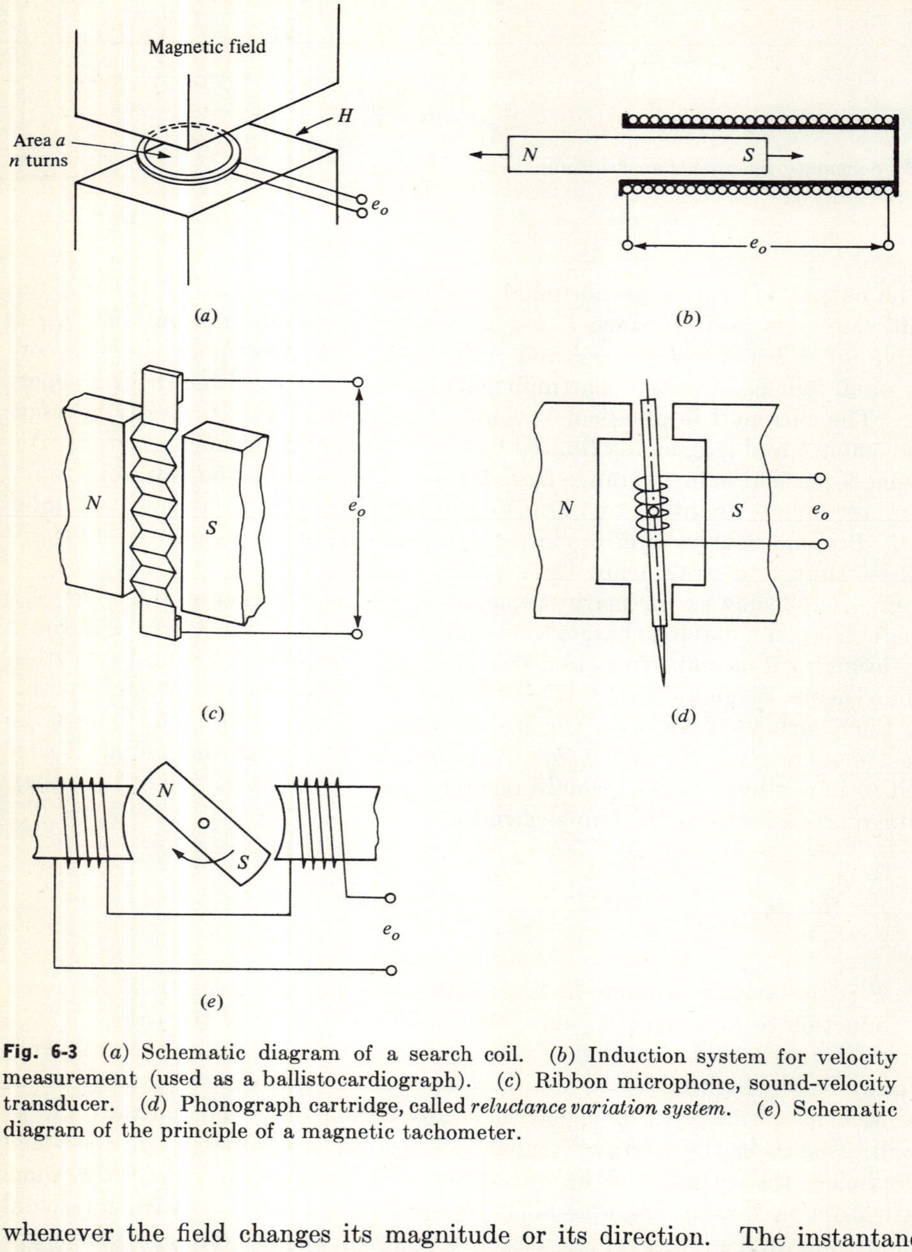

Fig. 6-3 (*a*) Schematic diagram of a search coil. (*b*) Induction system for velocity measurement (used as a ballistocardiograph). (*c*) Ribbon microphone, sound-velocity transducer. (*d*) Phonograph cartridge, called *reluctance variation system*. (*e*) Schematic diagram of the principle of a magnetic tachometer.

whenever the field changes its magnitude or its direction. The instantaneous output voltage e_o is proportional to the rate with which the field changes,

$$e_o = na\mu_0 \frac{dH}{dt} = na \frac{dB}{dt} = n \frac{d\Phi}{dt} \tag{6-2}$$

where $\Phi = aB = a\mu H$ is the total magnetic flux through the area a. If the field changes from $+H$ to zero ($\Delta H = H$) or from $+H$ to $-H$ ($\Delta H = 2H$), then such a total change is proportional to the time integral of e_o

$$\int_{t_0}^{t_1} e_o\,dt = na\mu\Delta H = na\Delta B = n\Delta\Phi$$

The time integral is independent of the velocity with which the field, the flux density, or the flux changes. Nevertheless, it may be necessary to accomplish such field variation in a short time since most instruments for the measurement of time integrals (e.g., ballistic galvanometers, RC integrators) introduce errors in the result unless the time of integration is considerably shorter than the time constants of these integrators.

Search coils for the measurement of magnetic fields and methods to cause field variations have been described extensively in the literature. The coils are frequently flat for insertion in narrow gaps and so small that the field is homogeneous over their area. A field variation can be accomplished by moving the coil in or out of the field, by rotating it, or by oscillating it in the field (but so that the oscillation causes a periodic increase or decrease of the flux through the coil, oscillation of a coil parallel to itself in a homogeneous field does not produce a variation of flux).

Several induction-type transducers are shown in Fig. 6-3*b* to *e*. These transducers furnish an output which is proportional to the velocities of the moving member. They can also be used for the measurement of displacements, either by integrating the output or in such cases where a known relationship exists between displacement and velocity (e.g., sinusoidal movement). Figure 6-3*b* shows a transducer which is frequently used for a small movement, such as in ballistocardiography (the measurement of the recoil movement of the human body caused by opposite movement of the body and the bloodstream). Figure 6-3*c* shows a microphone which contains a metal ribbon that is moved in the magnetic field by sound waves. The output voltage is proportional to the velocity of the ribbon. Figure 6-3*d* shows a schematic diagram of a phonograph cartridge which is often useful for velocity and displacement measurements in the laboratory. Figure 6-3*e* shows a tachometer, i.e., a transducer with an output which is proportional to the rotation velocity of the moving member.

A different type of induction transducer is shown schematically in Fig. 6-4. The current from a radio-frequency generator S is applied to a coil W and produces a magnetic field. The current depends on the applied voltage and the impedance of the coil. If a piece of metal M is brought into the magnetic field in the vicinity of the coil, the alternating field will induce eddy currents in the metal which, in turn, cause a magnetic field opposite to the field produced by the coil. This effect reduces the inductance of the coil and causes a change of the magnitude and the phase of the current. The effect increases as the metal object comes closer to the coil. The system can be used in different ways; for instance, as a proximity transducer with an electrical output that is a measure for the distance between the coil and the metal plate (useful range 10^{-5} to 10^{-2} in)

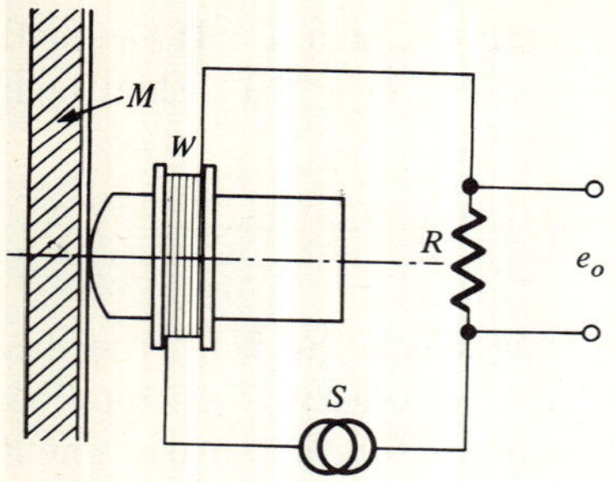

Fig. 6-4 Eddy-current transducer for proximity measurement, used as a metal locator.

or as a thickness gauge for a metallic or nonmetallic coating on a base metal, provided the coating has an electric conductivity that is substantially different from that of the base metal (e.g., paint on steel or silver on brass). The eddy-current transducer has also found application as a metal locator, for instance in archaeology, or for the detection of underground cables or landmines, or in surgery for locating metal particles in animal or human tissue.

Magnetic permeability If the flux density B in a doughnut-shaped coil (toroid, Fig. 6-5*a*) is measured, once when the coil is empty and once when it is filled with a core C of a material like iron, the experiments show a drastically different result: the flux density in the iron-filled toroid is many times higher than that of the empty coil. The ratio

$$\frac{\text{Flux density in the coil filled with material}}{\text{Flux density in the empty coil}} = \frac{B_m}{B} = \mu$$

where μ is the magnetic permeability of the material, a nondimensional quantity. Units and dimensions frequently used for permeabilities of the order of about 1 are:

The intrinsic induction B_i

$$B_i = B_m - B$$

that is the additional flux density resulting from the presence of the material within the coil

The magnetic susceptibility

$$\kappa = \mu - 1 = \frac{B_m - B}{B}$$

The specific magnetic susceptibility, i.e., the susceptibility per unit density ρ,

$$\chi = \frac{(\mu - 1)}{\rho}$$

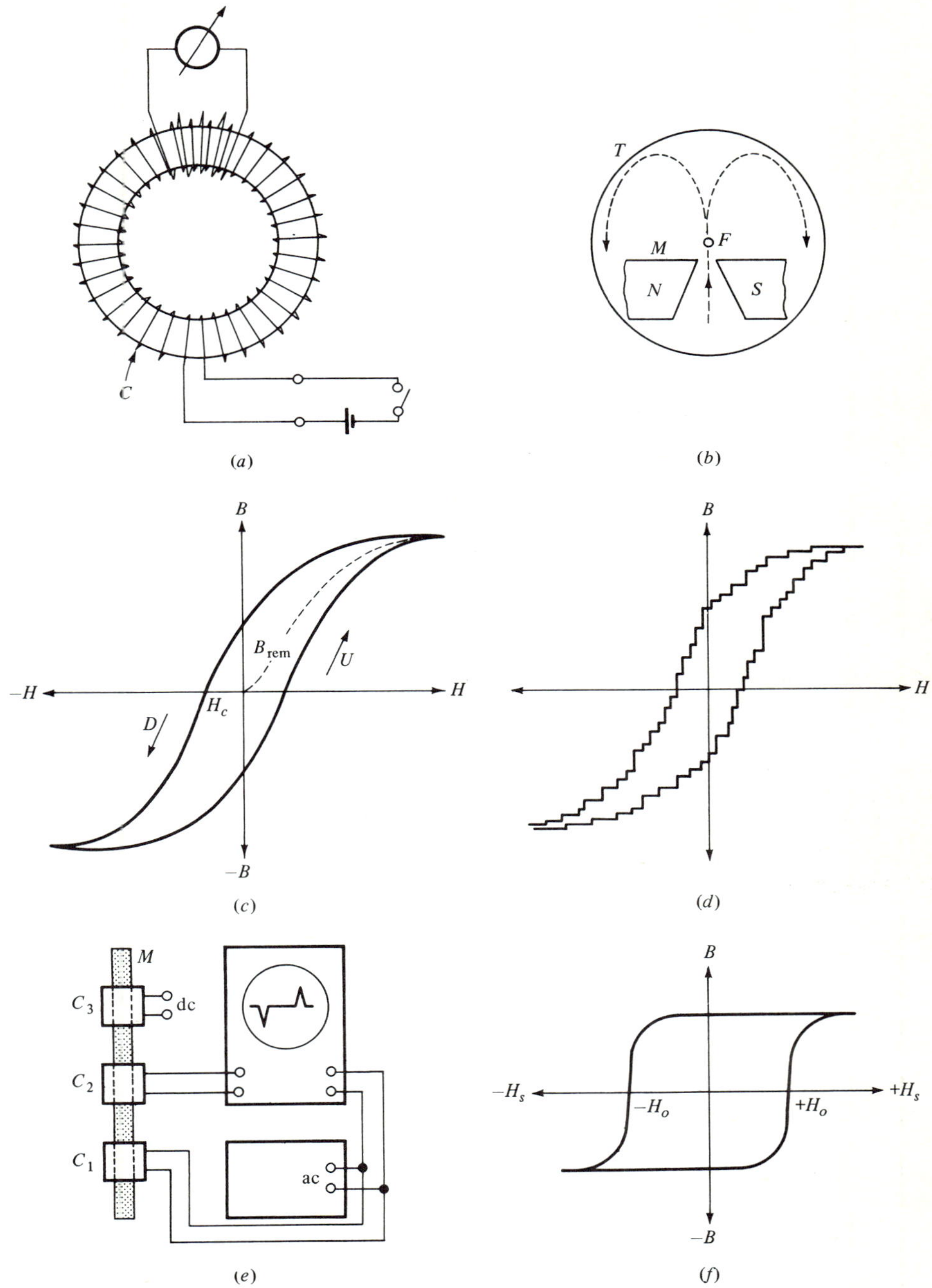

Fig. 6-5 (*a*) Toroid filled with a core *C* of a medium of the magnetic permeability μ; (*b*) magnetic-thermal transducer for the measurement of oxygen concentration; (*c*) cyclic magnetization characteristic of a ferromagnetic substance; (*d*) Barkhausen effect; (*e*) principle of the "peaking strip" system for the measurement of weak magnetic fields; (*f*) *HB* diagram.

All substances can roughly be divided into three groups:

1. *Diamagnetic substances,* substances with a permeability which is slightly smaller than 1 and constant, i.e., independent of the magnitude of the magnetizing field and also independent of the temperature.
2. *Paramagnetic substances,* with a permeability which is slightly greater than one, also independent of the magnetizing field but depending upon temperature.
3. *Ferromagnetic materials,* with a permeability which is several orders of magnitude greater than one and not constant but varying drastically with the applied field strength and with temperature.

Para- and ferromagnetic bodies in an inhomogeneous magnetic field are attracted toward the region of higher field strength; diamagnetic bodies are moved in the opposite direction. By suspending a paramagnetic or diamagnetic test sample in a strong and strongly inhomogeneous magnetic field and by measuring the force acting on such a sample, one can determine the magnetic permeability or susceptibility of the sample material.

The permeability of paramagnetic materials decreases with increased temperature in the form

$$x = \frac{C}{T}$$

where C is the Curie constant and T the temperature in degrees Kelvin. This relationship forms the basis of a temperature-measuring method at very low temperatures in the vicinity of the absolute zero point. A salt sample is exposed to the low temperature and its permeability is measured; the method is applicable at temperatures below about 4K.

The effect of permeability variation with temperature is applied in the oxygen-concentration transducer shown in Fig. 6-5*b*. The transducer consists of a tube T containing a heated filament F and a magnet M with a strongly inhomogeneous field. If oxygen (which is paramagnetic) is present in the transducer, the oxygen molecules will be attracted toward the region of highest field strength which is close to the filament. There the oxygen molecules will be heated and will lose their paramagnetic properties so that they will move out of the field into the upper part of the tube. They will be cooled by contact with the wall, and the process will be repeated. The "magnetic wind" caused by the oxygen will cool the wire and change its resistance considerably more so than cooling brought about by thermal conduction. Oxygen has the highest magnetic permeability of all gases (magnetic susceptibility of O_2 at 293K is $+3{,}449 \times 10^{-6}$ cgs. The only other strongly paramagnetic gas is nitrogen oxide (NO) with $+1{,}461 \times 10^{-6}$ cgs at 293K). In the absence of NO, the transducer output is a function of the oxygen concentration only and is practically independent of the presence of other gases in the mixture.

Also, the permeability of ferromagnetic materials decreases with increased temperature until a point is reached (the Curie point, varying for different materials from room temperature to several hundred degrees Celsius) where the ferromagnetism disappears suddenly and only a paramagnetic behavior remains. This phenomenon can be used for temperature measurement or measurement of other phenomena that cause a temperature variation, such as radiation.

Of particular interest for many instrumentation problems is the cyclic magnetization of ferromagnetic materials as shown in the hysteresis curve, Fig. 6-5*c*. An increase (arrow U) of the magnetizing field strength H causes an increase of the flux density B; there follows a region of saturation. Reducing the magnetizing field strength H into the negative range furnishes a second curve (arrow D) different from that of the upswing. For $H = 0$, a certain flux density still remains in the ferromagnetic material, the remanence B_{rem}. In fact, a demagnetizing field strength H_c (coercive force H_c) is needed to bring the flux density to zero. Different types of ferromagnetic materials show great variations in the shape and extent of the hysteresis curve. The area enclosed by the hysteresis curve is proportional to the energy lost in the cyclic magnetization process; it is converted into heat and raises the temperature of the ferromagnetic sample. The derivative of the B-H function, dB/dH, is the magnetic permeability μ. In general, μ rises, reaches a maximum value, and then declines.

If the instrument used for the observation of cyclic magnetization characteristics (Fig. 6-5*c*) has a high time resolution (cathode-ray oscilloscope), one observes that the characteristic does not follow a smooth curve, but exhibits small steps as shown in Fig. 6-5*d*. The occurrence of these steps is called the *Barkhausen effect*. The magnitude of the transitions depends to a large extent on the ferromagnetic material. The Barkhausen effect limits the signal-to-noise level of ferromagnetic elements.

The nonlinear relationship between magnetizing field and magnetic flux density finds numerous applications in the field of instrumentation. An example is the "peaking strip" magnetic transducer[1] shown in Fig. 6-5*e*. A probe M of a ferromagnetic material with an HB characteristic (see Fig. 6-5*f*) is surrounded by three coils. Coil C_1 is energized by an ac source which causes a magnetic field H varying periodically between $+H_s$ and $-H_s$.

Each time the magnetizing field passes through $-H_0$ or $+H_0$, the flux density changes sharply and induces in the coil C_2 a pulse which appears on the cathode-ray oscilloscope screen. The pulses in the setup described are symmetrical. If an external (dc) field is applied, the peaks will appear unsymmetrical because such a field will increase the magnetic ac field in one direction and decrease it in the opposite direction. Symmetry can be restored by applying a compensating dc field to the coil C_3. The compensating current is a measure of the applied (unknown) external field. The method is applicable for flux densities in a range from 0.1 to 500 Oe.

The permeability of some iron-alloy cores changes in response to dimensional changes, such as compression, expansion, or torque. This magnetic-elastic effect can cause the induction of voltages and, hence, can be used in mechanic-

[1] J. M. Kelly, *Rev. Sci. Instr.*, **22**:256 (1951).

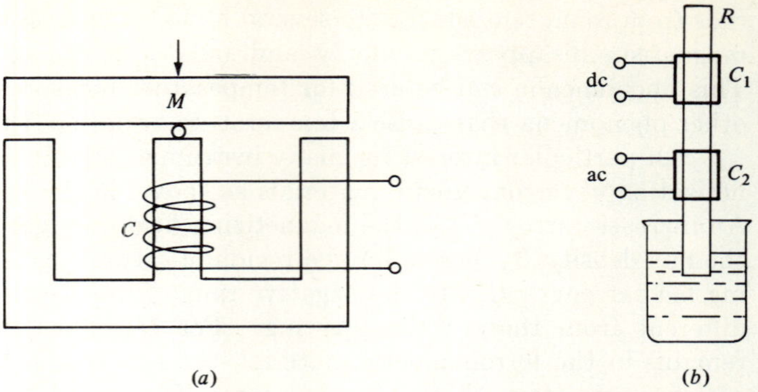

Fig. 6-6 (*a*) Schematic diagram of a magnetoelastic transducer; (*b*) magnetostrictive transducer for the generation of ultrasound.

electric transducers. Figure 6-6*a* illustrates a magnetoelastic accelerometer. Acceleration of the mass M causes a force and, hence, stress in the laminated core and changes its permeability. The change of permeability causes the induction of a voltage in the coil C surrounding the core. Since the force and the stress are proportional to the acceleration, that is, $f = m(d^2s/dt^2)$, and since the induced voltage is proportional to df/dt, the output is proportional to the third derivative of the displacement of the mass M. Difficulties may arise since the variation of the permeability is a nonlinear function of the applied stress and also changes with the magnetizing field, the temperature, and the composition of the alloy.

The reverse effect, a change of the dimensions of a body with magnetization, is called *magnetostriction*. The effect is most pronounced in certain iron-nickel alloys and even in these substances the dimensional change is rather small. If a rod (Fig. 6-6*b*) is premagnetized by a dc field (coil C_1) and an ac field is superimposed (coil C_2), it will periodically change the length of the rod. The effect will be maximum at resonance, i.e., if the ac frequency coincides with the lengthwise oscillation frequency of the rod. This frequency is, in general, at the high end of the audio-frequency range or in the ultrasonic range. Arrangements of the type shown in Fig. 6-6*b* can be used as powerful generators for ultrasonic oscillations which may cause effects like cavitation in liquids, the breaking up of particles such as bacterial cells, and the formation of stable emulsions of oil and water or even mercury and water. The same system can also be used as a tuned magnetoelastic transmitter for ultrasound.

A magnetic field may cause a change of the electric resistivity of a metal or a semiconductor. This magnetoresistive effect is particularly strong at low temperatures but extends for some metals, e.g., bismuth, into the range of room temperature. The resistance of a bismuth wire rises at low field strengths approximately with the square of the field strength. At field strengths of 10^5 A/m, it rises linearly with the field strength and can reach twice that measured at zero field strength. At very high field strength saturation occurs. A flat bifilar spiral of bismuth wire held between two sheets of mica can be used as a transducer for strong magnetic dc fields. The response characteristic of such a transducer depends strongly on the chemical composition (impurities) and the physical treatment of the wire. Resistance-temperature effects and thermoelectric effects are likely to introduce errors.

Important applications of the magnetization process are memory systems and tape recordings. A simple magnetic tape recorder is shown schematically in Fig. 6-7*a*. Its basic operation is as follows: a magnetic tape unrolls from supply reel *A* and passes roller *I*, magnetic recording head *R*, capstan *C*, which serves to assure constant tape speed, and take-up tension arm *S*. The magnetized tape is stored on take-up reel *B*.

The input signal to be recorded is applied to an encoder *E* where the information contained in the signal is converted to a form suitable for magnetic recording. For reproduction or playback, the output from a reproduce head is applied to a decoder *D*, where the signal is recovered in its original form. For simple direct recording of the incoming signal, the systems *E* and *D* are merely amplifiers with appropriate gains and frequency characteristics.

The magnetic tape generally consists of a plastic base, about 1,000 μin thick and uniformly coated with a layer of finely divided magnetic particles in a plastic binder. The magnetic layer is usually about 200 μin thick. Commercial tapes range from 0.25 to 2 in wide. Magnetic recordings can also be made on plastic chips or on magnetic disks or drums, especially when fast access to the recorded information is required.

A recording head is illustrated schematically in Fig. 6-7*b*. Essentially the head contains a nearly ring-shaped frame or core made up from thin laminations of a material of high magnetic permeability and high electric resistivity (to reduce eddy currents). The core has a gap of nonmagnetic material. A coil of wire is wound around the core; a current passing through the wire causes a magnetic flux in the core and in the gap. The magnetic tape moves past the gap with constant speed and the magnetic field produced in the gap enters the magnetic layer of that part of the tape adjacent to the gap. Particles of the magnetic tape are then magnetized and remain in a state of magnetization after the tape leaves the gap. Data analysis in instrumentation frequently requires multiple simultaneous recording on the same tape. For this purpose, recording and playback heads are frequently equipped with multiple recording systems which are separated by magnetic shields to reduce *crosstalk*. An illustration of a composite magnetic head is shown in Fig. 6-7*c*.

The requirements for the mechanical accuracy of recording heads, particularly for instrumentation recording (in contrast to audio application), are frequently very severe. Tolerances in the region of microinches are sometimes required, particularly if a tape is recorded on one machine and played back on another one.

The reproduce or playback head of a tape recorder is quite similar to the recording head. In fact, simple systems for voice or music reproduction use the

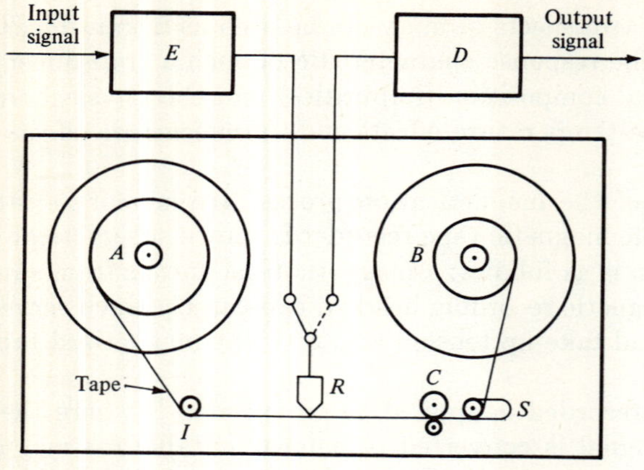

(a)

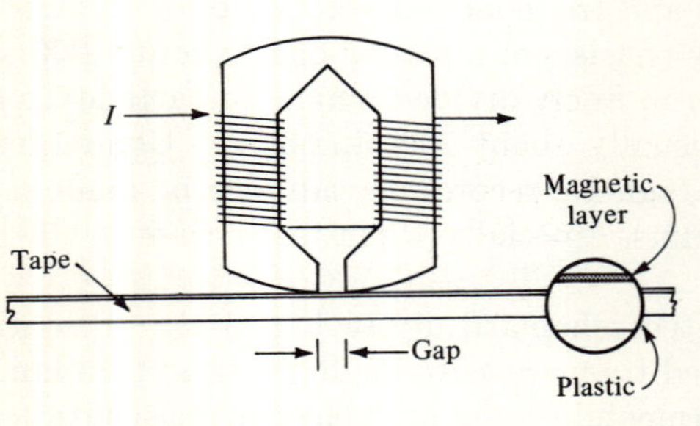

(b)

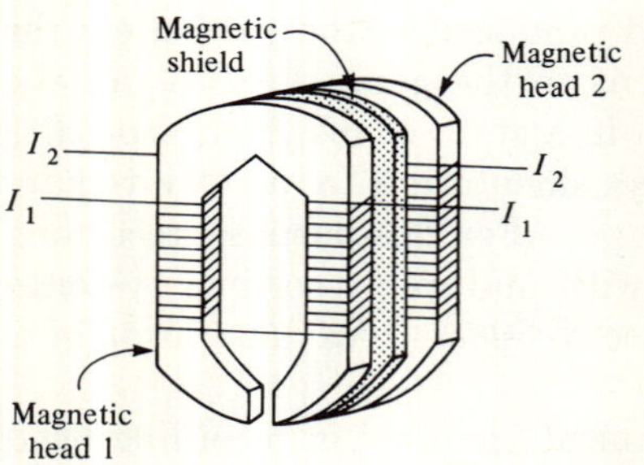

(c)

Fig. 6-7 (*a*) Basic diagram of a simple magnetic recorder and playback system; see text for explanation; (*b*) schematic diagram of recording or playback head with magnetic tape; (*c*) recording head for multiple recordings.

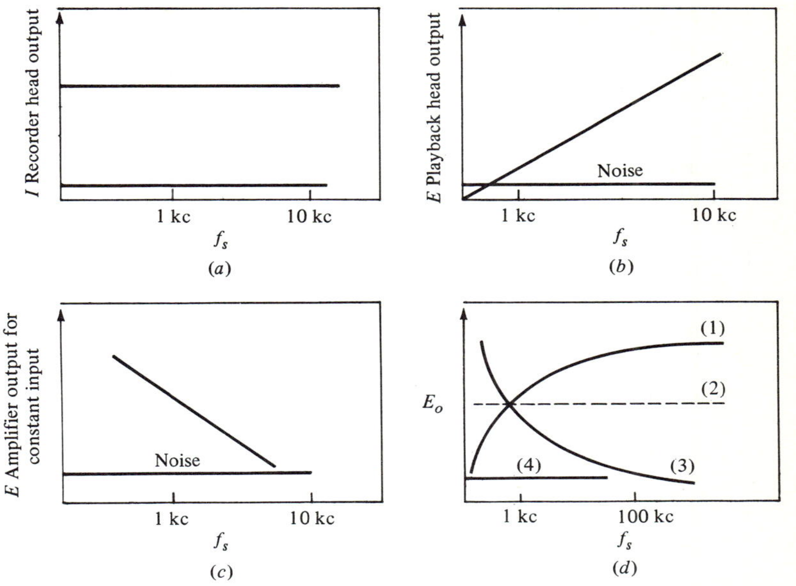

Fig. 6-8 Frequency characteristics of recording components. (*a*) Output current from recorder head versus frequency; (*b*) output voltage from playback head versus frequency; (*c*) amplifier with declining frequency characteristics; (*d*) equalization characteristics: (1) playback head, (2) amplifier, (3) combined effect of (1) and (2), (4) system noise characteristics.

same head for both. The process of the playback system is shown in Fig. 6-8. Magnetization of the tape causes a magnetic flux through the core of the playback head which varies with the varying state of magnetization of the tape. The magnetic flux through the core causes the induction of a voltage in the windings of the head. The voltage is proportional to the variations with time of the flux Φ,

$$e_o = k\frac{d\Phi}{dt} = k\frac{d}{dt}\Phi \sin 2\pi ft$$
$$= k2\pi f\Phi \cos 2\pi ft$$

The output voltage increases with increasing frequency (see Fig. 6-8*b*). Such a frequency characteristic is generally undesirable; many applications require a flat frequency characteristic. A flat characteristic can be obtained by applying the output from a playback head with rising characteristic (Fig. 6-8*b*) to an amplifier or decoder having an inverse (i.e., falling) characteristic (Fig. 6-8*c*). If properly matched, the amplifier will furnish an output which has a flat frequency characteristic (*playback equalization,* Fig. 6-8*d*).

The relationship between the current in the windings of the recording head and the magnetization of the tape is inherently nonlinear. If a sinusoidal current

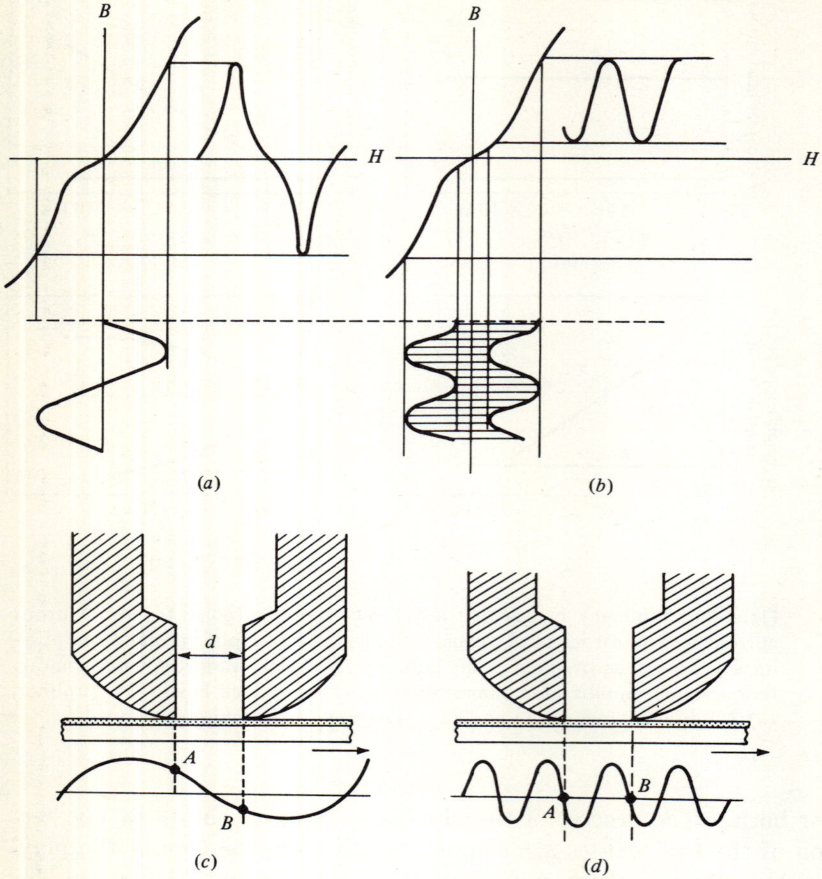

Fig. 6-9 *HB* diagram of a magnetization on tape. (*a*) Without bias, distorted output; (*b*) addition of an ac (high-frequency) bias, undistorted output; (*c*) limitation of frequency response; low-frequency, magnetized tape causes induction in playback head; (*d*) limitation of frequency response; high-frequency; magnetization *A* to *B* is zero and remains zero when the tape moves.

passes through the recording head, it will cause a distorted magnetization of the tape as shown in Fig. 6-9*a*. Now if an alternating current (*bias current*) that exceeds several times the signal amplitude and frequency is superimposed onto the signal, its effect will be an extinction of magnetization in the nonlinear region so that only a recording in the linear part of the transfer characteristic remains. This ac bias is of considerable importance; without it a linear operation of the magnetic tape recording would not be possible. The frequency and amplitude of the ac bias are not very critical. The frequency must be high enough so that any effect of the ac bias on the recording is eliminated. Customarily, the ac bias is of a frequency about 3.5 times the highest frequency for which the recorder is designed.

It is evident from Fig. 6-9*b* that at the low-frequency end of the characteristic the output signal from the playback head is so small that the signal is below the noise level. In other words, for direct recording, there exists a low-frequency limit below which successful operation of the magnetic recorder is not possible.

A limit of operation exists also at high frequencies. If the tape moves past the gap of a playback head with constant velocity and if the frequency of the input signal increases, a situation will occur like that illustrated in Fig. 6-9*c*. The magnetization of the tape, during the time the tape is opposite the gap of the playback head, changes from A to B. The average magnetization is zero and remains zero if the tape moves. This effect limits the useful frequency range of the recording. The effect can be reduced, and the frequency range can be expanded by increasing the tape speed or by decreasing the gap width. However, a reduction of the gap width reduces the magnetization of the tape and the output from the playback head, while an increase of the tape speed increases the wear between the head and the tape. For direct recording, the useful frequency range for a tape speed of 60 in/s and a gap width of 80 μin is from about 100 Hz to 250 kHz. The tape speed of commercial recorders is about 120 in/s. At that speed the highest recorded frequency is about 2 MHz. Recorders for color television can record in excess of 15 MHz at a tape speed of 1,560 in/s. Several methods exist to operate tape recorders with frequency modulation (see Chap. 7), pulse modulation (see Chap. 13), or with digital recording. The requirements for the precision of the speed regulation and of the start and stop accuracies are sometimes very severe.

The magnetic recording on a tape can be erased so that the tape can be used for a new recording. Simple erasing can be accomplished by moving the tape past a pole of a permanent magnet or past a sequence N-S-N-S, etc., of one or several magnets. The method is likely to leave some traces of a previous recording. Complete erasure can be accomplished by an erasure head supplied with alternating current. In general, the frequency is the same as that used for ac bias. The tape runs from the supply reel, over the erasure head and immediately afterwards over the recording head. Ac erasure systems can also be applied to tape in bulk while it is on the reel.

A further group of instruments, the magnetic amplifiers and the saturable reactors, are also based on nonlinear magnetization characteristics. A schematic diagram of a simple magnetic amplifier is shown in Fig. 6-10*a*. The amplifier contains a saturable reactor X_s, i.e., an inductance coil with a core made of a ferromagnetic alloy and carrying two coils L_1 and L_2. The load or power circuit consists of an ac source S in series with the coil L_2 (load winding), and a load, indicated by the load resistor R_L. The current I_L in the load circuit is controlled by the inductive reactance $2\pi f L_2$.

The inductance of L_2 depends on the effective permeability of the magnetic core which can be varied by variation of the magnetization of the core, i.e., by

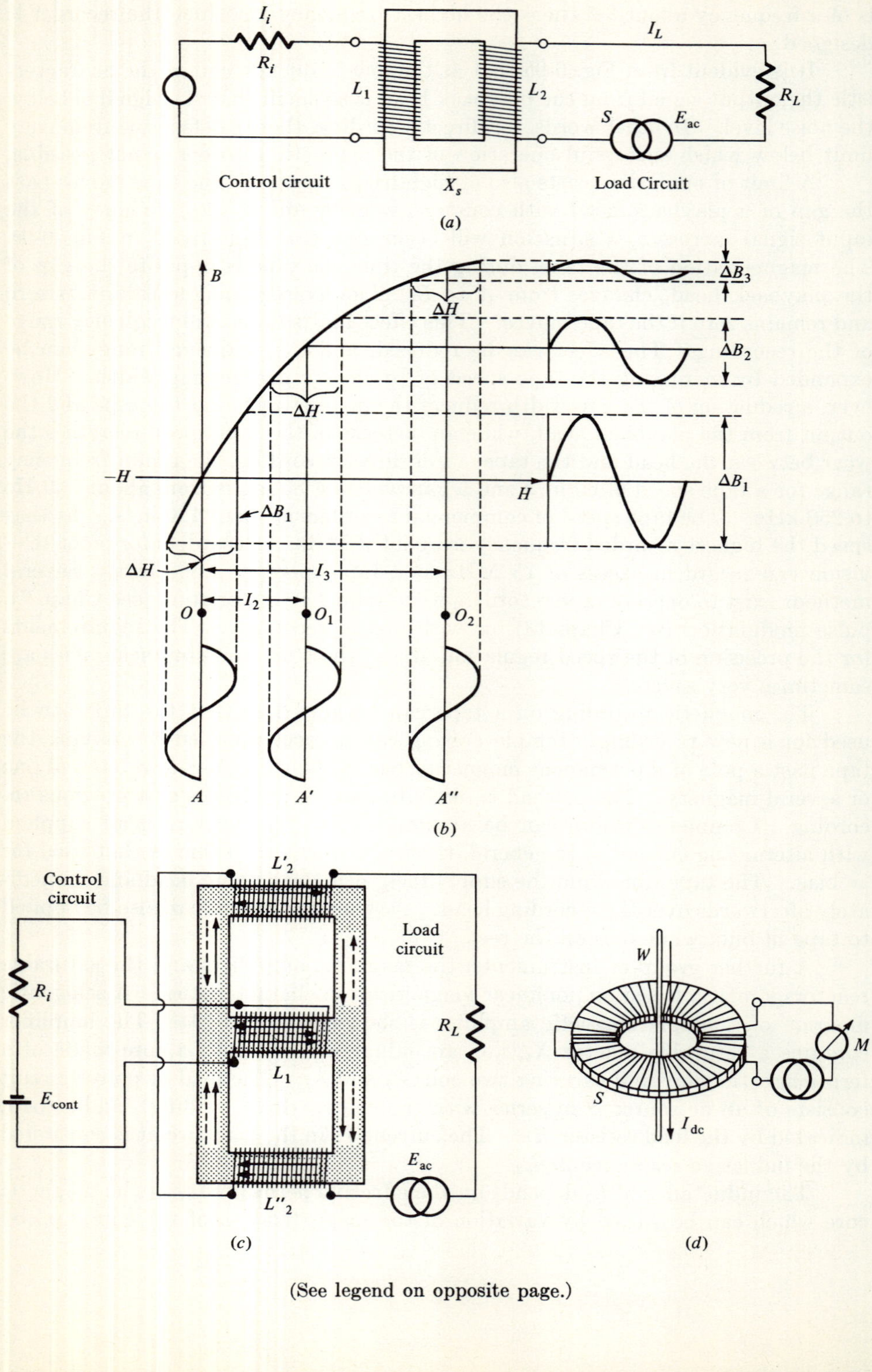

(See legend on opposite page.)

applying a dc current to the control winding, coil L_1. The effective permeability of the core, over a range of the HB characteristic that may be considered linear, is

$$\mu_{\text{eff}} = \frac{\Delta B}{\Delta H} \approx \frac{\Delta B}{\Delta(kI_L)}$$

Figure 6-10b shows how a dc control current I_2 or I_3 can change the effective permeability and, hence, the inductance of the load winding L_2. The figure shows a BH characteristic. An alternating field ΔH is applied from the load circuit; it causes a variation of the flux density ΔB which oscillates with line frequency about the point O. The slope of the characteristic at the point O is very high; therefore, the magnetic permeability is high and so is the reactance $2\pi fL_2$ and, therefore, the inductance L_2. Thus, the current I_L in the load circuit (Fig. 6-10a) is very small.

If now a dc current I_C (*control current*) is applied to the coil L_1, it will have the effect of shifting the operating point to O_1. At this point the slope of the characteristic $\Delta B_1/\Delta H$ is much smaller than before; the permeability of the core is decreased, and so are the inductive reactance and the inductance of L_2. Therefore, the current in the load circuit is now larger than before. The effect is even more pronounced if a controlling current I_3 is applied to L_1, leading to an operation about the point O_2.

The current in the load winding L_2 (Fig. 6-10a) will induce a voltage in the control winding L_1, which may have an undesirable effect on the circuit or the control source. The effect can be avoided by a balanced saturable reactor as illustrated in Fig. 6-10c. The winding L_2 is split into two windings L_2' and L_2'' which are wound on the outside legs of a three-legged transformer; the control winding is applied to the center leg. The direction of the magnetic flux lines generated by the load circuit is indicated by solid arrows, those generated by the control circuit by dotted arrows. The solid arrows cancel each other within coil L_1. There is no effect from the load circuit on the control circuit, but there is an effect from the control circuit on the load circuit, as is evident from the direction of the arrows in Fig. 6-10c.

Magnetic amplifiers, like other types of amplifiers, are commercially available as units (*operational magnetic amplifiers*) which only require connections to the power source, the load circuit, and the control circuit. Typically, they furnish an output of 5 V into a 1,000-Ω load; the input current is of the order of 1 μA; the frequency response is about 10 Hz; and the power requirement at 60 Hz is 2.5 VA at 118 V. Magnetic amplifiers are also available in substantially larger and more powerful units.

Fig. 6-10 (a) Magnetic amplifier, see text for explanation. (b) Characteristics of a magnetic amplifier. The output signals are proportional to ΔB_1, ΔB_2, and ΔB_3, which vary in response to the bias signals 0, I_2, I_3. (c) The effect of the load circuit on the control circuit is eliminated (see solid arrows in coil L_1). The influence of the control circuit on the load circuit is unaffected (see dotted arrows). (d) Magnetic amplifier used for the measurement of a dc current I in a conductor W, without contact between the wire and the structure S.

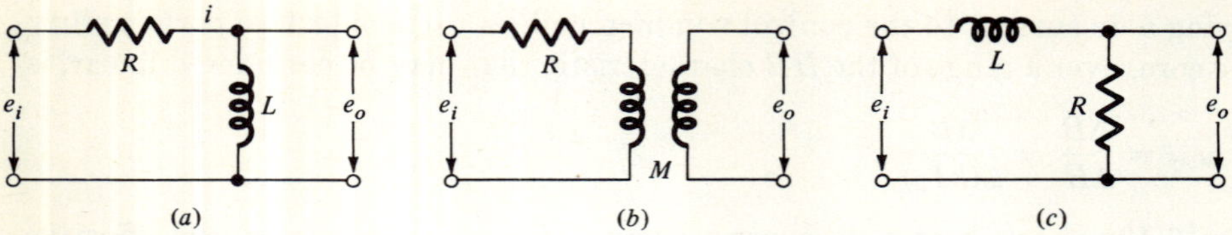

Fig. 6-11 (*a*) Inductive differentiating circuit; (*b*) differentiating circuit with mutual inductance; (*c*) inductive integrating system.

Saturable reactors can be used for the solution of a number of problems. An example is illustrated in Fig. 6-10*d*. The figure shows an instrument that measures the current passing through a wire without the need for cutting the wire or interrupting the current. Two semicircular cores of ferromagnetic material are applied to surround the wire or cable. (In general, the two cores are held in a pair of pliers (not shown), the pliers are opened, the cores are placed around the wire and a spring in the pliers presses the cores together.) The circuit is essentially that of Fig. 6-10*a* except that the magnetic field surrounding the wire is formed by one conductor only. This field causes a change of the magnetic permeability and, hence, a variation of the current in the meter M.

Inductance In Fig. 6-11*a* the current i passing through a coil L causes a magnetic field. Any increase or decrease of the current by the amount di causes a change of the magnetic field dH. Any change of a magnetic field causes an induced voltage $e_o \approx dH/dt$ in any conductor linked with the magnetic field, also in the coil itself. The induced voltage is

$$e_o = -L\frac{di}{dt} \qquad (6\text{-}3a)$$

A change of magnetizing current in one coil will induce a voltage in the windings of an adjacent second coil (Fig. 6-11*b*). The voltage is

$$e_o' = -M\frac{di}{dt} \qquad (6\text{-}3b)$$

The time integrals of the voltage are

$$\int_{t_0}^{t_1} e_o\,dt = L(I_1 - I_2) \qquad \text{or} \qquad \int_{t_0}^{t_1} e_o'\,dt = M(I_1 - I_2)$$

The proportionality factor L is the inductance of the coil, the factor M is the mutual inductance of the two coils; their dimensions are the same, namely 1 V·s/A = 1 H. The magnitude of L depends on the square of the number of turns n of the coil and the coil geometry, as well as the permeability μ of the medium within and around the coil:

$$L = \mu_0\mu n^2 G$$

where G is a geometric factor. For a long coil in air (length $l \gg$ diameter d), the inductance is

$$L = \frac{\mu_0 n^2 a}{l}$$

where μ_0 is the (universal) induction constant (1.256×10^{-6} V·s/A·m) and a the effective cross section of the coil (square meters). A henry is a large quantity which is practically obtained only in coils with ferromagnetic cores. Standard inductances are usually in the millihenry (mH) and microhenry (μH) range. Such inductances are frequently built in the form of a toroid as shown in Fig. 6-5*a*. The advantages of this shape are that it has a negligible magnetic stray field and that magnetic stray fields from the outside cause a negligible induction in the coil windings.

Equations (6-3*a*) and (6-3*b*) form the basis of the inductive differentiating and integrating circuits. The basic differentiating circuit is shown in Fig. 6-11*a*. A (time variable) voltage e_i is applied to the input. If the resistance R is large enough so that the current i in the circuit is determined primarily by the magnitude of R and if the resistance in the windings of the inductance is negligible, then the output voltage will be approximately

$$e_o = -L\frac{di}{dt} = -\frac{L}{R}\frac{de_i}{dt}$$

If L/R is small, the output voltage will be nearly proportional to the time derivative of the input voltage, but the output is small. An increase of L/R increases the output but also the error, i.e., the departure from the correct value of the derivative. The inductive differentiating method is of advantage where high efficiency (ratio e_o/e_i) but only moderate accuracy is required. A differentiating element with a mutual inductance M is illustrated in Fig. 6-11*b*. In analogy to the inductive differentiator, the output is

$$e_o = -M\frac{di}{dt} = -\frac{M}{R}\frac{de_i}{dt}$$

The condition for correct differentiation is that R is large, i.e., the large part of the applied input voltage e_i appears across R. However, in contrast to the previously described differentiator, no error is caused by the presence of a resistance in the primary windings of the transformer M as long as no current is drawn from the secondary side. Differentiators of the inductive and mutual inductive types operate best when supplied from a constant-current source.

Circuits containing inductance can also be used for integration. The integral form of Eq. (6-3*a*) is

$$i = \frac{1}{L}\int e\,dt$$

A voltage output, rather than a current output, can be obtained by inserting a resistance R in the circuit as shown in Fig. 6-11c. If R is small, that is, if $L/R \gg T$ (where T is the time of the slowest variation of e_i), the output is

$$e_o = \frac{R}{L} \int e_i \, dt$$

The inductive integrator is not frequently used. It requires a large inductance L which can be technically achieved only by inductors with ferromagnetic cores and which may introduce hysteresis and nonlinearity errors. However, at times such integrating circuits can be useful if an ohmic connection between the output terminals is required.

Another important consideration follows from Eq. (6-3a). In circuits containing a large inductance L, such as an electromagnet, the voltage which arises between the terminals can become dangerously high and can lead to destruction of the instruments if the rate of change of the current, di/dt, assumes large values, for instance, when the circuit is closed or, more so, when it is opened. It is advisable to close or open such circuits slowly, for instance, by continuous or stepwise increase or decrease of the current.

Inductive displacement transducers The inductance L of a coil depends on the number of turns, the permeability of the material within and surrounding the coil, and a geometric factor, as expressed above. This equation can be rewritten in the form

$$L = \frac{n^2}{\mathcal{R}}$$

where $\mathcal{R}$ is the reluctance of the magnetic circuit, a quantity analogous to the resistance in an electric circuit. In a magnetic circuit like that shown in Fig. 6-12a, the reluctance is composed of that in the iron core and that of the air gap. In many practical cases the reluctance is that of the air gap, the iron core contributing but little to it.

There exist, therefore, three possibilities to vary the inductance of a coil arrangement in response to a mechanical force or displacement: by changing the number of turns, the (effective) permeability of the core medium, or the geometry. Many types of inductive displacement transducers are based on geometric variations, like that shown in Fig. 6-12a, where the inductance L changes in response to a variation of width d of the air gap. Figure 6-12b is an example of a displacement transducer in which the applied mechanical force or displacement causes a change of the effective permeability.

Inductive transducers can be built so that the introduced variation causes a change of induction L or of mutual induction M. They can be built with or without ferromagnetic materials, and they can be constructed with single or double output. In the latter case, the subsequent element can be constructed to measure either the difference or the ratio of the two outputs.

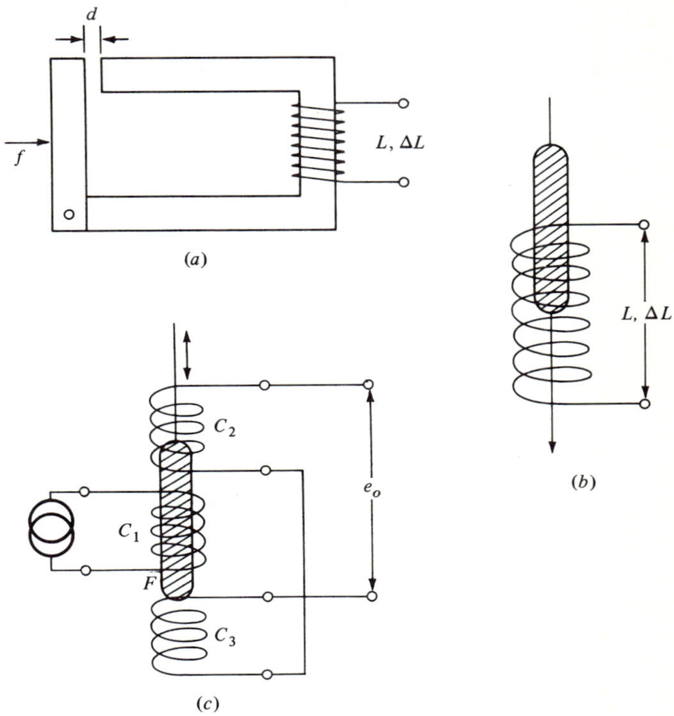

Fig. 6-12 (*a*) Reluctance variation mechanic-electric transducer; (*b*) mechanic-electric transducer based on permeability variation; (*c*) schematic diagram of linear variable differential transducer.

Differential output is particularly desirable if the change ΔL (or ΔM) caused by the applied mechanical force or the displacement is small or if the fractional variation $\Delta L/L$ is small. Figure 6-12*c* shows an example, the differential transformer. A primary coil C_1 is supplied from an ac generator; it induces voltages in the two coils C_2 and C_3. If the iron core F is in the center position, the induced voltages are equal and their difference is zero. Any variation of the position of F in the axial direction increases the voltage in one coil and decreases it in the other coil. The voltage difference appears at the output terminals. With the appropriate dimensions the output is a linear function of the core displacement over a wide range (*linear variable differential transformer*). Technical transducers, operated from a 6.3-V 60-Hz source, have sensitivities of the order of 20 mV/0.001 in in the linear part of their transfer function. Depending on their construction, their linear range may be as little as 0.005 in or as much as several inches. Displacements of less than 10^{-6} in have been measured. The mechanical axial force exercised by the magnetic field on the ferromagnetic core is proportional to the square of the current in the primary coil; it is zero if the core is in the center position and increases linearly with the displacement. For a commercial transducer operated at 60 Hz and for full displacement, the force is of the order of 0.3 g (300 dyn). That is about a hundred times more than the force requirement of a comparable capacitive transducer; but the output power of inductive transducers is, in general, larger than that of capacitive transducers. Miniature transducers of about 12-mm length and 3-mm diameter have been built for intracardial pressure measurements.

The shielding of magnetic stray fields or the protection from magnetic stray fields is not a simple matter. Three different physical mechanisms can be used to reduce the effect of magnetic stray fields on a sensitive structure or circuit,

1. Ferromagnetic shields
2. Eddy-current shields
3. Proper positioning of stray field sources with respect to stray field receiver

Ferromagnetic shields Only closed hollow shields are effective, such as spherical or cylindrical shields. The shielding effect depends on the geometry of the shield or the effective permeability μ of the shield material, and the thickness D of the shield. For instance, the attenuation (magnetic field strength outside the shield over that inside the shield) for a ferromagnetic sphere of the diameter R is approximately

$$S_s = \frac{2}{3} \mu_{\text{eff}} \frac{D}{R}$$

The attenuation of a cylindrical shield of a diameter R and of a length $l \geqq 4R$ is

$$S_c = \frac{1}{2} \mu_{\text{eff}} \frac{D}{R}$$

As previously stated, the effective permeability of a ferromagnetic material rises with increasing magnetization, reaches a maximum, and then declines. For a good magnetic shield, the maximum permeability should occur at the strength of the stray field. That means that small stray fields can be shielded satisfactorily only with Ni-Fe alloys which show high initial permeabilities, such as Mumetal or Permalloy, while at higher strength of the stray field materials like Permenorm are preferred. In general, the use of two dissimilar layers is recommended. Also, two thin layers of ferromagnetic material insulated from each other are better than one thick layer. "Netic" and "Co-Netic" foil[1] are frequently used.

Eddy-current shields are applicable for the attenuation of ac fields. The magnetic stray field causes eddy currents in a copper or aluminum plate. The eddy-current field has a polarity opposite that of the stray field so that the fields cancel each other. For a good shielding effect the shielded structure should be as small as possible and should have no slots and no holes. The shielding effect increases with frequency but decreases in the range beyond 10 to 100 MHz.

[1] Trade names of Magnetic Shield Division, Chicago, Ill.

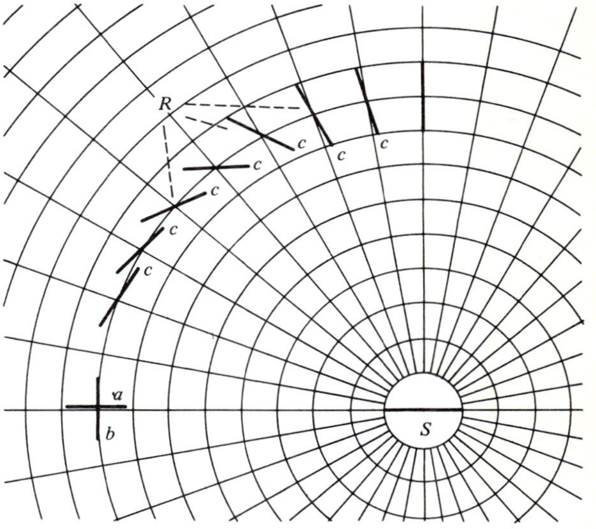

Fig. 6-13 Magnetic stray-field source S and receiver R positioned to reduce the influence of the magnetic stray field.

Positioning Figure 6-13 shows a stray field source S and a stray field receiver R, e.g., a transformer connected to the input of an amplifier in different positions. It is evident that the transformer in position a will be exposed to all stray-field lines in the same direction; the effect of the stray field is very high. In position b, the effect of the stray field on the receiver is cancelled, and the same occurs with receivers in positions c. In general, and in the absence of more than one stray-field source, one can rotate a receiver element into a position so that the influence of the stray field on the receiver is cancelled.[2]

REFERENCES

Steward, W. Earl: "Magnetic Recording Techniques," McGraw-Hill, New York, 1958.
Ettinger, George M.: "Magnetic Amplifiers," Methuen's Monographs on Physical Subjects, Methuen & Co., Ltd., London, and Wiley, New York, 1953.

[2] K. S. Lion, *Funk* (Germany), **28**:431 (1928).

7

Alternating Currents

INTRODUCTION

The elements described in this chapter furnish or convert electric signals in the form of periodically varying electric or magnetic quantities. The information may be contained in the voltage or current *level,* in the *reactance* or *impedance,* or in a *time function* of the signal, such as the frequency or phase of the periodically varying signal.

The simplest form of a periodic variation is the sinusoidal function; all other forms of continuous periodic variations can be synthesized from sine waves. The sinusoidal function is produced by the projection of a rotating vector, for instance a voltage vector $\mathbf{E}_p$ (Fig. 7-1*a*) on a vertical axis A-A. If the vector $\mathbf{E}_p$ rotates with the frequency f hertz (1 hertz = 1 cycle per second), i.e., with the angular velocity $\omega = 2\pi f$ rad/s (1 radian = 57.3°), then the instantaneous value of the voltage at any time t is

$$e = \mathbf{E}_p \sin (\omega t + \phi)^*$$

* Throughout this and the subsequent chapters, the following symbols are used: instantaneous values—e, i; all others, generally—E, I.

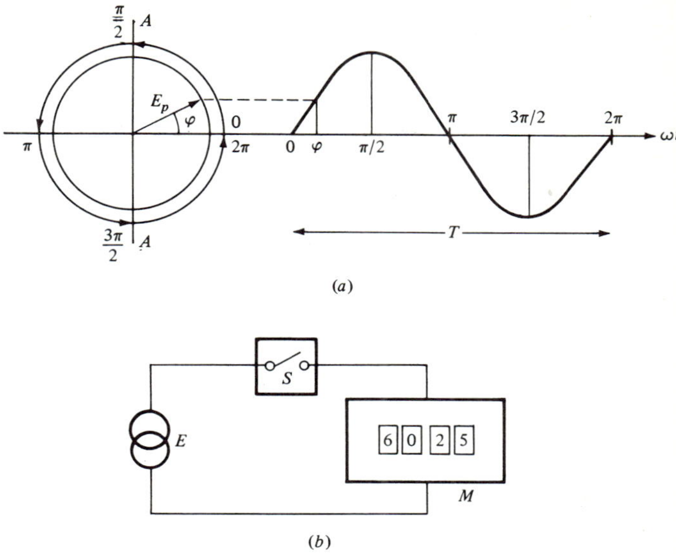

Fig. 7-1 (*a*) Sine function, basic form of a periodic alteration. (*b*) Digital frequency meter. *E*, source; *S*, electronic switch; *M*, digital counter.

where ϕ is the angle at $t = 0$. The instantaneous value of a voltage can be observed and measured with an oscilloscope that is sufficiently free of inertia so as to follow the voltage variation with negligible dynamic error.

FREQUENCY

The duration of a full period is by definition $T = 1/f$. In general, frequency can be measured to the nearest cycle. At frequencies of the order of 10^{10} Hz, frequency can be measured with an accuracy of better than one part in 10^{10}. Frequency is always measured by comparison with a calibrated frequency or time signal. An intrinsically simple method is illustrated in Fig. 7-1*b*. The output of the source E is applied to a scaler M via the very precise electronic time switch S. While switch S is closed for, say, 1 s, the scaler advances with each cycle of the applied voltage. At the end of 1 s the switch opens and the number of cycles per second can be read off the scaler. The requirements for accurate operation of the time switch S are quite severe. The switch frequently contains a highly stabilized oscillator of a frequency of, say, 1 MHz (10^6 cps) and a scaler unit which opens the main circuit after 10^6 cycles. An advantage of the method is the indication of the frequency in digital form. Also, the method is applicable over a wide frequency range. A disadvantage of the system, for certain applications, is the discontinuous indication of the result.

In the laboratory, the frequency is usually compared with that of a calibrated signal generator. A simple comparison method consists of applying the

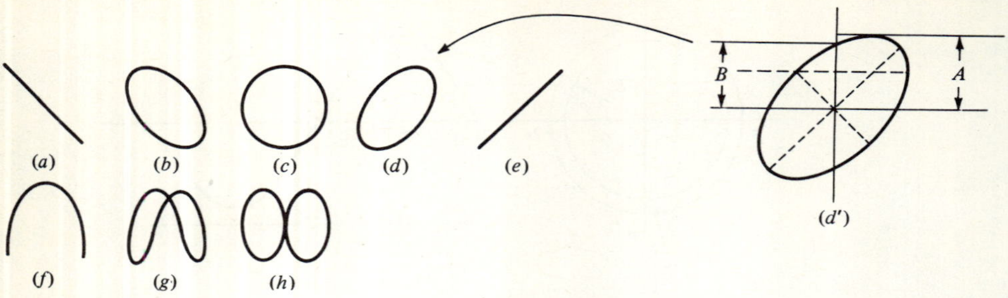

Fig. 7-2 (*a*) to (*e*) Frequency comparison with a cathode-ray oscilloscope; the voltages applied to the horizontal and vertical deflection plates are of the same frequency but of different phase. (*f*) to (*h*) Frequency comparison with a cathode-ray oscilloscope; the frequency applied to the vertical plates is twice that applied to the horizontal plates.

signal of the unknown frequency to the vertical deflection-producing plates of a cathode-ray oscilloscope and applying the output from the signal generator to the horizontal deflection-producing pair of plates. The frequency of the signal generator is continuously varied until a stationary figure appears on the oscilloscope screen. Figure 7-2 shows different configurations which arise when the signal generator frequency is either equal to or twice the size of the unknown frequency.

The measurement of frequencies from about 25 to 400 Hz can be carried out with a set of mechanical resonators (*vibrating reeds*) as illustrated in Fig. 7-3. The resonators consist of leaf springs of a magnetic material fastened at one end in a frame, while the other end can swing freely. Each reed is tuned to another

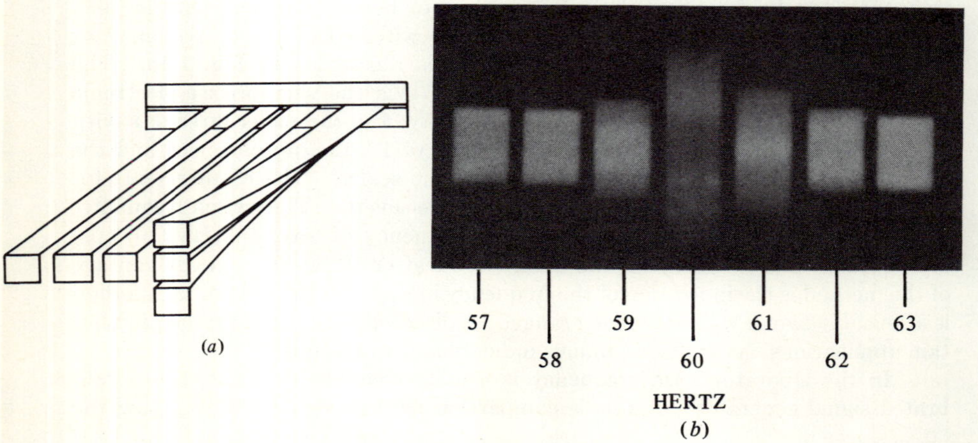

Fig. 7-3 (*a*) Frequency measurement by mechanical resonance; ac magnet (not shown) is below or above the vibrating reeds; (*b*) frequency measurement by mechanical resonance, appearance of vibration; the excitation frequency is 60 Hz.

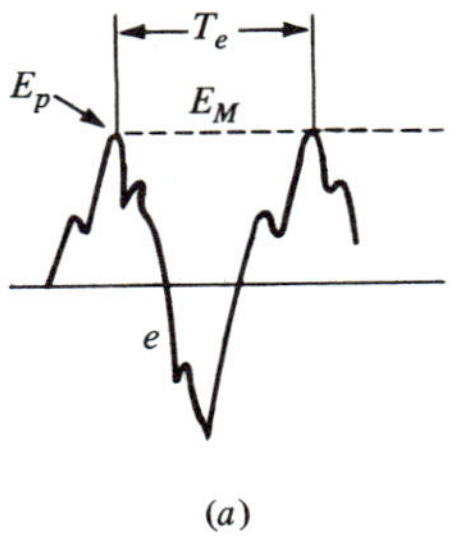

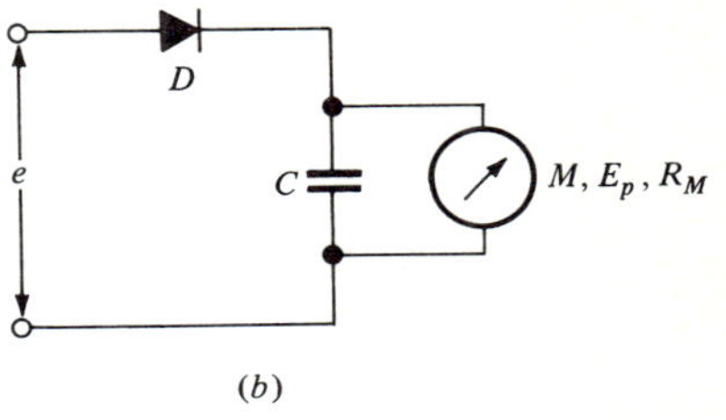

Fig. 7-4 Peak voltmeter. (*a*) Voltage-time function. (*b*) Circuit diagram: *e*, instantaneous value; E_p, peak value; E_M, voltage indicated by meter *M* in circuit (*b*).

frequency. The alternating current causes a magnetic field which attracts the leaf springs in the rhythm of the applied frequency (more accurately, with twice that frequency whenever the magnetic field reaches its maximum, twice per cycle, unless the springs or the magnet are polarized). The leaf spring with a resonant frequency equal to that of the alternating current shows a maximum amplitude. If the frequency of the alternating current is between that of two neighboring springs, the frequency can be estimated from the relative amplitudes of the spring excursions. Other frequency-measuring methods will be described below.

Extremely precise time signals and frequency signals in the radio- and audio-frequency range are broadcast by the radio stations of the National Bureau of Standards, WWV in Beltsville, Maryland, and WWVH in Hawaii. The emitted frequencies of 2.5, 5, 10, 15, 20, and 25 MHz are accurate within better than five parts in 10^{10} and may be used for comparison in any laboratory with a suitable radio receiver. Also transmitted by the same stations are audio signals of 600 and 440 Hz (musical A above middle C), as well as pulses of 5-ms and 1-s duration.

VOLTAGE AND CURRENT MEASUREMENT

Frequently the *peak* value of an alternating voltage or current is of interest, in particular in nonsinusoidal waveshapes. Peak voltages or peak-to-peak voltages can, of course, be read off the oscilloscope's voltage-time trace. The standard method for the measurement of peak voltages is illustrated in Fig. 7-4. The periodically alternating voltage *e* is applied to the input terminals of a circuit containing a diode *D* which permits the flow of current in one direction only. As long as the applied voltage on the left of the diode is larger than that on the right side, a current will flow and will charge the capacitor *C*. If the applied voltage recedes, the diode acts as an open switch and prevents a flow of current in the opposite direction which would discharge the capacitor. The result is that *C* gets charged to the peak value $\mathbf{E_p}$, and this voltage is measured with the voltmeter *M*. The resistance of the meter R_M must be large so that it does not

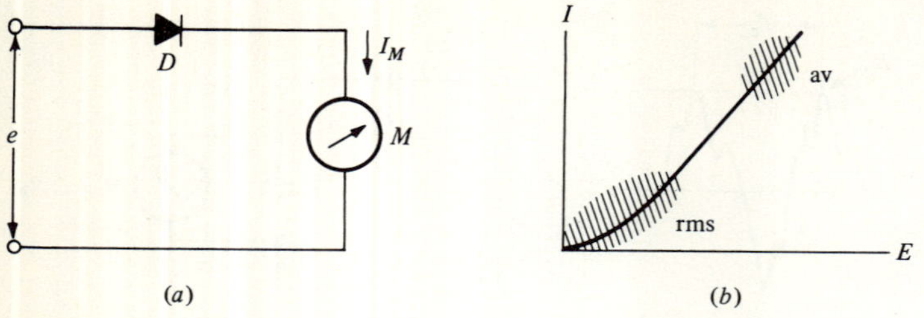

Fig. 7-5 System for the measurement of the average or rms voltage. (*a*) Circuit diagram. (*b*) *EI* characteristic of the rectifier diode *D*. If operated in the lower part of the characteristic, meter *M* measures rms value; if operated in the upper part, it measures average value.

appreciably discharge the capacitor during each cycle. In other terms: the condition for correct measurement of $\mathbf{E}_p$ is

$$R_M C > T_e$$

where T_e is the time of a complete period of the applied voltage. However, the time constant $R_M C$ must not be too large, so that the meter can follow the variations or recessions of the peak voltage (if such recessions occur) over a greater number of periods. If the time constant $R_M C$ approximates infinity, the meter will remain deflected until the capacitor gets discharged, either by its own parallel resistance or by a hand-operated switch. This type of circuit can be useful as a *crest voltmeter* to indicate the occasional occurrence of a large voltage peak. Very high peak voltages (>15 kV) can be measured with moderate accuracy by means of spark gaps as described in Chap. 8.

The *average* value of a symmetrical, periodically varying voltage is, of course, zero. The average voltage of a half-cycle $T/2$ of a symmetrical, periodically varying voltage is defined as

$$E_{av} = \frac{2}{T}\int_{t=0}^{t=T/2} e\,dt \tag{7-1}$$

For sinusoidal variations, the average voltage is

$$E_{av} = 0.637 \times \mathbf{E}_p \tag{7-1a}$$

The average voltage for any waveshape can be measured with the circuit shown in Fig. 7-5*a*. The meter *M* must have a high mechanical inertia so that it does not follow the instantaneous variation of the rectified current. The diode *D* must have a linear voltage-current characteristic. A diode has a characteristic like that shown in Fig. 7-5*b*, i.e., it exhibits nonlinearity for a small voltage across the diode. For correct measurement of the average voltage, it is essential to operate the diode in the upper part of its characteristic where the nonlinearity becomes negligible. Germanium diodes are preferable for this purpose since their characteristic is more nearly linear than that of silicon rectifiers. The

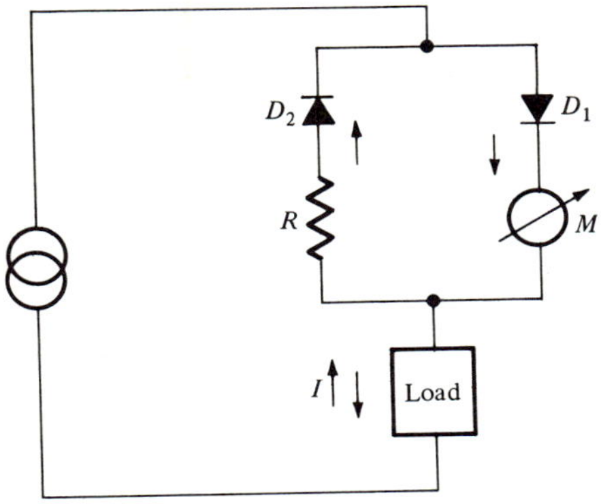

Fig. 7-6 System for the measurement of rms or average current. Current flow through meter M is unidirectional, but bidirectional ac current I flows through load.

polarity of the diode is unimportant if the applied alternating voltage is symmetrical; for unsymmetrical waveshapes, the value of the average voltage is, of course, different for both polarities.

The circuit in Fig. 7-6 is applicable for voltage measurements where the voltmeter current is so small as not to alter the potential distribution in the main circuit. This is a condition for any good voltmeter method. The circuit is, however, not applicable for *current* measurements since it has the effect of rectifying not only the current for the meter I_M but the main current as well. The circuit shown in Fig. 7-6 avoids the rectification of the main current I. In one polarity the current passes through diode D_1 and meter M; in the other polarity, the current can pass in the opposite direction through D_2 and R.

Of generally greater interest than the average voltage is the *root-mean-square value* of an alternating voltage, i.e., that dc voltage which has the same thermal effect as the alternating voltage. It is defined as

$$E_{\text{rms}} = \sqrt{\frac{1}{T}\int_0^T e^2\,dt} \tag{7-2}$$

and for a sinusoidal waveform, $E_{\text{rms}} = 0.707 \times \mathbf{E}_p$. To measure the rms value of a voltage or a current, the instantaneous voltage is first applied to a squaring element. The output from this element is proportional to the (instantaneous) square value of the input. A second element forms the time average; a third element forms an output proportional to the root of its input. If the latter element is a meter, the formation of the root is contained in its calibration. Many technical variations of this general scheme are available. A rectifier with a square characteristic can be used as the first element, but for the rms measurement the diode must be operated in the lower part of the EI characteristic, where approximately a quadratic relationship exists between voltage and current. Silicon diodes are preferable for this purpose. The time average is formed by a sufficiently inert meter. By calibrating the meter in rms units, it fulfills the functions of averaging and root forming elements. The greatest source of errors in such a system is usually in the rectifier device. A thermal device with

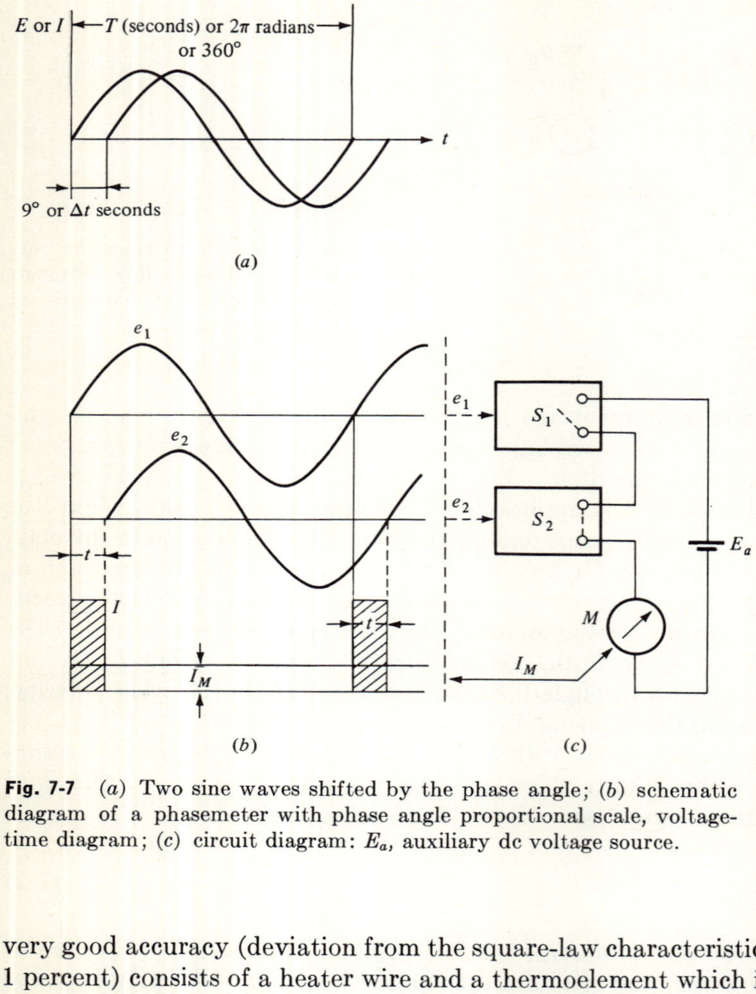

Fig. 7-7 (a) Two sine waves shifted by the phase angle; (b) schematic diagram of a phasemeter with phase angle proportional scale, voltage-time diagram; (c) circuit diagram: E_a, auxiliary dc voltage source.

very good accuracy (deviation from the square-law characteristic between 0.1 and 1 percent) consists of a heater wire and a thermoelement which is attached to the center of the wire. The entire system frequently is sealed in an evacuated vessel; the sensitivity of a system in vacuo is about ten times that of a system in air. The thermoelement output is proportional to the heater temperature and, therefore, proportional to the square of the applied voltage or current. Thermoconverters have been built for currents between 2 mA and 1 A; the output voltage is generally 5 to 25 mV full range. These converters exhibit a time lag of several seconds and they may be damaged by even moderate overload.

Electrostatic and electromagnetic meters with a mechanism causing a force or a displacement proportional to the square of the applied voltage or current will indicate the rms value directly on a suitably calibrated scale.

One difficulty with the measurement of alternating voltages and currents is that an acceptable standardization does not exist. Meters may measure peak values but are calibrated in rms values with the implied assumption that the system is to be used only for sinusoidal waveshapes.

PHASE

Figure 7-7*a* shows two periodic processes which occur at the same frequency or within the same duration of a cycle T, but which follow each other with a time delay. The phase difference or phase shift between both processes, i.e., the relative position of the two curves, can be expressed in time difference Δt but is more frequently expressed in angular degrees ϕ, the length of a whole cycle being 360° or 2π rad. Phase differences can be measured with high accuracy; therefore, elements which carry information by means of a variable phase angle are capable of high accuracy and sensitivity.

A simple but only moderately accurate method for the measurement of phase angles is the observation of two traces on a dual-beam oscilloscope. More accurate (but somewhat cumbersome) is the method of Lissajous' figures on a single-beam oscilloscope: one signal is applied to the horizontal deflection-producing plates, the other signal is applied to the vertical deflection-producing plates. Since both signals are of the same frequency, a stationary figure will appear on the screen of a form depending on the phase difference (see Fig. 7-2*a* to *h*). The phase angle is zero for Fig. 7-2*a*, 90° or $\pi/2$ in Fig. 7-2*c*, and 180° or π in Fig. 7-2*e*. In the intermediate positions the phase angle can be found from $\sin \varphi = \pm B/A$, where A is the ordinate maximum value and B is the value of the ordinate at the intersection with the Lissajous figure (see Fig. 7-2*d*).

A system that furnishes an output proportional to the phase angle is illustrated in the schematic diagram of Fig. 7-7*c*. The voltage e_1 is applied to an electronic switch S_1; whenever e_1 passes from a negative to a positive value, switch S_1 closes the output circuit. The voltage e_2 is applied to the electronic switch S_2, which closes when e_2 is negative and opens when e_2 changes to a positive value. The effect of the two switches is that a pulse arises in the output circuit of a duration t which is proportional to the phase difference between e_1 and e_2. The meter M in the output indicates the average current value I_M, which is also proportional to the phase shift between e_1 and e_2. Other methods for the determination of phase angles will be described later.

AC POWER

The power at dc was defined in Chap. 1 as the product of voltage and current. This holds true for the instantaneous values at ac, that is, $p = ei$. Since

$$e = \mathbf{E}_p \sin (\omega t + \varphi_1)$$

and

$$i = \mathbf{I}_p \sin (\omega t + \varphi_2)$$

it follows that the mean power is

$$P = \frac{1}{T} \int_0^T ei\, dt = EI \cos (\varphi_2 - \varphi_1) = EI \cos \varphi \tag{7-3}$$

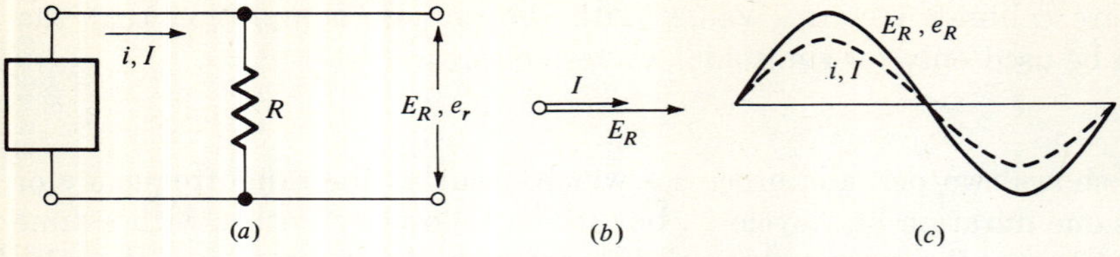

Fig. 7-8 Voltage-current relationship in a resistance. (*a*) Circuit; (*b*) vector diagram; (*c*) voltage- and current-time diagram.

The instantaneous power is rarely of interest. It can be measured with electrodynamic oscilloscopic systems or with analog multiplier systems which furnish an output that at any time is proportional to the product of both input signals e and i.

The mean power is usually measured with electrodynamic meters, i.e., systems containing a moving coil in the field of an electromagnet. If the current through the electromagnet is made proportional to e and that through the coil proportional to i, the average deflection of the coil will be proportional to $EI \cos \varphi$, where E and I are the rms values. Also electronic wattmeters are available in the form of phase-sensitive multipliers, which furnish an output signal proportional to the mean power transferred from one element to another.

RESISTANCE, REACTANCE, IMPEDANCE

If a resistance is connected in a circuit as shown in Fig. 7-8, the instantaneous voltage e_R arising between its terminals is proportional to the instantaneous current i,

$$e_R = Ri = R(\mathbf{I}_\mathbf{p} \sin \omega t)$$

It follows that the rms value of the resistive voltage E_R is also proportional to the rms value of the current I

$$E_R = RI$$

Voltage and current are in phase ($\varphi = 0$).

If a pure inductance L is connected in a circuit as shown in Fig. 7-9, the instantaneous voltage e_L arising between its output terminals is proportional to the negative time derivative of the current [Eq. (6-3*a*)]. If the current alternates sinusoidally, the instantaneous output voltage is

$$e_L = -L\frac{di}{dt} = -L\frac{d(\mathbf{I}_\mathbf{p} \sin \omega t)}{dt} = -\omega L\left[\mathbf{I}_\mathbf{p} \sin\left(\omega t + \frac{\pi}{2}\right)\right]$$

It follows that the rms value of the inductive voltage E_L is also proportional to the rms value of the current I

$$E_L = \omega LI = \mathbf{X}_\mathbf{L} I$$

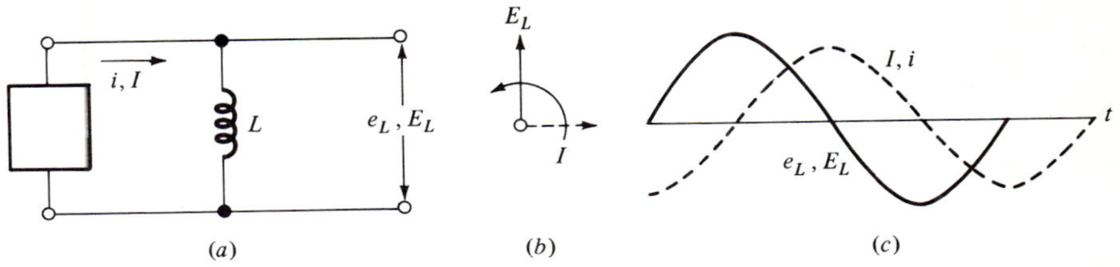

Fig. 7-9 Voltage-current relationship in an inductance. (*a*) Circuit; (*b*) vector diagram; (*c*) voltage- and current-time diagram.

where $\omega L = \mathbf{X_L}$ is the inductive reactance, in ohms. A phase shift of $\pi/2$ or 90 exists between the voltage and the current, the voltage vector leading.

If a pure capacitance C is connected in a circuit as shown in Fig. 7-10, the instantaneous current i_c in the circuit is proportional to the time derivative of the voltage [see Eq. (5-16)]. If the voltage alternates sinusoidally, the instantaneous current is

$$i_C = C\frac{de}{dt} = C\frac{d(\mathbf{E_p}\sin\omega t)}{dt} = \omega C\left[\mathbf{E_p}\sin\left(\omega t + \frac{\pi}{2}\right)\right]$$

It follows that the rms value of the current I_C is also proportional to the rms value of the capacitive voltage E_C

$$I_C = \omega C E_C \qquad \text{or} \qquad E_C = \frac{1}{\omega C}I = X_C I$$

where $1/\omega C = \mathbf{X_C}$ is the capacitive reactance, in ohms. A phase shift of $\pi/2$ or 90° exists between the voltage and the current, the current vector leading.

The described relationship between voltage, current, frequency, and phase angle of a reactance forms the basis for a number of methods of measuring a reactance and such physical phenomena which may cause a reactance variation. If nothing is known about the reactance, then at least its magnitude can be determined from voltage and current measurements: $\mathbf{X} = E/I$, where both the voltmeter and the ammeter must indicate the same parameters, such as peak

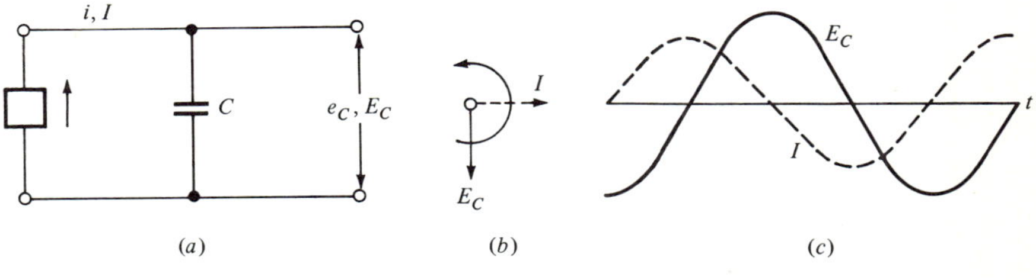

Fig. 7-10 Voltage-current relationship in a capacitive circuit. (*a*) Circuit; (*b*) vector diagram; (*c*) voltage- and current-time diagram.

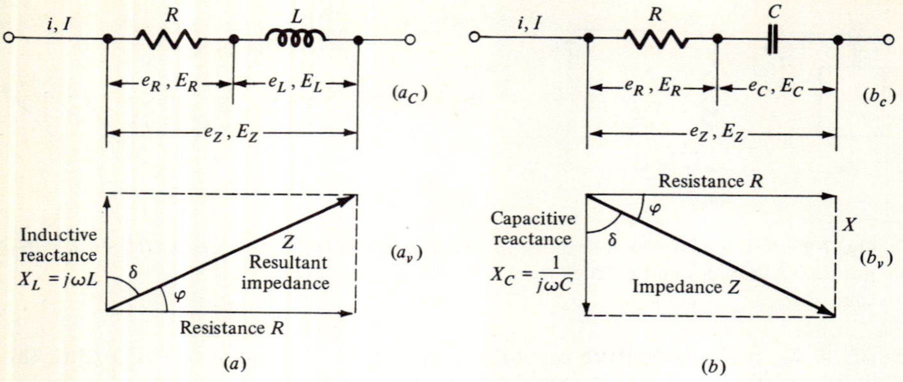

Fig. 7-11 (*a*) Resistive-inductive impedance. Circuit diagram above, vector diagram below. (*b*) Resistive-capacitive impedance. Circuit diagram above, vector diagram below.

values or rms values. Discrimination between inductive and capacitive reactance is more difficult since direct-indicating phasemeters are rarely available. The simplest method is by variation of the frequency; increase of the reactance with increased frequency indicates an inductive reactance.

The concept of pure reactances is, of course, an abstraction. Every inductor has a resistance which may be considered to be in series with the inductance. Every capacitor has an imperfect dielectric and, hence, may be considered a capacitance in parallel with a resistance. The combined effect of a reactance and a resistance is that of an impedance. A reactance is determined by magnitude and phase angle, i.e., it is a vector; therefore, reactances and resistances must be combined geometrically. Using the expressions of complex numbers, the impedance of a series combination of a resistance and a reactance is

$$\mathbf{Z}_\Omega = \mathbf{R} + j\mathbf{X}$$

where $j = \sqrt{-1}$. This is merely a symbolic analytical expression for the graphic procedure shown in Fig. 7-11*a* and *b*. The series combination of a resistance **R** and an inductive reactance ωL (Fig. 7-11*a*) is

$$\mathbf{Z} = \mathbf{R} + j\omega L$$

The series combination of a resistance with a capacitive reactance $1/\omega C$ (Fig. 7-11*b*) is

$$\mathbf{Z} = \mathbf{R} + \frac{1}{j\omega C}$$

Equally simple procedures for the graphic presentation of **Z** in parallel combinations are not available. The general method used to handle these problems is the conversion of the vectorial quantities **Z**, **R**, and **X** into corre-

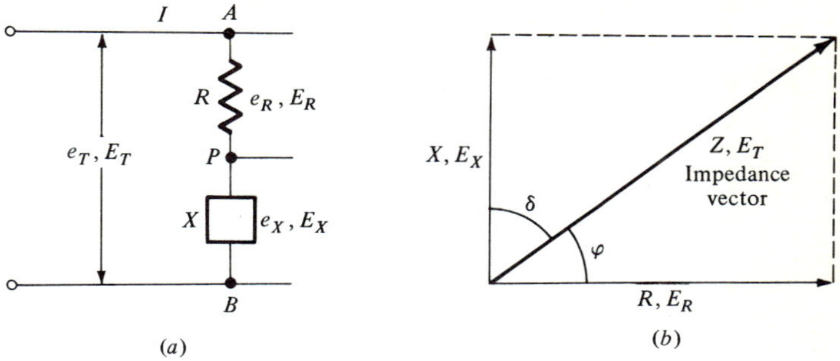

Fig. 7-12 Determination of an unknown impedance (R plus X) from voltage and current measurements. (a) Circuit; (b) vector diagram.

sponding reciprocal quantities: $\mathbf{Y} = 1/\mathbf{Z}$ (*admittance*), $\mathbf{G} = 1/\mathbf{R}$ (*conductance*), and $\mathbf{B} = 1/\mathbf{X}$ (*susceptance*). These new vectors are used in the same way as the former vectors in series combinations, but the analytical method is generally preferable.

The analytical expressions for the resistive-inductive parallel network are

$$\mathbf{Z} = \frac{\mathbf{R}j\omega L}{\mathbf{R} + j\omega L} = \mathbf{R}\omega L \frac{\omega L + j\mathbf{R}}{\mathbf{R}^2 + \omega^2 L^2}$$

and for the resistive-capacitive parallel networks

$$\mathbf{Z} = \frac{\mathbf{R}1/j\omega C}{\mathbf{R} + 1/j\omega C} = \frac{\mathbf{R} - j\omega C\mathbf{R}^2}{1 + \omega^2 C^2 \mathbf{R}^2}$$

The significance of the vectorial addition, for instance in a resistance-reactance series combination (see Fig. 7-12), is the following. The voltage (rms value) measured across the resistance $\mathbf{R}$ is $\mathbf{E_R}$; that across the reactance $\mathbf{X}$ is $\mathbf{E_X}$. The algebraic sum $(\mathbf{E_R} + \mathbf{E_X})$ is always greater than the measured total voltage E_T as evident from the vector presentation. Of course, the instantaneous voltages are added arithmetically, that is, $e_R + e_X = e_T$.

The magnitude of an impedance vector may be found experimentally from voltage and current measurements: $\mathbf{Z} = E/I$, as described above for reactances and as illustrated in Fig. 7-12. The direction of the impedance vector is expressed by the phase angle φ between the impedance vector $\mathbf{E_Z}$ and the resistive component $\mathbf{E_R}$ of this vector. If the resistance R is accessible, the voltages $\mathbf{E_R}$ and $\mathbf{E_X}$ can be measured, and the impedance vector and the phase angle φ can be found from triangulation. Usually $\mathbf{R}$ and $\mathbf{X}$ are not separately accessible, but in this case the direction of the impedance vector can be found since the resistive voltage $\mathbf{E_R}$ is in phase with the current I through the impedance. The direction or the phase angle of the impedance is, therefore, that between the voltage E_T applied to the impedance and the current I passing through the impedance. Triangulation methods or methods of determining impedances from phase-angle measurements are rarely very accurate. Bridge methods, as described below, are generally more elegant, more direct, and capable of achieving much higher accuracy.

The ratio of reactive to resistive component of an impedance, i.e.,

$$\frac{\mathbf{X}}{\mathbf{R}} = \tan \varphi = Q$$

(Q, called *storage factor*) has a special significance. The larger Q is, the more the impedance approaches the character of a pure reactance. Capacitances can be built with high Q values sometimes exceeding the value of 10^6. Inductances generally have Q values which do not exceed several hundred. The reciprocal of the storage factor Q is the dissipation factor

$$D = \frac{1}{Q} = \frac{\mathbf{R}}{\mathbf{X}} = \tan (90° - \varphi) = \tan \delta$$

(Fig. 7-12). The choice between Q or D in impedance calculations is a matter of convenience. For instance, the storage factor of a *parallel RC* network is

$$Q = \mathbf{R}\omega C$$

while the dissipation factor of a *series RC* network is

$$D = \mathbf{R}\omega C$$

Resistive-reactive networks can be used as filters (see Fig. 7-13). The magnitude of the reactance of a capacitance C is $\mathbf{X_c} = 1/\omega C$, that is, it diminishes

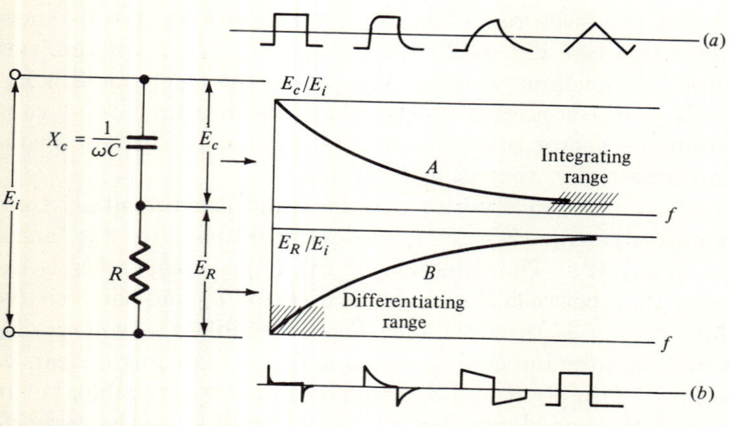

Fig. 7-13 Behavior of an RC network at different frequencies; left, circuit diagram; right, relative magnitudes of the output voltages, A across the capacitance and B across the resistance. Columns a and b show the appearance of an output signal at different frequencies if a square wave is applied at E_i.

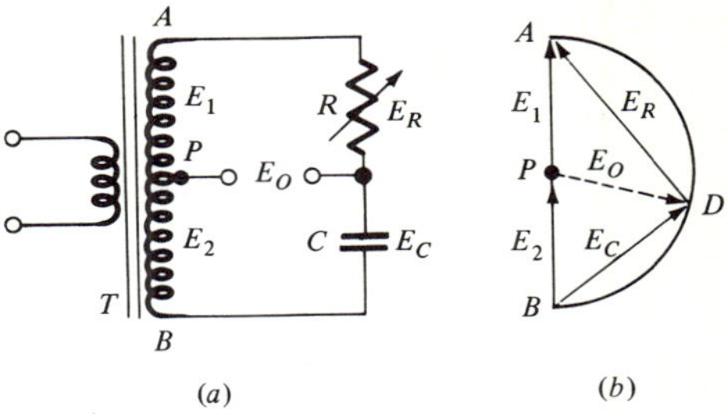

Fig. 7-14 Circuit which produces a voltage E_o that can be varied in phase from 0° to almost 180° by varying the resistance R. The magnitude of E_o is independent of the phase-angle setting.

with increasing frequency. The resistance **R** is, of course, frequency independent. The voltage across the capacitance is

$$E_c = E_i \frac{1/j\omega C}{R + 1/j\omega C} = E_i \frac{1}{RC}\left(\frac{1}{j\omega + 1/RC}\right)$$

At low frequency, when $\mathbf{X_C}$ is large, the larger fraction of the applied voltage E_i will appear across C as E_c; this partial voltage will diminish at higher frequencies. With E_c as the output, the network acts as a low-pass filter; its characteristic is shown in Fig. 7-13, curve A. The circuit is identical with the integrating circuit shown in Fig. 5-14b which integrates correctly for $E_o \ll E_i$, that is, for $R\omega C \gg 1$ or $\omega \gg 1/RC$. If a square wave is applied at E_i, it will emerge as a square wave at E_c if the frequency is low, but it will emerge in integrated form as shown in line a, i.e., as a triangular wave, if ω is large.

The voltage across the resistance is

$$\mathbf{E_R} = E_i \frac{1}{R + 1/j\omega C} = E_i \frac{j\omega RC}{j\omega RC + 1}$$

With sinusoidal input and at low frequency, the voltage $\mathbf{E_R}$ is small; it rises with increased frequency and approaches E_i asymptotically. Its characteristic is that of a high-pass filter as shown in Fig. 7-19b.

The circuit is identical with the differentiating circuit shown in Fig. 5-13, which differentiates correctly for $E_o \ll E_i$, that is, for $R\omega C \ll 1$ or $\omega \ll 1/RC$. If a square wave of low frequency is applied at E_i, the output $\mathbf{E_R}$ will be a short pulse of a width and height determined by the rise time of the square wave as explained in Chap. 5. If a square wave of high frequency is applied at E_i, it will emerge at $\mathbf{E_R}$ as a square wave as shown in line b.

Figure 7-14 shows a circuit which furnishes a voltage E_o that can be adjusted in phase from 0° to almost 180°. It consists of a transformer T with a center tap P. Between terminals AP and PB two equal voltages E_1 and E_2 exist that are in phase (Fig. 7-14b). Terminals A and B are connected by capacitor C and resistor R in series. Either resistor R or capacitor C should be made variable. The voltage across the resistor and that of the capacitor are 90° phase-shifted; the connecting point D is, therefore, on the semicircle drawn over AB. The output voltage E_o between P and D varies in position (phase angle) with any

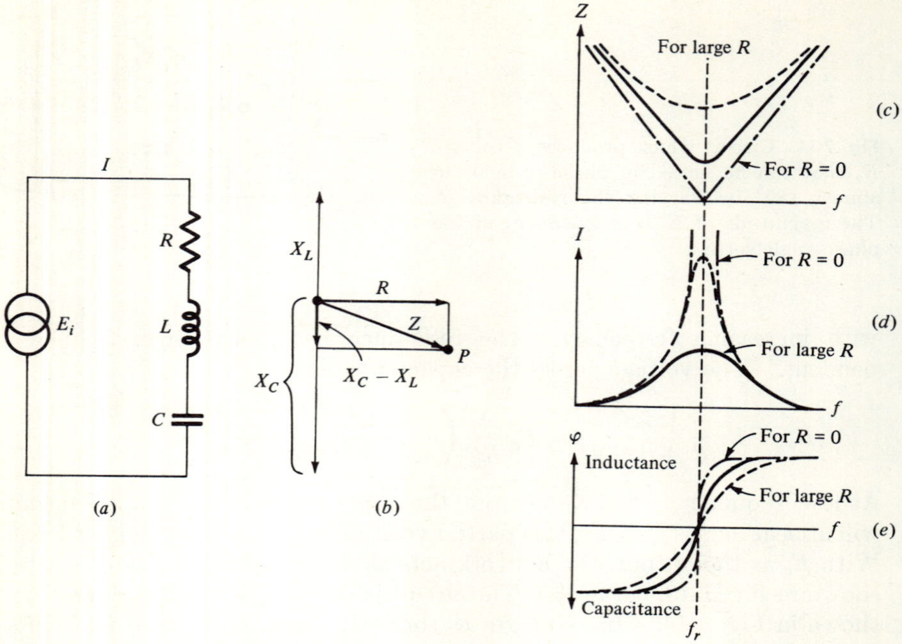

Fig. 7-15 Series resonance. (*a*) Circuit; (*b*) vector diagram; (*c*) impedance; (*d*) current; (*e*) the phase angle between applied voltage and circuit current, all plotted as function of the frequency of the applied voltage E_i.

variation of R or C, but E_o does not vary in magnitude. Because of the unavoidable resistive component of C and the reactive component of R, the phase angle cannot be changed by a full 180° (usually only by about 170° to 175°) although more complex circuits have been devised which permit a variation of the phase angle by 180° and more.

RESONANCE

Figure 7-15 shows a circuit containing in series a resistance $\mathbf{R}$, an inductance L, and a capacitance C. The total impedance is $\mathbf{Z} = \mathbf{R} + j\mathbf{X}$ (absolute value $|\mathbf{Z}| = \sqrt{\mathbf{R}^2 + \mathbf{X}^2}$), where $\mathbf{X} = \mathbf{X_L} - \mathbf{X_C}$. The vector diagram for this circuit is shown in Fig. 7-15*b*. If a constant voltage E_i is applied and if the frequency of this applied voltage increases, the inductive reactance $\mathbf{X_L} = \omega L$ will increase, while the capacitive reactance $\mathbf{X_C}$ will decrease. This means that in the vector diagram the end point P of the impedance vector moves upward and falls into the direction of the resistive component for

$$\omega L = \frac{1}{\omega C} \qquad \omega = \frac{1}{\sqrt{LC}} \qquad f_r = \frac{1}{2\pi\sqrt{LC}}$$

where f_r is the resonant frequency. At resonance the following phenomena occur in a series RLC circuit:

1. The total impedance $\mathbf{Z}$ of the circuit reaches a minimum, namely the value R; therefore, the current I reaches a maximum, the value E/R.
2. The circuit acts as a capacitive-resistive impedance at frequencies below resonance and as an inductive-resistive impedance above resonance. At resonance the phase angle is zero, and the circuit behaves like a pure resistance.
3. At resonance the voltages across the capacitance and the inductance reach their highest values, but these voltages are of opposite phase, i.e., they cancel each other. The voltage across L is $jI\omega L$, that across C is $I/j\omega C$, or if expressed by the storage factor Q of the circuit (where $Q = \omega L/R$ or $1/R\omega C$), $E_L = E_C = E_iQ$. The voltages across L or C can assume values that are many times larger than the applied voltage E_i; the series resonance circuit can be used for the generation of high voltages.

Figure 7-16*a* shows a parallel resonance circuit (*tank circuit*) consisting of an inductance L in series with a resistance R_L (R_L includes the dc resistance of the

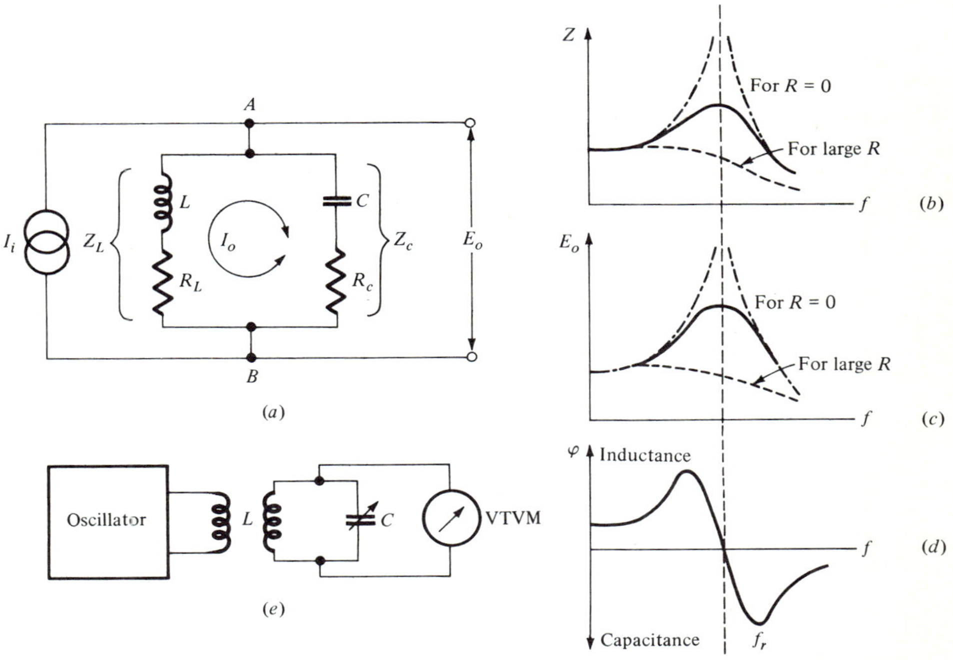

Fig. 7-16 (*a*) Parallel resonance circuit connected to a constant-current source I_i and synopsis of the physical processes in a parallel resonance circuit; (*b*) impedance Z between A and B; (*c*) voltage E_o between A and B; (*d*) phase angle between applied current I_i and voltage E_o all plotted as a function of the frequency of the applied current; (*e*) schematic diagram of a circuit for frequency analysis.

coil) and a capacitance C in series with a resistance R_C. The circuit is connected to a constant-current generator. The equivalent impedance of this circuit is

$$Z = \frac{Z_C Z_L}{Z_C + Z_L}$$

This circuit also shows extreme values at resonance, when $\omega L = 1/\omega C$. The resonant frequency is

$$f_r = \frac{1}{2\pi \sqrt{LC}}$$

which is the same frequency as for the series-resonance circuit.

The following phenomena occur in a parallel-resonance circuit:

1. The total impedance $\mathbf{Z}$ of the circuit reaches a maximum; therefore, if the parallel resonance circuit is supplied by a constant-current source, the voltage across the circuit reaches a maximum (Fig. 7-16*b* and *c*).
2. The circuit acts as a resistive network at very low frequencies; as the frequency rises, the circuit acts as an inductive-resistive impedance. Shortly before resonance the trend of the curve reverses, and at resonance the circuit acts as a pure resistance. At frequencies above resonance the circuit acts as a capacitive-resistive impedance as shown in Fig. 7-16*d*.
3. The current in the tank circuit consisting of L, R_L, C, R_C is small at all frequencies outside of resonance but increases steeply in the vicinity of the resonance frequency and reaches a maximum at resonance. The current at resonance is

$$I_o = I_i Q$$

where $Q = \omega L/(R_L + R_C) = 1/\omega C(R_L + R_C)$. The current I_o can reach values of from several tens to hundreds of amperes. This current flows in the tank circuit only (see arrow I_o in Fig. 7-16*a*). It results from a discharge of the capacitor C through the inductance L which causes a voltage in L which in turn charges C with opposite polarity. The process repeats with a periodicity given by C and L. The source I_i furnishes only the current to cover the losses in R_C and R_L.

Every inductor has a distributed capacitance and, hence, can be considered a parallel-resonance circuit. This may lead to the apparent paradox that an inductor operated at a frequency above resonance acts as a capacitor. Such effects are indeed observed and can be very disturbing in electronic oscillators where they frequently lead to undesirable "parasitic" oscillations.

Resonance circuits find numerous applications in instrumentation. If the circuit can be calibrated, it can be used for frequency measurements. An example is shown in Fig. 7-16*e* where a tunable-resonance circuit consisting of a

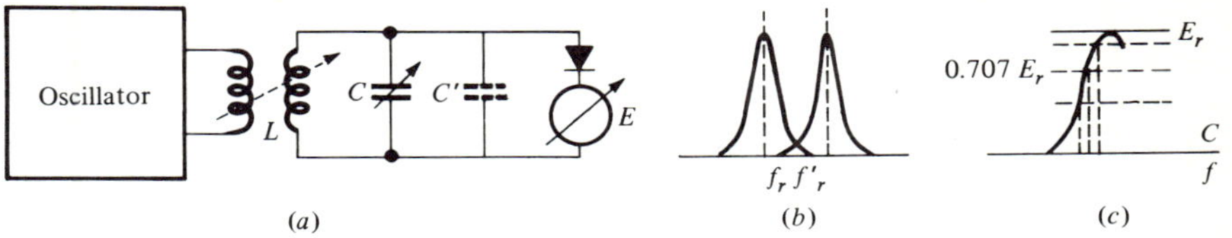

Fig. 7-17 Schematic diagram of a Q meter applicable to the measurement of the storage factor Q of the circuit or of the circuit components, such as L or C, as well as to the measurements of L, ΔL, C', ΔC, f, and related quantities. (*a*) Schematic of circuit; (*b*) resonance shift caused by the insertion of C'; (*c*) the method of "half-power point" for the observation of small shifts in frequency (see text).

fixed inductor L and a variable capacitor C is brought in the vicinity of an oscillator. The magnetic stray field from the oscillator coil induces voltages in the resonant circuit. These voltages can be measured, for instance, with an electronic voltmeter. The voltage reaches a maximum when the resonant frequency of the circuit equals the emitted frequency of the oscillator. If the oscillator emits a spectrum composed of different frequencies, it is possible by successive tuning of the resonance circuit to analyze the spectrum, i.e., to measure the frequency and the relative magnitude of every component of this spectrum.

The sharp rise of the resonance curve in high-Q circuits also permits very accurate and sensitive measurement of the inductance L, if C is known, or of the capacitance C, if L is known, or of any change ΔL or ΔC which may result from change of physical dimensions, displacement, force, or from change of material properties (such as magnetic permeability, dielectric constant, conductivity, or change in chemical composition or structure). An example is shown in Fig. 7-17. A resonance circuit containing the inductance L and the two capacitors C and C' is loosely coupled to the oscillator. The oscillator induces a voltage (hence, a current) in the resonance system. Because of the loose coupling, the reaction of the resonance circuit on the oscillator is negligible. The circuit is tuned to resonance; the resonance effect is indicated by the meter V. Any variation of the capacitance C' (caused in response to, say, a mechanical displacement of the capacitor plates) changes the resonant frequency. Two methods can be distinguished. Prior to the measurement, the circuit is tuned to resonance. The variation of C' causes a variation of the resonant frequency from f_r to f'_r (Fig. 7-17*b*); resonance can be reestablished through a variation of C by the amount ΔC. The change ΔC can be read off a calibrated dial. Alternatively, prior to measurement, the circuit is adjusted to a value below or above resonance so that the output power is one-half that at resonance; the voltage is then $1/\sqrt{2}$ of E_R or $0.707E_R$ as indicated in Fig. 7-17*c*. Any variation of C' causes an increase or decrease of the voltage across the resonance circuit which is measured with the electronic voltmeter V. The method is limited to small variations of C'. It is extremely sensitive particularly if the Q factor of the circuit and, therefore, the slope of the resonance curve, are high. The method requires a highly stabilized oscillator and very stable components in the resonance circuit.

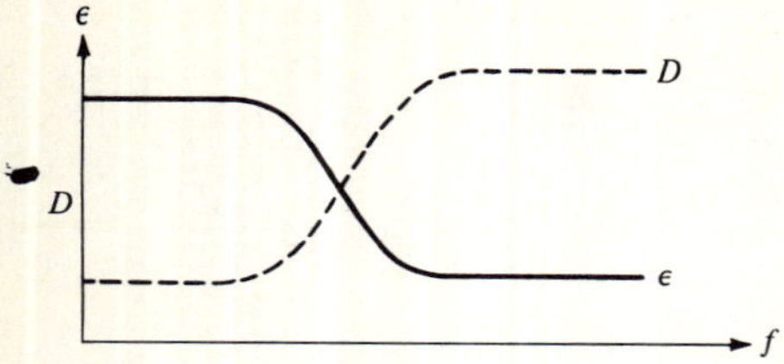

Fig. 7-18 Generalized diagram illustrating the variation of dielectric constant ε and the dissipation factor D of dielectric materials, plotted versus frequency.

The storage factor Q of the circuit determines the height of the resonance peak or the relative width of the resonance curve and can be measured. A Q meter (in principle the circuit shown in Fig. 7-17*a*) can be used to measure the Q value, the resistive, or the reactive value of circuit components. It also can be used to measure the dielectric constant and conductivity of substances brought between the plates of a capacitor. The measurement of the storage factor Q and of the capacitance C at different frequencies also permits the determination of the dipole moment of polar substances. A low concentration of a polar substance in a nonpolar medium is brought between the plates of a capacitor; the dielectric constant and the conductivity are measured at different frequencies. At low frequencies, the phase shift between the polarization of the dipoles and the buildup of the electric field in the capacitor is negligible but, as the frequency increases, the phase angle increases with the result that the dielectric constant diminishes to a smaller value as shown in the solid curve of Fig. 7-18. The frequency range in which the change of ε occurs is determined by the relaxation time of the dipole. In the same frequency range in which the dielectric constant diminishes, there is an increase of conductivity. As may be seen from the dotted curve of Fig. 7-18, these processes follow curves which have the character of an anomalous dispersion rather than of a resonance process. Therefore, no wavelength-specific effect in the sense of a maximum energy absorption at a particular frequency has ever been found in chemical or biological media, nor is such an effect expected.[1]

Inductances and capacitances can be used as low-pass or high-pass filters in a way similar to the resistance-capacitive networks illustrated in Fig. 7-13. Two examples are shown in Fig. 7-19. The low-pass filter of Fig. 7-19*a* attenuates but little any voltage of frequencies $f < 1/2\pi\sqrt{CL}$ applied to the input termi-

[1] See Paul Schupp, "Wiss. Veröff. Siemens Werke," vol. XVII, Springer, Berlin, 1938.

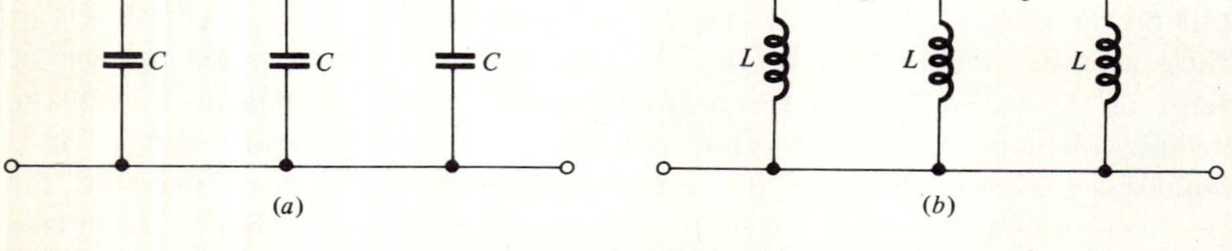

Fig. 7-19 *LC* filters. (*a*) Low-pass filter; (*b*) high-pass filter.

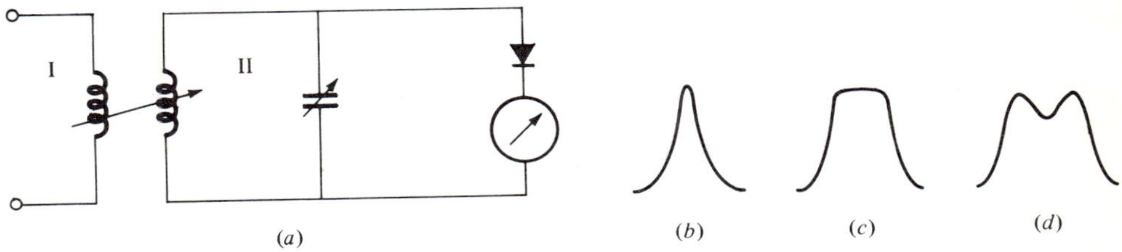

Fig. 7-20 Variation of the coupling between a primary driving system I and a secondary driven resonance system II. (*a*) Circuit; (*b*) resonance at loose coupling; (*c*) resonance at closer coupling; the resonance curve has a bandfilter character; (*d*) saddle-shaped curve for very close coupling.

nals, but attenuates strongly at frequencies $f > 1/2\pi\sqrt{CL}$. The high-pass filter of Fig. 7-19*b* has the reverse characteristic. The network shown in Fig. 7-20*a* has a transfer characteristic of the form of a resonance curve (Fig. 7-20*b*) for a loose coupling between the two coils. However, if the coupling is closer, there will be a reaction from the secondary system to the primary system. This effect is, of course, strongest at resonance and will cause a flattening of the resonance curve as shown in Fig. 7-20*c*. Such an arrangement acts as a bandpass filter; it permits the passage of a frequency band and cuts off rather sharply any frequencies outside of this band. The effect can be enhanced by staggering a number of stages. If the coupling between the two coils becomes too close, a transfer characteristic of the type shown in Fig. 7-20*d* (*saddle curve*) will ensue.

SUPERPOSITION OF ALTERNATING VOLTAGES AND CURRENTS

For two voltage sources e_1 and e_2 in series, two currents i_1 and i_2 in parallel, at dc, $e_T = e_1 + e_2$ and $i_T = i_1 + i_2$. The same relationships hold at ac as far as the instantaneous values are concerned, no matter what the frequencies or phase relationships of the two components are. The situation is shown in Fig. 7-21. The two components in Fig. 7-21*a* are of the same frequency and the same phase, and in Fig. 7-21*b* of the same frequency but of different phase. Since both sine waves are formed by the rotation of two vectors with the same angular velocity, the resulting wave can be obtained by the vectorial addition of the components $\mathbf{E}_1$ and $\mathbf{E}_2$ or $\mathbf{I}_1$ and $\mathbf{I}_2$, respectively.

The situation is different in Fig. 7-21*c*, where one vector $\mathbf{E}_2$ rotates with a frequency different from that of the other vector, $\mathbf{E}_1$. If the angular velocity or frequency of the second vector $\mathbf{E}_2$ is a multiple integer of that of the first vector, then a standing-wave figure will result which is repeated in every cycle of the lowest frequency (the *fundamental*, also called *first harmonic*). This case is illustrated in Figs. 7-21*c* and *d* which represent both the superposition of a fundamental and the third harmonic. The configurations of the resulting waveshapes are different for different phase shifts between the two components. Interestingly enough, for a number of applications, there is no difference between the effects of the resultant waveshapes in Fig. 7-21*c* and *d*. For instance, if two

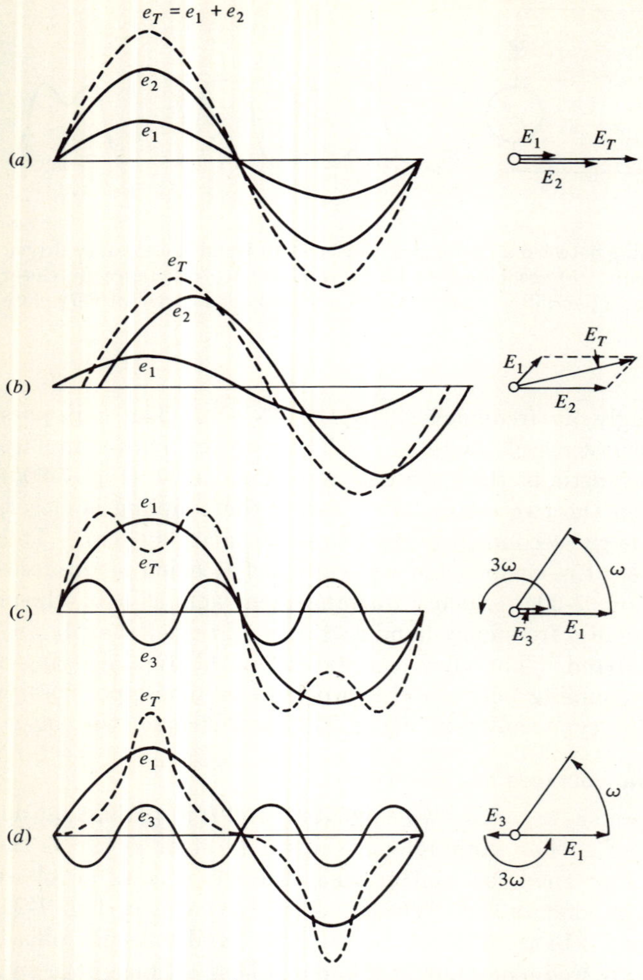

Fig. 7-21 Superimposition of sine-shaped voltages. The two components e_1 and e_2 are (*a*) the same frequency and the same phase; (*b*) same frequency, different phase; (*c*) different frequencies; (*d*) same two frequency components as in (*c*) but different phase. The instantaneous output voltages are e_T.

voltages with the same harmonic content are applied to a speaker, the human ear will not perceive any difference[1] due to the phase shift of the components; i.e., the ear acts as a harmonic analyzer, it is affected only by the frequencies of the components and their respective amplitudes. For many applications it is

[1] Differences may be perceived, however, because of secondary effects. For instance, the amplitude of the resultant in Fig. 7-21*d* is larger than in 7-21*c*. If the speaker or the ear has a nonlinear amplitude distortion, a difference will be perceived.

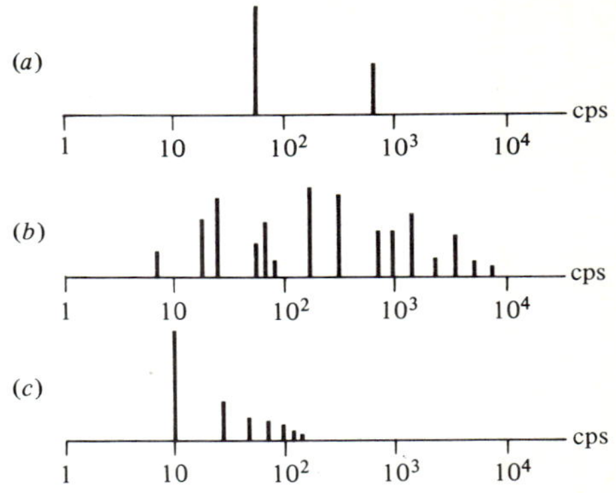

Fig. 7-22 Presentation of frequency spectra. (*a*) Spectrum of Fig. 7-21*c* or *d*, consisting of a 60- and a 180-cycle component; (*b*) spectrum of a violin tone; (*c*) spectrum of a 10-Hz square wave.

sufficient, therefore, to characterize a frequency mixture by a diagram (*Stumpf-diagram*) as shown in Fig. 7-22.

Harmonic analysis (*Fourier analysis*) is a means to discover hidden periodicities in amplitude-time functions. Its use is justified when there is a suspicion of the presence of a periodic function, in contrast to correlation analysis which is applied to evaluate the presence of a hidden relationship between two or more events which need not be of a periodic character. An experimental method for harmonic analysis is shown in Fig. 7-23*a*. The amplitude-time function, which may be given in the form of a graph, is first converted into an electric signal by an input transducer. This may be done by an optical curve tracer, a photoelectric device driven by a servosystem which follows the trace of the curve and causes an electric output signal analogous to the plotted amplitude-

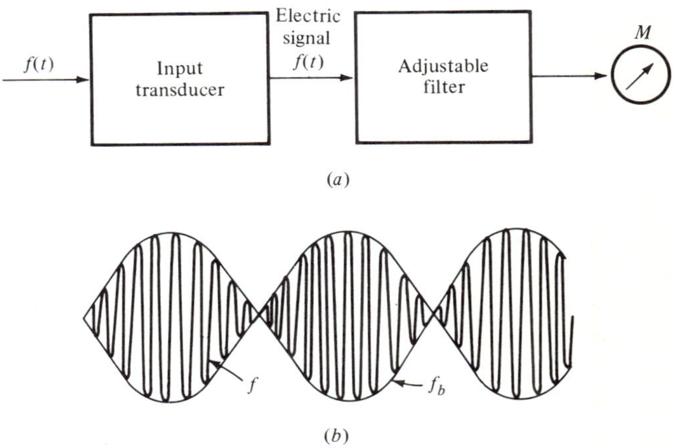

Fig. 7-23 (*a*) Schematic diagram of a frequency analyzer, see text for explanation. (*b*) Superimposition of a signal with a frequency f and another signal with a frequency F' close to f. The amplitudes (envelope curve) rise and fall with the beat frequency $f_b = \frac{1}{2}(f - f')$.

time function. The signal is then led through a tuned, adjustable filter which permits the passage of a narrow frequency range. The output amplitude is measured by a meter M. By running the signal through the system at different settings of the filter, the entire signal sequence can be investigated. Alternatively, a (sometimes considerable) number of filters and meters are used simultaneously. Adjustable filters in the subaudio- and audio-frequency range are difficult to build; therefore, the adjustable filter is frequently replaced by a fixed-frequency filter, and the frequency of the signal is made variable by superimposing an alternating current of variable frequency (see below).

In general, superimposed signals will not be multiple integers of a fundamental. The instantaneous value of two superimposed signals of the same amplitude can then be written

$$e = E \sin \omega_1 t + E \sin \omega_2 t$$
$$= 2E \sin \left(\frac{\omega_1 + \omega_2}{2} t \right) \cos \left(\frac{\omega_1 - \omega_2}{2} \right) t$$

i.e., a voltage alternating with the frequency $f = 1/2\pi[(\omega_1 + \omega_2)/2]$ and increasing and decreasing periodically with the difference or beat frequency $f_b = 1/2\pi[(\omega_1 - \omega_2)/2]$. The waveforms are shown in Fig. 7-23*b*. If ω_1 is 1 MHz (1,000,000 Hz) and ω_2 is 999,998 Hz, then the beat frequency will be 1 Hz, a magnitude which can be measured easily. The beat-frequency method offers an enormous sensitivity for the measurement of frequencies and, in further sequence, of such physical quantities that determine frequency, such as capacitance and inductance or the output from a transducer based on capacitance and inductance variations.

The superimposition method for the purpose of producing beat frequencies has other applications. Radio signals of different frequencies are picked up from an aerial. For detection and amplification, the received signal must be applied to a tuned amplifier with several stages (*bandpass filters*). Rather than adjusting each of these filters to whatever frequency is to be received, it is simpler to superimpose on the received signal another signal from a built-in (local) oscillator, mix the two frequencies so that an intermediate frequency f_i results, which is the same for each received frequency and which can be applied to the intermediate-frequency (IF) amplifier which now needs to be tuned for one frequency only. In a similar manner a harmonic analyzer can be built with a filter tuned to pass only one narrow frequency band which is the result of superimposing the unknown signal with a signal from a local oscillator.

The fact that all waveshapes of a periodically recurring function can be synthesized from sine functions of different frequencies, amplitudes, and phases has an important bearing on the dynamic characteristic of instrumentation elements and systems. If a periodic signal is applied to the input of a linear system (e.g., an amplifier), it will be reproduced faithfully at the output only if each frequency component contained in the signal is reproduced with the same gain. It is not necessary to investigate a system for every waveshape. The passage of

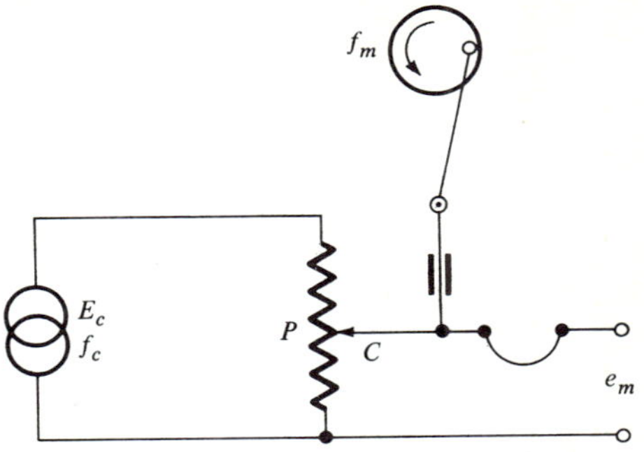

Fig. 7-24 Schematic diagram of amplitude modulation. The ac signal with the carrier frequency f_c is modulated in the rhythm of the modulation frequency f_m; signal e_m emerging at output terminals.

each frequency component with the same factor is sufficient to assure the faithful reproduction of a waveshape of any complexity which contains these frequency components.

The discussion above was concerned with the additive superimposition of alternating voltages and currents. A different, multiplicative, or *amplitude-modulation* type of superimposition is illustrated schematically in Fig. 7-24. A source E_C furnishes an ac voltage of the instantaneous value $e_c = E_c \sin \omega_c t$, where $f_c = \omega_c/2\pi$ is the carrier frequency. This signal is applied to a potentiometer P. The movable contact of the potentiometer is moved up and down in a sinusoidal manner with the modulation or signal frequency $f_s = \omega_s/2\pi$. Obviously, the movement of the sliding contact can alter the potentiometer output between two values: at one end, the output can be zero (not negative); at the other end, it can go to the fraction m of the maximum output. The fraction m, a figure between 0 and 1, is the modulation index. The instantaneous output voltage is

$$\begin{aligned} e_m &= E_c(1 + m \sin \omega_s t) \sin \omega_c t \\ &= E_c \sin \omega_c t + \tfrac{1}{2} m E_c \cos (\omega_c - \omega_s)t - \tfrac{1}{2} m E_c \cos (\omega_c + \omega_s)t \end{aligned}$$

The output voltage consists of three components with different frequencies: the original carrier wave $E_c \sin \omega_c t$ and two sidebands with the frequencies $f_c - f_s$ and $f_c + f_s$. It is evident that for the transfer of a modulated wave through a channel, a minimum bandwidth from $f_c - f_s$ to $f_c + f_s$ is required. The higher the signal frequency f_s is, the greater bandwidth required. Modulation is generally carried out with electronic means (see Chap. 13).

A schematic diagram of a *frequency-modulation* system is shown in Fig. 7-25*a*. The frequency f_c of the oscillator is determined by the resonance circuit LC; the capacitance of C is varied periodically with the signal or modulation frequency f_s, so that the instantaneous value of the frequency-modulated output is

$$e_0 = E_c \sin (\omega_c t + m_f \sin \omega_s t)$$

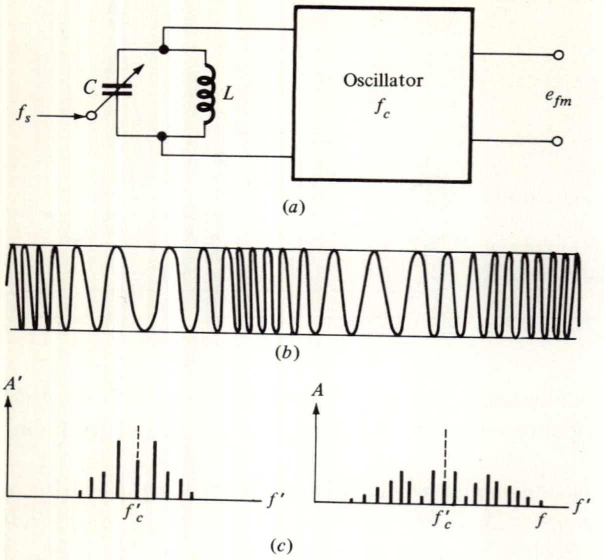

Fig. 7-25 Frequency modulation. (*a*) The oscillator frequency is determined by the resonance frequency of C and L; (*b*) a periodic variation of C causes an output voltage e_{fm}; (*c*) depending on the modulation index m (see text), different frequency spectra will result. A, amplitude; f, frequency; f_c, carrier frequency.

where m_f is the modulation index, a quantity analogous to the modulation index m in amplitude-modulation systems. An example of a frequency-modulated wave is shown in Fig. 7-25*b*. Such a wave represents a spectrum of the carrier wave and symmetrically grouped sidebands. The number and the relative amplitude of the sidebands depend not only on the signal frequency but also on the modulation index m_f. An example of a simple spectrum is shown in Fig. 7-25*c*.

One advantage of a modulation system, whether it be amplitude or frequency modulation, is the following: all signals appear in the form of alternating voltages and in a relatively narrow frequency band, even if the signal frequency is zero or constant for a long time, i.e., has a dc character. Dc signals can cause difficulties; dc amplifiers are more subject to drift than ac amplifiers. This is the reason that modulation systems are frequently used in spite of the fact that dc systems are, in principle, simpler.

The additive superimposition of dc and ac voltages or current poses no new problem as far as the instantaneous values are concerned. The total instantaneous voltage e_T, or the total instantaneous current i_T, is equal to the sum of their respective components. Not quite so trivial is the computation of the rms value of a superimposition of a dc plus an ac component. The rms value of an ac voltage was defined above as

$$E_{\text{rms}} = \sqrt{\frac{1}{T}\int_0^T e^2\,dt}$$

Replacing e by $e_{\text{dc}} + e_{\text{ac}}$,

$$E_{\text{rms}} = \sqrt{\frac{1}{T}\int (e_{\text{dc}} + e_{\text{ac}})^2\,dt}$$

which furnishes for the binomial the three terms

$$e_{\text{dc}}{}^2 + 2e_{\text{dc}}e_{\text{ac}} + e_{\text{ac}}{}^2$$

The second term disappears when integrated over a full period, so that

$$E_{\text{rms}} = \sqrt{E^2_{\text{ac,rms}} + E_{\text{dc}}{}^2}$$

i.e., the rms value of a dc voltage plus an rms ac voltage is equal to the *geometric* sum of these voltages.

AC BRIDGES

Figure 7-26*a* shows a Wheatstone bridge as described in Chap. 1, except that the dc source is replaced by an ac generator E_i and the galvanometer by an ac detector D. The balance condition is the same as that at dc, i.e., the detector will indicate zero voltage or current for

$$\frac{R_X}{R_S} = \frac{R_A}{R_B} \tag{7-3}$$

The potential of the point M will alternate in the rhythm of the applied ac voltage. The potential of N will alternate in the same rhythm, in the same phase,

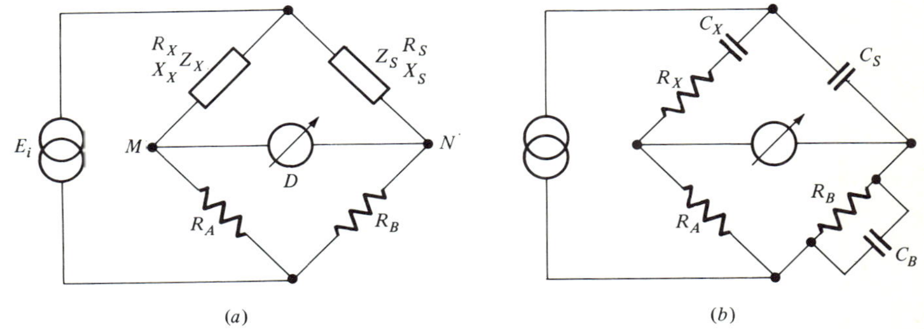

Fig. 7-26 (*a*) Generalized impedance bridge. For balance, the potentials of M and N must swing between equal limits and must be in phase. This requires two balancing conditions (see text). (*b*) Schering bridge. The series resistance-capacitance network in the X arm is balanced by a capacitance in the S arm and a capacitance C_B in parallel to the resistance in the B arm.

and (if the bridge is balanced) by the same amount, so that the potential difference between M and N is zero at all times. If the bridge is unbalanced, an ac voltage appears for $e_M \gtrless e_N$. The voltage disappears at balance for $e_M = e_N$ and reappears with 180° phase shift with further imbalance $e_M < e_N$.

If one of the bridge arms contains a reactance or an impedance, the potential difference between M and N will not be zero, even if the magnitude of the impedances would fulfill the bridge-balance condition, because the potentials of M and N are no longer in phase. In order to balance the bridge, at least one of the other three branches of the bridge must include a reactive component. Therefore, there are two conditions for the balance of an ac bridge: the magnitude and the phase of the impedances Z_X and Z_S must be balanced. This is equivalent to the conditions that (1) the reactive components must be balanced; e.g., for a balance in the S arm,

$$\frac{X_X}{X_S} = \frac{R_A}{R_B} \tag{7-4}$$

and (2) the resistive component must be balanced

$$\frac{R_X}{R_S} = \frac{R_A}{R_B} \tag{7-5}$$

Balancing can be accomplished by inserting appropriate impedances in any bridge arm. Since resistances, capacitances, and inductances are available for balancing and since these can be arranged in series or in parallel, a great number of different bridge arrangements can be built; however, only a few are practical, i.e., can be balanced by commercially available adjustable reactors or resistors.

The balancing of a bridge with two pure reactances (practically achievable with fair accuracy only for capacitors) requires, of course, only one adjustment. The bridge is balanced for

$$\frac{X_X}{X_S} = \frac{R_A}{R_B}$$

$$\frac{1/\omega C_X}{1/\omega C_S} = \frac{R_A}{R_B}$$

$$\frac{C_X}{C_S} = \frac{R_B}{R_A}$$

(Note sequence of indices A and B.) The situation is different in the series-resistance capacitance bridge. The unknown capacitance C_X with a series-resistive component embodied by R_X is balanced by a variable capacitor C_S and a variable resistor R_S. If balance is reached,

$$C_X = C_S\left(\frac{R_B}{R_A}\right)$$

and

$$R_X = R_S\left(\frac{R_A}{R_B}\right)$$

Balancing a bridge is not always a simple matter; the balance point is found by successive adjustment of C_S and R_S. Unless both components are balanced, there will be no zero indication but only a flat, ill-defined minimum. The fact that the adjustment of the reactive and the resistive component becomes interdependent may cause a condition whereby the successive adjustment may lead to successively greater unbalance (*sliding balance effect*). The effect is particularly pronounced if the storage factor Q of the measured impedance is low and if the balance of both components is accomplished in different bridge arms. A commercial bridge is available ("Ortho-Null" bridge, General Radio Company) in which both the reactive and the resistive components can be balanced simultaneously by means of dials which can be coupled mechanically.

In the Wien bridge the series-resistance capacitance in the X arm is balanced by a parallel resistance-capacitance network in the S arm. At balance

$$C_X = \frac{R_B(\omega^2 R_S^2 C_S^2 + 1)}{R_A \omega^2 C_S R_S^2}$$

and

$$R_X = \frac{R_A R_S}{R_B(\omega^2 R_S^2 C_S^2 + 1)}$$

The fact that ω appears in these equations indicates that the bridge is balanced for a given setting of C_S and R_S at only one frequency. If the supply voltage E_i contains higher harmonics and if the bridge is balanced for the fundamental frequency, then an output which contains all higher harmonics will appear at D. This has the disadvantage that for zero output, pure sinusoidal waveshapes are needed for the operation of the Wien bridge. But it offers as an advantage the possibility of using the Wien bridge as a filter for the separation of higher harmonics from the fundamental. The Wien bridge can, therefore, be used to measure distortion and intermodulation, such as in amplifiers which exhibit a nonlinear transfer function. Wien bridges can also be used as frequency-determining elements in electronic oscillators.

A further example of an ac bridge, the Schering bridge, is shown in Fig. 7-26*b*. The unknown capacitance C_X is balanced by the capacitance C_S; the unknown loss resistance R_X by a capacitance C_B in the opposite bridge arm. At balance

$$C_X = C_S\left(\frac{R_B}{R_A}\right) \tag{7-6}$$

and

$$R_X = R_A\left(\frac{C_B}{C_S}\right) \tag{7-7}$$

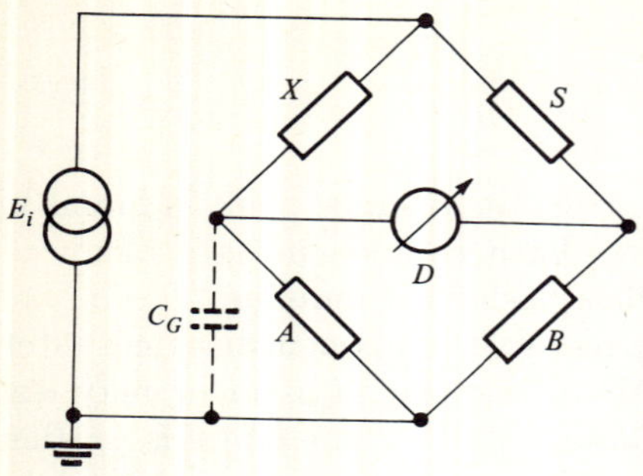

Fig. 7-27 Grounding in ac bridges. One terminal of the source E_i is grounded. The line-voltage-operated detector (e.g., cathode-ray oscilloscope) is grounded through the unavoidable capacitance C_G; an error is introduced since C_G is in parallel with the A arm.

This is a useful bridge for the measurement of the dielectric constants and losses of dielectrics in capacitors, particularly at high voltages in the vicinity of the breakdown voltage. Figure 7-26*b* and Eqs. (7-6) and (7-7) are correct under the assumption that the capacitor C_S is free of losses which, of course, is only an approximation. But even if C_S has losses, they can be balanced by a further capacitor in parallel with R_A (not shown).

Bridges with inductances operate in a manner analogous to the capacitive bridges. One of the advantages of the ac bridge circuits is that they furnish simultaneous information of the resistive and reactive components of an impedance, which sometimes can be interpreted as the resistivity and the dielectric constant of a substance, i.e., of the status of mobility and of polarization of charges. However, bridge measurements furnish information about the complex impedance only with lumped reactance and resistance not about the detailed composition or the actual structure of an impedance. More information about the detailed structure of an impedance can sometimes be obtained from measurements at various frequencies.

The technical difficulties in operating a bridge increase, of course, with demands for higher accuracy. Such difficulties stem primarily from stray fields, which necessitate adequate shielding and grounding, and from circuit residuals.

If one terminal of the *source* is grounded, then the detector cannot be grounded, otherwise there would be a short circuit across one bridge arm. Even if the source or the detector is grounded through a capacitance,[1] such as C_G in Fig. 7-27, an error would arise, since the capacitance C_G would appear in parallel with the A arm of the bridge. A way out of this difficulty is to use either a "floating" detector or generator (technically unsatisfactory) or a shielded transformer. Such a transformer can be inserted either between the source and the bridge (preferable because of better impedance matching) or between the bridge and the detector. Although the shielded transformer does not completely avoid the appearance of capacitances across the bridge arms, these capacitances are small and, hence, are of little influence on the bridge balance. Also, they are well defined so that their influence on the bridge balance can be compensated.

[1] All line-voltage-operated instruments have substantial capacitance to ground, even if they contain a line-to-power supply transformer.

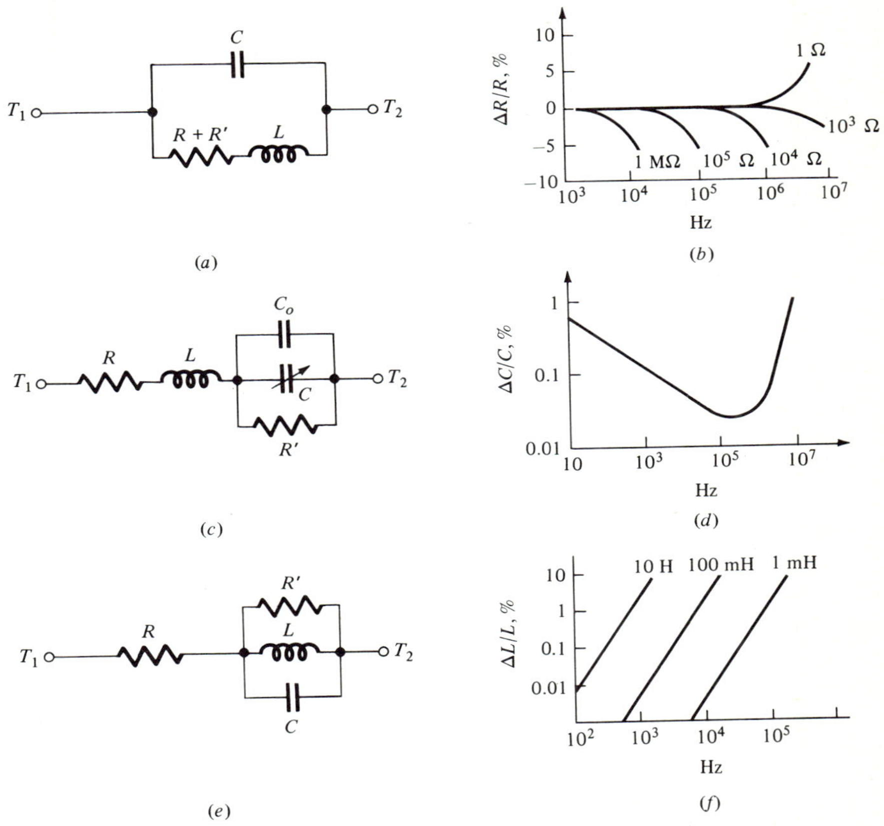

Fig. 7-28 Circuit element residuals. (*a*) Equivalent circuit of a resistor; (*b*) the circuit residuals lead to an error in calibration at higher frequencies; (*c*) equivalent circuit of a capacitor with (*d*) the calibration characteristic; (*e*) equivalent circuit of an inductor; (*f*) frequency characteristics.

Any circuit element exhibits undesirable "residual" reactance or resistance which may influence the balance conditions of a bridge or the operation of an electric or electronic system. An example is the equivalent circuit of a resistor (Fig. 7-28*a*), consisting of a fixed resistance component R; a variable resistance component R' (e.g., representing skin effect, the effect that at higher frequencies current conduction takes place only in the outer layers of a wire); an inductance L in the wire, which may be reduced by appropriate methods of winding, but which can never be completely avoided; and capacitance C between the terminals. The residual components may be negligible at dc and low frequencies. At high frequencies, they cause a variation of the calibration and of the phase angle. The percentage change in calibration for different frequencies is shown in Fig. 7-28*b*. Similar equivalent circuits and their effects on the calibration are presented in Fig. 7-28*c* and *d* for a capacitor and in Fig. 7-28*e* and *f* for an inductor.

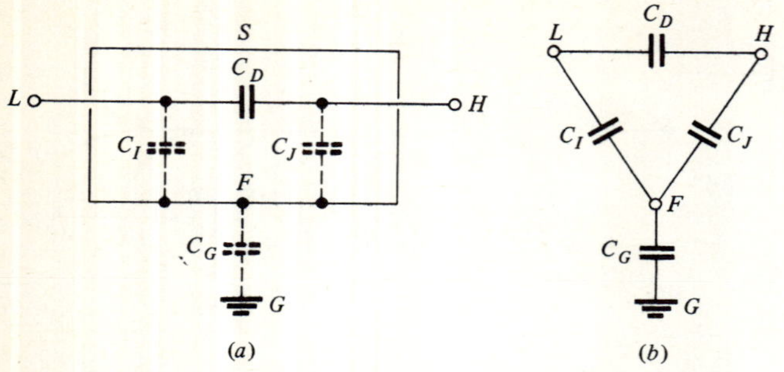

Fig. 7-29 (*a*) Another equivalent circuit of a capacitor C_D in a shielded container *S*, insulated from ground *G*; (*b*) the same equivalent circuit redrawn.

A simplified equivalent circuit of a capacitor in a shielded container *S* is shown in Fig. 7-29. Located between the two terminals *L* and *H* is the direct capacitance C_D, which is the capacitance to be measured. Residual capacitances C_I and C_J exist between each terminal and the shield; these capacitances are generally not alike. The terminal with the higher capacitance to ground (i.e., the terminal which develops a lower potential to ground if exposed to an electric stray field) is called the *low terminal*, the other is the *high terminal*. Furthermore, a capacitance C_G exists between shield *S* and ground *G*.

If such a three-terminal capacitor is inserted in a bridge, and if C_D is large compared with C_I, C_J, and C_G, these residual capacitances may be neglected. However, even if C_I, C_J, and C_G are of the same order of magnitude as C_D, there are means of separating them or avoiding their effects on the bridge balance. For instance, in Fig. 7-30, the capacitor is inserted in the *X* arm. Now if point *F* is connected to ground (dotted line), the capacitor C_G is shorted and its influence on the bridge balance is eliminated. Capacitance C_J is in effect connected between points *O* and *P*, i.e., it is in the bridge diagonal, in parallel with the source and, therefore, without influence on the bridge balance. Capacitance C_I appears in parallel to the *A* arm and causes an error in the determination of the

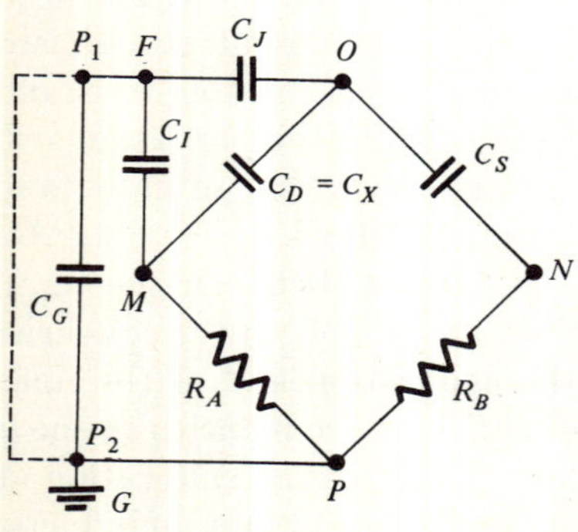

Fig. 7-30 The capacitor of Fig. 7-29*b* is inserted in a bridge. By connecting point *F* or P_1 with P_2, the influence of C_G and C_J on the bridge balance are eliminated. The influence of C_1 is frequently negligible. The source (omitted in Figs. 7-30 and 7-31) is connected between *O* and *P*, the detector between *M* and *N*.

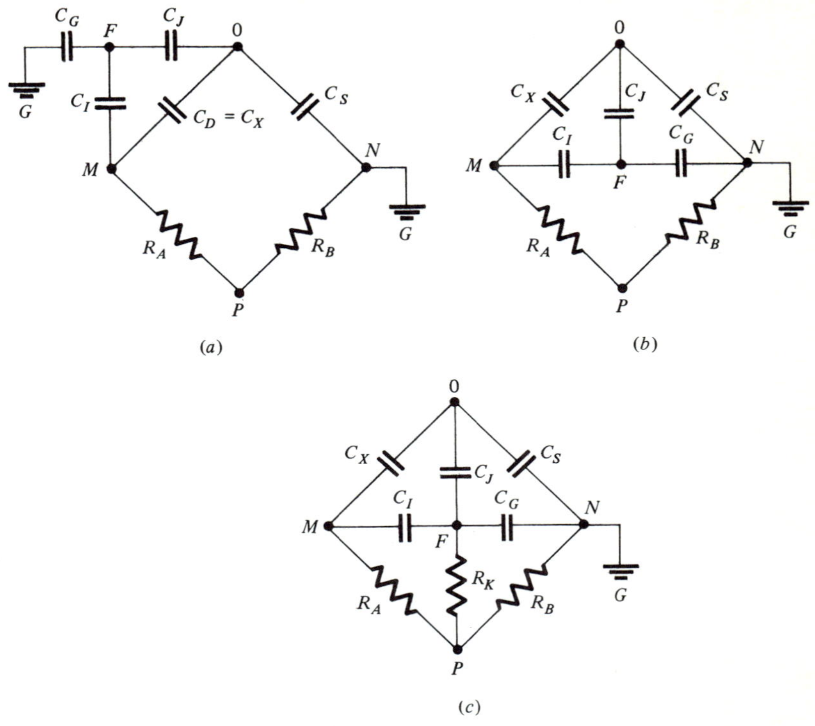

Fig. 7-31 (*a*) The same circuit as Fig. 7-30, but point N is grounded. (*b*) The same circuit redrawn. The influence of the residual capacitances is completely eliminated by the insertion and balancing of R_K in (*c*) so that the potentials of M, F, and N are alike (*Wagner ground*). The supply source is to be connected to points O and P, the detector between N and F and, after balancing, between N and M. Repeated balancing is required.

phase angle, but not of the magnitude, of C_D. The circuit of Fig. 7-32 is, of course, only one of several ways of grounding a bridge and reducing the effect of residual reactance or resistances.

There is also the possibility of completely balancing the effect of the residual capacitance. Figure 7-31*a* shows a circuit similar to that in Fig. 7-30 before making the connecting P_1–P_2; the ground connection is now made to point N. Figure 7-31*b* shows the same circuit redrawn. If point F has the same potential as points M and N, then the influence of the residual capacitances would be completely eliminated. This is done in Fig. 7-31*c* by inserting R_K in the circuit and balancing it so that

$$\frac{Z_X}{Z_S} = \frac{Z_A}{Z_B} = \frac{Z_J}{Z_K} \tag{7-8}$$

or

$$\frac{Z_X}{Z_A} = \frac{Z_S}{Z_B} = \frac{Z_I}{Z_G} \tag{7-9}$$

In practice, the detector is connected between N and M and the bridge arms X, S, A, and B are balanced. The detector is then connected between N and F and the bridge arms J, S, K, and B are balanced. The process is repeated until the equilibrium conditions, Eqs. (7-8) or (7-9), are met.

RECTIFIER BRIDGES

The complexities of a bridge circuit can be avoided in applications where the exact values of both the resistive and reactive components of an impedance need not be measured. For instance, if a transducer is inserted in the bridge arm X which changes capacitance under the influence of a displacement or a force (Chap. 5), it is generally acceptable to disregard the resistive component of the capacitor and to measure only the capacitance C_X either by balancing the bridge by means of a capacitor C_S or by using the deflection method, i.e., by establishing the calibration characteristic $E_0 = fC_x$ and measuring E_o.

The circuit shown in Fig. 7-32 solves both the phase-shift and the grounding problem. In a simplified way its operation may be described as follows: The source E_i furnishes an ac voltage. During one half-cycle, the diode D_1 is conductive and the capacitor C_1 is charged to the peak value of the source voltage E_i. During the other half-cycle, C_1 discharges through resistance R_1 and meter M. The meter current I_1, hence the meter deflection, is proportional to the capacitance C_1. During the same half-cycle during which the capacitor C_1 discharges through R_1 and the meter, the capacitor C_2 is being charged with a polarity opposite that of C_1 (because of the direction of D_2). During the following half-cycle the capacitor C_2 is discharged through R_2 and meter M, causing a current I_2. Under a balanced condition, that is, if $C_1 = C_2$ and $R_1 = R_2$, the currents I_1 and I_2 are alike. The meter will then read zero, provided it is sufficiently inert to indicate the average current over a full cycle. Under unbalanced conditions, the meter deflection is proportional to C_1–C_2.

One can show that the sensitivity of the system reaches a flat maximum if the time constants C_1R_1 and C_2R_2 are of the same order of magnitude as the period of the source voltage E_i. In the same region, the output current I_1–I_2

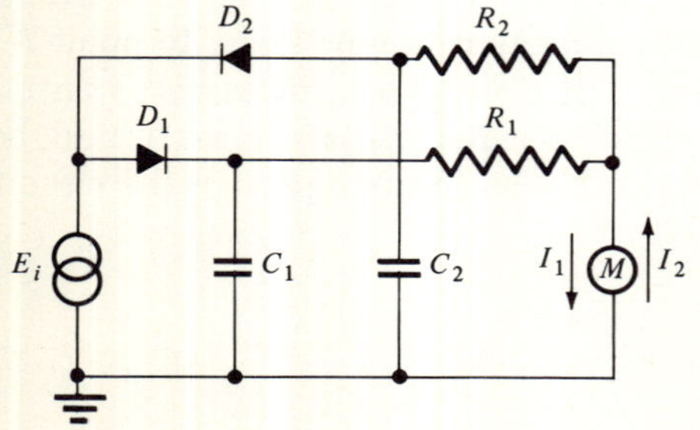

Fig. 7-32 Twin-T bridge with diodes D_1 and D_2. E_i, input voltage; M, meter; C_1, unknown capacitor; C_2, comparison capacitor; the source E_i, the capacitors C_1 and C_2, and the meter M are grounded.

is practically independent of the frequency of E_i. The circuit has the further advantage that the source, the transducer capacitor, the balancing capacitor, and the output meter are all connected to ground.[1]

REFERENCES

Michels, W. C.: "Electrical Measurements and Their Applications," Van Nostrand, 1957.
Stout, Melville B.: "Basic Electrical Measurements," Prentice-Hall, 1950.
Easton, Ivan G.: "Impedance Bridges Assembled from Laboratory Parts," General Radio Co., Cambridge, Mass., 1953.

[1] K. S. Lion, Nonlinear Twin-T Network for Capacitive Transducers, *Rev. Sci. Instr.*, **35**:353–359 (1964). K. S. Lion, U.S. Patent No. 3,012,129.

8

Elements Based on the Transport of Charges in Gases

INTRODUCTION

The preceding chapters have dealt essentially with linear circuit elements, i.e., with elements that exhibit a linear relationship between applied voltage and resulting current as shown in Fig. 2-2. The resistance is expressed by the slope of the characteristic

$$R = E/I = \frac{1}{\tan \varphi} = \frac{\Delta E}{\Delta I}$$

R is constant, i.e., independent of the applied voltage; resistances and impedances are linear elements.

The more general case of a nonlinear voltage-current characteristic N is illustrated in Fig. 8-1a. The slope, hence the resistance, varies with the applied voltage. If a voltage E_p is applied, a current I_p will flow. The resistance of the circuit element, if operated at the point P, is $R_p = E_p/I_p$ and is expressed by the slope ($1/\tan \varphi$) of the dotted line R_p connecting P with O.

If an element is operated at the point P and the applied voltage is varied by the incremental amount ΔE, a current variation ΔI will ensue. In general, the ratio $\Delta E/\Delta I = 1/\tan \varphi_d = r_d$ will be different from R_p. It may be smaller or it may be larger, depending on the curvature of the EI characteristic N at the point P. The quantity r_d, the incremental resistance (also called *dynamic* or *differential resistance*), has a physical reality and can be measured, for instance, with the circuit shown in Fig. 8-1*b*. The circuit contains a dc source E_p, which determines the location of the operating point P in Fig. 8-1*a*, and an ac source E_Δ, which causes a small symmetrical variation ΔE about the operating point P. Both voltages superimposed are applied to the nonlinear element N. The currents I_p and ΔI are measured from the voltage drop across the (small) resistance r_s. The dc meter is a moving-coil system; it measures the dc voltage drop only. The ac meter measures only the ac voltage drop across r_s since capacitor C blocks the direct current from the meter. Both meters may have a scale calibrated in ohms so that the value of the resistance R_p can be read off the dc meter; that of the incremental resistance r_d can be read off the ac meter.

The ac output voltage arising between terminals A and B in Fig. 8-1*b* is proportional to the *slope* of the EI characteristic of the nonlinear element N at any selected point P. The circuit acts as a differentiating circuit. In contrast to all formerly described differentiating circuits which furnish an output proportional to the time derivatives de/dt or di/dt, this circuit furnishes an output proportional to di/de. If e and i are functions of other variables x and y, the output will be proportional to dy/dx.

Figure 8-2*a* shows two nonlinear elements N_1 and N_2 with different characteristics. If these two elements are connected in series and a voltage E_T is applied (see lower part of Fig. 8-2*b*), it is not immediately apparent what the resulting current I_T is and what partial voltages E_1 and E_2 appear across each

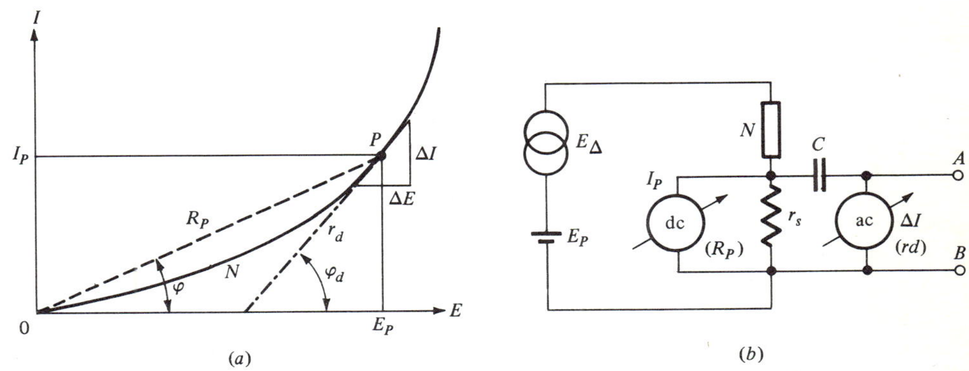

Fig. 8-1 (*a*) Characteristic N of a nonlinear circuit element. The slope is indicative for the resistance R_p, the slope r_d for the incremental or dynamic resistance r_d. (*b*) Circuit for the measurement of R_p and r_d, also used as differentiating circuit. The output between terminals A and B is proportional to di/de.

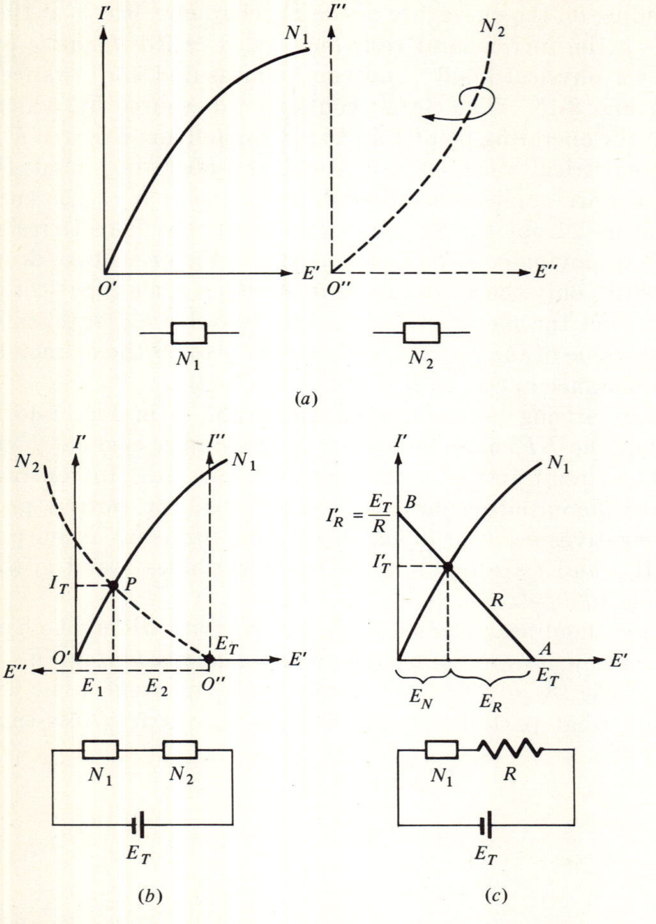

Fig. 8-2 (*a*) Two nonlinear circuit elements N_1 and N_2 and their *IE* characteristics; (*b*) series connection of two nonlinear elements N_1 and N_2 and their superimposed characteristics; (*c*) series connection of a nonlinear element N_1 and a resistor *R*, characteristic and load line.

element. The graphic method shown in Fig. 8-2*b* permits the determination of these quantities.

This graph is constructed in the following way: first, the graph E'-I' for N_1 is drawn (Fig. 8-2*b*). Then, on the abscissa, the point E_T is found (Fig. 8-3*b*); the distance O to E_T corresponds to the total applied voltage. Furthermore, from E_T in the reverse direction (from right to left) the graph E''-I'' is drawn. Both curves intersect at point P. The projection from point P onto the I axis furnishes the current I_T in the circuit; the projection from point P onto the E axis furnishes the partial voltages E_1 across N_1 and E_2 across N_2. The con-

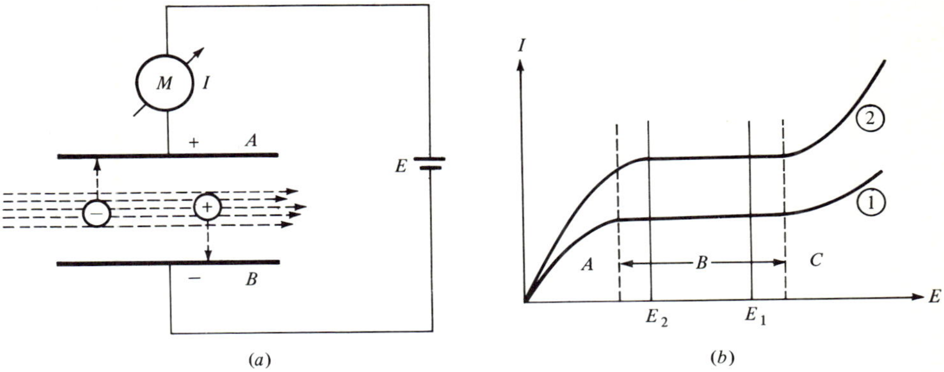

Fig. 8-3 (*a*) Schematic diagram of an ionization chamber. (*b*) Two voltage-current characteristics of an ionization chamber for (1) low-intensity and (2) high-intensity incident radiation. *A*, region of operation for very intense incident radiation; *B*, normal operation range, saturation region; *C*, collision ionization range.

struction of the graph is particularly simple for the series connection of a nonlinear element N_1 with a linear element, such as the resistance R shown in Fig. 8-2*c*, because the resistance characteristic is a straight "load line" between point A with coordinates $E' = E_T$ and $I' = 0$ and point B with coordinates $E' = 0$, and $I'_R = E_T/R$.

Elements based on the transport of charges in gases, vacua, electrolytes, and semiconductors are frequently nonlinear because the current causes a depletion or an additional supply of carriers of electricity.

NON-SELF-SUSTAINING GAS DISCHARGES

Ionization chambers A schematic diagram of an ionization chamber is shown in Fig. 8-3*a*. The chamber consists of two plates, A and B, which are insulated from each other. A voltage source E causes an electric field between the plates. If charged particles of either polarity are present between the plates, they will be separated. They will move toward the plates and form a current pulse in the external circuit. If charges are continuously introduced into or produced in the space between the plates, they will form a continuous current.

The mere movement of the charged particles (not their arrival at the electrodes) constitutes a current. Movement of positive charges toward the cathode causes a current in the same direction as the movement of negative charges toward the anode. The current-voltage characteristic of an ionization chamber is shown in Fig. 8-3*b*. For small applied voltages (region A), the current is small because the charges move slowly and some of them will recombine with charges of opposite polarity so that they will not continue to participate in the conduction process. As the applied voltage is raised, the particles will move faster, and the current will increase until a condition is reached (region B, plateau, saturation range) where all ions are collected at the electrodes before they have a chance to recombine. The length of the plateau depends on the nature of the gas in the chamber and on the ionization density. If the voltage is further increased, new ions and electrons will be formed by collision ionization, and the current will rise again (region C). For the measurement of ionizing radiation, the ionization chamber is generally operated in region B where the current is independent of the applied voltage.

If a steady flux N_0 of ionizing particles arrives each second in the sensitive volume of the chamber, and in the absence of recombination, the ionization current is

$$I = N_0 Slpe$$

where S is the specific ionization of the gas, i.e., the number of ions formed per centimeter of length and per atmosphere of pressure; l is the range of the ionizing particles or the active length of the chamber, whichever is shorter; p is the pressure in the chamber; and e is the charge of an electron.

Ionization chambers are usually operated at atmospheric pressure. Operation at higher pressure is sometimes required to increase the absorption (stopping power) for incident high-energy particles.

The current can be measured by means of a sensitive galvanometer M as shown in Fig. 8-3a. Alternatively, a resistor is inserted in the circuit, in place of the meter, and the voltage drop across this resistance is amplified and measured. A capacitor can also be connected in parallel with the ion chamber or the ion chamber can be used as a capacitor, which is initially charged to a voltage E_1 and exposed to an ionizing radiation. The ion current will discharge the capacitor, so that after a time t the voltage will be diminished to the value E_2. The (average) ion current can then be found from

$$I = \frac{Q}{t} = \frac{C(E_1 - E_2)}{t}$$

provided that E_1 and E_2 are both within region B in Fig. 8-3b.

Ionization chambers with instantaneous read-out can be used for the measurement of ionizing radiation to the limit of about 1 μr/h (1 microroentgen per hour); integrating systems, like that based on capacitive discharge, are capable of much higher sensitivity.

If the incident ionizing radiation is kept constant, the ion and electron current in the chamber will depend only on the pressure and the specific ionization of the gas. The ionization chamber with a built-in radioactive source can be used for the measurement of pressure (vacuum) or for gas analysis.

A schematic diagram of an ionization vacuum gauge is shown in Fig. 8-4a. It contains a radioactive source A which emits α particles and causes ionization of the gas. The number of ions formed is proportional to the gas pressure as long as the range of the α-rays exceeds the dimensions of the chamber. Increase of pressure reduces the range of the α particles. If the α particles are stopped in

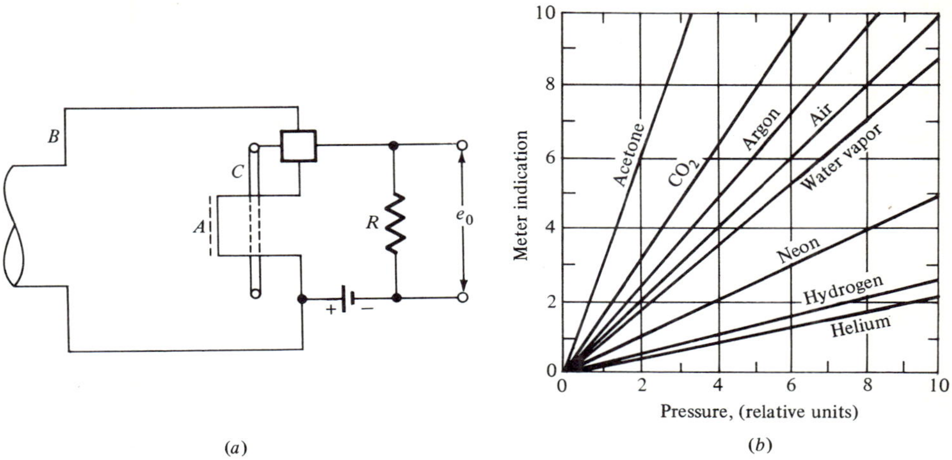

Fig. 8-4 (*a*) Schematic diagram of an ionization vacuum gauge. *A*, ionizing radiation source; *B*, housing; *C*, collector. The ionization current passes the resistor *R* and causes an output signal e_o. (*b*) Calibration characteristic of an ionization vacuum gauge for different gases.

the gas before they reach the end of the gauge, then the number of ions formed reaches an upper limit, i.e., the ion current does not increase with a further increase of pressure. Practical ionization gauges can be used for pressure measurements from 10^3 to 10^{-4} torr. At pressures lower than 10^{-4} torr, x-rays produced by the impact of particles on the walls of the gauge cause photoemission and an electron current which is indistinguishable from the ion current.

The ions are collected at an electrode *C*; the ion current is of the order of 10^{-9} to 10^{-13} A. The transfer characteristic is linear; the slope of the characteristic is different for different gases as shown in Fig. 8-4*b*. The radioactive decay of the radiation source limits the lifetime of the gauge; practical gauges have a lifetime of about five years.

Because of the different reaction to different gases (Fig. 8-4*b*), ionization gauges can be used for leak detection in vacuum systems. A gauge is connected to a vacuum system which is continuously evacuated by a pump. Air enters the system through a leak; the pressure within the system comes to an equilibrium, which is indicated by the gauge. An atmosphere of a test gas (hydrogen or helium) is then brought in the vicinity of a suspected leak. If a leak exists, the test gas will enter the system, and the change of gas composition will be indicated by the ionization gauge. This is only one of several methods available for leak detection in vacuum systems.

Atmospheric air normally contains ions of both polarities, ranging in mass from single molecules to large aggregations of molecules, like dust particles (*Langevin ions*). The composition, charge, and concentration of these ions vary widely and are influenced by atmospheric electrical conditions (thunderstorm, ion clouds), the presence of radioactive materials, cosmic radiation,

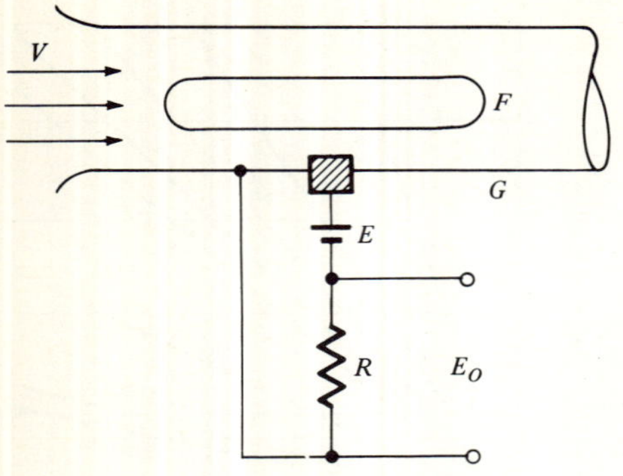

Fig. 8-5 Gerdien ion counter. Air containing ions flows through the cylinder G with the velocity V. The ions are attracted by the center electrode F; they form a current through the resistor R and an output signal E_o.

smoke, electric discharges, x-rays or ultraviolet radiation, and the breaking up of water droplets (*waterfall electricity*). The presence and concentration of such ions can be measured, e.g., with the Gerdien counter shown schematically in Fig. 8-5. The counter consists of an external, grounded cylinder G and an internal electrode F. Air containing ions is drawn through the system with the flow velocity V. A voltage E is applied to operate the counter in the plateau range. The ions are attracted toward the center electrode and form a current

$$I = nqV$$

where n is the ion concentration and q is the charge of the ions. Positive and negative ions are measured separately by applying either a negative or positive potential E to the center electrode F.

If the counter is operated at voltages below the saturation voltage E_s, only the lighter and, therefore, more mobile ions are drawn to the center electrode. The current then depends on the ion mobility [i.e., the velocity of an ion when exposed to a field strength of unity; dimension of mobility (m/s)/(V/m)]. It can be used for the measurement of ion mobility and for an analysis of different ions contained in a gas.

A cloud of ions formed in air or other gases by a high-voltage spark discharge will persist for several seconds before all ions are recombined into neutral molecules. If the airstream moves, the ion cloud will follow the air without influencing the flow pattern. This phenomenon forms the basis of the transportation-of-charge method for the measurement of gas-flow velocity. Ions are produced by a spark discharge at a point within a gas duct. At a time Δt later and at a distance d downstream from the spark gap, the ions are detected. The flow velocity can be found from

$$v = \frac{d}{\Delta t}$$

The method measures the average stream velocity of the gas volume in and around the ion cloud; it is applicable to high-flow velocities above 20 mi/h. At

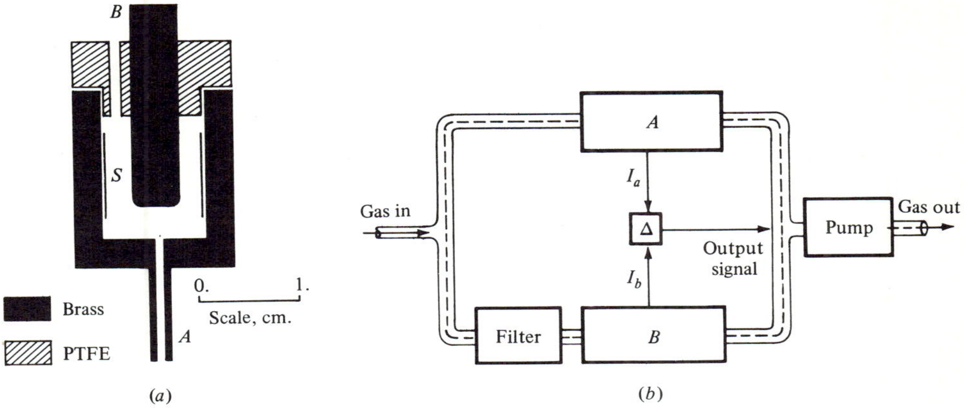

Fig. 8-6 (*a*) Ionization gas detector (cross-section detector) formed by the two electrodes *A* and *B*, which are insulated from each other. Ionization is caused by the radioactive foil source *S*. (*From J. E. Lovelock, Analytical Chemistry,* **33**:162–178 (1961). (*b*) Gas-specific (CO_2-sensitive) ionization detector (*Lion Research Corporation*). The gas to be analyzed for CO_2 reaches the ionization chamber *A* directly, the ionization chamber *B* through a CO_2 absorbing filter. The output signal is proportional to the CO_2 concentration of the gas.

smaller velocities the diffusion and recombination of the ions reduce the local definition of the ion cloud and cause errors.

Ionization systems can be used for the detection of an unknown gas, the test gas, in a carrier gas of known composition (e.g., pure helium), and for the measurement of the concentration of a test gas in an inert carrier. Ionization detectors are based on the differential ionization of the test gas molecules and those of the carrier gas, or on variations of ion recombinations or of ion mobility, or on exchange of charges caused by the presence of a test gas in the ionization chamber.

An ionization detector consists, in principle, of a chamber with two electrodes as shown in the schematic diagram Fig. 8-3*a*. It contains a sealed source of α or β radiation and is generally operated at atmospheric pressure. Other ionization detectors use a hydrogen flame in air as the source of ionization. Means are provided for the passage of the carrier gas and of a volatile derivative of the test substance (gas or vapor) through the detector chamber.

Figure 8-6*a* shows a cross-section ionization detector. If the ionization cross-section Q of the test gas molecules differs significantly from that of the carrier gas, e.g., if the test gas is a polyatomic gas of high density while the carrier gas is monoatomic and of low density, then the test gas molecules will be ionized in preference to the carrier gas and can be measured electrically. In the ideal case (but physically not realizable) in which only the molecules of the test gas are ionized and none of the carrier gas, it would be possible to detect single molecules. In practice, the sensitivity of the method is limited by the ionization efficiency and by the always present background current caused by the ionization of some carrier gas molecules.

The current, i.e., the concentration of ion pairs formed in the cross-section ionization detector, is

$$i = \frac{KP}{RT} \Sigma xQ$$

where x = molar fraction
Q = cross section of the components of the gas mixture
K = geometric factor
P = pressure
R = gas constant
T = temperature

The detector can be used for the measurement of all gases and vapors up to 100 percent and down to 0.01 percent volume concentration. The detector is most frequently used with hydrogen or helium as a carrier gas, but almost any gas can be used. The applied voltage is usually between 300 and 1,000 V.

A variety of ionization detectors has been described based on different physical effects and varying in operation, chamber geometry, range, sensitivity, and suitability for different test substances. Ionization detectors frequently are used as detectors in gas chromatography. The sensitivity of the ionization detector is potentially higher than that of any other physical method of gas analysis.

Most ionization detectors are unspecific, i.e., they furnish information about the amount or concentration of a test gas, but not about the type or nature of the gas. Several methods have been described to obtain information about the concentration of a specific test gas. An example is shown in Fig. 8-6*b*. The incoming gas stream is divided; one part passes through the ionization chamber A, the other part passes through a filter which absorbs a specific gas, such as CO_2. The gas then passes through the ionization chamber B. The electric output signal is proportional to the difference of the ionization currents I_a and I_b and furnishes information as to the concentration of CO_2 only, regardless (within limits) of other gases that may be present in the incoming gas stream.

Amplifying gas-filled counters If the voltage applied to an ionization chamber is raised, the current rises as indicated in region C of Figs. 8-3*b* or 8-7*a*. In this region the counter is less suitable for the measurement of continuous incident radiation because of instability, but lends itself to the measurement of single ionizing events and to the counting of ionizing particles.

The counter frequently is built in the form shown in Fig. 8-7*b*. It consists of a cathode in the form of a cylindrical metal or metallized container and a thin central wire as anode. An incident-ionizing particle causes a discharge in the counter which leads to the formation of an output pulse of the magnitude

$$\Delta E = kAn$$

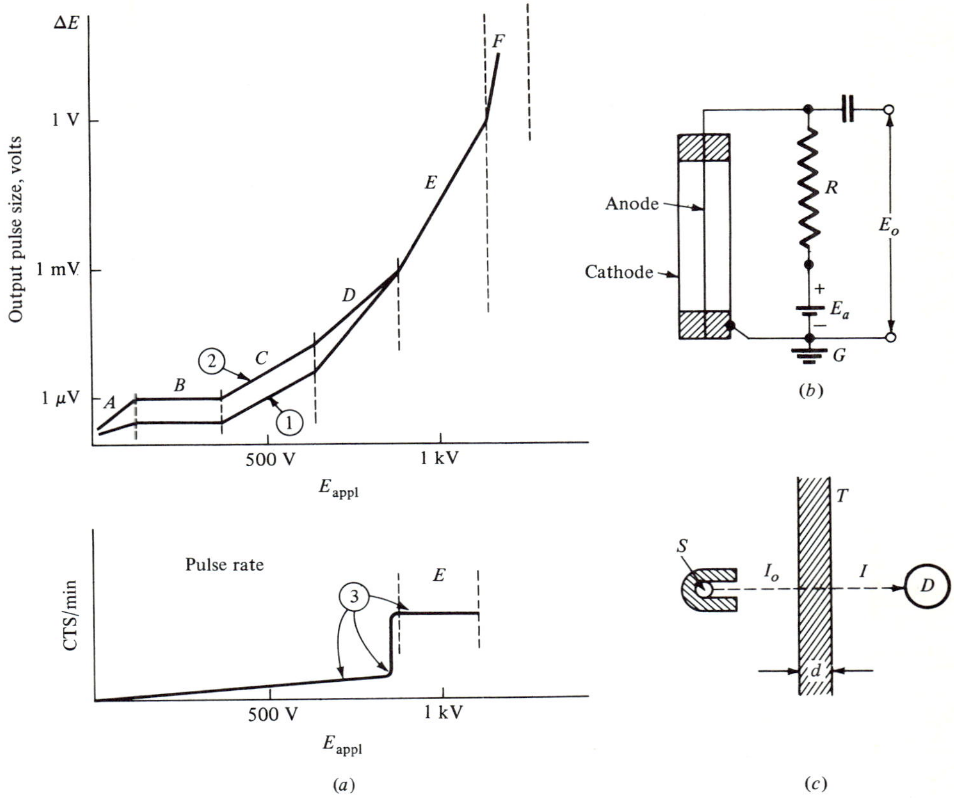

Fig. 8-7 (*a*) Operating ranges of gas-filled counters: (1) for low-incident ionization intensity and (2) for high intensity. Region *A* rarely used; *B*, normal operating range for ion chambers; *C*, range of proportional counter; *D*, region of limited proportionality; *E*, Geiger-counter range; *F*, region of self-sustained discharge. (*b*) Schematic diagram of a Geiger counter, see text for explanation. (*c*) Schematic diagram of beta or gamma gauge. *S*, source; *T*, test object of thickness *d*; *D*, detector. Incident-beam intensity, I_o; emerging-beam intensity, I.

where k = a constant depending on the kind of gas filling and the geometry of the counter
A = the gas amplification factor
n = the number of ion pairs formed by the incident-ionizing particle

In the range *B*, where the counter operates as an ionization chamber, the amplification factor is unity; no amplification takes place. Operated at elevated supply voltages in the range *C*, the amplification factor increases with the applied voltage and can reach values between 1 and 10^4. In this range, the output signal ΔE is proportional to the number n of initially formed particles. For instance, an incident β particle of a certain energy may cause one ion pair (curve 1 in Fig. 8-7*a*); an incident α particle may cause many hundred ion pairs (curve 2). Throughout the region *C*, the output remains proportional to the number of ion pairs initially formed; the counter acts as a proportional counter.

At higher voltages, there follows a range *D* of limited proportionality which is poorly understood and in general is not of practical interest.

In all regions *A* to *F* except region *B*, the pulse sizes increase with increased voltage applied to the counter. In all regions except region *E*, the pulse rate (i.e., the number of pulses per unit time for a given, constant primary ionization) increases with increased voltage applied to the counter. Only in region *E* is the counting rate independent of the applied voltage (*Geiger plateau*). In the region *E* all pulses are of the same size regardless of the primary ionization, indicating that in this range a new mechanism sets in which determines the pulse size (*Geiger counter range*). The process which leads to the development of these large pulses is the following.

An incident-ionizing particle causes ionization of one or several molecules, i.e., the formation of at least one electron and a positive ion. The positive ion moves toward the cathode, the electron with much higher velocity (because of its small mass) toward the anode. The electron accumulates kinetic energy in the electric field and causes further ionization by collision and the formation of a discharge avalanche. A second process now enters into the discharge mechanism: some of the ions and electrons recombine, a process which leads to the emission of light. The light causes new electrons and ions by photoionization which takes place throughout the counter, but only in the vicinity of the central electrode is the electric field high enough to separate electrons and ions. Therefore, a gas discharge spreads along the center anode in both directions until the entire anode is covered with ions. Hence, the current reaches a maximum which is independent of the primary ionization.

The discharge ceases when the positive-ion sheath surrounding the anode moves toward the cathode and virtually increases the effective anode radius, thus decreasing the field strength in the vicinity of the anode. Eventually the field strength falls to a value below the minimum sustaining voltage.

When ions reach the cathode, they may liberate new electrons which may reignite new discharges. To prevent this effect, a quenching mechanism must be provided. Quenching can be accomplished by the insertion of a large resistance in the circuit which increases the time constant, i.e., the recharging time of the counter so that all ions are swept out of the counter before the voltage reaches the level necessary for the discharge mechanism. Of course, an increase of the time constant carries with it an undesirable effect: The counter has a relatively long recovery time; if an ionizing particle arrives in the counter before the end of the recovery time, it will not be counted. Geiger counters are not applicable for counting rates higher than 10^4 counts per second. Another means for quenching new undesirable discharges initiated by the impact of positive ions on the cathode consists of the admixture of a polyatomic gas (e.g., alcohol) or a halogen. These gases cause a gradual depletion of the potential energy of the ion so that the ions arriving at the cathode have too little energy to cause a new avalanche.

Beta and gamma gauges An application of a nuclear radiation detector for the noncontact measurement of the thickness of a test object is shown in Fig. 8-7*c*. The gauge consists of a source *S*, usually a radioactive isotope in a shielded container, which emits radiation; the radiation is attenuated in the test object *T* and measured with a radiation detector *D*. If the source emits beta or gamma radiation of homogeneous energy and if the intensity of the incident radiation is I_o, then the intensity of the radiation after penetration of the test object is

$$I = I_o e^{-\mu d} \tag{8-1}$$

where μ = the absorption coefficient
d = the thickness of the test object

If μ is known, the thickness *d* or its variations can be measured. If *d* is kept constant, variations of μ can be measured, such as those produced in binary liquids by the change of one of the components.

For a given material, μ is proportional to the density ρ (mass m/volume V) of a substance. The exponent in Eq. (8-1) is, therefore,

$$\mu d = \text{const. } \rho d = \text{const. } \frac{md}{V} = \text{const. } \frac{m}{a}$$

i.e., the absorption of test material depends on the mass *m* per unit area *a* (dimension, grams per square centimeter) of the absorber.

The absorption coefficient is different for different types of radiation and changes with the radiation energy. γ-rays and high-voltage x-rays are very penetrating; therefore, gamma gauges are suitable for the investigation of heavy metals and thick test objects. Beta gauges are customarily used for light metals, metal foils, paper, or plastic layers with a gauging range between 1 and about 1,200 mg/cm^2. Also α-rays can be used, but alpha gauges are more difficult to use than beta and gamma gauges, because the absorption of α-rays does not follow an exponential law. α-rays are absorbed in very thin layers; therefore, alpha gauges are useful only for the thickness measurement of very thin films.

SELF-SUSTAINED GAS DISCHARGES

The different forms of gas discharges described above are not self-sustaining, i.e., they require the injection or formation of charged particles in the discharge space by an external source, such as a nuclear or ultraviolet radiation source. Gas discharges can become self-sustaining when new charged particles, electrons or ions, are formed by the discharge mechanism, and when the number of ions produced in the discharge balances or exceeds the number of ions lost by recombination or other causes.

A vessel for the observation of a self-sustaining discharge is illustrated in Fig. 8-8*a*, which shows a tube with two disk-shaped electrodes, a cathode *C*, and an anode *A*. An adjustable voltage E_a is applied; a resistor *R* is provided to limit the current. The voltage between the electrodes *C* and *A* is $E_t(E_a - RI)$. The pressure in the tube can be varied.

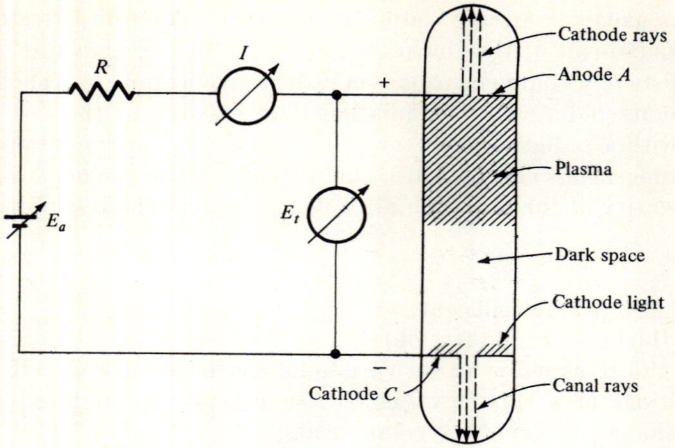

(a)

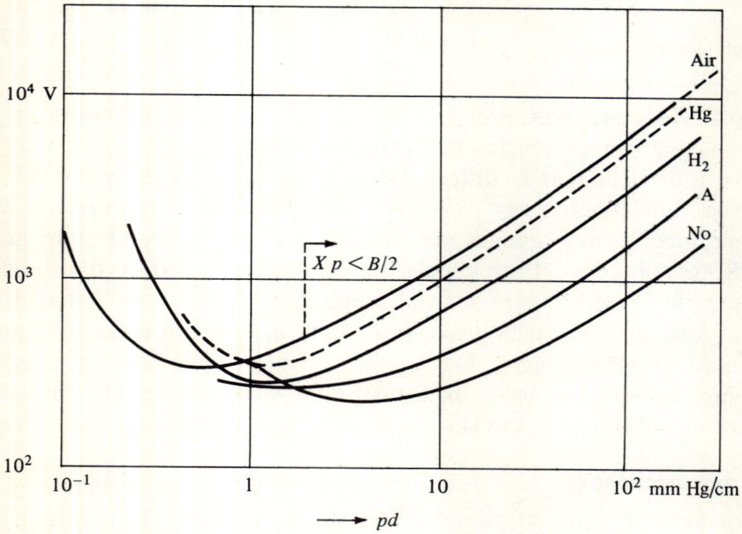

(b)

Fig. 8-8 (*a*) Schematic diagram of gas-discharge tube for the illustration of different discharge forms; (*b*) starting potential E_s as a function of the reduced electrode distance *pd* in several gases. [*From A. von Engel, Ionized Gases, Oxford Clarendon Press*, 1955 (*by permission*).]

If the pressure is between atmospheric pressure and about 1 torr and if the voltage E_t is of the order of several volts, the tube still acts as an ionization chamber; a non-self-sustaining discharge will take place and a current of the order of 10^{-12} A will flow as described above. The electrons and ions may be caused by cosmic radiation or by the natural radioactivity of the walls of the vessel.

As the source voltage E_a is raised so that the voltage E_t reaches a critical value E_s (*starting or firing voltage*), a luminous discharge appears suddenly (*region of normal glow discharge*). The current I rises abruptly to the order of 10^{-3} A and because of the voltage drop across R, the voltage across the tube E_t drops to a smaller value E_o (*operating voltage*).

Several luminous regions can be distinguished in the glow discharge. The most important one is the cathode glow: the cathode surface is partly covered with a luminous sheath. A different luminous column is observed in front of the anode (plasma) and extends throughout the larger part of the tube. Between both luminous regions are one or several dark spaces.

With increased applied voltage E_a, and thus E_t, the light-covered area of the cathode increases and the current I increases sharply with the result that the voltage drop across R increases further and the voltage E_t remains practically constant (*region of normal cathode fall*). Once the entire cathode area is covered with light, the current rises less steeply at E_b (*region of abnormal cathode fall*). If the applied voltage is raised still further at E_{cl}, a new region of instability is reached; the current rises again and the discharge assumes the character of an arc.

If the tube is evacuated to about 10^{-2} torr and if the applied voltage is raised to several kilovolts, an ion beam can be observed in front of the cathode. If there is a canal in the cathode or if the cathode is made from a wire screen, the ion beam will continue into the field free space beyond the cathode and will form a *canal ray*, the forerunner of the particle accelerator or mass spectrograph.

If the tube is further evacuated, the dark space will extend further until it fills the space between the cathode and the anode. If there is a canal in the anode, a beam of *cathode rays* will be observed in the space beyond the anode which causes fluorescence of the glass wall or of a luminous screen brought in that part of the tube.

The starting or firing voltage E_s depends on the type of gas in the gap and the material of the electrodes. For a given gas-electrode combination, the starting potential also depends on the product of gas pressure p and distance d in a manner shown in Fig. 8-8*b*. The occurrence of a minimum starting voltage is explained as follows: the discharge is initiated by an electron injected from the outside. The number of molecules in the gap is proportional to the product pd, where p is the gas pressure within the tube and d is the distance between the electrodes. At low pressure, the mean free path of an electron is large; few electrons can collide with the gas molecules so that the chances for ionization by collision are small. At a higher pressure, the mean free path of the electrons is small and few electrons can acquire sufficient energy over a mean free path to ionize. A similar argument would apply for a variation of the distance d. Hence, a condition for the most favorable (minimum) breakdown potential exists. The minimum breakdown voltage for air between iron electrodes is 330 V and occurs at a value of $pd = 0.57$ torr·cm. The minimum breakdown voltage in inhomogeneous fields follows a similar function; i.e., it shows a minimum for a given value of pd, but also depends on the polarity of the setup. For instance, for a point-to-plane electrode system, the positive point-negative plane arrangement has a lower sparking potential than that with opposite polarity. Tubes with two electrodes of different shape show a rectifying effect if operated with alternating current.

Glow-discharge tubes A voltage-current characteristic of a glow-discharge tube is shown in Fig. 8-9. As long as the applied voltage E_a remains below the starting voltage E_s, the tube acts almost like an open circuit; the current is practically zero so that the voltage drop across the resistance R_s is negligible and the voltage E_t between the tube terminals is equal to the applied voltage E_a.

As soon as the voltage E_t reaches the region of the starting or firing voltage E_s of the glow tube, the current rises suddenly as explained above. The firing voltage is constant for a given tube; its value depends on the type of gas within the tube, the gas pressure, the material of the electrode surface, primarily that of the cathode, and the tube geometry. The firing voltage of commercial glow lamps may be as low as 60 V or as high as several hundred volts. The current after firing may be from several microamperes to about 100 mA; its value depends on the tube parameters but also on the magnitude of the applied voltage and the series resistance in the circuit.

When the tube fires, the rise of current causes a voltage drop across the resistance R_s and thus a reduction of the terminal voltage E_t. If after firing or after operation of the tube at any point along the characteristic the voltage is reduced, the glow will persist until the voltage across the tube reaches a lower

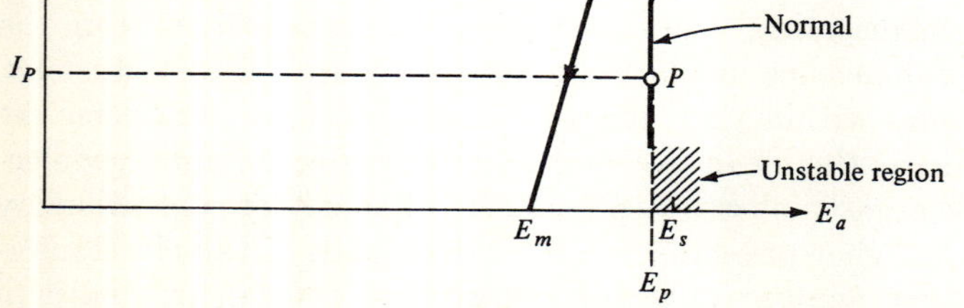

Fig. 8-9 Rising and falling characteristic of a glow discharge. Insert, circuit diagram.

value E_m, the minimum sustaining voltage, which is usually 2 to 20 V lower than the starting voltage E_s.

The equivalent circuit of a glow lamp operated in the region of the normal cathode-fall characteristic is that of a voltage source E_m in series with a resistance R_i and bypassed by a capacitor C. The current can be found from

$$I = \frac{E_a - E_m}{R_s + R_i} \tag{8-2}$$

The starting voltage depends slightly on external circumstances which facilitate the formation of discharge-initiating electrons in the discharge space. Some glow lamps contain a trace of a radioactive substance to facilitate the onset of a discharge. Illumination of the cathode may cause the emission of photoelectrons, primarily from the cathode. Electric fields in the vicinity of the tube, in particular those of radio frequency, may cause a local increase of field strength. If the tube was ionized and then extinguished, some long-living charge carriers which facilitate the formation of a discharge may still be present in the tube.

Since the formation of a discharge depends on the presence of initiating charged carriers, there is frequently a time delay between the application of a voltage to the tube and the initiation of a discharge, which may last from 25 μs to several seconds. The time delay decreases if the applied voltage exceeds the starting voltage of the tube (overvoltage or excess voltage). The deionization time, i.e., the time after disconnecting the supply voltage when practically all charged carriers are recombined, is of the order of 50 ms.

The glow is primarily observed as a luminous sheath covering the cathode partly if the tube is operated within the region of the normal cathode fall, or covering it completely if operated in the region of the abnormal cathode fall. The plasma usually is reduced to a small luminous cloud in the vicinity of the anode.

If glow lamps are operated within the rating of the manufacturer, their life is usually in excess of 20,000 h continuous operation. The application of higher-than-rated voltages and currents diminishes the lifetime drastically. The discharge causes sputtering of the electrode metal on the glass wall; the glass wall gets black and the light output is diminished. Sputtering also reduces the gas pressure (gas is absorbed by the sputtered metal), and the firing voltage increases. The greatest aging effect takes place during the first hundred hours of operation.

Glow tubes frequently are used as signal lights to indicate the presence or absence of a voltage or as read-out devices in digital counters. They are reliable since they do not burn out like incandescent bulbs, but they gradually get darker, indicating the need for renewal. They can be used for voltages and currents varying over a wide range if a current-limiting resistor is connected in series with the glow lamp. Glow tubes can be operated with direct current or with alternating current at power, audio, or radio frequency.

Figure 8-10*a* shows a glow tube used as a digital read-out device for numerical or character indication; a circuit diagram of the tube is shown in Fig. 8-10*b*.

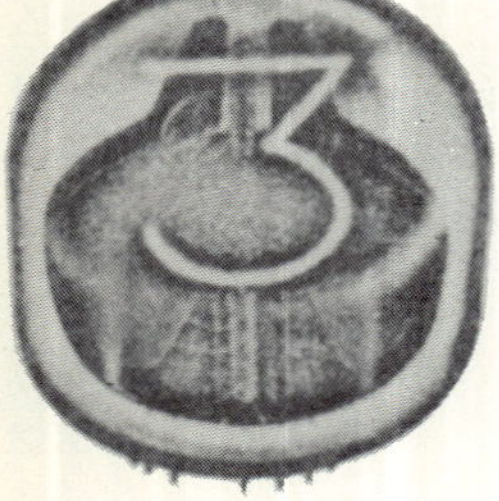

(a)

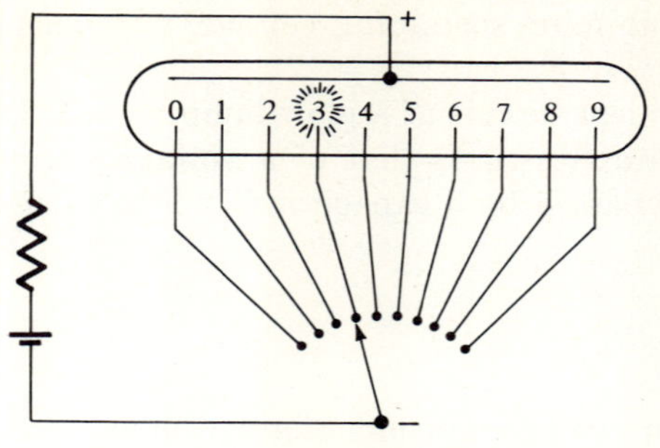

(b)

Fig. 8-10 (*a*) Glow-tube numerical indicator. ("Nixie tube," trade name of The Burroughs Company); (*b*) circuit diagram for the operation of a glow-tube numerical indicator.

The tube contains an anode and ten cathodes. The cathodes have the shapes of figures, for instance, 0 to 9. The application of a negative potential to any one cathode (e.g., cathode 3 in Fig. 8-10*b*) causes the cathode to light up. The life expectancy of these tubes under dynamic operating conditions (when all cathodes are used for equal times) may be more than 200,000 h. The power consumption is small (e.g., 170 V, 1.5 mA; 250 mW). The cathodes can be prebiased so that only a fraction of the operating voltage of 170 V is needed to initiate the glow of one cathode.

APPLICATIONS OF ELECTRICAL GAS DISCHARGES

An indicator for a critical voltage interval, frequently used in inexpensive recording or audio equipment, is shown in Fig. 8-11*a*. If the voltage applied to terminals A' and B' is lower than the starting voltages of tube I, none of the glow lamps will light up. Increase of the voltage beyond a critical level will cause tube I to light up. Increase of the applied voltage above a second critical level will cause tubes I and II to light up. The circuit can be adapted for different critical levels by means of biasing supplies.

The use of a glow tube as a trigger or switching circuit is illustrated in Fig. 8-11*b*. An auxiliary dc source E_a and a voltage divider R_1, R_2 holds the anode A of the glow tube at a potential below the starting voltage; the cathode C is connected to ground via the resistor R_3. As long as the glow tube is nonconductive, no output signal appears between the output terminals T_o, T'_o. If a pulse is applied to the input terminals T_i and T'_i, the glow tube becomes conductive and an output signal appears.

Glow tubes can be used in relaxation oscillators. An example is shown in Fig. 8-11*c*. An auxiliary voltage E_a charges capacitor C through resistor R with a time constant determined by the product RC. As soon as the voltage across C reaches the value of the glow-lamp starting voltage E_s, the glow tube becomes

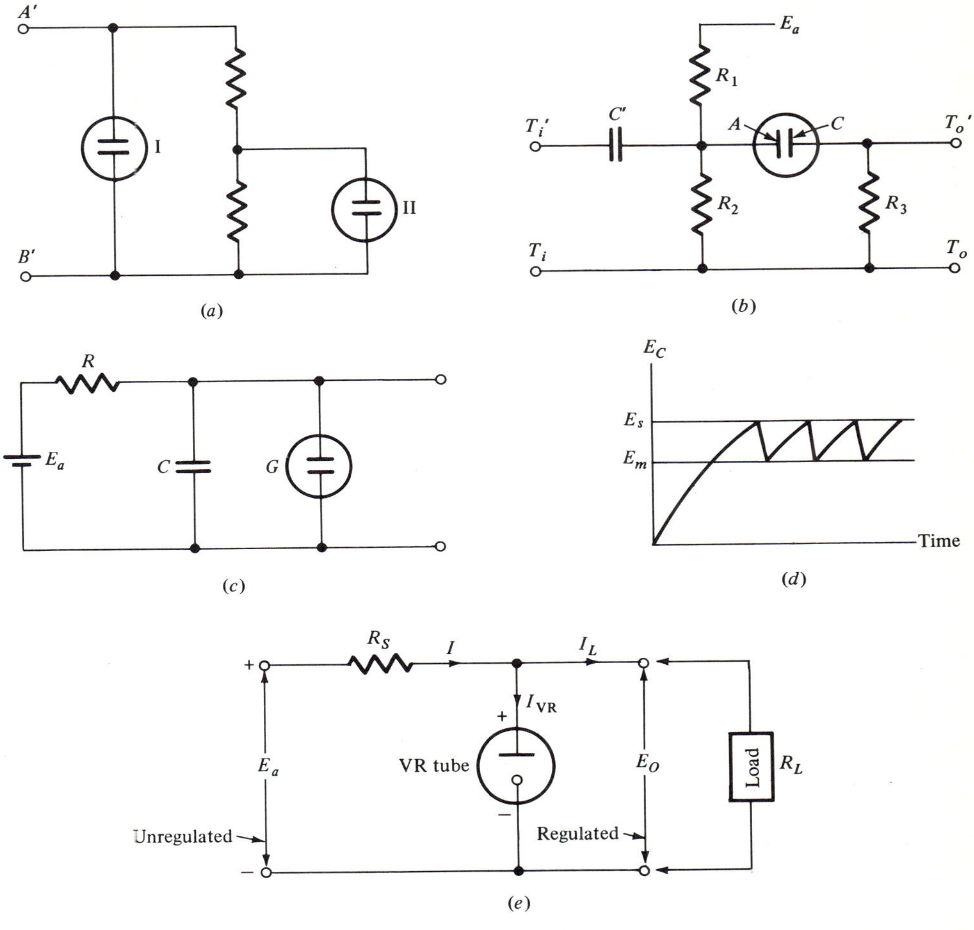

Fig. 8-11 (*a*) Glow lamp I indicates a first critical level; glow lamp II, a second higher critical level; (*b*) trigger or switching circuits with glow lamps, see text for explanation; (*c*) glow-lamp relaxation oscillator circuit; (*d*) voltage-time diagram; (*e*) glow-discharge voltage-regulator circuit.

conductive and the capacitor discharges through the glow tube. If the voltage across C reaches the value of the minimum sustaining voltage E_m, the glow tube extinguishes and the charge-discharge cycle repeats. The waveshape of the output has a sawtooth character as illustrated in Fig. 8-11*d*.

Gas-discharge voltage-regulator (VR) tubes are glow tubes with a large cathode surface area. A voltage-regulator (VR) circuit is shown in Fig. 8-11*e*. The tube is operated in the region of the normal cathode fall. In this region the incremental resistance of the tube $r_d = dE/dI$ is very small, of the order of 50 to 500 Ω. As a practical rule, the input voltage E_a should be at least 50 percent higher than the voltage across the VR tube. The larger the input voltage E_a and the larger the series resistance R_s, the better the voltage regulation will be.

If the load current changes by the amount ΔI_L, the output voltage will change by

$$\Delta E_o = \Delta I_o r_d$$

The value $r_d = dE/dI$ of the VR tube can be found from the slope of the EI characteristic.

If the input voltage varies by the amount ΔE_i, the output will vary by

$$\Delta E_o = \frac{r_d}{R_s} \Delta E_i$$

The regulation can be improved if in place of R_s a nonlinear circuit element with a large differential resistance is used, such as a ballast tube, a pentode, or a transistor.

The output voltage is determined by the tube construction and cannot be varied. Tubes (even of different voltage ratings) can be operated in series in order to obtain a regulated voltage which is the sum of the rated voltages of the single tubes. For such arrangements each tube should be bypassed by a resistor (order of a megohm) to secure satisfactory starting of all tubes.

If the load current diminishes, the current to the VR tube increases. The maximum tolerable variation of the load current is limited by the maximum current rating of the VR tube. The VR-tube current can be increased by operating VR tubes in parallel; a resistance (about 100 Ω) should be connected in series with each VR tube.

Arc-discharge tubes If the voltage applied to a gas discharge and the pressure in the tube are increased, the discharge form changes abruptly and the current increases drastically. The discharge assumes the character of an arc discharge which can lead to the destruction of the tube. The arc discharge is characterized by a thermoelectric emission of electrons from the cathode. If the cathode can be heated by a separate source and if the resistance in the circuit is correctly chosen, the tube current can be limited and controlled. Figure 8-12*a* shows a circuit used for this operation. For a given cathode temperature, the current rises rapidly as soon as the applied voltage exceeds the level E_i, the ionization voltage (order of 15 V). The current is limited by the resistance R_L in the external circuit,

$$I_{\max} = \frac{E_a - E_i}{R_L}$$

Highly luminous plasma fills most of the space between cathode and anode. Gas-type diodes are used as highly efficient rectifiers in ac-dc supplies. In many cases they are superior to high-vacuum diodes; however, the sudden rise and fall of the tube voltage in each cycle can cause parasitic high-frequency oscillations in associated circuits. Mercury-filled arc rectifiers for very large power usually have a mercury pool as a cathode. An applied high voltage initiates an arc between the anode and a narrow region of the pool surface which becomes sufficiently hot to emit electrons.

Thyratrons The onset of a gas discharge in a tube with a heated filament can be controlled by a grid, i.e., a wire or screen electrode surrounding the cathode. The function of the grid is as follows: electrons are emitted by the heated cathode and are accelerated by the positive anode. Their velocity increases until they cause further ionization by collision. The presence of a negatively charged grid surrounding the cathode (i.e., negative with respect to the cathode) counteracts to some extent the influence of the positive anode. If the negative potential of the grid is reduced or if the positive potential of the anode is raised, the tube will fire and carry a current. Once an arc discharge takes place in a thermionic gas-filled triode, the grid is no longer effective in controlling the current. Only if a large resistor is present in the plate circuit so that the ion density in the arc is small, is it possible to extinguish a thyratron by the sudden application of a strongly negative potential to the grid. The circuit diagram of a thyratron is shown in Fig. 8-12*b*.

The grid potential E_g at which a thyratron fires depends on the applied anode or plate voltage E_a; the higher the positive plate potential, the more negative is the grid voltage at which discharge sets in. This relationship is expressed in Fig. 8-12*c*, which shows whether or not firing of a thyratron occurs with a given combination of an applied voltage $+E_a$ and a grid voltage $-E_g$. There are three regions of operation. Region A indicates the combinations of parameters where no firing takes place. For instance, if the applied voltage is $+E_a'$ and the grid voltage is $-E_g'$, the intersection point P is in the field A and the tube carries no current. Region C indicates the combination of parameters where firing occurs. Between these fields is a region B where firing is likely to occur; it is not sharply defined but depends on the time function of the applied signal, on temperature, or on other physical circumstances. The discharge can be terminated by temporarily interrupting the plate circuit by opening the switch S in Fig. 8-12*b*, or by applying an ac source in place of the dc source E_a; the discharge ceases during the negative half-cycle.

Two principal circuits are available for the control of thyratrons, the dc and the ac phase-shift methods. The dc method is illustrated in Fig. 8-12*d*. An ac voltage is applied as plate supply and the dc voltage applied between grid and cathode is such that the thyratron does not fire. If the negative dc grid voltage is reduced, firing will occur at the peak ac value, and a current will flow during 50 percent of the positive half-cycle. If the negative grid voltage is further reduced, firing will set in at a point t before the ac plate voltage reaches its peak value and current can flow during 50 to 100 percent of the positive half-cycle.

The ac phase-shift control method is illustrated in Fig. 8-12*g*. A phase-shifting network permits the application of an ac voltage to the grid. As soon

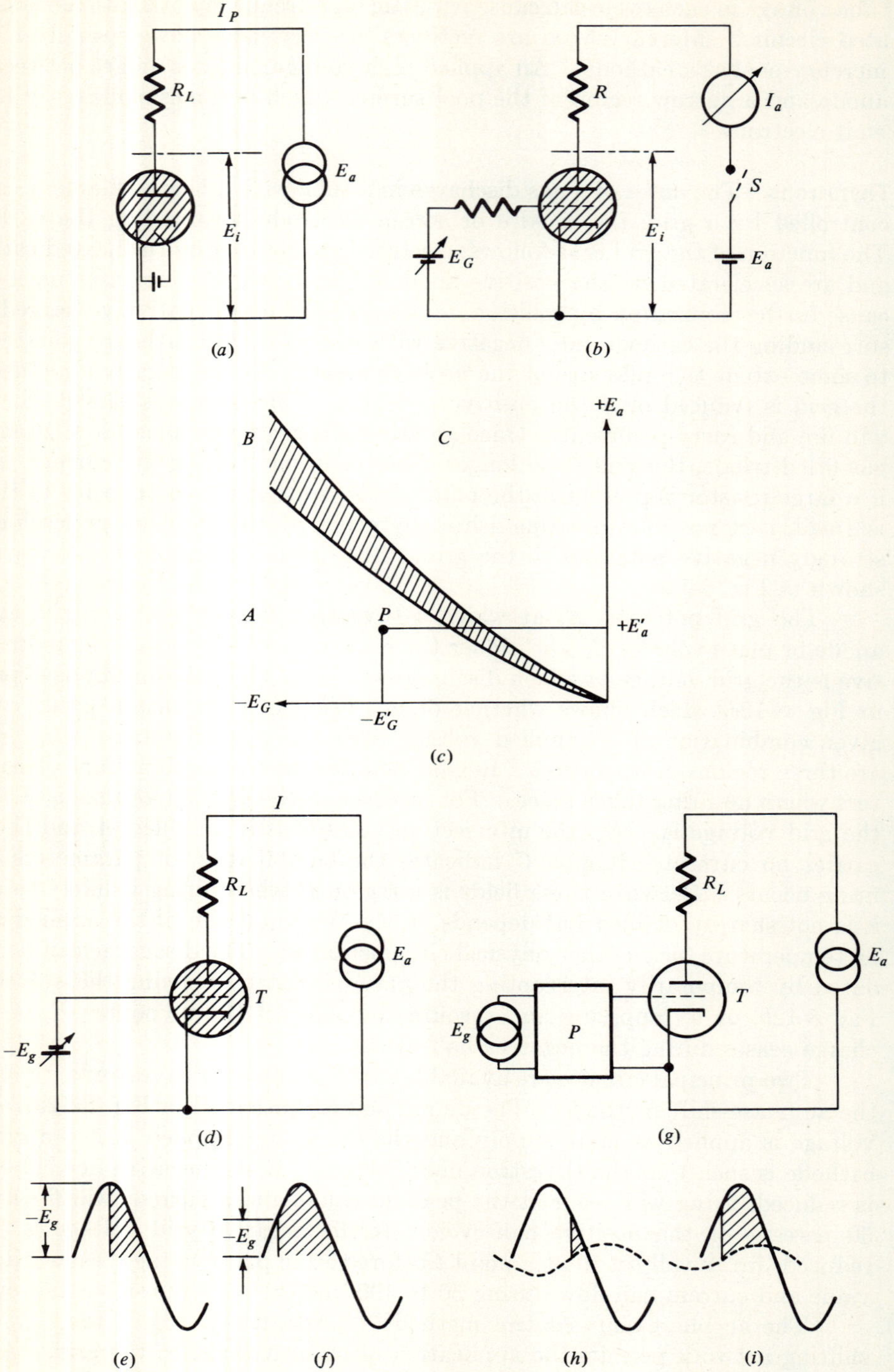

(See legend on opposite page.)

as this voltage reaches the firing potential the tube will fire and can thus be controlled from zero to 100 percent of the positive half-cycle.

Thyratrons can be used for average currents of several amperes, for peak currents of the order of 1,000 A (*Ignitrons*), and for voltages of several hundred volts. Depending on the construction, they will withstand peak inverse voltages (voltages applied in reverse polarity that will not cause a voltage breakdown) of several thousand volts. Their speed of response or maximum repetition rate of operation is determined by the deionization time or recombination time of the ions. This time is of the order of 10^{-4} s in mercury- or argon-filled tubes and about 10^{-5} s in hydrogen-filled thyratrons.

Arc discharges have found numerous applications as light sources in air or gases at atmospheric or reduced pressure. One of these sources, the electronic flash tube, is shown in Fig. 8-13*a*. The tube contains two electrodes and is filled with a gas (frequently xenon, but also air, nitrogen, or argon) at a pressure of about 70 to 200 torr. The flash-tube circuit contains a dc voltage source E_a of the order of several thousand volts which charges a capacitor C through a resistor R. As soon as a critical voltage level across C is reached, the tube fires and a discharge takes place. More frequently, the capacitor is charged to a voltage below the self-starting level. A high-voltage pulse (order of 20 kV) is then applied to an external trigger electrode T; the pulse causes an ionization within the tube and initiates the main discharge from the capacitor through the tube. Since the resistance of the ionized gas discharge within the tube is between a fraction of an ohm to several ohms, a very high current of the order of a hundred amperes and a correspondingly very high light intensity of short duration will occur. Peak powers in the megawatt region have been observed. The discharge time ranges from less than a microsecond to about 20 ms. Short flashes are obtained with short gap tubes, high voltages, and small capacitances. The efficiency of the flash tube is of the order of 35 lm·s/W·s. The voltage-time and the current-time relationships of a flash tube are shown in Fig. 8-13*b*.

A time delay, of the order of 0.5 to 50 μs, exists between the application of the trigger pulse and the development of high-intensity light. The delay time depends on the type of gas and the gas pressure within the tube and is least for xenon-filled tubes at low pressure. The decay of current in an RC circuit is an exponential function so that an exponential decay of light intensity is expected. However, it is found that light (called *afterglow*) continues after the current ceases

Fig. 8-12 (*a*) Simple arc diode-rectifier circuit. E_i, potential difference which develops between cathode and plate (order of 15 V). (*b*) Circuit for the operation of a thyratron, see text for explanation. (*c*) Three ranges of grid voltage-plate voltage combination. Range A, no discharge will take place; range B, discharge *may* take place, depending on physical circumstances; range C, discharge *will* take place. (*d*) Circuit for the control of a thyratron by a variation of the dc grid voltage. (*e*) Current flow (shaded area) for a grid voltage E_g causing firing at the peak value of the ac plate voltage. (*f*) Current flow for a further positive dc grid voltage. (*g*) Circuit for the control of a thyratron by a variation of the phase angle between applied ac plate voltage E_a and ac grid voltage E_g. P, adjustable phase shifter (such as shown in Fig. 7-4). (*h*) and (*i*) Current flow with different phase angles.

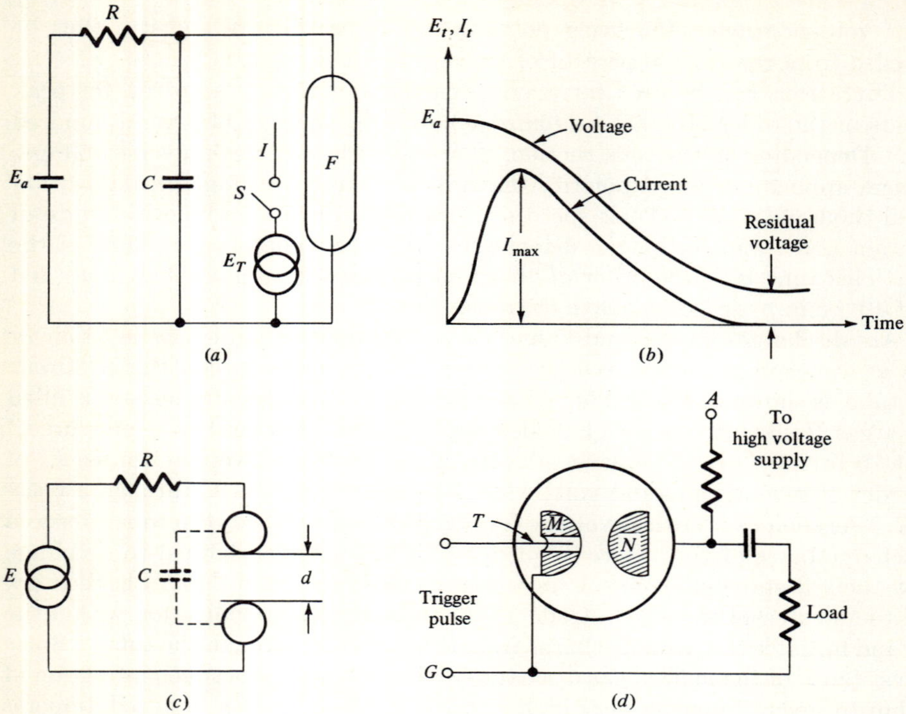

Fig. 8-13 (*a*) Flash discharge tube. E_a, applied voltage of several thousand volts, R, C, charging network; E_T, S, T trigger voltage source (approx. 20,000-V pulse), switch and trigger electrode. (*b*) Voltage-time and current-time characteristic of a flash discharge tube in a circuit as shown in Fig. 8-20*a*. (*c*) Spark-gap circuit for the measurement of a high voltage E. R, current-limiting series resistance; C, interelectrode capacitance; d, distance between spheres. (*d*) Triggered spark-gap. The triggering pulse is applied between the electrode T and ground, the high voltage between A and ground. M and N are the spark-gap electrodes.

to exist. The afterglow is probably caused by recombination of ions on the walls of the tube.

Flash tubes are frequently used for stroboscopic illumination of oscillating or rotating elements. The repetition rate of the strobotron tube is limited by the deionization time in the tube and the unavoidable discharge-lengthening resistance and inductance in the associated circuit. The upper frequency limit of commercial strobotrons is of the order of 20,000 Hz, but discharge rates as high as 10^5/s have been obtained.

The discharge light from a xenon-filled glass flash tube covers a spectral range from approximately 300 nm to 1.5 μm; that from a quartz tube extends into the ultraviolet range to 180 nm. The light approximates the color of sunlight. The maximum light output is in the infrared region, between 0.8 and 1.1 μm with a peak emission at 0.95 μm.

Spark discharges are observed if a high voltage is applied to two electrodes in a gas at atmospheric or higher pressures. The discharge originates at a region of highest field strength, e.g., in the vicinity of sharp points or at the place of closest proximity of the electrodes. The initial discharge form has the appearance of a weakly luminous, branch-like figure (*corona discharge*). If the applied voltage is sufficiently high, the corona discharge develops into an electric spark which is in many respects similar to an arc discharge. The current may be very high, of the order of 1,000 A, and the duration very short, in the microsecond region. The usually brilliant light emission contains spectral lines which are characteristic of the gas between the electrodes and of the material of the electrodes. The minimum voltage necessary to cause a spark is a function of the electrode distance.

Spark gaps used under standardized conditions (geometry of the electrodes, composition, pressure, and temperature of the gas) can be used for high-voltage measurements. The two electrodes of the gap, usually in the form of two polished metal spheres, are connected to the terminals of the high-voltage source E in Fig. 8-13*c*. The distance d is small compared to the radius of the sphere. A series resistance R of several megohms is provided to limit the current. The distance d between the spheres is diminished until breakdown occurs. The distance is measured and the unknown voltage (*peak value*) is found from empirical tables. The accuracy is of the order of several percent.

At dc, the magnitude of the resistor R does not influence the result or the accuracy of the measurement since before the onset of the discharge the current and, hence, the voltage drop across R, are negligible; once the spark is initiated, the discharge is largely that of the capacitance C between the two electrodes. At ac, a capacitive current flowing through C may cause a voltage drop across R and may give rise to a considerable error.

Spark discharges, like other gas-discharge forms, are initiated by the injection of an electron from an external source, such as a radioactive source, cosmic radiation, or photoelectric effect.* If a voltage is applied to two electrodes, a discharge will not occur immediately but with a statistically varying time delay. The average delay time depends on the applied voltage; it diminishes drastically if the applied voltage is raised. A spark gap, either the plane-parallel-plate type or the point-to-plane type, can be operated at a critical voltage so that it responds to the arrival of a photon or high-energy ionizing particle and, hence, acts as a counter. The advantages of spark counters over Geiger counters are a high output and a direct visual indication of a discharge which makes possible the observation or photography of an ionizing particle. Spark chambers consisting of an array of spark counters can be used as a means to trace the path of a high-energy particle.

A further advantage of a spark counter over a Geiger counter is its small critical volume and, hence, its low background and its high geometric definition.

* (Spark counters) K. S. Lion and G. F. Vanderschmidt, *J. Opt. Soc. Am.*, **45**(12):1024–1027 (1955).

For instance, a point-to-plane counter will react only if an ionizing event occurs within a small cone-shaped space extending from the point to the plane.

Spark discharges can be triggered to act as relays furnishing an output from 10^3 to 10^5 A. An example is shown in Fig. 8-13*d*. A high-voltage supply is applied between points A and G (ground). The voltage is not high enough to cause a breakdown of the spark gap between the electrodes M and N. If a pulse is applied between the electrode M and the trigger electrode T, a discharge will take place with a time delay of 10^{-9} to 10^{-7} s.

REFERENCES

Rossi, Bruno B., and Ham H. Staub: "Ionization Chambers and Counters, Experimental Techniques," McGraw-Hill, New York, 1949.

Price, William J.: "Nuclear Radiation Detection," McGraw-Hill, New York, 1964.

Korff, Serge A.: "Electron and Nuclear Counters," Van Nostrand, New York, 1949.

Wilkinson, D. H.: "Ionization Chambers and Counters," Cambridge University Press, 1950.

von Engel, A.: "Ionized Gases," Clarendon Press, Oxford, 1955.

Lovelock, J. E.: Ionization Methods for the Analysis of Gases and Vapors, *Anal. Chem.*, **33**(2):162–178 (1961).

9

Elements Based on Transportation of Charges in Vacua

If a vessel is evacuated from atmospheric pressure (roughly 10^3 torr*) to a pressure of 10^{-6} torr, then the mean free path of the remaining molecules can approach or even exceed the dimensions of the vessel depending, of course, on the size of the vessel. Electrons or ions brought into, or produced in, the evacuated vessel and accelerated by electric or magnetic fields will move essentially undisturbed by the remaining molecules.

Instrumentation elements based on the acceleration and the transportation of charges in vacua deal with two kinds of carriers, ions and free electrons. Ions are used primarily in mass spectrographs or particle accelerators; electrons are used in photoelectric and thermionic tubes. The following presentation deals with the controlled movement of electrons.

Electrons can be obtained from metallic conductors by imparting sufficient energy to the electrons so that they can overcome the forces which keep them bound to the metal lattice. There are primarily two methods of imparting energy to the electrons, leading to photoelectric emission and thermal emission.

Photoelectric cells A cell for photoelectric emission is illustrated in Fig. 9-1. The cell contains a larger electrode C and a smaller electrode A in an evacuated

* 1 torr = 1 mm Hg.

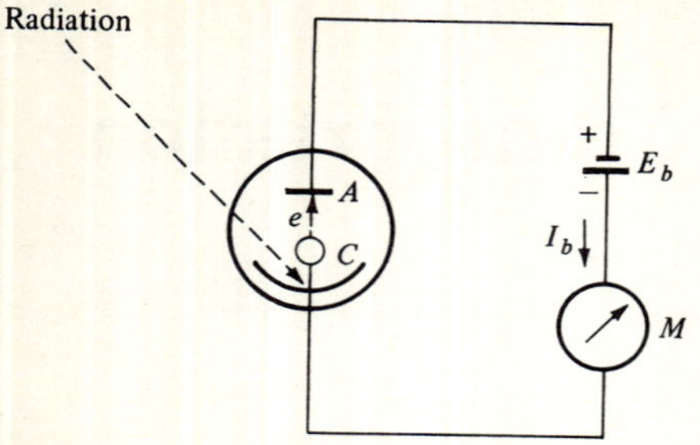

Fig. 9-1 Basic-circuit diagram for the operation of a photoemissive cell using a galvanometer M; anode, A; cathode, C.

vessel. Light radiation strikes electrode C and interacts with the electrons in the metal whereby the entire energy of a light quantum hc/λ is converted into the kinetic energy W of an electron.

$$W = \frac{hc}{\lambda}$$

where h = Planck's constant, 6.6×10^{-34} W·s²*
c = propagation velocity of light, 3×10^{8} m/s
λ = wavelength of incident light radiation

Electrons with sufficient energy to overcome the forces which bind them to the metal will be emitted. In the emission process an electron will lose a part of its energy $e\varphi$ so that the kinetic energy of the emitted electron is

$$W' = \frac{hc}{\lambda} - e\varphi$$

where e = charge of the electron, 1.6×10^{19} A·s
φ = work function of the metal, V

The kinetic energy of the electron becomes zero for

$$\frac{hc}{\lambda_0} = e\varphi \tag{9-1}$$

$$\lambda_0 = \frac{hc}{e\varphi} \tag{9-2}$$

where λ_0 is the threshold wavelength for a metal surface with the work function φ. If the wavelength of the incident radiation is larger than λ_0, there will be no electron emission. The threshold wavelength of commercial photoelectric emitters is, in general, between 600 and 1,200 nm, depending on the work function of the surface layer of the electrode C. The alkali metals, particularly alloys combining several alkali metals, have a very low work function and are used, therefore, in the construction of photoelectric cathodes.

Closely connected with the work function is the quantum yield of photocathodes, i.e., the number of electrons emitted for a given number of absorbed

* Planck's constant, 6.652×10^{27} erg·s converted to watt seconds is 6.6×10^{34} W·s.

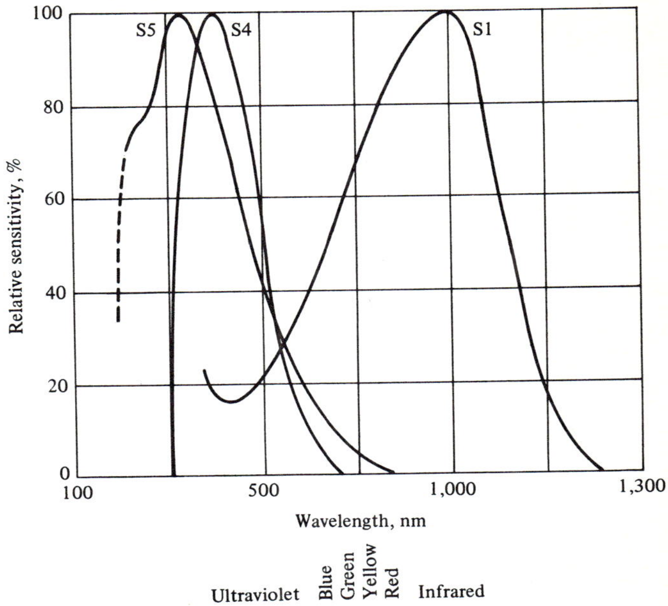

Fig. 9-2 Spectral sensitivity of three different photoemissive cathode surfaces.

quanta. The quantum yield of pure metal surfaces is of the order of 0.001 (one electron per thousand absorbed quanta); that of commercial photocathodes is around 0.05; quantum yields as high as 0.3 have been observed. The quantum yield is zero at the threshold wavelength and increases with decreasing wavelength (*normal photoelectric effect*). In some materials, notably in the presence of hydrogen or oxygen, a narrow spectral region is observed where the emission increases considerably above that of the normal photoelectric effect. This selective photoelectric effect has its origin in a thin surface layer. In technical photoelectric cells, the normal and the selective photoelectric effects are superimposed. The spectral-response characteristic of a photoelectric cell is influenced, in the short wavelength (ultraviolet) region of the spectrum, by radiation absorption in the glass envelope. Some spectral sensitivity curves of commercial photoelectric cells are shown in Fig. 9-2.

An auxiliary voltage E_b is applied between cathode C and anode A as shown in Fig. 9-1. The emitted electrons move toward the anode and form a current I_b which can be read on the meter, such as a galvanometer, or which causes a voltage drop E_o across a load resistor connected in series with the photoelectric cell. The output voltage E_o can be amplified and modified in subsequent stages.

A voltage-current characteristic of a vacuum photoelectric cell is shown in Fig. 9-3*a*. Without the application of a positive potential to the anode, the photoelectrons form a cloud in the vicinity of the cathode; the resulting negative space charge reduces the emission of further electrons. The current I_b is

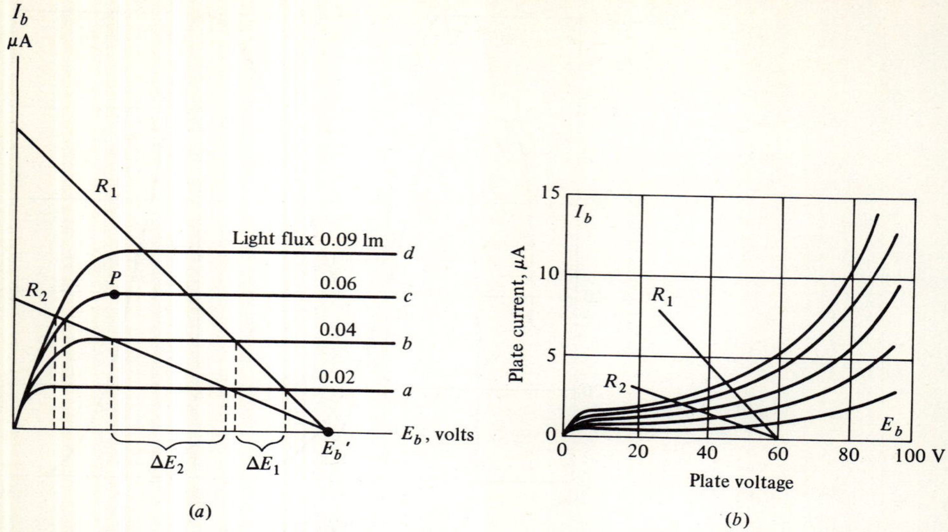

Fig. 9-3 (*a*) Voltage-current diagram of photoemissive vacuum cell. R_1, load line for a small resistance, R_2 for a larger resistance connected in series with a cell. (*b*) Voltage-current characteristics of a gas-filled photoemissive cell.

very small; it is due primarily to the kinetic energy of those electrons that can overcome the decelerating forces caused by the negative space charge. As the applied voltage E_b increases, the effect of the space charge on the emission current diminishes, and the current rises. With further increase of the applied voltage, the current reaches a saturation value. It is then limited only by the emission and, therefore, proportional to the incident light flux as shown in Fig. 9-3*a*, curves *a* to *d*.

Two load lines are drawn in Fig. 9-3*a*, one for the smaller resistance R_1 and one for the larger resistance R_2. If the light flux changes from 0.02 to 0.04 lm, the voltage across R_1 changes by ΔE_1, that across R_2 by ΔE_2. It is evident that under these conditions the larger resistance furnishes a larger output voltage variation. However, if the light flux increases to 0.04 or 0.06 lm, the load line for R_2 proceeds into a region of nonlinearity, while the voltage across R_1 still increases linearly with the light flux. The optimum resistance value can be found by considering the expected variation of the light flux, the maximum voltage E_b' that can safely be applied to the cell and the tolerable nonlinearity. For instance, if a light flux variation between zero and 0.06 lm is expected and if the maximum voltage applicable to the photoelectric cell is E_b' (usually given by the tube manufacturer), and if linearity is essential, a resistance represented by a load line passing through the points E_b' and P would be an acceptable compromise.

If the resistance in the circuit is small, the photoelectric current is proportional to the light flux Φ,

$$I_b = S\Phi$$

where S is the sensitivity of the photoelectric cell. The light flux Φ can be expressed in physical terms, watts, or in photometric terms, lumens, based on the sensitivity characteristic of the human eye. Correspondingly, the sensitivity S can be expressed either in amperes per watt (*radiant sensitivity*) or in amperes per lumens (*luminous sensitivity*). It is customary to compare photoelectric cells by means of a light source of standardized spectral distribution; for instance, an incandescent lamp with a tungsten filament operated at a temperature of 2870K has been selected for this purpose. On the basis of this source, vacuum photoelectric cells have a luminous sensitivity of 10 to 100 μA/lm, or a radiant sensitivity of 0.002 to 0.1 A/W. However, the sensitivity defined is of limited significance. Different photoelectric cells vary widely in their spectral response, and the measured light may have any spectral composition. Where high sensitivity is essential, the spectral sensitivity curve of the cell (Fig. 9-2) should be matched to the spectral composition of the incident light.

The sensitivity of a photoelectric cell can be enhanced by gas amplification. A gas-filled cell generally contains an inert gas at a pressure of about 1 torr. The emitted electrons are accelerated by the electric field and cause ionization and thus further electrons by collision. This process may provide a gain over the response of a vacuum cell of about 5 to 10. Voltage-current characteristics of a gas-filled photoelectric cell are shown in Fig. 9-3*b*. If the applied voltage E_b exceeds a critical value (depending on tube geometry, gas filling, and illumination), a glow discharge may set in which can damage the cathode. A resistor should always be connected in series with the gas-filled cell to limit the current in case of an accidental overvoltage.

Gas-filled photoelectric cells are not as stable over long periods of time as vacuum cells, and their transfer function is not as linear. Also, they exhibit a time lag in response to a light pulse which has the effect of reducing the response to modulated or chopped light at frequencies above about 10^4 Hz.

A small current can be observed in photoelectric cells even in the absence of radiation. This *dark current* which is of the order of 10^{-8} A in vacuum cells and about 10^{-7} A in gas-filled cells has its origin in

1. Thermionic emission from the cathode
2. Leakage current over the glass wall
3. Current caused by ion impact on the cathode (primarily in gas-filled tubes)
4. Field emission from the cathode

Thermionic emission takes place even at room temperature of the cathode because of its low work function. It can be reduced by cooling the photoelectric cell to the temperature of liquid nitrogen, but frequently a simpler way is to separate it from the photoemissive current by light modulation. This method is illustrated in Fig. 9-4. A light beam L, on its way to the cathode, passes through a rotating shutter S (chopper) which causes a pulse-modulated illumination of the cathode. The current output I_b consists of the steady (dc) value I_t, resulting from thermionic emission, and a superimposed pulse (ac) of the magni-

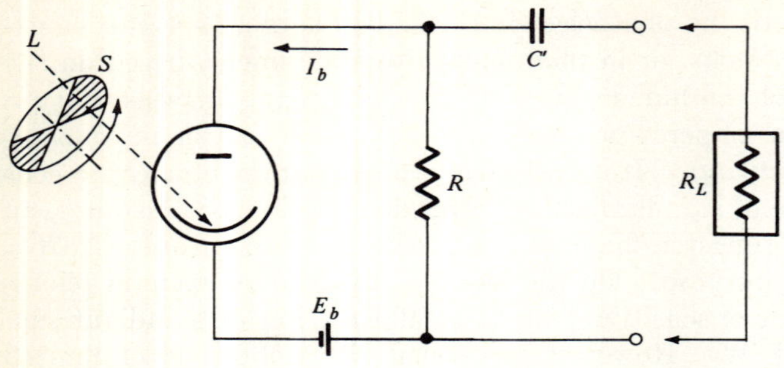

(a)

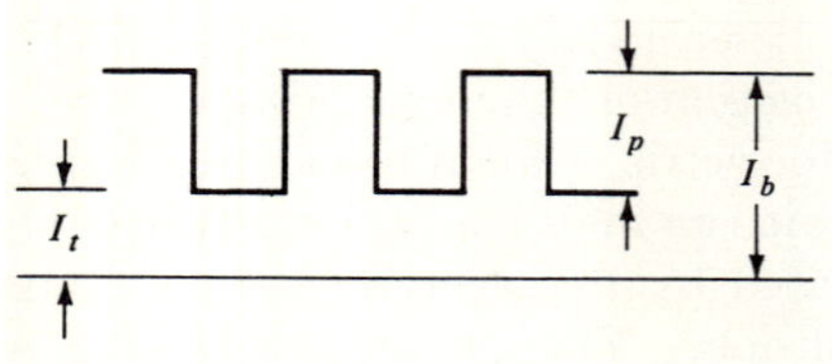

(b)

Fig. 9-4 (*a*) Circuit for the separation of a photoelectric current from a dark current (I_c) by means of a light chopper S. (*b*) I_t diagram; I_b, total current; I_p, photoelectric current; I_d, dark current.

tude I_p, resulting from photoelectric emission. The dc and the ac components are separated by a capacitor C'; only the photoelectric component appears at the output terminals. The influence of leakage current over the output glass wall can be eliminated by a guard ring (see Fig. 4-12) or with the above-mentioned light-modulation method.

Currents caused by the impact of positive ions on the cathode and field emission from the cathode are generally not of practical importance; they can be reduced by operating the photoelectric cells at low plate voltages.

The output from a photoelectric cell illuminated by a constant light flux is not a pure dc voltage but fluctuates randomly about an average value. The fluctuation of the output voltage, i.e., the *shot noise*, can be considered an ac voltage with a uniform distribution over the frequency spectrum. However, the measurable part of this ac voltage is limited by the bandwidth Δf of the associated equipment. The mean-square value of the current is

$$\bar{I}_s^2 = 3.2 \times 10^{-19}\, I_o \Delta f$$

where I_o is the average emission current of the cell. If a load resistance R is connected in the circuit, the noise output voltage caused by shot effect is

$$\bar{E}_s^2 = 3.2 \times 10^{-19}\, I_o R^2 \Delta f$$

In addition to the shot effect the Johnson effect caused by the thermal agitation of the electrons in a resistor also contributes to a noise voltage by the amount

$$\bar{E}_{\text{Th}}^2 = 4HTR\Delta f$$

(see Chap. 2). At room temperature

$$\bar{E}_{\text{Th}}^2 = 1.6 \times 10^{-20} R\Delta f$$

The total noise voltage, comprising shot effect and thermal agitation effect, is

$$\bar{E}_{N,\text{rms}} = \sqrt{3.2 \times 10^{-19} R(I_oR - 0.05)\Delta f}$$

Photomultipliers A schematic diagram of a photomultiplier is shown in Fig. 9-5*a*. Light strikes a photocathode C and liberates electrons. The electrons are accelerated by a voltage E_1 and focused on an electrode, the dynode D_1,

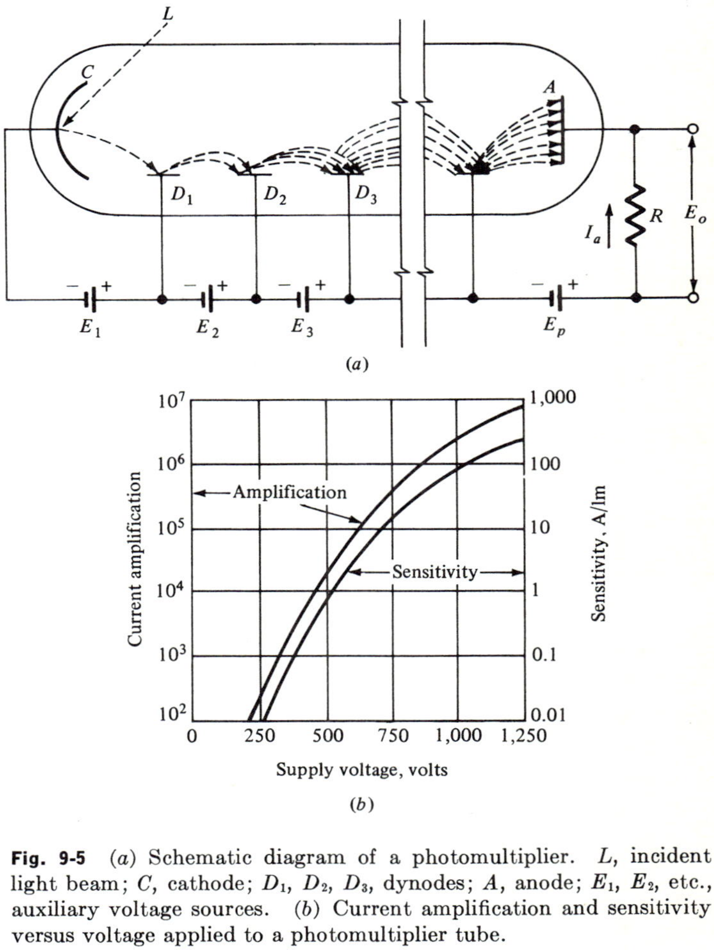

Fig. 9-5 (*a*) Schematic diagram of a photomultiplier. L, incident light beam; C, cathode; D_1, D_2, D_3, dynodes; A, anode; E_1, E_2, etc., auxiliary voltage sources. (*b*) Current amplification and sensitivity versus voltage applied to a photomultiplier tube.

where each incident electron causes the emission of several secondary electrons. The same process is repeated at the dynodes D_2, D_3, ... The electrons from the last dynode stage are collected by a positive anode A, and the current I_a is measured. If the gain of each stage (number of electrons formed by secondary emission for each primary electron) is g and if n dynode stages are used, the total amplification is

$$A_o = g^n$$

The value of g varies with the voltage between successive dynodes and with the surface composition and the geometry of the dynodes from a value of 0.5 to about 10. The number of stages n of commercial photomultipliers is between 9 and 14, and the resulting gain A_o between 10^5 and more than 10^7. The sensitivity of a photomultiplier exceeds that of a photoelectric cell by several orders of magnitude. Generally it is about 20 A/lm, compared with 100 μA/lm for the vacuum photoelectric cell. Sensitivities of photomultipliers as high as 2×10^4 A/lm have been reported. Since the maximum permissible current at the output of a photomultiplier for continuous operation is of the order of 10^{-5} A, the photomultiplier is to be used for the measurement of a light flux less than 10^{-6} lm. With special precautions, it can be used for an incident light flux as small as 10^{-16} lm, i.e., about one light quantum per second. The spectral response of photomultipliers is the same as that of photoelectric cells.

The gain per stage and, hence, the amplification and the sensitivity of the photomultiplier increase almost in an exponential fashion with the voltage applied per stage as shown in Fig. 9-5*b*. On the one hand, this fact offers the possibility of controlling the sensitivity over a wide range by varying the supply voltage; on the other hand, the drastic change in sensitivity with the supply-voltage variation demands that the supply voltage be highly stabilized if constant sensitivity for continuous operation is required.

There are several modes of operation for a photomultiplier. In the range from about 10^{-5} to 10^{-8} lm, the photomultiplier can be used as shown in the general circuit diagram Fig. 9-5*a*. Either a meter is inserted in the plate circuit or a resistance R ($\sim$1 MΩ) is inserted. The voltage drop E_o across this resistance is amplified in further dc amplifier stages. For an incident light flux between 10^{-8} and 10^{-11} lm, a dc amplifier in the output is usually not sufficiently stable, and separation of the photoelectric current from the dark current of the photomultiplier (order of 10^{-7} A at room temperature) is required. Therefore, light modulation and replacement of the dc amplifier by an ac amplifier are needed. In the range below 10^{-11} lm, cooling of the photomultiplier usually is required to reduce thermionic dark current and the noise component of this current. At light levels below 10^{-14} lm, only counting of the amplified pulses caused by the arrival of single quanta at the cathode furnishes results. Operation of the photomultiplier at high light levels (in excess of 10^{-5} lm) can cause damage which may be temporary (*fatigue*); the photomultiplier will recover if left in darkness for several hours. If exposed to very high levels of illumination, the photomultiplier may be permanently damaged.

The response time of a photomultiplier is of the order of 10^{-8} s, i.e., the transit time of an electron through the photomultiplier. Very short light pulses of less than 10^{-9} s duration show a time dispersion effect: the amplified output is a somewhat broadened pulse. The effect is caused by differences of the initial velocity of electrons emitted at the cathode and the difference of the path length of electrons within the photomultiplier.

Noise, i.e., a random fluctuation of the output, arises in a photomultiplier as in any other electronic tube. In the photomultiplier the noise is composed of the random variation of the photoemissive current i_p and the unavoidable thermionic emissive current i_t. The rms current measured at the anode is

$$\sqrt{\overline{I^2}} = A_o \sqrt{2e\Delta f(i_t + i_p)} \qquad (9\text{-}3)$$

where A_o is the photomultiplier gain and Δf is the bandwidth of the detecting system. The thermal emission from the cathode i_t is about 10^{-14} A at room temperature. It is evident that the reduction of i_t, by operating the tube at low temperature, is one of the most effective means to reduce the noise level and, hence, to improve the signal-to-noise level at low illumination.

Instead of using a photomultiplier one can, of course, use a photoelectric cell with subsequent amplifier. A comparison shows that, as far as the signal-to-noise level is concerned, the photomultiplier is superior at low levels of illumination, but that at higher levels of illumination, when the photoelectric current from the cathode reaches the order of 10^{-6} A, the signal-to-noise ratio is almost the same for both systems.

Thermal emission A thermionic diode is illustrated in Fig. 9-6a; it contains a cathode C and an anode A in an evacuated vessel. The cathode can be heated by

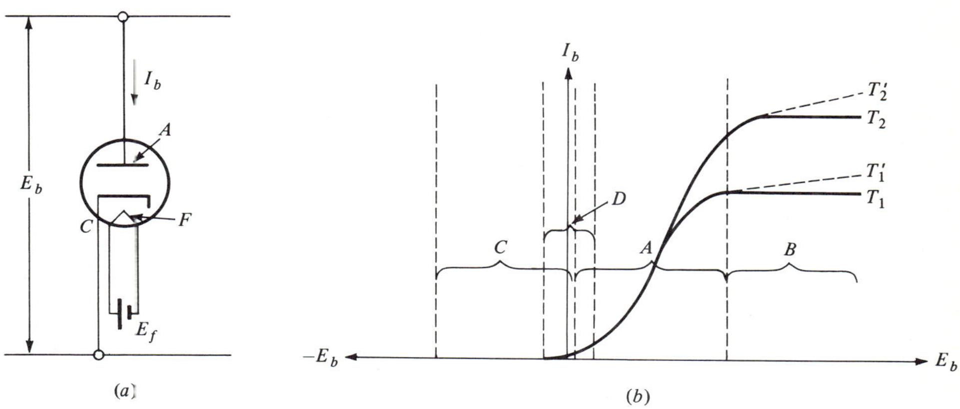

Fig. 9-6 (a) High-vacuum diode. A, anode; C, cathode heated by a filament F; E_f, supply voltage for filament. (b) Current I_b in a high-vacuum diode versus applied voltage E_b. Region A, space-charge-limited operation; B, temperature-limited range; C, blocking direction for inverse applied voltage $-E_b$; D, logarithmic region of operation.

a filament F energized by the auxiliary-voltage source E_f. Increase of the temperature of the cathode leads to an increased energy of the molecules in the crystal lattice of the cathode and, in further sequence, to an increase of the mean electron energy, also to a widening of the energy distribution curve of the electrons. Because of the increased energy, some electrons will eventually overcome the binding forces of the metal and will be emitted. The electron-current density (number of emitted electrons per surface area of the cathode) rises steeply in an exponential function with the cathode temperature.

The emitted electrons will form a cloud (a negative space charge) around the cathode which prevents further electron emission. If the electrode A is made positive with respect to the cathode, electrons from the space charge will move toward the anode and form a plate current I_b. If the applied voltage is further increased, the space charge will disappear, and the current will be limited by the emission process, i.e., by the temperature of the cathode.

Figure 9-6*b* shows the voltage-current characteristic of a vacuum diode for two different cathode temperatures T_1 and T_2. In the region A, the current is limited by the space charge. The current rises with the applied voltage E_b as

$$I_b = KE_b^{3/2}$$

where K is a constant depending on the electrode geometry. In the region B, the current is limited by the temperature of the cathode and does not rise with a further increase of the applied plate voltage (*saturation effect*). The effect can be demonstrated only with low-emitting tungsten cathodes. Tubes with cathodes covered with a material of low work function, which emit a much higher current at normal operating temperatures, show a further increase of emission with increased plate voltage, as indicated by the dotted curves T_1' and T_2'. In the reverse direction (region C), where the potential of the anode is negative with respect to that of the cathode, there is only a small, usually negligible, current stemming from insulation leakage, or from photoemission at the anode, or from thermal emission at the anode if the anode is heated by electron impact during a previous conduction cycle or by thermal radiation. Diodes are operated, in general, only in regions A and C. The forward resistance dE/dI of vacuum diodes in the space-charge-limited region is low, of the order of a few ohms; in the temperature-limited region, it is very high, several megohms.

In region D, which overlaps regions C and A, the current is small and rises exponentially with the applied voltage. If the current is varied (if the source is a constant-current generator) as shown in Fig. 9-7*a*, the voltage across the diode measured with an electronic voltmeter with high input impedance is

$$E_o = k \log I_i \tag{9-4}$$

Equation (9-4) is correct over a large range of current variation of several orders of magnitude as shown in Fig. 9-7*b*. A circuit like that shown in Fig. 9-7*a* can be used as a logarithm-forming system. Its use results in a scale of the output meter like that shown in Fig. 2-3*b*.

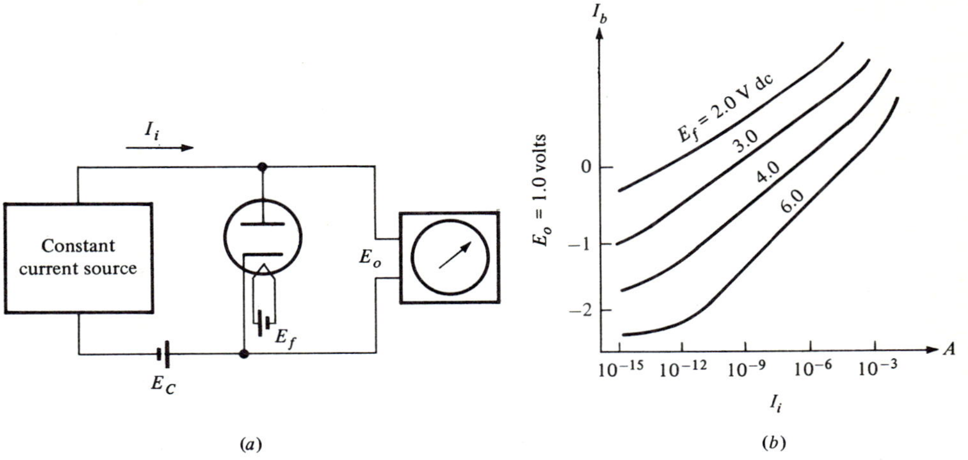

Fig. 9-7 (a) Operation of a diode in the logarithmic region; (b) *IE* characteristics in the logarithmic region.

If a resistance R_L is connected in series with a diode (see Fig. 9-8a), the resulting current in the circuit as well as the partial voltages across the resistance R and the diode D can be found by the same graphic construction of a load line as shown above in connection with photoelectric cells (Fig. 9-3). An example is shown in Fig. 9-8b for three different values of the supply voltage E_1, E_2, and E_3. The projections from a point P of the intersection of the diode characteristic $I_b = f(E_b)$ and the load line $I = (E - E_b)/R$ furnish the desired information.

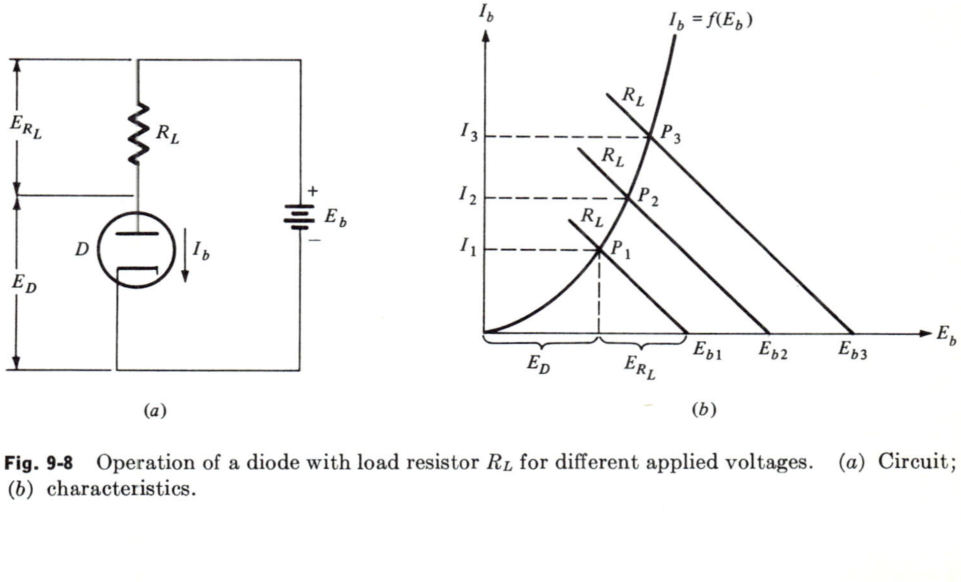

Fig. 9-8 Operation of a diode with load resistor R_L for different applied voltages. (a) Circuit; (b) characteristics.

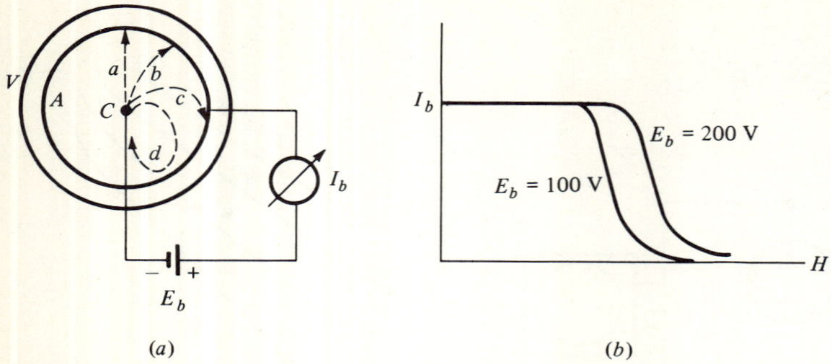

Fig. 9-9 Magnetron. (*a*) Circuit diagram; (*b*) current magnetic field characteristics.

Controlled flow of electrons in a vacuum The flow of electrons from the cathode to the anode can be controlled by magnetic and electric fields. An example for control by magnetic fields, the magnetron, is shown in Fig. 9-9*a*. In an evacuated vessel V, electrons are emitted from a cathode C which extends perpendicular to the drawing plane and is heated by an auxiliary source (not shown). The cathode is surrounded by a cylindrical anode A. The cathode is connected to the negative terminal and the anode to the positive terminal of a voltage source E_b of several hundred volts. Without the application of a magnetic field, the electrons move radially (arrow a) from C to A thus causing a current I_b in the external circuit. Application of a magnetic field in a direction into the drawing surface causes a deflection of the electron beam (arrows b, c) in the form of a circle with a radius of curvature which diminishes with increased magnetic field strength. Eventually, the electrons are bent to a circle without reaching the anode (arrow d); the current I_b then decreases as indicated in Fig. 9-9*b*. The magnetron is used extensively for the generation of oscillations of very high frequencies.

Triodes The control of an electron current by electric fields is illustrated in Fig. 9-10*a*. The cathode C of a triode T is connected to a grounded bar O which for the following discussion will serve as the reference level. The anode or plate P is kept at a level of several hundred volts (e.g., +300 V) positive with respect to O, by means of a voltage source E_b. The potential of the grid electrode G may be a few volts positive or a few volts negative with respect to O. There are then two electrodes, P and G, which affect the flow of electrons. Of these, the grid G is closer to the cathode and, therefore, is more effective than the anode. The relative effect of these two electrodes on the electron current can be measured. The plate may be at a potential E_{b1}, the grid at a potential E_{g1}; under these conditions the electron current (and, thus, the current in the external circuit) may be I_{b1}. Now if the grid potential is raised by ΔE_g (e.g., 1 V), the current will rise to I_{b2}. In order to bring the current back to the value I_{b1}, the plate potential

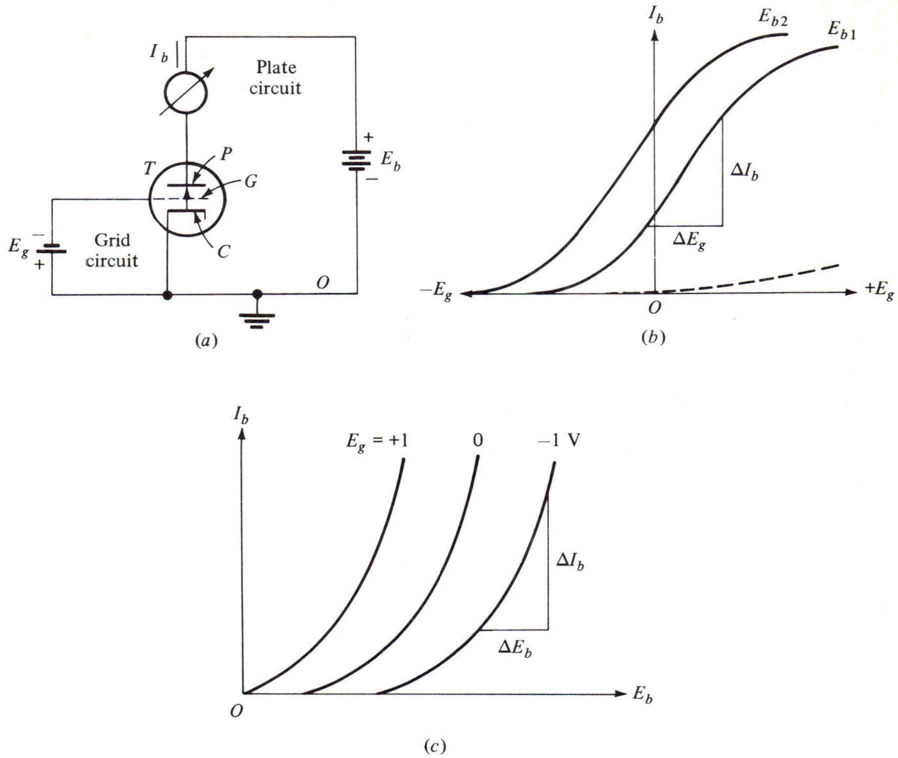

Fig. 9-10 (*a*) Triode circuit; see text for description; (*b*) grid-voltage–plate-current characteristics of a triode; (*c*) plate-voltage–plate-current characteristics of a triode.

must be reduced by the amount ΔE_b, which generally is considerably more than 1 V (e.g., 75 V). The ratio

$$\frac{\Delta E_b}{\Delta E_g} \text{ (for } I_b = \text{const.)} = \mu \tag{9-5}$$

The quantity μ is the amplification factor of the triode. (In the numerical example $\mu = 75$.) The amplification factor is not identical with the gain of a triode as used in an amplifier. The gain can approach the value of the amplification factor under certain conditions which, however, are rarely met.

If the plate voltage E_b is kept constant and the grid voltage is varied from a highly negative value through zero to a positive value, the current I_b will rise as indicated in Fig. 9-10*b*. The slope of the curve, i.e., the ratio

$$\frac{\Delta I_b}{\Delta E_g} \text{ (for } E_b = \text{const.)} = g_m \tag{9-6}$$

is the *mutual transconductance of the triode*, a somewhat unfortunate expression based on the formalism that the ratio I/E has the dimensions of a conductance.

The meaning of g_m is the variation of current in one circuit (the plate circuit) caused by the variation of the voltage in another circuit (the grid circuit). The constant g_m has the dimension ampere per volt; the technical literature uses the unit mho (1/Ω) from which the micromho (1 μA/V) is derived.

If the grid voltage E_g is kept constant and the plate voltage E_b is varied, the plate current will change as illustrated in Fig. 9-10*c*. The reciprocal slope of the E_bI_b characteristic, i.e., the ratio

$$\frac{\Delta E_b}{\Delta I_b} \text{ (for } E_g = \text{const.)} = r_p \tag{9-7}$$

is the plate resistance of the triode. The plate resistance is an incremental resistance as described at the beginning of Chap. 8. Inspection of Eqs. (9-5) to (9-7) shows that the parameters μ, g_m, and r_p are related. If two of these parameters are known, the third can be found from

$$g_m r_p = \mu \qquad \frac{\Delta I_b}{\Delta E_g}\frac{\Delta E_b}{\Delta I_b} = \frac{\Delta E_b}{\Delta E_g}$$

Numerical values for the amplification factor μ in triodes vary from 10 to about 100; for the mutual transconductance g_m from 10^3 to 5×10^3 μA/V, and for the plate resistance from 2,000 to about 100,000 Ω.

Tetrodes Considerably higher values of the amplification factor μ and of the plate resistance r_p are obtained by the insertion of a second grid, the screen grid *S*, between the control grid and the plate. The screen grid is usually kept at a fixed positive potential E_{C2} somewhat lower than that of the plate. The screen grid has the effect that a constant and positive potential is established in the space between the control grid and the plate. The plate potential itself, though always positive, would not serve the same purpose. If a load resistance R_L is inserted in the plate circuit, the time-varying plate current i_b, controlled by a time-varying signal applied to the control grid, will produce a time-varying voltage drop R_Li_b across the load resistance. The result is that the electron-accelerating potential of the plate E_b' is less than E_b; it is $E_b - R_Li_b$. Therefore, the time-varying potential of the plate counteracts to a certain extent the controlling action of the control grid, an effect which is considerably reduced by the steady positive potential of the screen grid.

A second, equally important effect of the screen grid, is its action as an electrostatic shield between the plate and the control grid. The shielding action reduces the undesirable influence of a time-varying plate potential on the control grid. The plate-to-grid influence is expressed by the grid-to-plate capacitance C_{gp}, which is about 2 to 10 pF in a triode but diminishes to about 0.005 to 0.2 pF when a screen grid is inserted between grid and plate. The grid-to-plate capacitance, as evident from the foregoing explanation, cannot be considered a simple two-terminal capacitance.

An unexpected secondary effect reduces the usefulness of the tetrode: electrons falling onto the plate cause the emission of secondary electrons which move to the screen grid if the potential of the plate drops below that of the screen grid. The resulting E_bI_b characteristic has a region of negative slope which leads to instable operation.

Pentodes The instability effect can be avoided by the insertion of another grid, the suppressor grid D (Fig. 9-11*a*) which is usually connected to the cathode and suppresses the flow of secondary electrons from the plate to any other electrode. A circuit diagram of a pentode with its associated supply sources as well as a family of E_bI_b characteristics is shown in Fig. 9-11*a* and *b*. Average numerical values for the parameters of a pentode are: for the amplification factor μ, from several hundred to several thousand; for the mutual transconductance g_m, from 1,000 to about 10,000 μA/V; for the plate resistance r_p, from about 0.5 to 2 MΩ. Pentodes are normally suitable for large output impedances and a high ratio of output to input voltage. A special group of pentodes, the beam-power tubes, is particularly suitable at the output stages of amplifiers where a high power level is required. Beam-power tubes have plate resistances of the order of 10,000 to 100,000 Ω.

With increased plate voltage the plate current of a pentode rises first and then remains almost constant (Fig. 9-11*b*). The shape of these characteristics shows that the pentode is most suitable for the construction of a constant-current source more so than a simple large resistor as in the constant-current source shown in Fig. 2-1. The plate resistance of the pentode $r_p = \delta e_b/\delta i_b$ is very large; in addition, the current level can be varied within wide limits by setting the grid voltage E_g of the pentode.

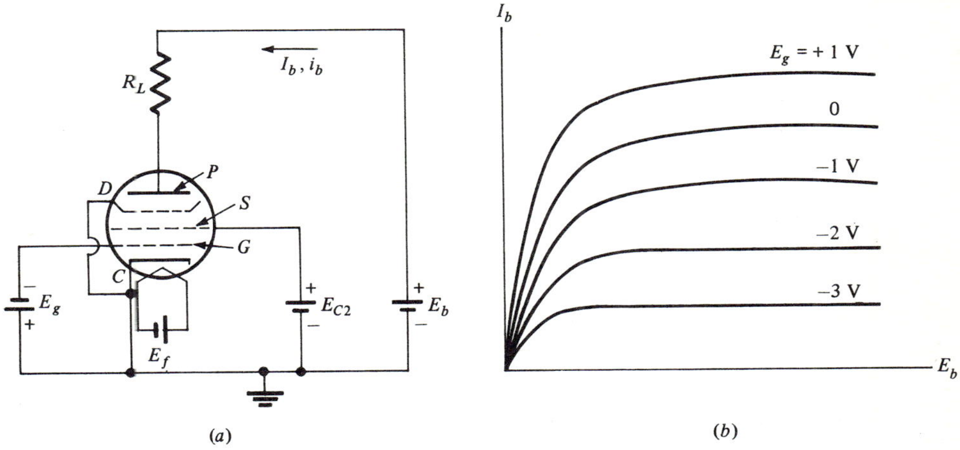

Fig. 9-11 (*a*) Pentode with suppressor grid D; (*b*) I_bE_b characteristics for different values of control grid voltages E_g.

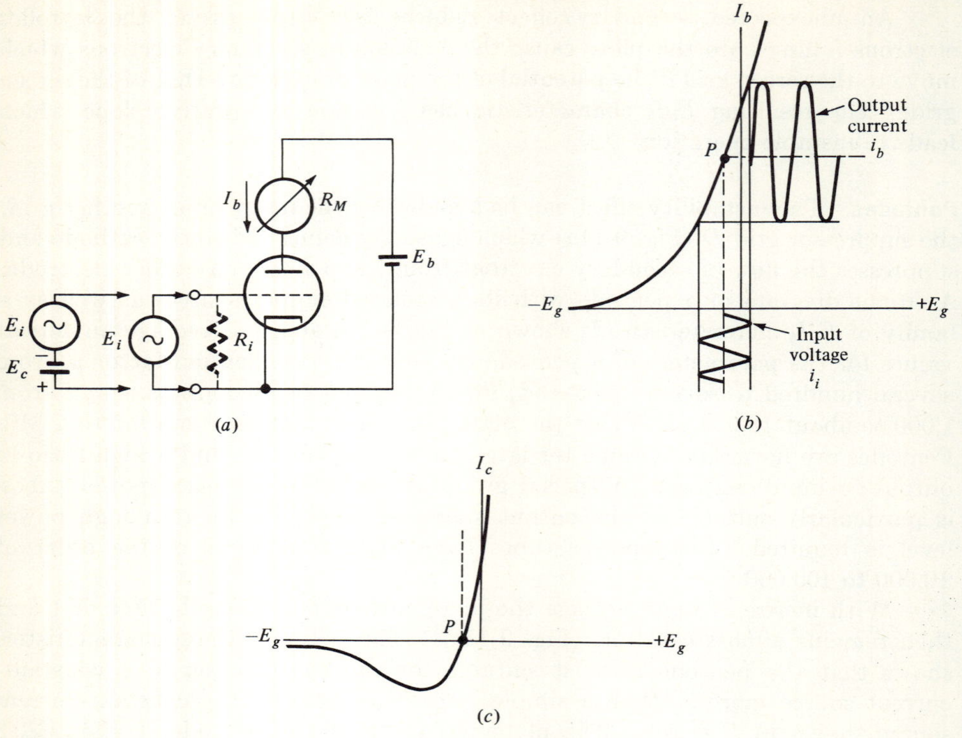

Fig. 9-12 Grid circuits in triodes. (*a*) Circuit diagram; (*b*) input voltage–output current characteristics; (*c*) grid current versus grid voltage, see text for explanation.

Grid circuits Figure 9-12*a* shows a triode circuit with a meter inserted in the plate circuit. The resistance of the meter may be small so that the current I_b is determined primarily by the tube and by the applied voltage E_b. The circuit serves as a vacuum-tube voltmeter. If a signal voltage is applied between grid and cathode, the current through the meter can be found from the I_bE_g characteristic shown in Fig. 9-12*b*. A variation of the signal voltage by the amount $\pm\Delta E_g$ will cause a variation of the current through the meter by the amount $\pm\Delta I_b$. If the signal E_i were applied to the meter directly, the power required from the signal source would be E_i^2/R_m where R_m is of the order of 10^2 or 10^3 Ω. However, if the signal is applied to the input of the tube, the power requirements are E_i^2/R_i, where R_i, the input resistance of the tube, is of the order of 10^6 to 10^{16} Ω. Thus, the loading of the source caused by a vacuum-tube voltmeter is many orders of magnitude less than that caused by direct connection to the meter; the tube acts as a power amplifier.

The input impedance of a vacuum tube is not constant, the grid current–grid voltage characteristic is nonlinear, as shown in Fig. 9-12*c*. The main component of the grid current is a current of electrons from the cathode to the grid, if the grid is positive with respect to the cathode; minor sources of the grid current are emission from the grid (if the grid is heated by thermal radiation), ion current, and leakage current. Figure 9-12*c* shows that the grid current is small if the grid is kept at a negative potential in the vicinity of the operating point P. This can be accomplished by the use of a grid-bias source E_c as indicated in Fig. 9-12*a*. The signal is superimposed on the fixed-bias voltage. The selection of an appropriate bias is not always a simple matter. The operating point P may come to lie in a range of the I_bE_g characteristic where the slope is small so that the resulting sensitivity is low, or in a curved range of the characteristic, so that nonlinear distortion occurs. A compromise between these different requirements frequently is needed.

Special tubes (*electrometer tubes*) with extremely low grid current are available; but in many cases tubes with a well-insulated grid connection (*acorn tubes*) operated at reduced filament and plate voltages are satisfactory. Reduction of the loading of the preceding element can also be accomplished by feedback methods (see below). A simple method to check for grid current is to insert a large resistance R_i (several megohms) into the grid connection as shown in Fig. 9-12*a*. If the grid current I_g is negligible, opening or closing switch S will have no effect on the meter M. If the grid current is not negligible, it will produce a voltage drop I_gR which will cause a deflection ΔI_b of the meter. Since $\Delta I_b = g_mI_gR$, the value of I_g can be found.

Plate circuits A load resistance R_L is usually inserted in the plate circuit as shown in Fig. 9-13 for a triode. The plate current I_b, the voltage E_p across the tube, and the voltage E_{R_L} across the load resistor may be found from the load-line construction as in previous similar cases. A variation with time of the signal e_i applied to the grid causes a variation of the potential at the point P; an increase of the grid potential leads to a decrease of the potential at the plate; the variational plate voltage e_b and the grid voltage e_g are 180° phase-shifted.

The output voltage across the load resistance is

$$e_o = -e_g\mu \frac{R_L}{r_p + R_L}$$

The voltage gain of an amplifier stage is, therefore,

$$G = -\mu \frac{R_L}{r_p + R_L}$$

The art of designing a good amplifier is to find an acceptable compromise between high gain and low distortion. Amplitude distortion is caused by the asymmetry and nonlinearity of tube characteristics. The operating point P should be located in a region of the diagram where I_bE_b characteristics are approximately linear and equidistant.

The maximum voltage $E_{b,\max}$ that can be applied to a vacuum tube is determined by breakdown considerations; the maximum current $I_{b,\max}$ by the

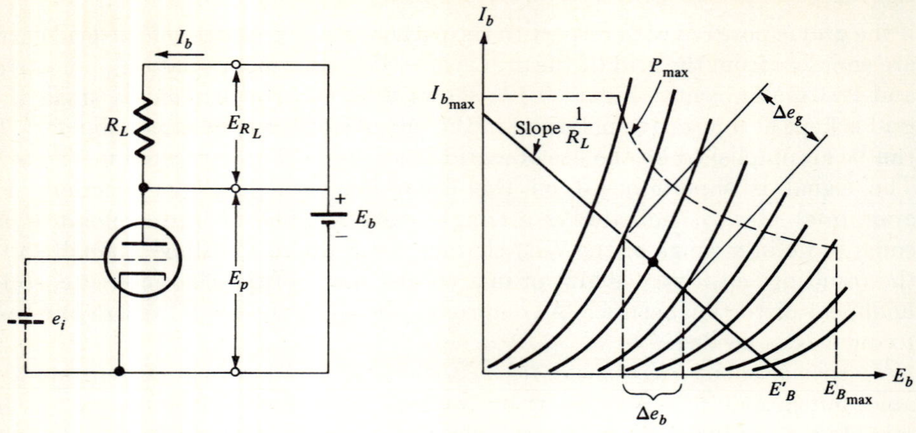

(a) (b)

Fig. 9-13 Operation of a triode. (*a*) Circuit diagram; (*b*) I_bE_b characteristics; the maximum permissible values of plate current I_b, plate voltage E_b, and power dissipation P are indicated (dashed lines).

electron emission from the cathode; the maximum power P_{max}, i.e., the product $(E_bI_b)_{max}$ by the dissipation of the plate which heats up if the electrons fall on it and transmit their kinetic energy to the plate. These resulting limits of operation of a tube are entered in Fig. 9-13*b*.

Noise The current in an electron tube fluctuates randomly by a small amount about its average value (*shot effect*). The plate current can be considered to be composed of a dc component I_b, an ac component, and the noise current I_n, with equal energy distribution over the entire frequency spectrum. The rms value of the noise current in a diode operated in the temperature-limited region (region B in Fig. 9-12) is

$$I_{n,\text{rms}} = 0.564 \times 10^{-9}\, I_b \Delta f \qquad (9\text{-}8)$$

Diodes operated in the saturation region and followed by an amplifier frequently are used as noise generators. If a diode or a triode is operated in the space-charge-limited region, the noise current is reduced to two-thirds of the value computed from Eq. (9-8). It is sometimes convenient to express the noise current in the plate circuit of a triode by the equivalent noise voltage applied to the grid of the triode (since $E_{gn} = I_n/g_m$) or, one step further, by the equivalent resistance inserted between grid and cathode which causes (by thermal agitation, Johnson effect) the grid noise voltage E_{gn} and the noise current I_n in the tube. The equivalent resistance, which at 300°K causes the same thermal noise as the shot effect in a space-charge-operated triode, is

$$R_{\text{equ}} = \frac{3}{g_m}$$

The magnitude of the shot noise for a normal triode and for a bandwidth of the detector of 10 kHz is of the order of 1 μV. The noise level in a pentode is about three times as large as in a triode. This noise, primarily that arising at the input stage, limits the usefulness of an amplifier since it is superimposed on the signal and amplified in subsequent stages.

Tubes are classified according to the number of systems they contain, such as diodes, triodes, tetrodes, or pentodes. There are also tubes containing, in one envelope, several tube systems, such as duodiodes, duotriodes, diode plus triode, diode plus pentode, etc. Tubes are also classified according to heater construction and rating. There are indirectly heated tubes, in which the heater is electrically insulated from the cathode and heat is transferred from the filament to the cathode by thermal conduction. There are directly heated tubes containing a filament which acts simultaneously as heater and as cathode. Indirectly heated tubes afford greater freedom in circuit design since several tubes can be heated from the same source (usually ac) while their cathodes can be at different potentials; directly heated tubes warm up faster. The filament voltage rating of most tubes is 6.3 V. Tubes are classified according to the material of the envelope (glass, metal, or ceramic), according to size of the envelope (standard, miniature, subminiature size), and according to the size of the socket and the number of socket prongs.

REFERENCE

Receiving Tube Manual, *Technical Series* RC-24, RCA Corporation, Harrison, N.J.

10
Elements Based on Transportation of Charges in Electrolytes

BASIC CONCEPTS

The classical experiments in the field of physical chemistry have demonstrated that the molecules of a salt, an acid, or a base in an aqueous solution are dissociated into ions. An NaCl crystal contains Na^+ and Cl^- ions bound within the regular array of the crystal lattice; if dissolved in water, the positive Na^+ ion becomes a stable entity in the electrolyte as does the negative Cl^- ion.

The charged ions in the electrolyte exert electrostatic forces on each other. The forces depend on the molar concentration C_i and on the valence number or ionic charge z (charge in terms of numbers of electrons) of each ion species i in the electrolyte. A measure of the interaction of the dissociated ions is the ionic strength μ

$$\mu = \tfrac{1}{2}\Sigma C_i z_i^2$$

When an electric field is applied to an electrolyte, as illustrated in Fig. 10-1, a force will act on the ions. This force will accelerate the ions and move them through the solution against the decelerating frictional forces exercised by viscous drag of the ions in the solution. The voltage E' measured with an electrometer or potentiometer between the electrodes T_1 and T_2 (i.e., without drawing a

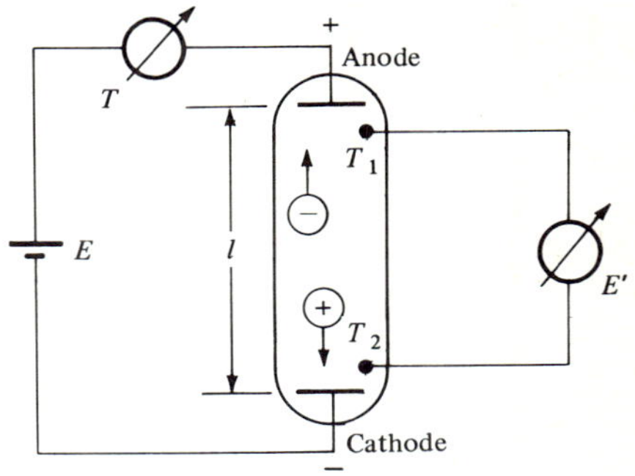

Fig. 10-1 Circuit for the demonstration of the validity of Ohm's law in electrolytic conductors.

current from these electrodes) is proportional to the current I and follows Ohm's law. However, the current I is not proportional to the applied voltage E because of polarization phenomena occurring at the surface of or in the vicinity of the current-carrying cathode and anode.

The current, i.e., the transport of charges through an electrolytic conductor, is always accompanied by a transportation of matter. The positive ions or cations (e.g., metal ions, hydrogen ions, or the positive ions of radicals) migrate to the cathode; the negative ions or anions migrate to the anode. The ions may be plated as neutral molecules at the electrodes, or they may react with the electrodes or with the solvent.

The total current is the sum of the currents contributed by the positive ions and the negative ions

$$I = \frac{(nzev)_+ + (nzev)_-}{l} \tag{10-1}$$

where n = the number of ions of one polarity
ze = the charge of an ion
v = its velocity
l = the distance between anode and cathode

The velocity v of an ion depends on the applied field strength E/l and on the ion mobility u; i.e., the velocity of an ion in a field of 1 V/cm [dimension of u: (cm/s)/(V/cm) or cm²/(V·s)]

$$v = \frac{E}{l} \tag{10-2}$$

The ion mobility can be measured. The small hydrogen ion H^+ has the relatively highest mobility, 32×10^{-4} cm²/(V·s, in highly dilutent aqueous solution; the hydroxyl ion OH^-, a mobility of 17.4×10^{-4} cm²/V·s; the mobility of other, larger ions is of the order of 3 to 6×10^{-4} cm²/(V·s). The mobility of each ion species is in the first approximation independent of the mobilities of other ionic species in the same solutions. The mobility u is independent of the applied field strengths below 10^5 or 10^6 V/cm and independent of the frequency

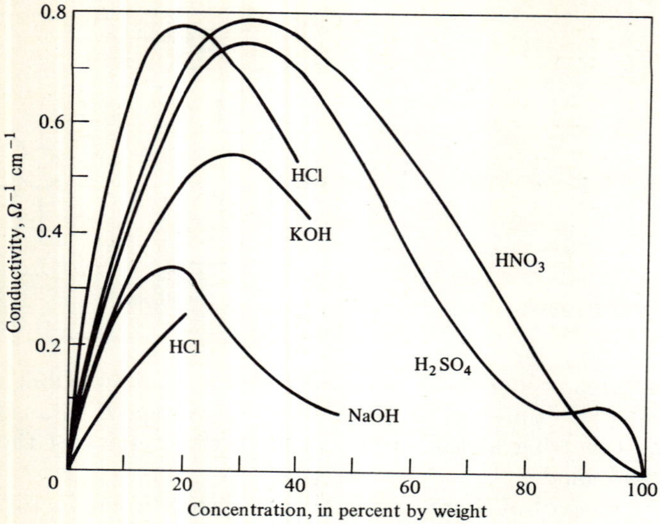

Fig. 10-2 Conductivity of electrolytes versus concentration of the solute.

below 10^7 Hz. At higher field strengths and at higher frequencies the mobility of the ions increases.

CONDUCTIMETRY

The higher the mobility of the ions and the higher their concentration, the higher will be the conductivity of an electrolyte. The conductivity κ of an electrolyte can be derived from Eqs. (10-1) and (10-2)

$$\kappa = \frac{I}{E}\frac{l}{a} = (N'_v zeu)_+ + (N'_v zeu)_-$$

where $N'_v = n/V = n/al$ is the concentration of a certain ion species, i.e., the number n of ions in the volume $V = al$ of the electrolyte. In the simple case of a cylindrical electrolytic column, a is the cross-sectional area of the column.

The concentration of ions in a solution is not identical with the total concentration of the solute but depends on both the number of ions released per molecule dissociating and the degree of dissociation. Measurement of the conductivity of electrolytes furnishes information only about the *ion* concentration. There is no simple relationship between conductivity and the total concentration of the solute except at very low concentrations. Values for the conductivity as a function of concentration are taken from empirical tables. Figure 10-2 shows examples for conductivity versus concentration of several electrolytes.

In highly diluted, strong electrolytes the conductivity is approximately proportional to the molecular concentration

$$\kappa = C\Lambda$$

where Λ is the equivalent conductivity of a solute.

The measurement of the electrolytic conductivity furnishes information about the total concentration of dissociated ions in a solution, but the method is unspecific, i.e., it does not furnish information about the types or species of ions present. The conductimetric method is suitable for concentration measurements when the type of the dissolved ion is known, or in certain chemical-process-control applications when the type of solute is constant and only the concentration varies. The method frequently is used for the measurement of water purity. Very pure water has a conductivity of $3.82 \times 10^{-8}\ \Omega^{-1}\ \text{cm}^{-1}$ (theoretical value); the conductivity increases by one to several orders of magnitude in the presence of traces of dissociable salts.

Different conductimetric cells are shown in Fig. 10-3. The resistance between the terminals of a cell is

$$R = \frac{1}{\kappa}\frac{l}{a} = \frac{C}{\kappa}$$

where l = the effective distance between the electrodes
a = the effective cross-sectional area

These two parameters are hard to measure; it is common practice, therefore, to fill the cell with a substance of known conductivity κ_c and determine the cell constant C from the conductivity κ_c and the measured resistance R_c.

The resistance is usually measured with a bridge as shown in Fig. 10-4*a*. If the capacitance of the cell is large, it may be balanced by a variable capacitor C_s in an adjacent bridge arm. An electronic oscillator or, for technical measurements, the 60-Hz line with a step-down transformer furnishes the ac voltage for the operation of the bridge. As a detector, a pair of headphones, an oscilloscope, a magic-eye detector, a sensitive ac meter, or an amplifier with subsequent rectifier and meter may be used. The conductivity increases strongly with temperature (1 to 9 percent/C; in general about 2 percent/C) so that thermostatic control is required for most measurements. Electrolytic conductivities can be measured in the laboratory with an error of 0.005 percent. The error of industrial measuring systems is, in general, of the order of 1 percent.

Errors in the measurement of conductivity may be caused by polarization. The passage of a direct current through an electrolyte causes electrolysis (i.e., a change of concentration of one ion species near one electrode, or a plating of the electrodes) so that a potential difference, the polarization voltage, appears between the two electrodes. Electrolysis may also cause the formation of gas bubbles at the electrodes and lead to a change of the effective surface area, hence, cross section a of the electrodes, or it may cause a significant change of the composition of the electrolyte. To avoid these effects, resistance measurements

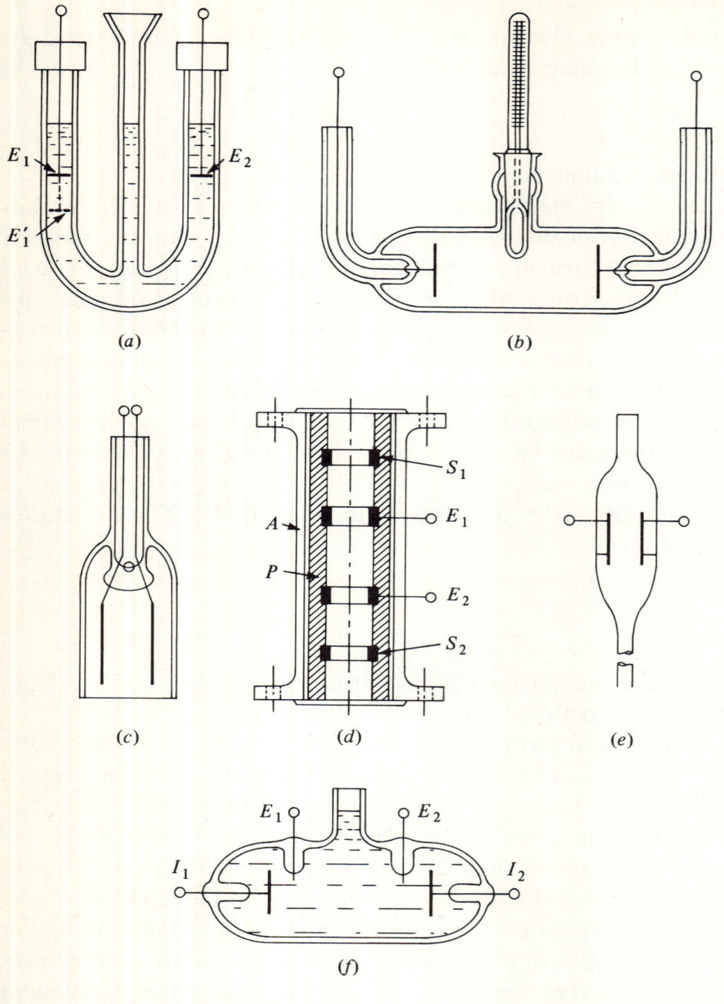

Fig. 10-3 Different forms of conductimetric cells. (*a*) Laboratory-type cell with adjustable electrodes E_1 and E_2. (*b*) Laboratory-type cell with fixed electrodes. (*c*) Dip cell. (*d*) "Flow-through cell" with metal housing *A*, insulator *P*, electrodes E_1 and E_2, and shield electrodes S_1 and S_2. (*e*) Pipet-type cell. (*f*) Cell with separate current electrodes I_1, I_2 and voltage electrodes E_1 and E_2.

are usually carried out with a small alternating current. A chemical alteration of the electrodes during one half-cycle is then reversed in the next half-cycle and the concentration of the electrolyte remains unaltered. The effect depends on the frequency; ordinarily a frequency of the order of 1,000 Hz will yield adequate results. Polarization and the formation of gas bubbles can be reduced by reducing the current density at the electrodes (large area, covering the electrodes with platinum black).

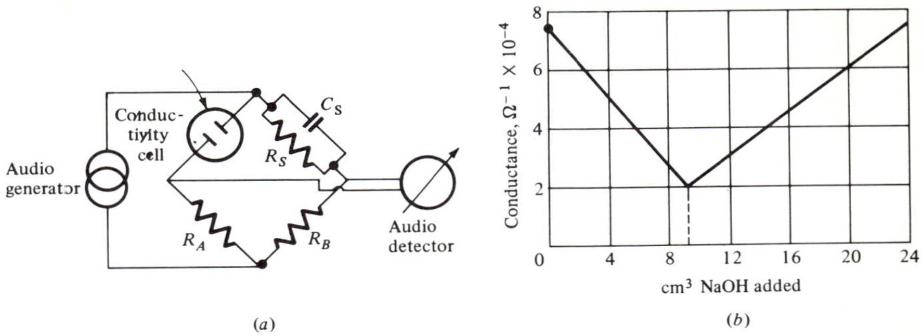

Fig. 10-4 (*a*) Bridge for the resistance measurement of electrolytic conductors; (*b*) conductimetric titration curves. (*From J. H. Harley and S. E. Wiberley, Instrumental Analysis, Wiley, New York,* 1954, *by permission.*)

The conductimetric method can be used to follow the course of a titration (Fig. 10-4*b*). For instance, if an HCl solution is titrated with a solution of NaOH, the conductivity is first reduced because the highly mobile H^+ ion reacts with the OH^- ion and is replaced by the less mobile Na^+ ion. After all H^+ ions have reacted, the addition of more NaOH solution introduces an excess of OH ions and the conductivity rises again. The titration end point can be found from the break point of the characteristic.

Transducers containing a layer of an electrolytic conductor in an appropriate binder may be used for the measurement of humidity in air or other gases. A schematic diagram of such a transducer is shown in Fig. 10-5. The transducer consists of an insulating carrier with two conductors and a layer of a hygroscopic salt, such as lithium chloride, calcium chloride, or phosphorous pentoxide between them. If water from the surrounding gas atmosphere is absorbed in the hygroscopic salt layer, the ions will move more freely and the resistance of the salt layer decreases drastically (e.g., from 10^8 to 10^3 Ω, depending on the concentration of water in the gas). The resistance of the humidity cell can be measured; the humidity can be found from an empirical calibration curve. Some transducers are useful over a limited range of relative humidity only (e.g., from 20 to 35 percent relative humidity); the resistance depends strongly on temperature.

The variation of electrolytic conductivity with the concentration of the solute is also the basis of the tracer method for the measurement of flow velocity

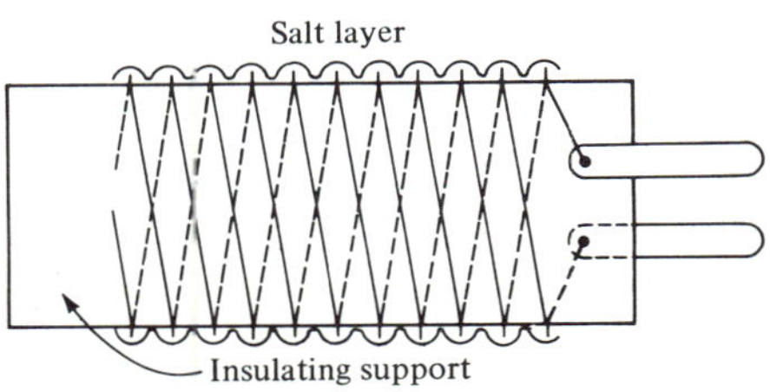

Fig. 10-5 Electrolytic humidity transducer.

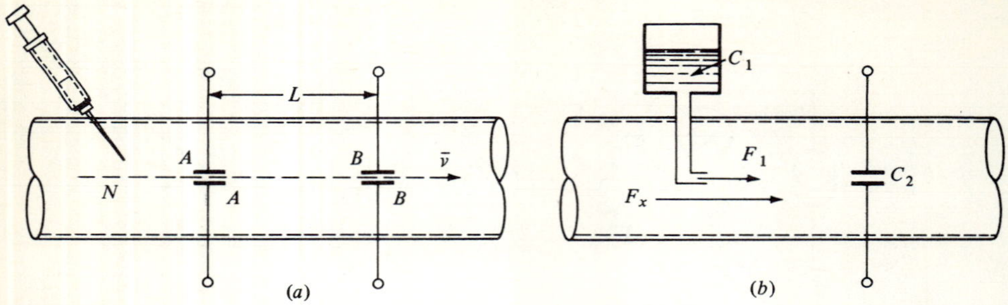

Fig. 10-6 Electrolytic-flow velocity transducer. (*a*) Injection method; (*b*) dilution method.

(see Fig. 10-6*a*). A liquid flows through an isolating duct with the average linear velocity v. A salt solution is injected into the liquid at a point N, causing a cloud of electrolyte to travel with the liquid; the conductivity of the cloud should be markedly higher than that of the liquid. The cloud of electrolyte causes a change of current, first between the two electrodes AA and, at a time t afterward, between the electrodes BB. The distance between AA and BB is L. The flow velocity can be found from $v = L/t$.

The method is suitable for discontinuous measurements only. An electrolytic method for continuous measurements of volume flow velocity (in cubic centimeters per second) is illustrated in Fig. 10-6*b*. An electrolyte of the concentration C_1 flows continuously with the volume flow rate F_1 into a liquid flowing with the volume flow rate F_x. After mixing, the concentration is C_2. The unknown flow rate is

$$F_x = F_1 \frac{C_1}{C_2}$$

The resistance of an electrolytic conductor can be varied easily by a variation of the length l of the current path or by a variation of the cross-sectional area a. An example of a displacement transducer, based on a variation of a and l, is shown in Fig. 10-7*a*. Similar transducers have been used for respiratory studies. The tube is wrapped under moderate tension around the human or animal chest; any expansion of the chest causes an increase of the length and a decrease of the cross-sectional area. The resulting resistance variation can be measured. Transducers of this type are neither accurate nor linear, but they are frequently satisfactory for experiments where the presence or absence of a signal or the signal timing is of interest.

A transducer for the measurement of liquid levels and level variations (e.g., water waves) is illustrated in Fig. 10-7*b*. It consists of two wires A and B. The resistance between the two wires varies in inverse proportion to the height h of the liquid level.

Figure 10-7*c* shows a schematic diagram of the Coulter counter for the counting of blood cells. It consists of two chambers A and B with an electrode

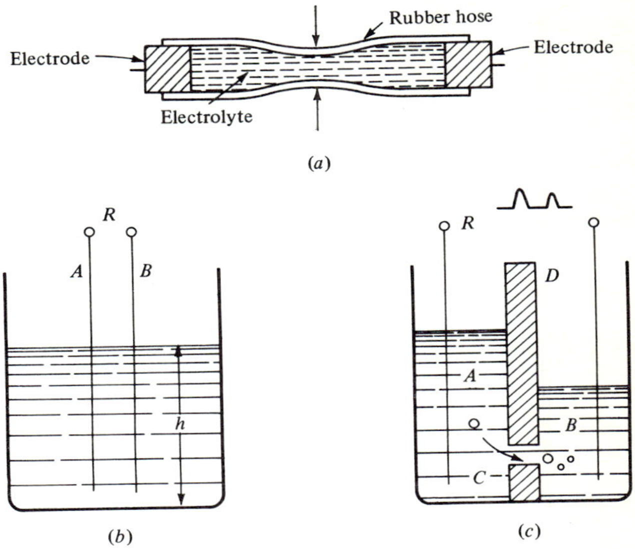

Fig. 10-7 (*a*) Electrolytic displacement transducer; (*b*) electrolytic liquid-level transducer; *A* and *B*, wire electrodes; (*c*) schematic diagram of transducer for counting of blood cells (Coulter counter).

in each chamber. A nonconductive wall *D* with a narrow canal *C* separates the two chambers. Both chambers are filled with diluted blood; the blood is forced to flow from chamber *A* to *B*. The resistance between the electrodes is primarily that of the liquid column in the canal *C*. Whenever a blood cell (which is surrounded by a nonconducting lipoid layer) enters the canal, it causes a reduction of the conductive cross-sectional area of the canal and an increase of the resistance between the electrodes which is recorded and counted. Different sizes of blood cells cause different heights of the output signal which can be sorted and counted separately by a pulse-height analyzer.

POTENTIOMETRY

If a bar M of an *n*-valent metal is immersed in a solution of its ions (e.g., silver in silver nitrate solution, $n = 1$), the following reaction takes place

$$\mathrm{M} \leftrightharpoons \mathrm{M}^{+n} + ne$$

Either the metal M is oxidized, i.e., it loses an electron and goes into solution as a positive M^{+n} ion, leaving the metal bar negatively charged with respect to the solution; or metal ions from the solution are reduced, i.e., deposited on the bar where they take up an electron, making the bar positively charged. The reaction proceeds until an equilibrium is reached; the entire process requires an infinitesimal amount of ion transfer. The sign and magnitude of the potential difference between the metal and the solution (the potential of the "half-cell") depend on the chemical nature of the metal and the concentration of the M^+ ion in the solution. Thus, a measurement of the potential difference E_o can furnish information about the constituents of a solution and their concentration.

The measurement of the potential difference between the electrode M (indicator electrode) and the solution requires the presence of a second electrode R (reference electrode). The potential E_o is the sum of the potentials of both half-cells, i.e.,

$$E_o = E_{M/\text{solution}} + E_{\text{solution}/R} \tag{10-3}$$

The only information of interest in potentiometry is the potential and its variations arising at the indicator electrode $E_{M/\text{solution}}$. Knowledge of the potential at the reference electrode $E_{\text{solution}/R}$ is, in principle, not important, provided it is constant.

The half-cell potential E between an electrode and the surrounding electrolyte is expressed by the Nernst equation

$$E = E^\circ + \frac{RT}{nF} \ln C_{M^+} \tag{10-4}$$

where E° = standard potential for the reaction (see below)
R = gas constant
T = temperature in degrees Kelvin
F = Faraday constant
C_{M^+} = concentration (more accurately the activity) of the cations in solution

At 25°C and with the use of Briggs' logarithms, rather than natural logarithms, the Nernst equation becomes

$$E = E^\circ - \frac{0.059}{n} \log C_{M^+} \tag{10-5}$$

For a concentration $C_{M^+} = 1$, Eq. (10-4) becomes

$$E = E^\circ$$

i.e., the half-cell potential is equal to E° (*E standard*) which is characteristic for a specific reaction taking place at one electrode in an electrolyte at a molar concentration of unity. However, the standard potential is still not measurable since the potential E_o measured between the electrodes is the sum of two unknown potentials [Eq. (10-3)]. By arbitrarily assigning the value $E^\circ = 0$ to the potential of a normal hydrogen electrode (NHE) in a solution of unit activity, at any temperature all other half-cell potentials can be related to this value.

At concentrations higher than 10^{-3} M, the measured half-cell potential differs from that computed from the Nernst equation because it depends on the activity rather than on the molar concentration. The activity is

$$a = \gamma C$$

The activity coefficient γ is unity for highly diluted solutions. Its value depends on the concentration, the ionic strength, and the temperature of the solution. In general, γ is determined experimentally.

Only cells with electrochemically reversible half-cell reactions furnish reproducible results. A criterion for reversibility is the reaction of a half-cell to a small imposed current in either direction. If the reaction proceeds with equal facility in both directions and if the current causes only a small change of the half-cell potential, the reaction is reversible. A reversible reaction does not exist if the solution reacts spontaneously with the electrodes or if a reaction occurs in the presence of a current, such as the formation of a gas, oxidation, or passivation.

If two different valence forms of the same species are in the solution (e.g., Fe^{++} and Fe^{3+}), the higher valence may take up electrons from the electrode (may be reduced) or the lower valence form may furnish electrons to the electrode (may be oxidized).

If an external source is applied to the electrode and is varied, then generally a current will flow and either one or the other event will take place. If no current flows (e.g., if the electrode is floating), then the electrode will assume a potential determined by the types of ion species present and their relative activities (concentrations).

$$E = E^{\circ}_{Fe^{3+}/Fe^{++}} + \frac{RT}{nF} \ln \frac{a(Fe^{3+})}{a(Fe^{++})} \tag{10-6}$$

E is called the *Redox potential*.

Potentiometry is used frequently for the measurement of hydrogen-ion concentration in electrolytes. In very pure water a small fraction of the water molecules is dissociated into positive H ions and negative OH ions; the concentration of both ion species is alike.

$$[C_H] = [C_{OH}] = 10^{-7} \text{ mol/l at } 25°\text{C}$$

The product of both concentrations $[C_H] \times [C_{OH}]$ remains constant if the concentration of one of these ion species is increased. For instance, if $C_H{}^+$ is raised from 10^{-7} to 10^{-2} mol/l by the addition of an acid, the concentration of OH ions changes from 10^{-7} to 10^{-12} mol/l.

The concentration of hydrogen ions is commonly expressed by the exponent pH rather than by the concentration C_H. Both values are related by

$$C_H = 10^{-\text{pH}} \qquad \text{or} \qquad \text{pH} = \log \frac{1}{C_H}$$

For example, for pure water

$$C_H = 10^{-7} \text{ mol/l}$$
$$\text{pH} = \log 10^7 = 7$$

The problem of potentiometry is to find an indicator electrode responsive to an ion species and its concentration in the solution and a reference electrode

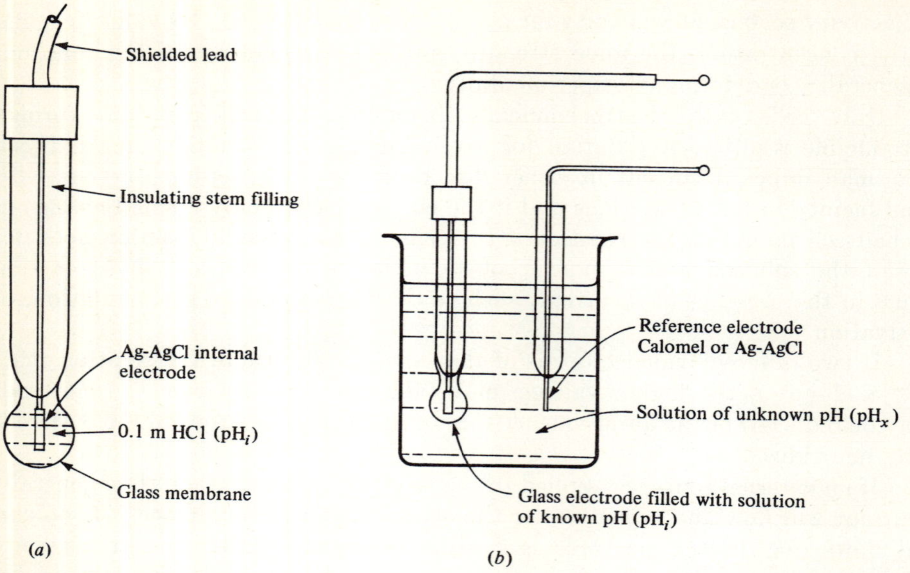

Fig. 10-8 (*a*) Diagram of a glass electrode, see text for explanation; (*b*) diagram of a potentiometric pH meter cell with a glass electrode and a reference electrode.

with a potential that is unvarying, i.e., independent of the variation of concentration of the same ion species in the electrolyte.

No electrode by itself is an indicator electrode or a reference electrode. For instance, the glass electrode is an indicator electrode for hydrogen ions; its potential in an electrolyte depends on the hydrogen-ion concentration of the electrolyte. Used in a solution with constant hydrogen-ion concentration, the glass electrode can be used as a reference electrode. The silver–silver-chloride electrode is uninfluenced by the hydrogen-ion concentration in the electrolyte; it can serve as a reference electrode for pH measurements. However, if the hydrogen-ion concentration of the solution is kept constant and the chloride-ion concentration changes, the potential of the silver–silver-chloride electrode will vary while the half-cell potential of the glass electrode remains constant.

Essentially two different kinds of electrodes are used: (1) metals, or gases adsorbed on metal, in contact with their own ions or (2) metals in contact with a sparingly soluble salt of the metal and a soluble salt with the same anion.

An example of a metal-gas electrode is the normal hydrogen electrode (NHE). It consists of a strip of platinum covered with electrolytically deposited platinum black (to increase its surface area). It is immersed in a solution and hydrogen is bubbled over its surface. The hydrogen is adsorbed at the surface. The potential of the half-cell is

$$E = -0.0591 \text{ pH}$$

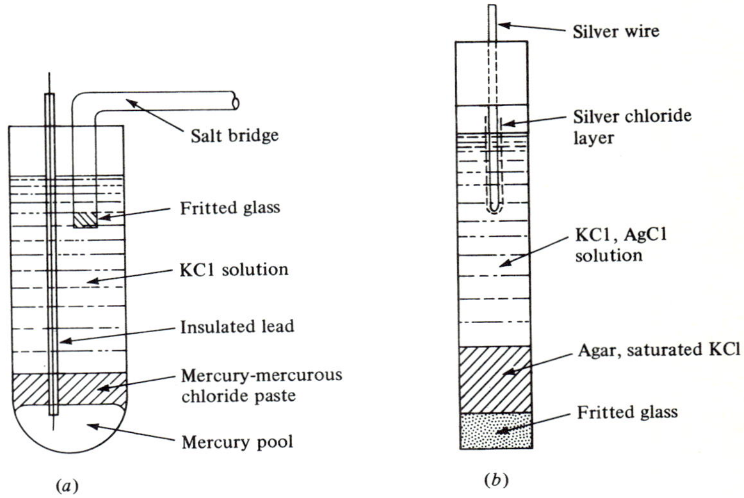

Fig. 10-9 (*a*) Construction of a calomel electrode; (*b*) construction of a silver–silver-chloride electrode.

Although all standard potentials are expressed with reference to the NHE, they are usually experimentally determined against a more convenient reference electrode (e.g., calomel).

The most commonly used indicator electrode for pH measurements is the glass electrode. This is a membrane electrode and depends on a different principle from those discussed above. Its construction is shown in Fig. 10-8*a*. It consists of a thin glass membrane which forms the end of a tube. The tube is filled with a buffer solution of known hydrogen-ion concentration pH_i, usually a 0.1 M HCl solution. An internal reference electrode (silver–silver chloride or calomel; see below) is inserted in the glass tube. Another electrode, either silver–silver chloride or calomel, is used as an external reference electrode in the solution with unknown hydrogen-ion concentration (see Fig. 10-8*b*).

The entire cell can be represented by

Reference electrode/	Solution i/glass/	Solution x/	reference electrode
Internal	pH_i	pH_x	External

and the measured cell potential at 25°C is

$$E_o = E_g^\circ + 0.0591\ (pH_x - pH_i)$$

Even for $pH_x = pH_i$, an asymmetry potential E_g° remains, so that empirical calibration of the glass electrode in a medium of known pH is required.

The physical mechanism of the glass electrode is complex and is based on an ion-exchange mechanism at the two glass/solution interfaces. In effect, it acts as if it were a membrane permeable only to hydrogen ions.

The resistance of the glass electrode is very high, of the order of 10^7 to $10^9\ \Omega$,

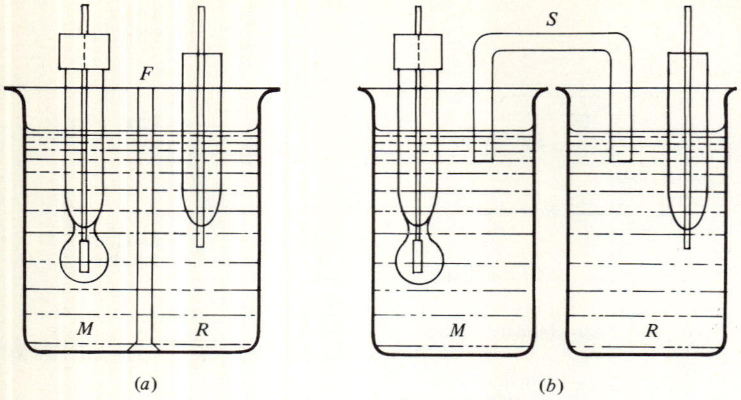

Fig. 10-10 Separation of a measuring half-cell M and a reference half-cell R. (*a*) Separation by a permeable membrane F (fritted glass disk); (*b*) separation by a salt bridge S.

so that for the measurement of the cell potential only systems with a very high input impedance can be used. The glass electrode can be used for measurements over a wide range of pH values, from about 0.5 to 13.5. Special types of glass are needed for measurements above a pH value of 10.

Two different types of reference electrodes frequently are used, the calomel electrode and the silver–silver-chloride electrode. The saturated calomel electrode is shown in Fig. 10-9*a*. It consists of a mercury pool, a layer of calomel paste (mixture of roughly equal weights of mercurous chloride and potassium chloride), and a column of saturated KCl solution. The half-cell reaction is

$$Hg_2Cl_2 + 2e = 2Hg + 2Cl^-$$

The silver–silver-chloride electrode is shown in Fig. 10-9*b*. It consists of a silver wire, covered with a layer of silver chloride, in a saturated solution of potassium chloride and silver chloride. The half-cell reaction is

$$AgCl + e^- = Ag + Cl^-$$

The half-cell potential of the reference electrode can be rendered independent of the variation of the electrolyte surrounding the indicator electrode by separating the electrolytes. Two methods are shown in Fig. 10-10*a* and *b*. In Fig. 10-10*a* the electrolytes are separated by a permeable membrane (e.g., fritted glass disk) and in Fig. 10-10*b* they are separated by a salt bridge consisting of a salt solution with added agar or gelatin. In each case a potential difference will arise at the boundary (junction potential E_j), so that the output voltage measured between the electrodes is

$$E_o = E_M + E_j + E_R$$

The junction potential is caused by an unequal diffusion rate of ions across the boundary between the solutions. The junction potential is less than ± 2 mV with the salt bridge (Fig. 10-10*b*) if the concentration of the electrolytes on both sides is less than 0.1 N, if the salt concentration in the bridge is high, and if the salt consists of a cation and anion with nearly equal mobility (e.g., KCl, NH_4NO_3).

Potentiometric measurements frequently are used for the determination of a titration end point; the output from a potentiometric cell rises or falls sharply as the titration end point is reached.

ELECTRODEPOSITION

Figure 10-11*a* shows an electrolytic cell with two electrodes. Electrode A is silver–silver chloride as described above and as shown in Fig. 10-9*b*, electrode C consists of platinum. The electrolyte is a solution of potassium chloride containing a small amount of copper chloride. If a voltage E is applied and increased from zero, no current will be observed until a critical value E_B is reached.

In the region between zero and E_B, ions will move toward the electrodes to form a double layer of a potential difference equal to the applied voltage and of opposite polarity. Therefore, there will be no electric field within the solution and no current, i.e., no further transportation of ions. As soon as the applied voltage exceeds the voltage E_B, a current will flow, i.e., ions will migrate through the solution and will react at the electrodes. In the example of Fig. 10-11*a* copper will be plated on the cathode and silver chloride formed at the anode. The process will lead to a depletion of copper in the solution. The slope of the voltage-current characteristic is determined by the conductance of the bulk solution. The potential between both electrodes is the sum of the anode and the cathode potentials. Of these, only the cathode potential is of interest, not the anode

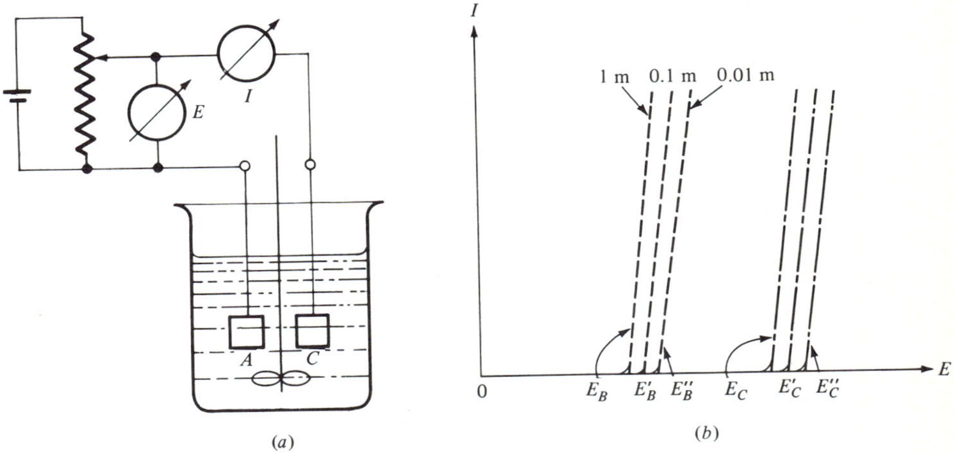

Fig. 10-11 (*a*) Schematic diagram of an apparatus for electrodeposition and coulometry. (*b*) Voltage-current characteristics of a coulometric cell containing two species of ions, *b* and *c*. Electrolysis takes place when the applied voltage E reaches the values E_b, E_b', E_b'' (for different concentrations of ion species *b*), and again when E reaches the values E_c, E_c', E_c'' for another ion species *c*.

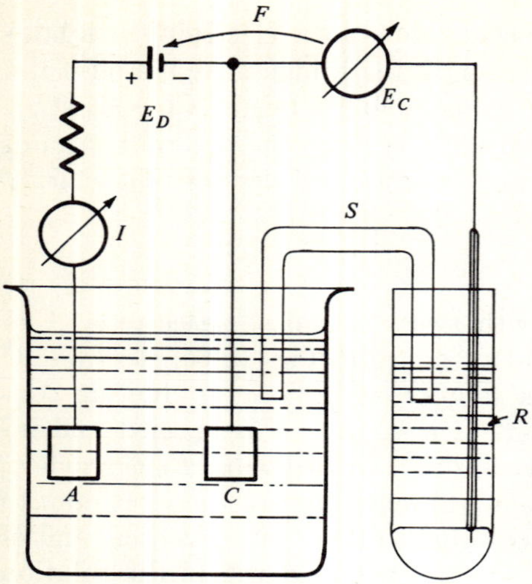

Fig. 10-12 Coulometric cell for controlled cathode potential operation. *A*, anode; *C*, cathode; source E_D; *R*, reference electrode connected to cathode half-cell by means of salt bridge *S*; *F*, feedback loop (schematic).

potential, provided the anode potential is constant. The anode potential is constant here because of the high concentration of potassium chloride.

The critical value of the cathode potential at which the reduction of the metal ions begins to take place depends on the nature of the ion and its concentration as shown in Fig. 10-11*b*.

If two or more species are in a solution, it is possible to deposit them selectively on the cathode, provided their cathode potentials are sufficiently far apart. As long as the applied potential is greater than E_B in Fig. 10-11*b*, but less than E_C, only the ion species *B* with the cathode potential E_B will undergo electrolysis, and only this species will become depleted. If the amount or concentration of another component *C* in the solution is to be determined, component *B* must first be quantitatively electrolyzed.

In practice, a sufficiently high voltage is applied to both electrodes and the current is adjusted to a convenient value. Both voltage and current are measured. The solution is well stirred. The current will decline when the ion species in the solution is depleted and deposited on the cathode. The deposited metal can be determined by weighing or it can be dissolved from the platinum electrode and analyzed further. If the deposition potentials of two ion species in a solution are not well separated, it is possible that at the end of the deposition period of the first metal a deposition of the second metal will set in and no complete separation will occur. This difficulty can be reduced considerably by controlled potential electrolysis as shown schematically in Fig. 10-12. An adjustable voltage source E_d is applied to the two electrodes *A* and *C* in an electrodeposition cell and causes electrodeposition of a metal at cathode *C*. The potential of the cathode half-cell is measured with the help of reference electrode *R*. Any variation of the cathode potential E_c is fed back to the source (for instance, by manual control) to keep the cathode half-cell potential constant.

COULOMETRY

The quantity M, in moles, electrolytically deposited on one electrode, is proportional to the total electric charge Q, in ampere seconds or coulombs, passing through the circuit during electrolysis

$$M = kQ = kIt = k\int i\,dt \tag{10-6}$$

where

$$k = \frac{1}{zF} = \frac{1}{zNe}$$

and z = valence, i.e., number of elementary charges per ion
F = Faraday constant, 9.65×10^4 A · s/mol
N = Avogadro's constant, 6.03×10^{23} molecules/mol
e = charge of the electron, 1.6×10^{-19} A · s

The current is either constant in time ($Q = It$) or varies with time (instantaneous value i and $Q = \int i\,dt$).

The relationship indicated in Eq. (10-6) offers the possibility for determining the quantity M of an ion species that is electrolytically deposited or liberated at an electrode by electrical means, i.e., by measuring Q rather than by weighing the deposit or by volumetric measurements.

Coulometry can be carried out with constant current or with constant voltage. The constant-current method has the advantage that, once the current has been adjusted, only the elapsed time of current flow need be measured. The constant-voltage method in coulometry permits accurate setting of the half-cell potential and, therefore, good separation of different ion species, but it requires the measurement of the current-time integral. This integration may be carried out graphically by recording the current as a function of time and determining the area under the curve. Alternatively, the current-time integral may be formed by an electromechanical integrator, such as a fluxmeter, an ampere-hour meter, or a rate servomechanism. There are also coulometric systems for integration (described below) in which the weight, volume, or concentration of an electrolytically developed substance is used as a measure for the current-time integral.

Coulometric titrations Electrolysis can be used for the generation of active titrants, in particular, those which are normally unstable or difficult to handle (e.g., bromine). An inactive form of the titrant (e.g., a bromide) is added to the solution and is oxidized electrically to bromine. The oxidized or active form reacts quantitatively with the species to be determined (e.g., sulfite). The end point of the titration can be determined optically or electrically. If the electrical oxidation is 100 percent efficient, i.e., if only the inactive form of the added titrant reacts at the anode, then the quantity of charge which passes through the cell is proportional to the quantity of the species to be determined.

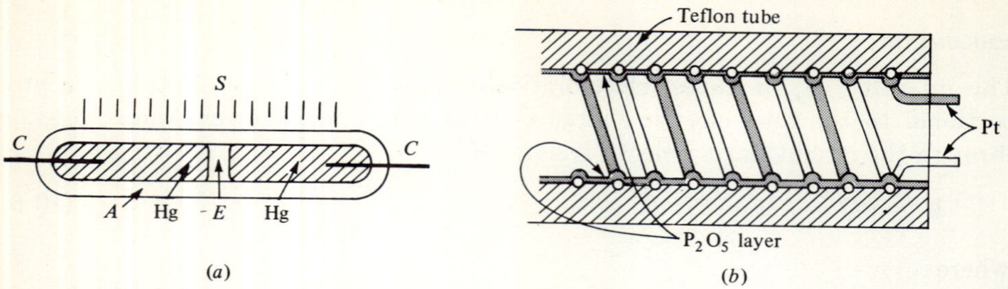

Fig. 10-13 (*a*) Coulometric mercury integrator for current-time integration or for use as elapsed-time meter; (*b*) coulometric humidity transducer for the range from 1 to 1,000 ppm of humidity in gases.

Coulometric integrators The coulometric method can be used for the integration of current signals over a wide time range, from several minutes to months, where other types of integrators cannot be used.

A very accurate (although now infrequently used) integrator is the silver coulometer which consists of a platinum crucible as cathode and a silver cylinder as anode. The crucible is filled with 1 *M* silver nitrate. The amount of silver deposited on the platinum is determined by weighing.

A volumetric integrator contains two electrodes in an aqueous electrolyte. Hydrogen is liberated at the cathode and oxygen at the anode. The liberated gas volume can be read off a burette scale. It is reduced to standard pressure and temperature and the current integral is calculated from the reduced gas volume.

The system illustrated in Fig. 10-13*a* contains two mercury electrodes and a gap filled with an aqueous electrolyte (solution of potassium mercury iodide). The applied current causes mercury ions to be transported across the gap so that the gap moves; its position can be read on a scale. Systems of this type can be used for integration of currents, such as ampere-hour meters, in the range from -5 to $+5$ mA dc (larger currents by the use of shunts, however, the cell resistance increases with decreasing current). The system can be used with constant current as an *elapsed-time meter* with a range from two to several thousand hours.

The coulometric method furnishes one of the most precise means for the measurement of humidity in gases. A schematic diagram of a humidity transducer based on electrolysis is shown in Fig. 10-13*b*. Two platinum wires are helically wound inside a small Teflon tube. The gap between the wires is bridged by a thin layer of phosphorus pentoxide (P_2O_5). A gas stream flows continuously through the tube; the humidity in the gas is absorbed by the P_2O_5 layer. A dc voltage larger than the polarization voltage (>2 V) is applied to the wires and causes electrolytic decomposition of the water absorbed by the electrolytic layer. The system comes to an equilibrium when the rate of water absorbed in the electrolytic layer is equal to that electrolytically decomposed into gaseous H_2 and O_2. The current is then

$$I = kcv_m$$

where k = a constant (0.0066)
c = volume concentration of water in the gas, in ppm
v = mass flow velocity of the gas mixture

The system can measure very small concentrations of humidity, of the order of 1 to about 1,000 ppm.

VOLTAMETRIC METHODS

Figure 10-14*a* shows an arrangement in which one electrode, the working electrode, is small so that electrolysis occurs within a small volume, and only a small fraction of the ions in the solution are electrolyzed. The current-voltage characteristic of such an arrangement is illustrated in Fig. 10-14*b*. At low values of the applied voltage ($E_{\text{applied}} < E_D$) the current is practically zero. As the applied voltage increases beyond the cathode potential E_D, the current rises as shown in Fig. 10-14*b*. If the voltage is increased further, most ions of the species being electrolyzed will be deposited or plated onto the electrode; the space in the vicinity of the electrode will then become depleted of such ions. The current is determined by the rate at which new ions move into the depleted space. New ions will arrive from the bulk solution under the influence of the existing *potential* gradient and of the *concentration* gradient in the depleted space. The potential gradient can be reduced to a practically insignificant level by strongly

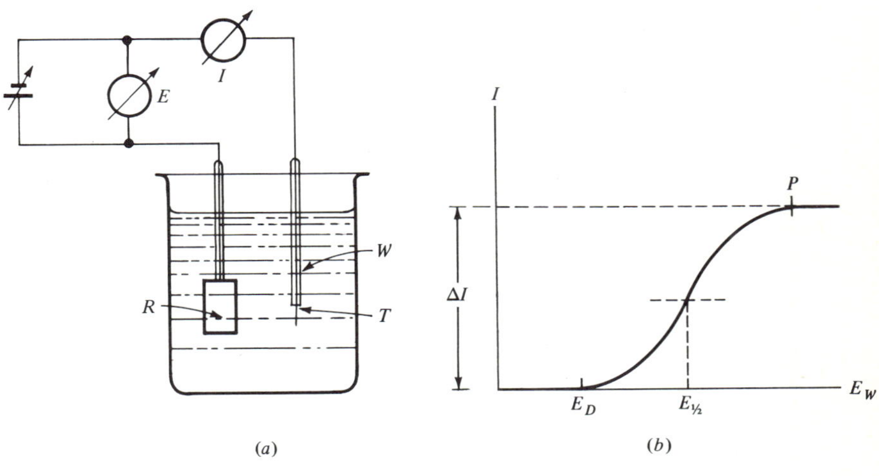

Fig. 10-14 (*a*) Schematic diagram of a polarographic cell. E, voltmeter; I, ammeter; R, reference electrode; W, working electrode; T, metal tip. (*b*) Voltage-current characteristic of a polarographic cell.

increasing the conductivity of the electrolyte. The addition of a large concentration of an electrolyte which does not enter into a reaction at the working electrode, i.e., which has a cathode potential substantially higher than E_D, increases the conductivity and, hence, reduces the electric field in the electrolyte. With the potential gradient thus practically eliminated (and in the absence of a convection current in the solution), the current depends on the concentration gradient of the reacting ion species in the space between the bulk solution and the electrode or, in other words, the current is limited by the concentration of the reacting ion species in the bulk solution. A further increase of the applied voltage will have no effect on the current; the current reaches a plateau of the value

$$I = \frac{DnF}{d}(C - C_0)$$

where D = diffusion coefficient
n = number of electrons required to discharge one ion
F = Faraday constant, 9.65×10^4 A · s/mol
d = thickness of the depleted region (linear gradient assumed)
C = concentration of the ion in the bulk solution
C_0 = concentration of that species at the electrode surface

Since C_0 is practically zero, the current is proportional to the concentration of the reacting ion species in solution.

Polarography The voltage-current characteristic shown in Fig. 10-14*b* furnishes qualitative as well as quantitative information regarding the reacting ion species in the solution. The *type* of the ion species reacting at the working electrode is indicated by the half-wave potential $E_{1/2}$, i.e., the potential at the inflection point of the EI characteristic. The *concentration* of the ion species is indicated by the rise of current ΔI. The polarographic method is also applicable if more than one species is in solution.

A simplified diagram of a polarographic apparatus is shown in Fig. 10-15. An unpolarizable electrode serves as a reference electrode. It may be used either in the same vessel as that containing the working electrode or in its own electrolyte which is connected by a salt bridge to the solution to be analyzed.

The usefulness of a specific working electrode is determined by the range of potential over which it is inert, by its surface area, and by its shape. Mercury and platinum are most commonly used. The platinum electrode has a high oxygen overpotential, i.e., the reaction with oxygen at its surface occurs at a much higher potential than the standard potential would predict. The platinum electrode can be used at positive potentials as high as +0.9 V versus the NHE in aqueous solutions. The mercury electrode has a high hydrogen overpotential and can be used at negative potentials as high as −1.8 V versus the NHE in aqueous solutions.

Of particular practical importance is the dropping mercury electrode (Fig. 10-15). Mercury flows from a reservoir through a small (0.05 mm) capillary

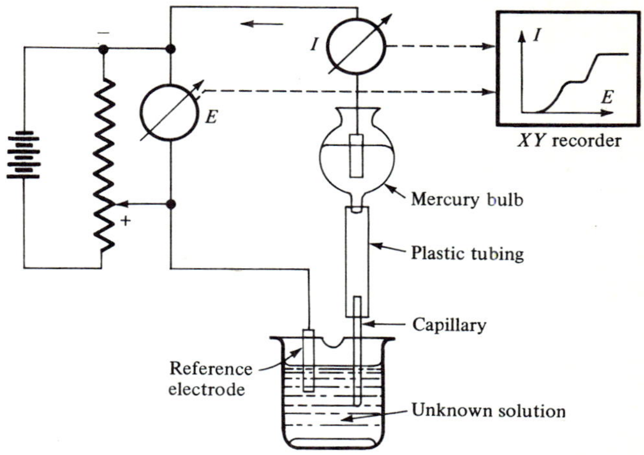

Fig. 10-15 Polarograph with dropping mercury electrode and x-y recorder.

and forms at the end of the capillary a small, approximately spherical drop which grows until its weight causes it to break off. The dropping mercury electrode has several advantages: the mercury surface is continuously renewed, the shape of the drop is easily reproducible, and the surface area may easily be calculated. The alternating growth and break off of the mercury drop causes a fluctuation in the EI characteristic which is superimposed on the average characteristic and which can be eliminated by the use of filters or average-forming networks. Figure 10-16 shows a standard and an idealized polarographic diagram.

The relationship between the current and the concentration of a given species for the dropping mercury electrode is (Ilkovič equation)

$$i_d = 607nD^{1/2}\,cm^{2/3}\,t^{1/6}$$

where i_d = average current over the period of one drop
n = number of electrons transferred to discharge one ion
D = diffusion coefficient of the species
c = concentration, millimol/l
m = mercury flow rate, mg/s
t = period of one drop (usually about 2 to 5 s)

Modern practice tends to use maximum current at drop fall rather than average current. This changes the factor 607 in the Ilkovič equation to 708 and eliminates the need for filters or averaging networks. The half-wave potential of the species being electrolyzed, for a given supporting electrolyte, is related to the standard potential by

$$E_{1/2} = E^0 - \frac{0.059}{n}\log\left(\frac{D_{\text{ox.}}}{D_{\text{red}}}\right)^{1/2} + \Delta E \qquad \text{at } 25°\text{C}$$

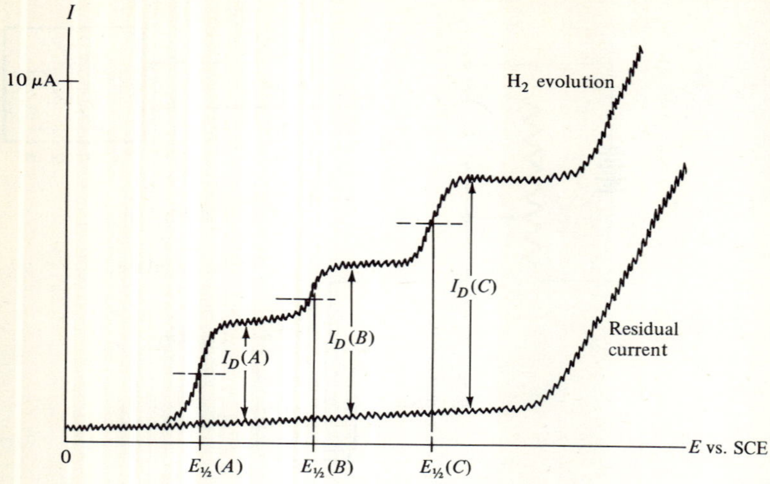

Fig. 10-16 Polarographic characteristic. (*a*) $E_{1/2}$, (*b*) $E_{1/2}$, (*c*) $E_{1/2}$; half-wave potentials of the ion species *A*, *B*, and *C*, rise of current indicative for the concentration of each ion species.

The symbol D is the diffusion coefficient of the reduced or the oxidized state. ΔE is a constant for a given system; it depends mainly on the ratio of the activity coefficients of the oxidized and reduced forms. Its value is usually small and can be determined empirically.

The polarographic method can be used for the analysis of many elements, organic and inorganic compounds in concentrations of 10^{-2} to 10^{-5} M and in amounts as small as 0.01 ml, either single or in mixtures. The method can be applied with modifications for the determination of titration end points (*amperometry*) and to studies of hydrolysis, solubility, complex formation, kinetics, dissociation constants, diffusion coefficients, and many other phenomena.

Clark oxygen electrode A number of polarographic systems have been developed that can be used to monitor the concentration of specific elements. One of the more frequently used systems is the Clark oxygen electrode which measures the partial pressure of oxygen in either gas or liquid samples. The electrode is basically a polarographic cell utilizing a solid platinum electrode. A gas permeable membrane of polyethylene or Mylar separates the platinum cathode from the sample. The construction of the electrode assembly is shown in Fig. 10-17. A constant dc voltage (about 0.6 V) is applied between a silver–silver-chloride half-cell and the platinum cathode. The voltage must be of sufficient magnitude so that any oxygen molecule diffusing to the cathode surface is immediately reduced to water.

The hydrogen ions are supplied by a buffer system such as potassium mono- and dihydrogen phosphates in equimolar concentrations, dissolved in the

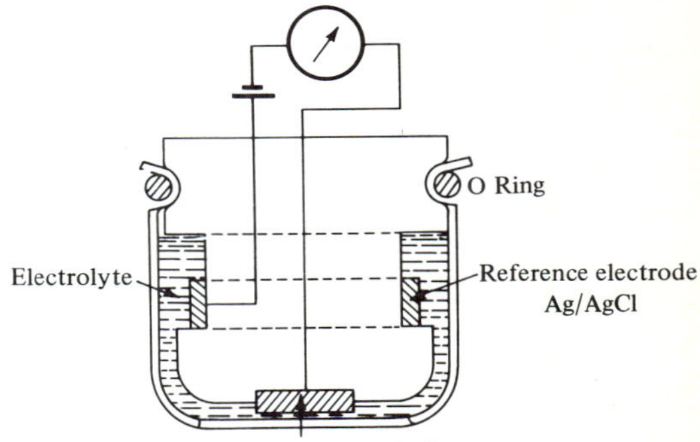

Fig. 10-17 Clark oxygen electrode, see text for explanation.

thin film of electrolyte solution covering the cathode and connecting the platinum cathode with the reference electrode. Reduction of one molecule of oxygen gives rise to a current which produces the following reaction at the reference electrode.

$$4Ag + 4Cl^- = 4AgCl + 4e^-$$

In addition to the buffer system, the electrolyte also contains a high concentration of potassium chloride to maintain the chloride ion at a nearly constant level. The current passing between the cathode and the reference electrode also passes through the external electrical circuit and is measured by the meter shown in Fig. 10-17. At the potential used, oxygen diffusing to the electrode through the membrane and the electrolyte film is immediately reduced. The current is, therefore, limited by the diffusion rate which in turn is governed by the concentration difference across the membrane-electrolyte pair: $C - C_0$. Because $C_0 = 0$, the current is directly proportional to C, the concentration of oxygen in the sample.

Typical electrodes with areas of about 0.5 cm^2 and polyethylene membranes of about 0.02-cm thickness give currents of about 10 μA in air or air-saturated water. The useful linear range of measurement extends down to about 0.02 atm, being limited by residual currents at the cathode due to trace amounts of reducible impurities in the electrolyte. The time response to reach the steady state depends on the membrane thickness and is about 30 s to 1 min for typical electrodes.

In use, the Clark electrode is calibrated using air as the most convenient standard. Since the diffusion coefficient is a strong function of temperature, standardization must be performed at the same temperature as the sample or an empirical temperature correction must be applied. When used in sample solutions containing suspended or colloidal matter, the calibration must be repeated at rather frequent intervals since deposition of solid material on the membrane surface changes the effective membrane thickness. Other gases which are capable of diffusing through the membrane and also of being reduced at the cathode (such as chlorine, sulfur dioxide, and the oxides of nitrogen) must be absent from the sample.

Clark electrodes have found many applications in such widely diverse fields as physiological measurements in blood, monitoring of oxygen in sewage and waste disposal plants, and control of oxygen levels in industrial process gas streams.

A Clark electrode with a suitable electrolyte and reversed polarity can be used to measure SO_2 by oxidation to sulfate.

REFERENCES

Ewing, G. W.: "Instrumental Methods of Chemical Analysis," McGraw-Hill, New York, 1954.

Lingane, James J.: "Electroanalytical Chemistry," Interscience, New York, 1958.

Considine, D. M. (ed.): "Process Instrument and Controls Handbook," McGraw-Hill, New York, 1957.

Willard, H. H., L. L. Merritt, Jr., and J. A. Dean: "Instrumental Methods of Analysis," 3d ed., Van Nostrand, 1958.

Strobel, H. A.: "Chemical Instrumentation," Addison-Wesley, 1960.

Bair, Edward J.: "Introduction to Chemical Instrumentation," McGraw-Hill, New York, 1962.

11
Elements Based on Transportation of Charges in Semiconductors

INTRODUCTION

The structure of semiconductors and the mechanism of current conduction in semiconducting materials have much in common with the corresponding properties in metals. Both metals and semiconductors exist in the form of crystals. However, metals are used normally in the form of polycrystals, i.e., they consist of a great number of randomly oriented small crystallites, while the semiconductors which are of interest in connection with control and amplification are made from single crystals. The conduction process in both metals and semiconductors is that of a transport of electronic charges rather than of ions. The ions remain stationary in the crystal lattice. Therefore, the electric current in metals or semiconductors is not connected with a transport of matter as in electrolytes and does not cause a chemical alteration within the crystal.

The current-conduction process in metal differs from that in semiconductors primarily in the following ways. The packing density of electrons (i.e., their number per unit volume) *in metals* is large and fixed; it is approximately equal to the packing density of metal atoms. Increase of temperature of metallic conductors reduces the conductivity, not because of a change of the number of electrons participating in the conduction process, but because an increase of temperature causes a decrease in the mobility of the electrons.

By contrast, *in semiconductors* the packing density and number of electrons, or carrier of electricity in general, are several orders of magnitude smaller than that in metals. In very pure and very perfect semiconductor single crystals at low temperature, the conductivity (*intrinsic conductivity*) is so small that the crystal behaves almost like an insulator. The resistivity of very pure germanium at room temperature is of the order of 10 Ω·cm, that of silicon of the order of 10^5 Ω·cm. An increase of temperature in pure crystals generally increases the number of available carriers of electricity. The number of carriers also increases by several orders of magnitude in the presence of impurities (*extrinsic conductivity*).

CONDUCTION MECHANISM

The mechanism of current conduction in semiconductors may be explained as follows. The elements in group IV of the periodic table, such as carbon (diamond), silicon, or germanium, have four electrons in their outer orbit or valence shell. The atoms are held in the crystal lattice in a three-dimensional array. The atoms are bound by interaction with the electrons, an electron sharing an orbit with an adjacent atom. This kind of bond, the covalent bond, is characteristic for all elements in group IV of the periodic system.

IMPURITY CONDUCTION—DOPING

The electrons in the crystal lattice are relatively loosely bound, i.e., they require only a small amount of energy to break loose and become free-conduction electrons. In many cases the amount of thermal energy available to the electrons at increased temperature of the crystal is sufficient to cause conduction, i.e., to produce a free (movable) electron and a positively charged (locally bound) ion. However, a concentration of conduction electrons obtained in this manner is insufficient for technical applications. Much higher concentrations of current carriers can be obtained by adding a controlled small amount of impurities, such as atoms from the third or fifth group of the periodic table. Such adding of impurities is called *doping* of the basic semiconductor material.

If a four-valent semiconductor (germanium, silicon) is doped with a small amount of a five-valent material (one impurity atom in 10^4 to 10^2 silicon atoms), such as phosphorus P, arsenic A, or antimony Sb, an atom of the five-valent element will replace an atom of the four-valent element within the crystal lattice. It is evident that of the five electrons in the valence shell, only four are bound; the fifth or excess electron is free to leave the parent atom. When it leaves, the five-valent atom remains positively charged; it becomes an ion which remains fixed in position. The five-valent atom is called the *donor* element. The donor causes conduction by negatively charged excess carriers, i.e., electrons. This type of conduction is called *n-type conduction*.

The *n*-type conduction is not the only possible conduction mechanism in semiconductors. A conductive crystal can be formed from a four-valent-atom crystal lattice with an added three-valent atom from the third group of the periodic table, such as boron B, gallium Ga, or indium I. It is evident that there is one electron deficiency or *positively charged hole* into which an electron may move. If an electron moves into this hole, another hole arises into which another electron will move, and so on. An observer with the hypothetical ability to see these electron deficiencies will get the impression that the positive deficiency, or the positive hole, moves in a direction opposite to that of the electron. This type of conduction mechanism is called *positive hole* or *p-type conduction*. The impurity atom which forms a deficiency conduction is called the *acceptor*. The total current is the sum of the currents carried by n- and p-type conduction.

Even though donor impurities cause predominantly n-type conduction and acceptor impurities cause p-type conduction, there are always carriers of the opposite polarity present, although in low concentration. Their presence results from intrinsic conductivity, imperfections of the crystal, or thermal liberation of carriers. These carriers are then called *minority carriers* in contrast to the much larger concentration of the desirable type of majority carriers. Both n- and p-type carriers may recombine and neutralize each other, or they may be caught in traps and cease to participate in the current transport. Carriers may also recombine with the stationary ion and form neutral molecules. In the absence of an external electric field, the movement of the carriers is at random (*diffusion current*). If an electric field is applied, a current component (*drift*) in the direction of the field will arise and be superimposed on the diffusion current.

DIODES

Figure 11-1*a* shows a drawing of an n-type semiconductor on the left. The positive ions are shown to be held in the fixed array of the crystal lattice. The negative carriers, i.e., the free electrons, are indicated by small black dots randomly distributed throughout the crystal. On the right is a p-type semiconductor. Here the negative ions are stationary in the crystal lattice, and the free positive carriers or holes are expressed by small white dots. If these two semiconductors are brought in contact (Fig. 11-1*b*), the free electrons (black dots) will migrate toward the junction where they will combine with the positive holes (white dots), which also will migrate toward the junction until the space charge caused by the *stationary* ions (+ circles and − circles) sets up a field which prevents further free carriers from moving. Therefore, a thin region exists between the semiconductors which is starved or depleted of free charges. This region is not a complete insulator because there are always free carriers from intrinsic conduction or thermal origin. However, the resistance of this *depletion layer* is several orders of magnitude larger than the resistance of the bulk of the semiconductors. Therefore, the depletion layer practically determines the current in the circuit. Any means which permits a variation of the thickness of the depletion layer can be used to control the current in a p-n junction and in an external circuit connected to the p-n junction. In Fig. 11-1*c*, a voltage is applied in the blocking direction, i.e., with such a polarity that the external field attracts the free carriers toward the electrodes and away from the depletion region, thus widening the depletion region and causing an increase of its resistance and blocking the passage of current. With the field applied in the opposite direction

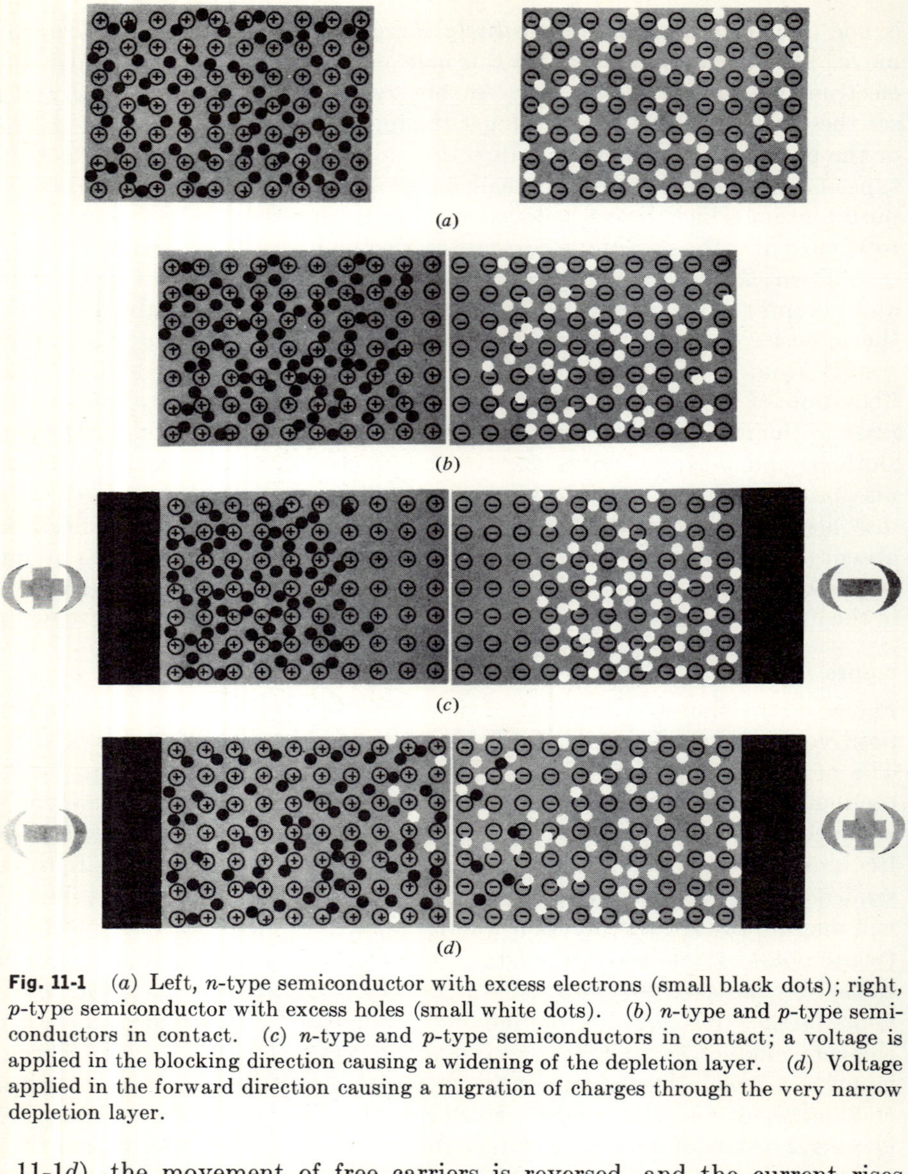

Fig. 11-1 (*a*) Left, *n*-type semiconductor with excess electrons (small black dots); right, *p*-type semiconductor with excess holes (small white dots). (*b*) *n*-type and *p*-type semiconductors in contact. (*c*) *n*-type and *p*-type semiconductors in contact; a voltage is applied in the blocking direction causing a widening of the depletion layer. (*d*) Voltage applied in the forward direction causing a migration of charges through the very narrow depletion layer.

(Fig. 11-1*d*), the movement of free carriers is reversed, and the current rises steeply with the applied voltage.

Voltage-current characteristics of a germanium and a silicon diode are shown in Fig. 11-2. In the forward direction the germanium diode shows a steeper increase of current at lower voltages than the silicon diode; in this range the germanium diode is more suitable for linear rectification than the silicon diode (see Chap. 7). In the voltage range between about 0.1 to 0.6 V, the current in both semiconductors rises logarithmically with the applied voltage.

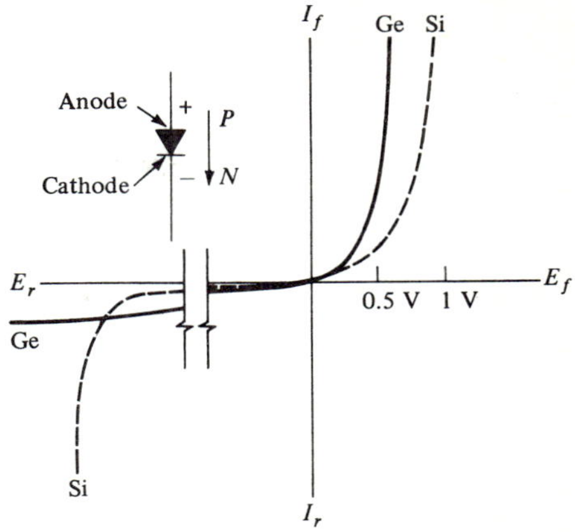

Fig. 11-2 Voltage-current characteristics of a germanium and a silicon diode.

The logarithmic characteristic may extend over 4 to 10 decades. Diodes operated in that range can serve as linear-to-logarithmic converters in analog computers.

The depletion layer with the two adjacent semiconducting layers forms a capacitor. Its capacitance changes with the thickness of the depletion layer and, therefore, with the applied voltage. This effect can be used advantageously for the control of a capacitance by means of an applied voltage, such as the tuning of a resonance circuit by electrical rather than mechanical means. On the other hand, the presence of a capacitor can limit the use of a rectifier at high frequency since the junction capacitance represents an undesirable path for the alternating unrectified current.

The generally accepted symbol of a diode is an arrow against a flat electrode as shown in Fig. 11-2. The direction of the arrow is that of the flow of conventional current, i.e., of positive charges. Current *passes* the diode if the arrow, or *p-conducting* electrode, is positive; the arrow is then the anode. In the blocking direction, the polarity of the voltage applied to the diode is reversed; the diode is reverse-biased.

Metallic diodes Metallic diodes, selenium or copper oxide rectifiers, have been used in the past but generally are superseded by semiconductors of the germanium and silicon type. Semiconductor diodes have a 10^2 to 10^3 times higher ratio of forward current to reverse (*leakage*) current than metallic diodes. They can be operated at higher current density, and they exhibit lower capacitance than the metallic diodes. The capacitance of metallic diodes is astonishingly high: for selenium rectifiers the capacitance is of the order of 0.01 F/cm^2, for copper oxide rectifiers of the order of 0.2 F/cm^2.

Zener diodes If the concentration of impurity levels, i.e., of donors and acceptors in both the n- and the p-type materials, is increased over the level in ordinary diodes, then a breakdown in the reverse direction is facilitated as shown in Fig.

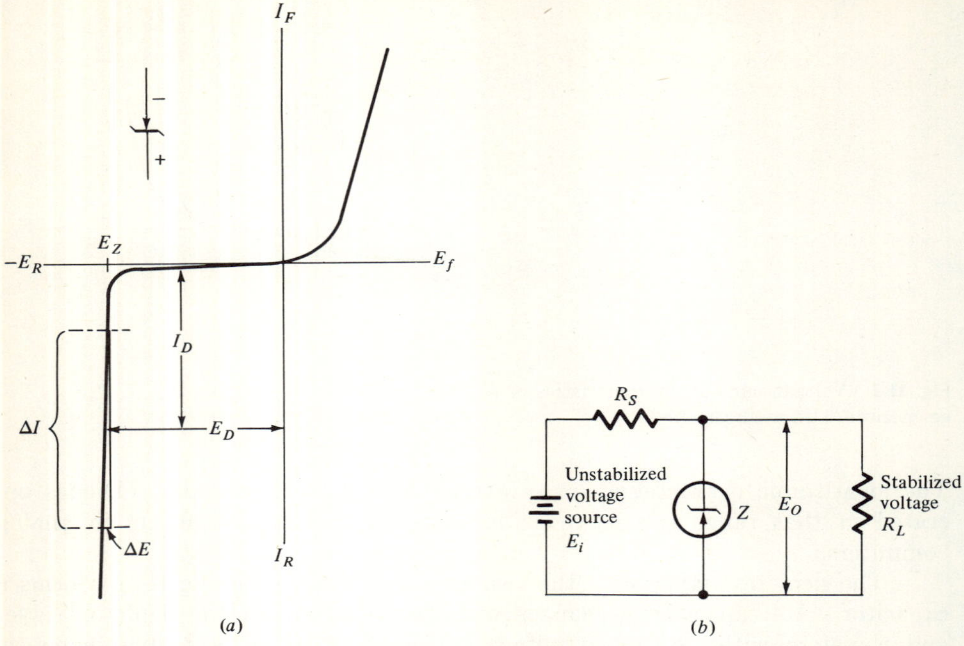

Fig. 11-3 (*a*) Voltage-current characteristic of a zener diode; (*b*) circuit diagram of a simple zener-diode–stabilized-voltage source.

11-3*a*. Therefore, if a small voltage is applied to a semiconductor diode in the reverse or blocking direction, the current passing the diode junction is only that due to leakage and is very small. If the voltage is increased, a critical point E_Z will be reached where a breakdown occurs and where the current rises steeply in the negative direction. This breakdown may be due to an avalanche effect: the electrons moving across the depletion region acquire sufficient kinetic energy to cause ionization by collision. The breakdown may also be initiated by tunnel effect (*zener breakdown*). The breakdown effect is reversible if means are provided to limit the breakdown current to a safe value. The breakdown occurs in a very short time after the application of a voltage. In silicon diodes the time delay between voltage and current is in the microsecond region. If operated in the zener breakdown region, the diode can be used for voltage stabilization of a supply source, current limiters, clippers, clamps, and other circuits (see Chap. 13).

The stabilization mechanism is similar to that of a gaseous voltage-regulator tube as described in connection with Fig. 8-9. A simple circuit is shown in Fig. 11-3*b*. The unstabilized voltage E_i, which should be 10 to 50 percent larger than the stabilized output, is applied to a resistance R_S and a zener diode Z in series. Any variation of the input voltage by the amount ΔE_i will cause a variation of the output voltage by

$$\Delta E_o = \frac{r_d}{R_S} \Delta E_i$$

where $r_d = \Delta E/\Delta I$ is the incremental (*dynamic*) resistance of the zener diode. Because of the steep rise of the zener-diode current with a variation of voltage in the operating range, r_d is very small and the output voltage variation ΔE_o is considerably less than the variation of the input ΔE_i. A variation ΔR_L of the load resistance R_L causes a variation of the output voltage E_o of

$$\Delta E_o = E_o \frac{\Delta R_L}{R_L} \frac{r_d}{R_L}$$

(assuming that $\Delta R_L \ll R_L$). In this equation, $\Delta R_L/R_L$ is the fractional variation of the load and r_d/R_L can be called the regulation coefficient, again a very small quantity.

Zener diodes are rated according to the voltage at which breakdown occurs (from two to several hundred volts) and to the maximum power they can safely dissipate (from milliwatts to hundreds of watts).

Tunnel diodes If the concentration of donors and acceptors on both sides of the *p-n* junction is still further increased over that in zener diodes, the characteristic changes for a typical germanium diode (see Fig. 11-4*a*). In the reverse-bias direction A, breakdown occurs at 0-V bias, and the current increases linearly in the negative direction with increased negative-bias voltage. In the forward-bias direction B and for voltages less than 100 mV, the current increases in a positive direction. The current in the forward direction reaches a peak value I_p at an applied voltage of the order of 100 mV, then diminishes to a lower value C. The current reaches a minimum I_V for an applied voltage E_V of about 300 mV and then rises again, like that of a normal diode D. In the region C the current diminishes with increased voltage, i.e., the diode in this range has a negative resistance. The effect is caused by tunneling of the electrons: the electrons can move through a potential threshold of an energy level higher than that of the electrons if the region of the threshold is narrow and the field strength is high.

Negative resistances are inherently unstable but can be used in the construction of oscillators and switching devices. An example of a tunnel-diode oscillator is shown in Fig. 11-4*b*. The supply voltage E_{BB} is applied via a voltage divider R_1 and R_2 to a tunnel diode. The primary reason for this voltage divider and the capacitor C_B is the establishment of a low-impedance path for the radio frequency in close proximity to the tunnel diode in order to avoid parasitic (undesirable) oscillations of the connecting leads between the voltage source and the tunnel diode. The diode is connected with an LC tank circuit, which is brought to oscillation by the tunnel diode. Circuits of this kind find application in small transmitters used frequently for radio telemetering of data from animals or from the inside of the human body. The outlines of such a transmitter can be made so small that the entire transmitter can be attached, ingested, or surgically implanted. If a quantity to be measured inside the body can be made to change the frequency of the oscillator (e.g., temperature, heart-beat frequency,

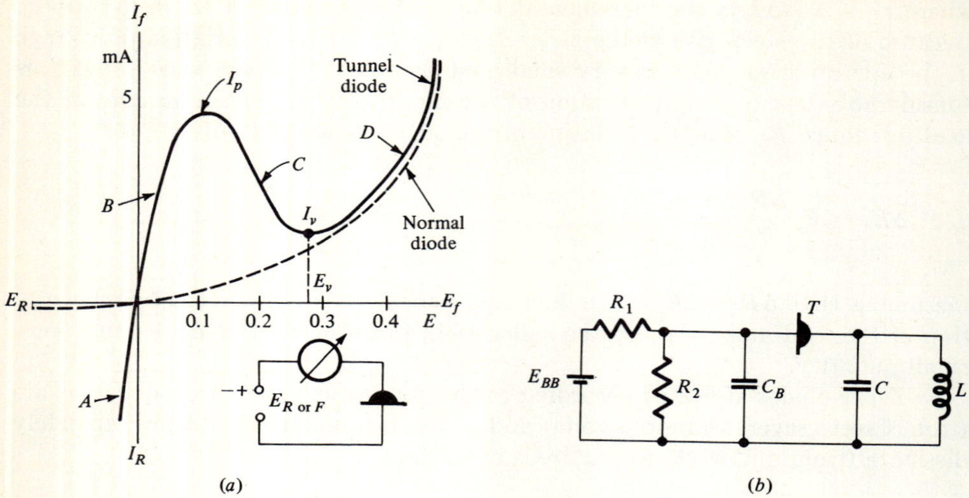

Fig. 11-4 (*a*) Voltage-current characteristic of a tunnel diode; insert, circuit diagram; (*b*) tunnel-diode oscillator, see text for explanation.

etc.), then such frequency-modulated signals can be detected by a radio receiver outside the body and at a considerable distance allowing unrestrained movement of the experimental animal or human subject.

Transistors A junction transistor consists essentially of three semiconducting layers in either the sequence *p-n-p* or *n-p-n*. The middle layer is very thin compared with both outside layers. All three layers are commonly made from one slab of a single crystal; the doping is accomplished by thermal diffusion methods.

A *p-n-p* transistor is schematically illustrated in Fig. 11-5 and operates as

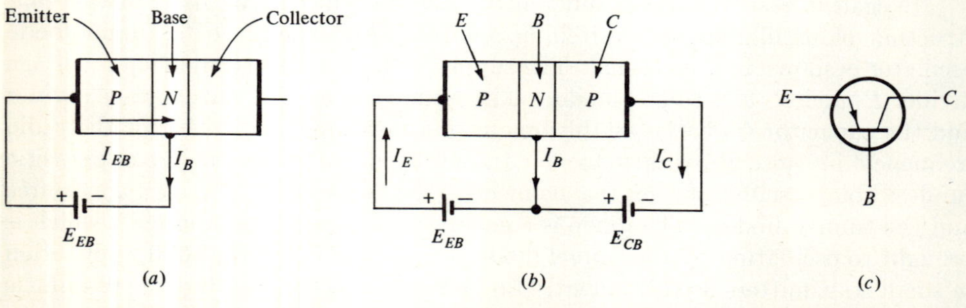

Fig. 11-5 (*a*) Schematic diagram of the emitter-base part of a *p-n* diode; (*b*) diagram of a *p-n-p* transistor in a common-base circuit; (*c*) symbol of a *p-n-p* transistor, the arrow shows the direction of the positive (hole) carriers.

follows. A voltage source E_{EB} is applied in the forward direction, i.e., so as to make the emitter-base diode conduct. Majority carriers (positive holes) move in the direction of the arrow, from the emitter to the base. A small fraction of the positive-hole current will combine with the n-carriers (electrons) in the n-type base layer and form a current, the base current I_B. Then a new effect occurs: because of the thinness of the base layer, some of the holes injected into the base region will move through the base layer and enter the collector region. The collector is connected to the negative pole of a voltage supply E_{CB} (Fig. 11-5*b*). From the viewpoint of a rectifier action at the base-collector interface, this is a reverse bias. However, the negative potential applied to the collector terminal has the effect of attracting the positive carriers; the carriers form a current I_C by combining with the electrons from the voltage supply. The analogous operation of an n-p-n transistor is illustrated in Fig. 11-6. The operation is almost identical with that of a p-n-p transistor, but the polarities of the supply voltages and those of the carriers (hence, the direction of the currents) are reversed.

A change of the emitter current causes a change of the collector current. This effect is the basis of the control or amplifying mechanism of a transistor. In the common-base circuits shown in Figs. 11-5*b* and 11-6*a*, the emitter current I_E is divided into one fraction $I_C = \alpha I_E$ passing into the collector region and forming the collector current, and another fraction $I_B = (1 - \alpha)I_E$ forming the base current. The numerical value of α varies between 0.90 and 0.99, depending on the transistor type and the operating conditions. Thus, the collector current is smaller than the emitter current; the total forward-current gain for this circuit, $I_C/I_E = \alpha$, is less than unity. Nonetheless, a voltage gain and a power gain can be realized with this circuit if a large load resistance is inserted in the collector circuit.

There are three basic circuits for the operation of transistor amplifiers, the common-base (described above and redrawn in Fig. 11-7*a*), the common-emitter, and the common-collector circuit.

An n-p-n transistor connected in a common-emitter configuration is shown

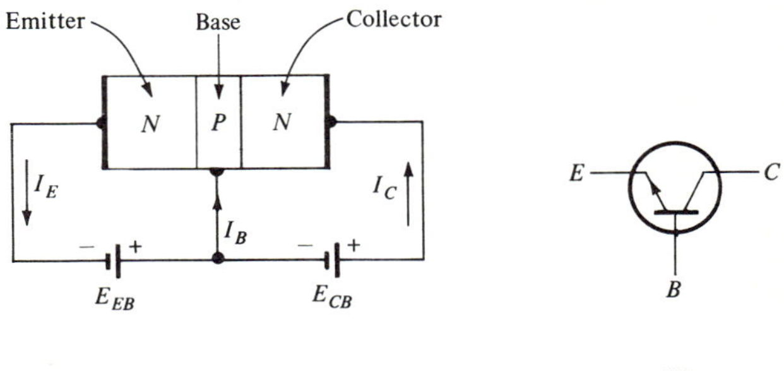

Fig. 11-6 (*a*) Schematic diagram of an n-p-n transistor, common-base circuit; (*b*) symbol of an n-p-n transistor.

Circuit	Common base	Common emitter	Common collector
Diagram	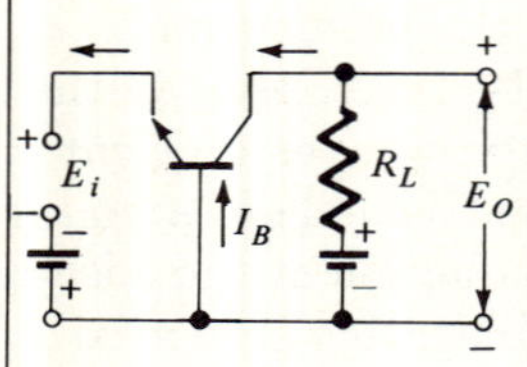 (a)	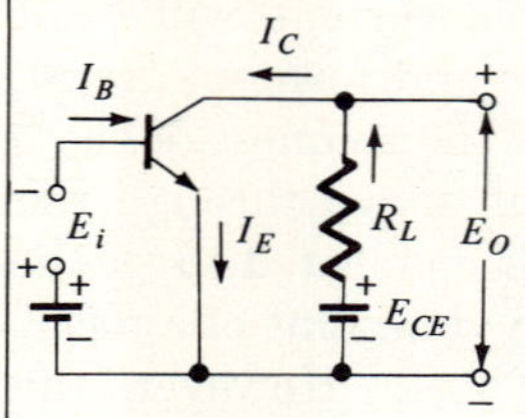 (b)	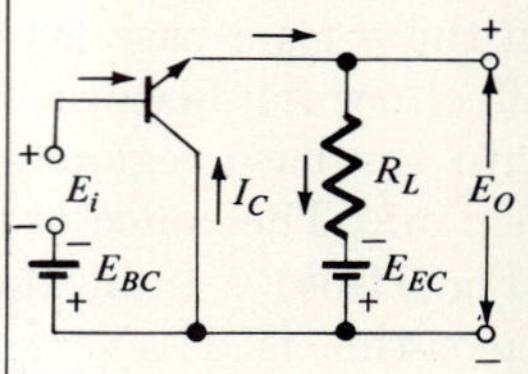 (c)
Power gain	Yes	Yes (highest)	Yes
Voltage gain	Yes	Yes	No (less than unity)
Current gain	No (less than unity)	Yes	Yes
Input impedance	Lowest (≈ 50 Ω)	Intermediate (≈ 1 kΩ)	Highest (≈ 300 kΩ)
Output impedance	Highest (≈ 1 MΩ)	Intermediate (≈ 50 kΩ)	Lowest (≈ 300 Ω)
Phase inversion	No	Yes	No

Fig. 11-7 Synopsis of the three basic transistor circuits.

in Fig. 11-7*b*. The signal E_i is fed into the base. The forward-current gain is

$$\frac{I_c}{I_b} = \frac{\alpha I_E}{(1-\alpha)I_E} = \frac{\alpha}{(1-\alpha)} \simeq \beta$$

The quantity β is the static dc current gain of a transistor in a common-emitter circuit. The current gain is considerably larger than that of a common-base circuit with the same transistor. Assuming a transistor with a value of $\alpha = 0.98$, the current gain of a common-emitter amplifier is

$$\beta = \frac{\alpha}{1-\alpha} = \frac{0.98}{0.02} = 49$$

β is not a constant quantity; it varies in a nonlinear fashion with the applied voltage. It frequently is replaced by the quantity h_{fe}, the forward-current transfer ratio for small signals, that is $\Delta i_C/\Delta i_B$, while h_{FE} is the static forward-current transfer ratio. The instantaneous output current under short-circuit condition is

$$\Delta i_C = h_{fe}\Delta i_B$$

while the dc output current is

$$I_C = h_{FE} I_B$$

The third configuration, the common-collector circuit, is shown in Fig. 11-7*c*. The circuit furnishes a current gain and a power gain, but the voltage gain is less than unity. The circuit is used primarily as an impedance changer; its input impedance is high; its output impedance is low.

Almost all parameters of a transistor change with temperature. The most critical changes are

1. The leakage current from collector to base (I_{CBO}, measured between base and collector with the emitter terminal open)
2. The leakage current from collector to emitter (I_{CEO}, measured between emitter and collector with the base terminal open)
3. The input resistance
4. The quantity β
5. The base emitter voltage E_{BE}

Leakage currents in germanium transistors are typically 100 to 1,000 times greater than in silicon transistors. For this reason, bias stability considerations in germanium-transistor circuits are concerned mainly with the influence of I_{CBO} and I_{CEO}. Variations of E_{BE} and β are of importance in both silicon and germanium transistors. If a transistor is operated so that it develops a high temperature at high collector current, its temperature will rise causing an even higher collector current. The higher collector current will cause increased heat dissipation and will lead to a thermal runaway effect which eventually destroys the transistor. Thermal runaway can be controlled by limiting the collector current to a safe value and, for power transistors, by providing adequate cooling with heat sinks. Transistor amplifiers are described in Chap. 12.

Field-effect transistors An illustration and a schematic diagram of a unipolar field-effect transistor (FET) are shown in Fig. 11-8. The transistor consists essentially of a semiconductor bar of either the p or n type having on both ends ohmic (nonrectifying) contacts called source S and drain D. The resistance of such a bar is determined by its length, its effective cross-sectional area, and its conductivity. If the bar is n-type conducting and a supply voltage is applied with the polarity as indicated in Fig. 11-8*a* and *b*, electrons will move from the source to the drain forming a current in the circuit. Between the source and the drain is a p-type electrode, perpendicular to the direction of the current flow. If a voltage is applied between the source and the gate, with the indicated reverse-bias polarity, a layer of n-type carriers will form a negative space charge around the gate electrode. This negative space charge has the effect of reducing the effective cross section of the bar. The electron flow is reduced, and the higher the applied gate voltage, the lower will be the current I_D in the circuit. The operation of a field-effect transistor is evident from Fig. 11-8*c*. For a given applied gate voltage E_G (e.g., 0 V) the drain current rises with increased drain voltage from zero until a value E_p, *the pinchoff voltage*, is reached. With further increase of the drain voltage, the drain current remains practically constant

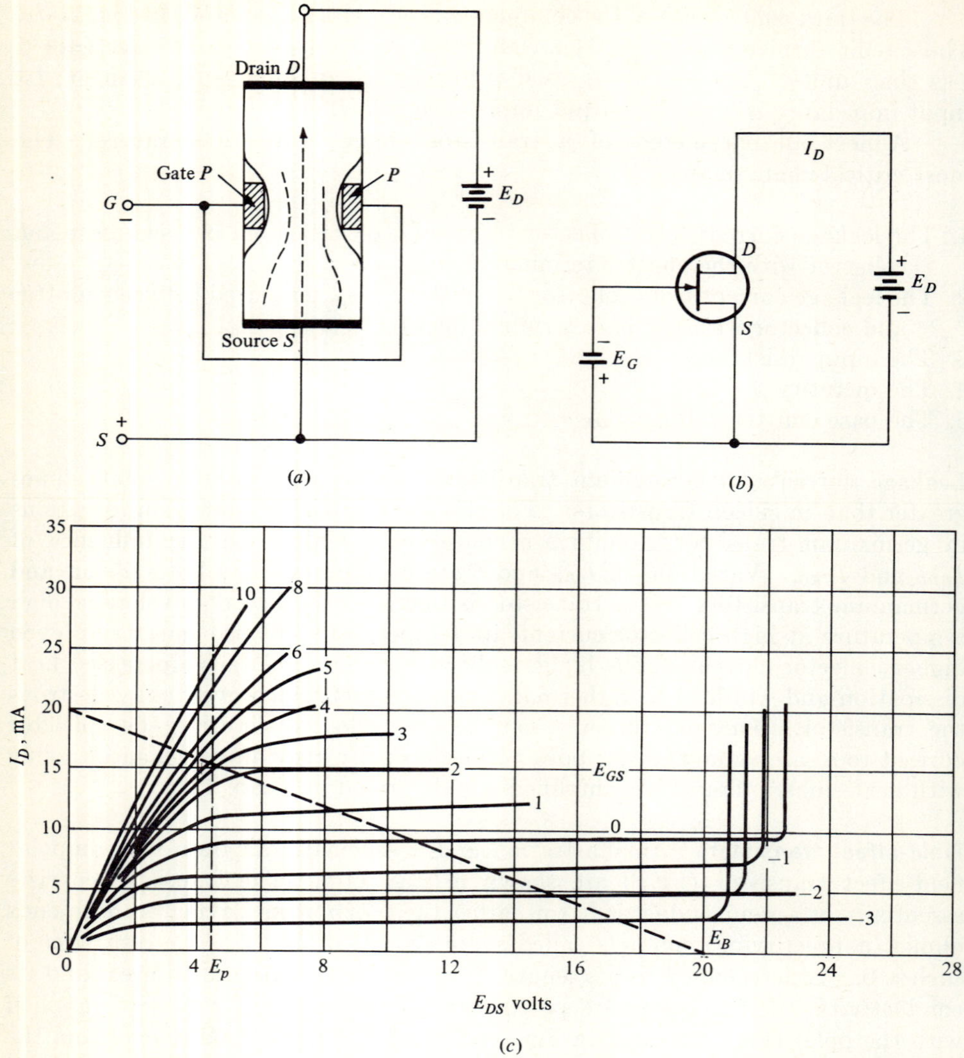

Fig. 11-8 (*a*) Schematic diagram of a unipolar field-effect transistor; (*b*) symbolic presentation; (*c*) characteristics and load line of a field-effect-transistor amplifier.

until at a second value of drain voltage, the breakdown voltage E_B is reached. If a load resistor is inserted in the drain circuit, the load line for this resistor can be constructed in the usual way (see Fig. 11-8*c*). The field-effect transistor has the outstanding characteristic that its input impedance (impedance between *G* and *S*) is high (typically 10^{12} Ω); i.e., its input current is low. The input resistance is relatively insensitive to temperature changes and is independent of the gate-bias conditions. The input capacitance is small (typically 4 pF).

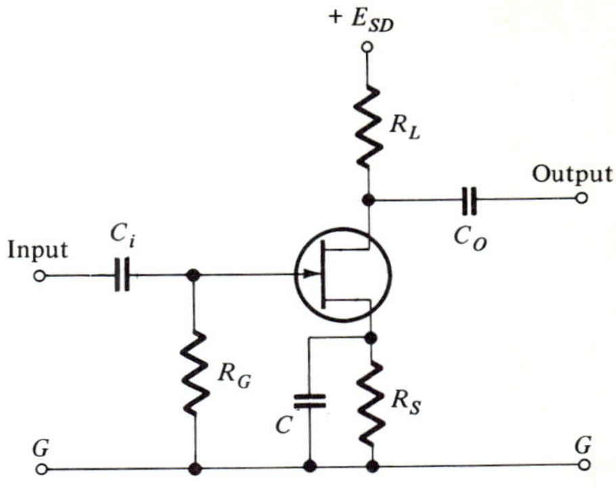

Fig. 11-9 Amplifier circuit for ac signals using a field-effect transistor.

The transconductance $\Delta I_D/\Delta E_G$ varies between 200 μA/V and 10 mA/V. Some field-effect transistors are built so that the source and the drain connections are interchangeable; i.e., they are symmetrical so that they can be used for symmetrical clipping action. An amplifier for ac signals based on field-effect transistors is illustrated in Fig. 11-9.

Because of its very high input impedance, the field-effect transistor can be used for many applications where ordinary junction transistors are unsatis-

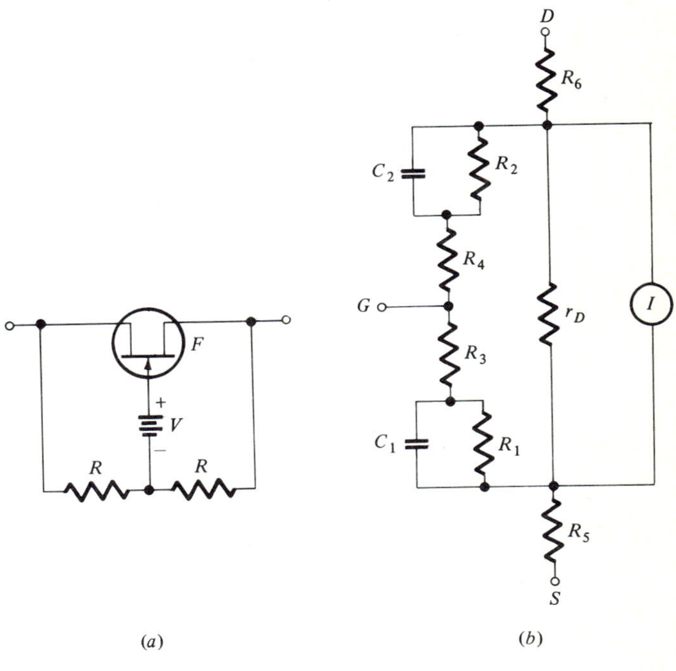

Fig. 11-10 (*a*) Circuit diagram of a voltage-controlled current source using a field-effect transistor; (*b*) equivalent circuit diagram of a field-effect transistor.

factory. Its power gain is high. No gate bias is required for small signals. It can be operated in the region below and above the zero bias. In the region above the pinchoff voltage, it can serve as a voltage-controlled current source as illustrated in Fig. 11-10*a*. In the region below pinchoff, the field-effect transistor can be used as a voltage-variable resistor. In the region of the pinchoff voltage, it can be used as a nonlinear circuit element for mixers or modulators. If operated in the region above the pinchoff voltage and below the breakdown voltage, it can be used as an amplifier with a wide frequency range. It can also be used for the construction of oscillators.

Operation at high frequency is limited only by the capacitance in the input circuit. An equivalent circuit diagram which shows the behavior of the field-effect transistor is provided by Fig. 11-10*b*.

METAL-OXIDE SEMICONDUCTOR TRANSISTORS

The metal-oxide semiconductor (MOS) is a modification of the field-effect transistor. It consists of a body of *n*-type silicon with two *p*-type electrodes, the source and the drain, as shown in Fig. 11-11*a*. On the surface of the body, between the source and the drain, is deposited a thin layer (1,000 AU) of silicon dioxide, a highly insulating substance. A metal electrode is deposited over the oxide layer and forms the gate contact.

If a voltage E_{SD} is applied between source and drain, no current will flow, since in either direction of the applied voltage, one of the *p-n* junctions will always be reverse-biased. If the gate is charged negatively with respect to the source, positive holes will enter from the source into the channel, i.e., the region between both *p*-type electrodes, until a surface layer of the channel has changed from *n*- to *p*-type conduction. At this point there is no more *p-n* junction between drain, channel, and source, and a small current will flow through the surface layer of the channel. As the negative gate voltage increases, the con-

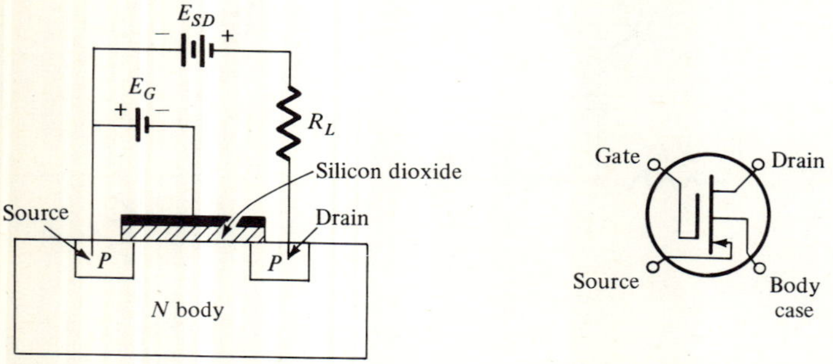

(*a*) (*b*)

Fig. 11-11 (*a*) Schematic diagram of a metal-oxide semiconductor field-effect transistor; (*b*) socket connections.

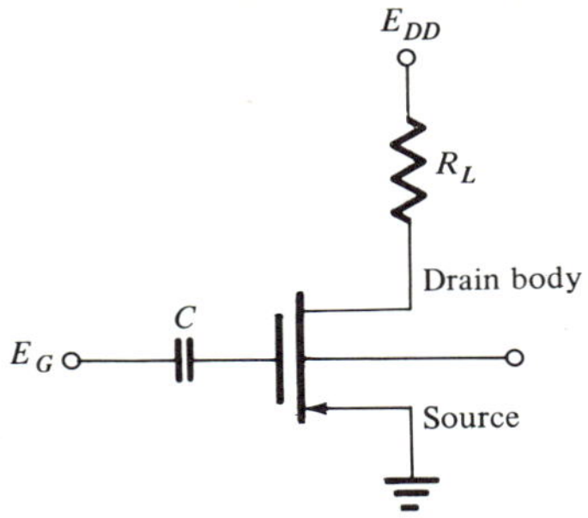

Fig. 11-12 Schematic circuit diagram of an ac amplifier using a metal-oxide semiconductor field-effect transistor.

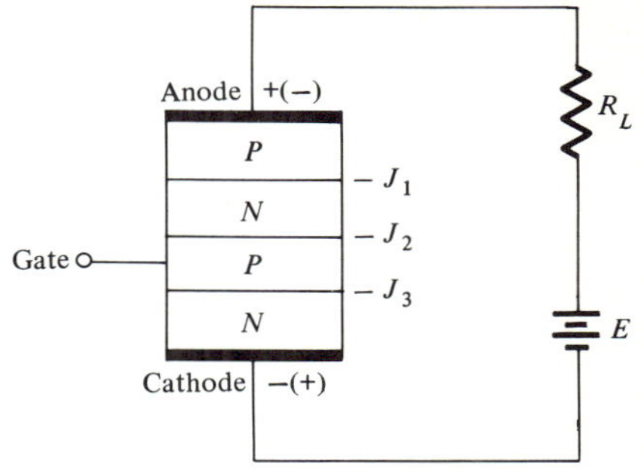

Fig. 11-13 Schematic diagram of a silicon-controlled rectifier.

ductive layer deepens, and the current from the source to the drain increases in a way similar to that of a field-effect transistor.

The gate-input resistance R_{GS} is very high, of the order of 10^{14} Ω. The source-drain resistance in the off-bias condition is of the order of 10^{10} Ω, and in the on condition of the order of 200 to 300 Ω. The input capacitance between the gate metal electrode and the body is typically 4 pF. A simple amplifier circuit diagram is shown in Fig. 11-12.

Four-layer semiconducting devices Semiconductors made with four layers of the configuration *n-p-n-p* have physical characteristics which make them suitable for switching operations or for the control of ac power. Because of their similarity to thyratrons, i.e., gas-filled switching tubes as described in Chap. 8, they frequently are called *thyristors*. A schematic diagram of a thyristor, the silicon-controlled rectifier (SCR), is shown in Fig. 11-13. If the anode is made negative with respect to the cathode, the junctions J_1 and J_3 are reverse-biased so that no current passes. If the anode is made positive with respect to the cathode (forward-bias direction), the junction J_2 is reverse-biased and will block the current. However, external means can be used to forward-bias the junction J_2 so that a current will pass through the *n-p-n-p* structure and through an external circuit connected to it.

The operation of an SCR can be illustrated by the equivalent circuit of Fig. 11-14*a* and *b*. The four-layer structure is presented as a combination of two transistors: an *n-p-n* transistor 1 with emitter e_1, base b_1, and collector c_1 and a *p-n-p* transistor 2 with emitter e_2, base b_2, and collector c_2. A gate electrode G permits the connection of an external circuit to the base b_1. If a gate current I_G flows into the base b_1, it causes the emitter e_1 to inject electrons with the result that the collector current I_{c1} increases

$$I_{c1} = \beta_1 I_{b1}$$

I_{c1} is also the base current I_{b2} for *p-n-p* transistor 2. An increase of I_{b2} causes

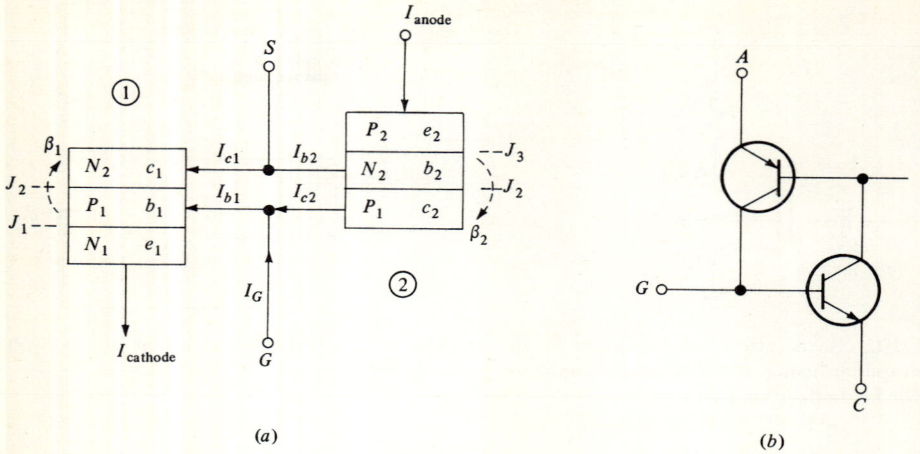

Fig. 11-14 (*a*) Equivalent circuit of a silicon-controlled rectifier; (*b*) same circuit using two transistors.

an injection of holes by the emitter e_2 and an increase of the collector current I_{c2}

$$I_{c2} = \beta_2 I_{b2}$$

The increase of I_{c2} has the effect of further increasing I_{b1} until saturation is reached, i.e., until the region P_1 is saturated with holes and region N_2 with electrons. The junction J_2 is then forward-biased and since J_1 and J_3 are also forward-biased, a high current can flow from the anode to the cathode. This occurs when $(\beta_1 n\text{-}p\text{-}n)(\beta_2 p\text{-}n\text{-}p) \geq 1$. (It is important that β_1 and β_2 are not constant quantities but that both increase as the collector-junction voltage increases and approaches the collector-junction breakdown voltage.) Once triggered, the anode current I_{anode} $(= I_{\text{cathode}})$ remains high, regardless of the presence or absence of the gate current I_G, and is limited primarily by the applied voltage and by the resistance in the external circuit. Once the SCR is in a state of conduction, the gate circuit does not exercise any control. To reduce the current to zero, the applied voltage must be removed or inverted. Such a procedure reduces the concentration of carriers and brings about the condition where the gate can be used again for further control.

A circuit diagram (Fig. 11-15) illustrates the operation of an SCR. The SCR in series with a load resistance R_L is connected to a supply source E_{CC}. Another supply source is applied between gate G and ground. Closing switch S_1 does not cause the flow of current in the circuit. However, when switch S_2 is also closed and the gate current supply is sufficiently high to cause saturation and breakdown of the junction J_2, then the current through the SCR rises fast to a finite value. Opening switch S_2 will have no further influence on this current. The current ceases if switch S_1 is opened, if only for a short duration.

n-p-n-p systems have also been developed in which all four semiconductor

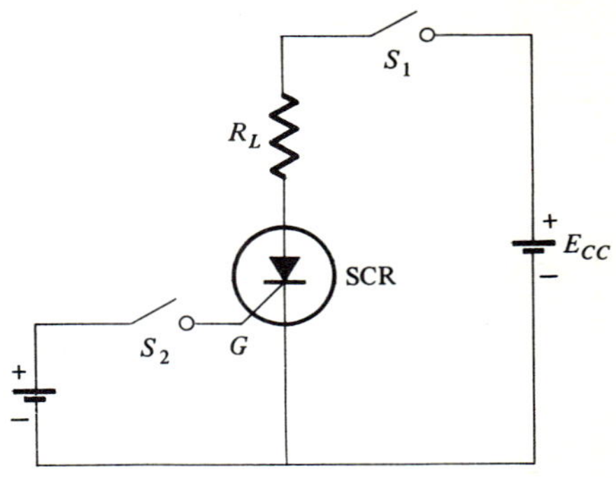

Fig. 11-15 Circuit illustrating the operation of a silicon-controlled rectifier.

layers are accessible: silicon-controlled switches (SCS) or gate-controlled switches (GCS). An example is shown in Fig. 11-16. Some of these switches permit the initiation of as well as the blocking of an anode-to-cathode current by the application of a pulse to the cathode or to the anode gate. For instance, in the circuit of Fig. 11-16 the current flow can be initiated by the application of a positive pulse to gate G_1 and can be blocked off by the application of a positive pulse to gate G_2 or of a negative pulse to gate G_1.

A *p-n-p* device may be triggered, even though the voltage applied to the gate electrode is below the triggering value, if a direct current is applied to the anode at a high rate, i.e., with a fast rise time. This rate effect is due to internal capacitance within the *n-p-n-p* device which causes an initiation of a discharge by intrinsic feedback. The rate effect need not be considered for power control devices operated at line frequency. The rise time at power frequencies is generally too low to cause a rate effect within the SCR.

A number of circuits have been devised which permit the electrical triggering (in contrast to manual or mechanical) of a discharge through a gate-controlled rectifier or switch circuit as well as the proper timing and shaping of a trigger

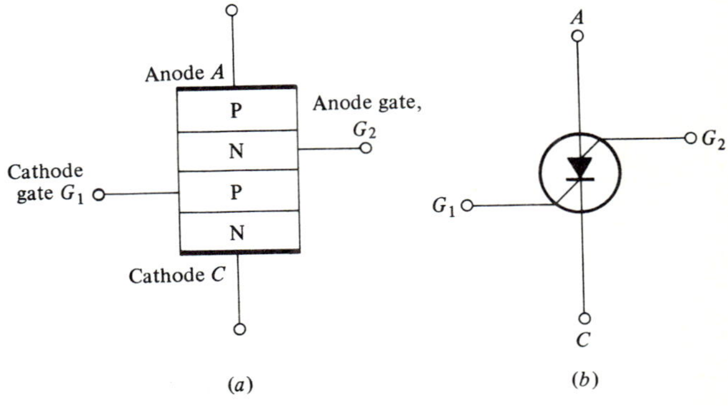

Fig. 11-16 (*a*) Silicon-controlled switch in which all four layers are accessible; (*b*) same in symbolic presentation.

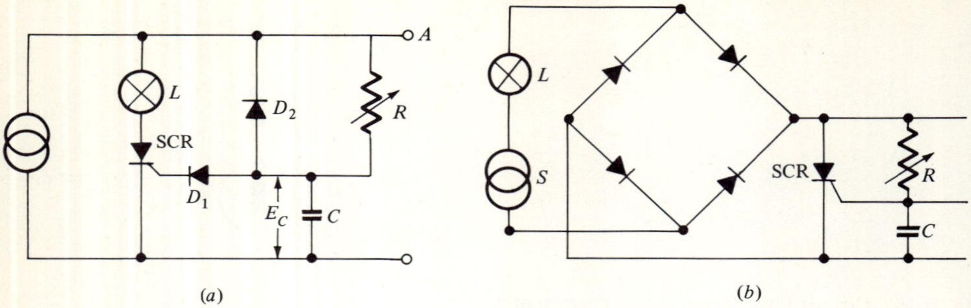

Fig. 11-17 (*a*) Circuit diagram for the control of electric power; half-wave operation, i.e., 0 to 50 percent; (*b*) circuit for the electrical control of power, 0 to 100 percent.

pulse. A very simple circuit for the control of ac power for heating control or dimming lights is shown in Fig. 11-17*a*. The light bulb L in series with an SCR is connected across the power-line feeders A and B. Also connected across the same feeders is a variable resistance R in series with a capacitor C. If A is positive, the capacitor C is charged through the variable resistance R with the time constant RC. The voltage E_C across the capacitor is applied to the gate of the SCR through diode D_1 and causes the SCR to fire when E_C reaches the proper potential. The time when this potential is reached and, thus, the duration of conduction of the SCR are controlled by the setting of R. In the negative half-cycle the SCR blocks the current and the light source extinguishes. The capacitor is charged to the negative peak value through diode D_2. Diode D_1 prevents damage to the SCR resulting from the application of the high negative potential across capacitor C. The circuit shown in Fig. 11-17*a* is operative only in the positive half-cycle and furnishes a regulation between 0 and 50 percent of the power of lamp L. A circuit capable of a control from 0 to 100 percent is shown in Fig. 11-17*b*. Bidirectional control can also be accomplished with gate-controlled, full-wave ac silicon switches (Triacs, a trade name of the RCA Corporation).

Different circuits have been described which cause termination of a discharge through an *n-p-n-p* system by electrical means. In principle, most of these circuits reduce the applied anode voltage below the level at which a discharge may be maintained (see, for instance, Fig. 11-17*a* and *b*). Some circuits introduce pulses that can be injected in different parts of the circuit and momentarily disrupt the current through a GCS, thus terminating or blocking the discharge.

SOLID-STATE PHOTOELECTRIC DEVICES

Figure 11-18 illustrates the general scheme of a solid-state photoelectric device. Optical radiation, x-rays, and certain corpuscular radiations fall on a solid-state semiconductor and are absorbed. The process of energy absorption produces

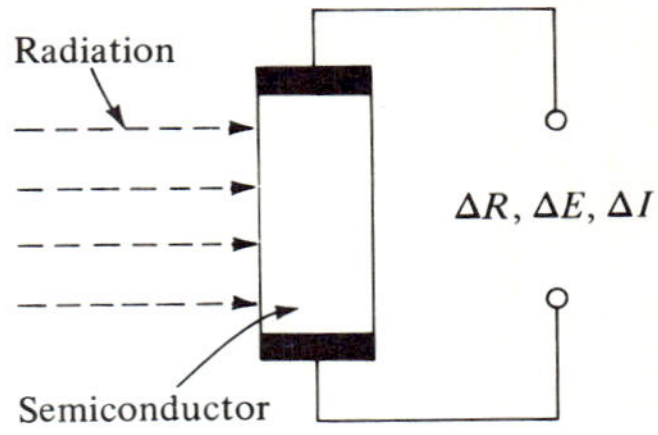

Fig. 11-18 Schematic diagram of solid-state photoelectric cell.

movable charges or a change of mobility of the charge carriers in the semiconductor. Of practical importance in the near-ultraviolet, visible, and infrared regions of the spectrum are two types of cells: photoconductive and junction photocells.

Photoconductive cells These are also called *photoresistive cells.* A cutaway view of a typical photoconductive cell is shown in Fig. 11-19*a*. A photoconductive material, such as activated polycrystalline cadmium sulfide or cadmium sulfoselenide, is deposited on a ceramic substrate. Electrodes of tin or indium are applied by evaporation onto the photoconductive layer, usually in an interdigitating pattern. The electrodes make ohmic (nonrectifying) contact with the photoconductive layer. The basic circuit for the use of a photoconductive cell is shown in Fig. 11-19*b*.

If the cell is kept in darkness, its resistance (dark resistance) can be as high as 10^{13} Ω. If the cell is illuminated, its resistance diminishes as shown in the transfer characteristics of Fig. 11-20. The resistance depends, of course, on the physical character of the photoconductive layer as well as on the dimensions of the cell and the electrode configuration. The current is proportional to the applied steady-state (dc) voltage up to about 1 mA, depending on the cell type. For a $\frac{1}{2}$-in-diameter cell and an applied voltage of 12 V dc illuminated with 1-fc

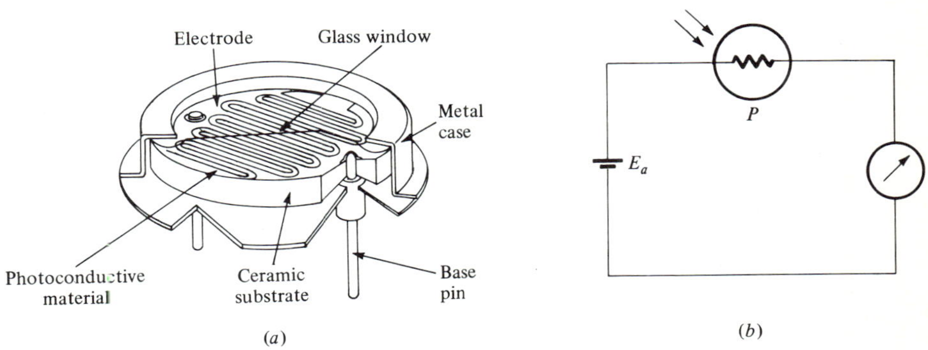

Fig. 11-19 (*a*) Cutaway view of a typical photoconductive cell. (*b*) Basic circuit diagram for the operation of a photoconductive cell; E_a, auxiliary or bias voltage; *P*, photoconductive cell; I_o, output meter.

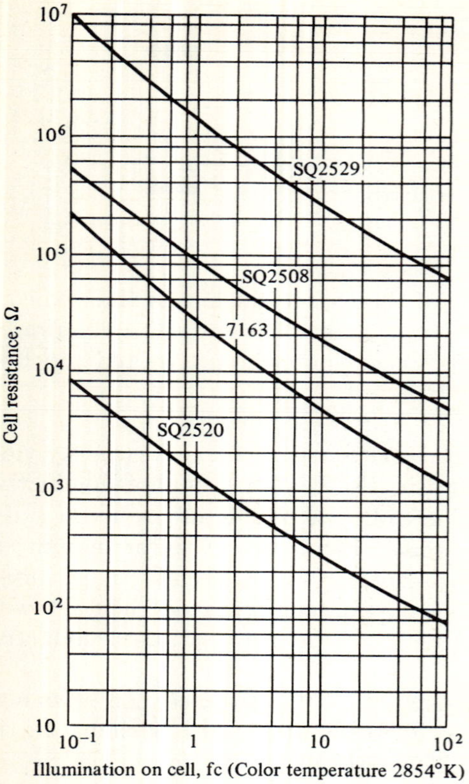

Fig. 11-20 Graph illustrating the variation of resistance of four different photoconductive cells as a function of their illumination. [*From RCA catalog (by permission).*]

incandescent light of a color temperature of 2870K (standard color temperature), the current is of the order of 1 to 10 mA.

The photocurrent does not follow instantaneously a variation of the illumination but rises gradually after the illumination is applied and decays gradually after the illumination is removed. Both the buildup and the decay-time constant depend on the material of the photoconductor and the level of illumination. Typical rise and decay characteristics are shown in Fig. 11-21*a* and *b*.

Temperature has little influence on the performance and sensitivity of a photoconductor as long as the illumination is of the order of 10 fc and more, but it has a great influence at low-light levels where the sensitivity increases strongly at lower temperature.

The spectral response characteristics of two different commercial cells are shown in Fig. 11-22. There is almost no response to radiation of a wavelength shorter than 300 nm. Cadmium sulfide cells have a peak response in the green part of the spectrum at 510 nm and can be used in the near-infrared region up to

---

Fig. 11-21 (*a*) Typical rise characteristic of a cadmium sulfide photoconductive cell for different illuminations; *A*, 10 fc; *B*, 1 fc; *C*, 01 fc; *D*, 0.01 fc. (*b*) Typical decay characteristics, same parameters.

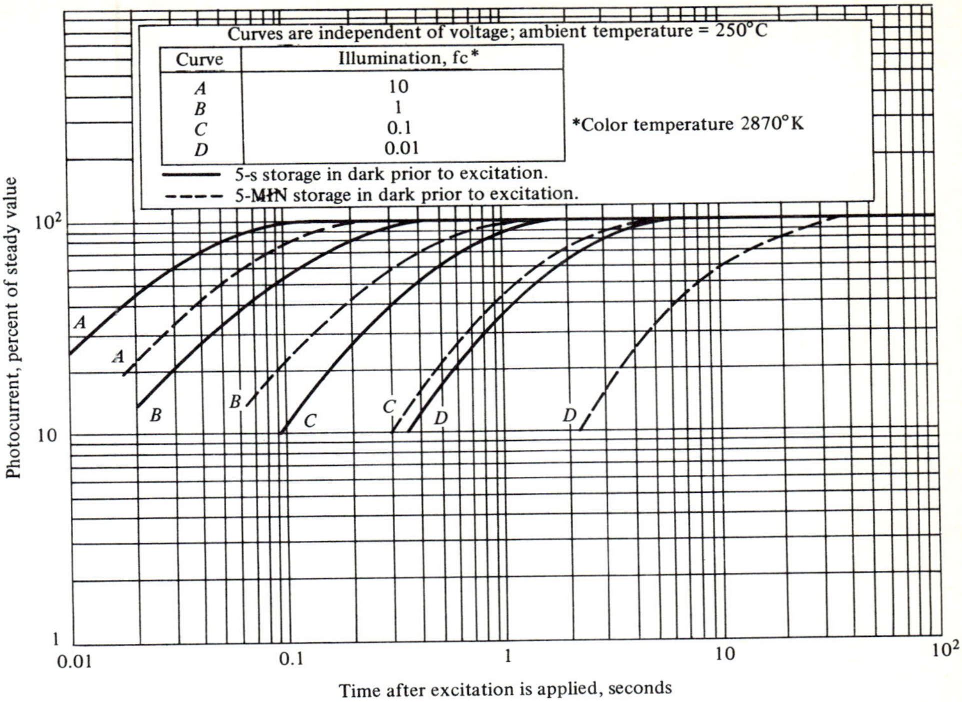

Curves are independent of voltage; ambient temperature = 250°C
Curve
Illumination, fc*
A 10
B 1
C 0.1
D 0.01
*Color temperature 2870°K
5-s storage in dark prior to excitation.
5-MIN storage in dark prior to excitation.
Photocurrent, percent of steady value
10^2
10
1
0.01
0.1
1
10
10^2
A
A
B
B
C
C
D
D
Time after excitation is applied, seconds

(*a*)

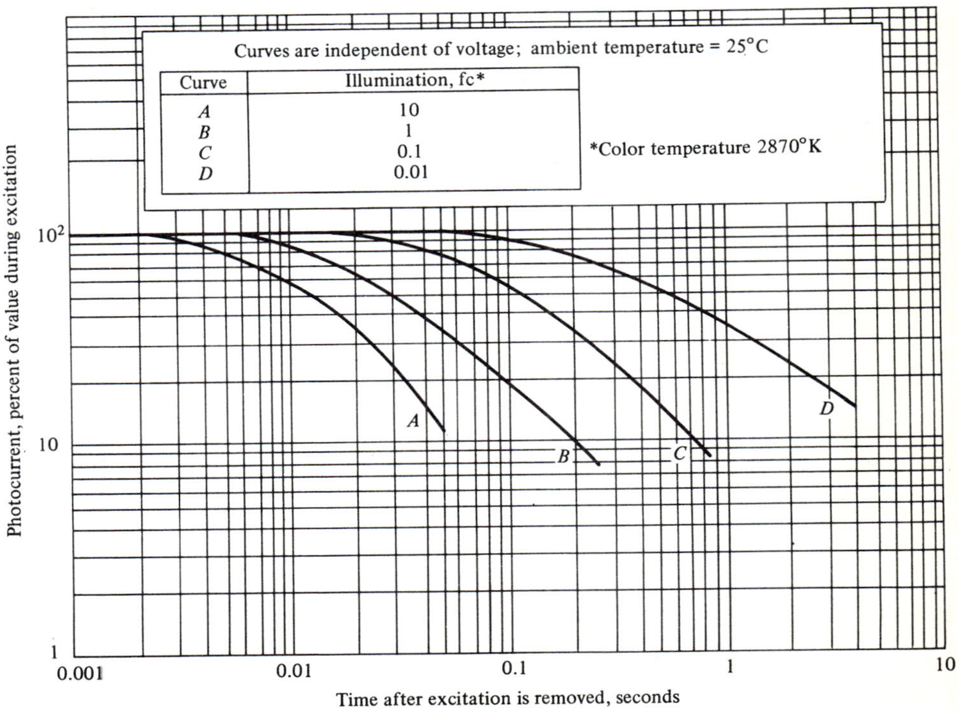

Curves are independent of voltage; ambient temperature = 25°C
Curve
Illumination, fc*
A 10
B 1
C 0.1
D 0.01
*Color temperature 2870°K
Photocurrent, percent of value during excitation
10^2
10
1
0.001
0.01
0.1
1
10
A
B
C
D
Time after excitation is removed, seconds

(*b*)

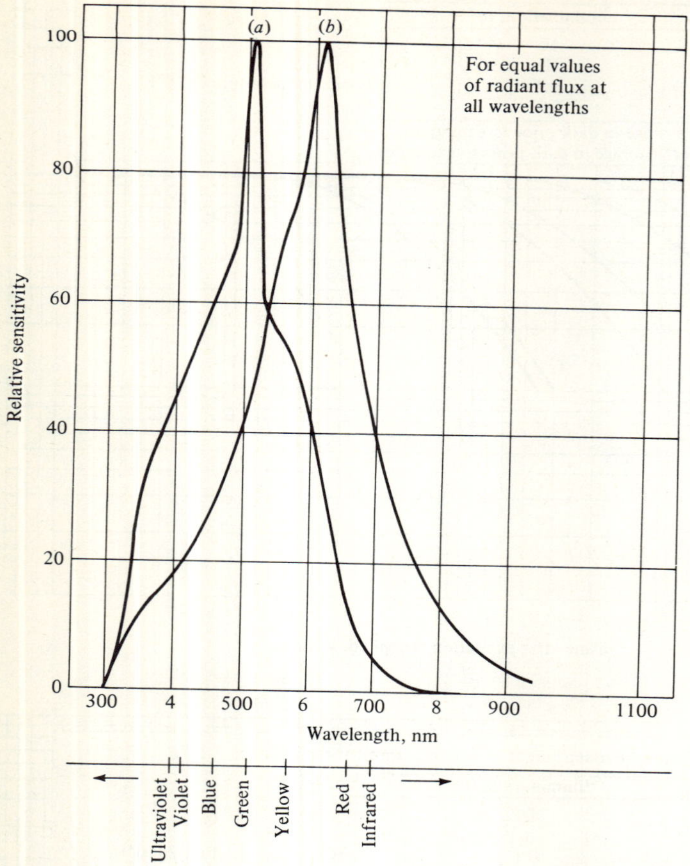

Fig. 11-22 Spectral response characteristics of photoconductive cells. (*a*) Cadmium sulfide cell; (*b*) cadmium sulfoselenide cell.

about 750 nm. The maximum response of cadmium sulfoselenide is in the yellow-orange at 615 nm; it can be used in the infrared region up to about 1,000 nm.

The infrared response of some solid-state photoconductive cells can be extended considerably by cooling the cell. For instance, a gold-doped germanium single crystal that is cooled to liquid nitrogen temperature can be used as a photoconductive detector and is sensitive to infrared radiation over a wavelength range from 1.0 to 9.0 nm.

Photojunction cells These are also called photodiodes, phototransistors, photovoltaic cells, and semiconductor photocells—the terminology in the literature varies widely. All these photocell systems have in common a construction comprising an n- and a p-type layer. The system has the electrical characteristics

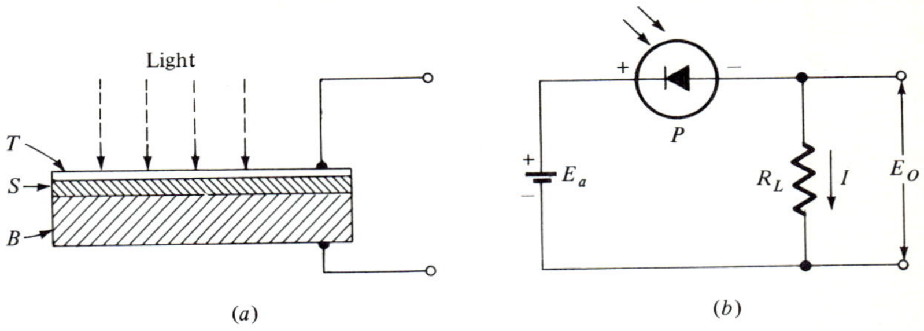

Fig. 11-23 (*a*) Schematic diagram of a photojunction cell; (*b*) basic circuit diagram for the operation of a photojunction cell (photodiode).

of a rectifier. Radiation is directed to the vicinity of the *p-n* junction and causes a current.

An example of a simple photojunction cell is shown in Fig. 11-23*a*. The cell consists of three layers: a base plate *B*, a semiconductive layer *S* usually made from silicon or germanium, and a transparent metal layer *T*. Depending on the construction and the processing, an active junction arises between layers

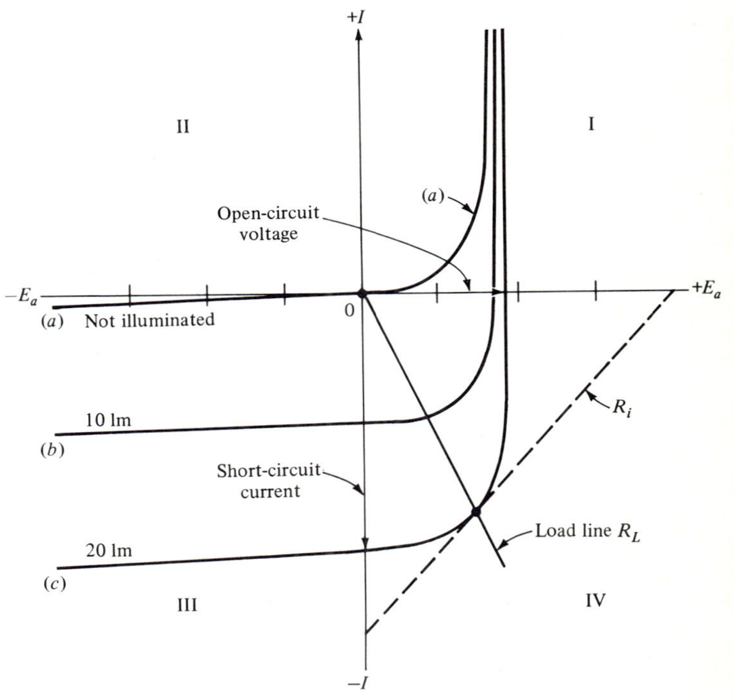

Fig. 11-24 Voltage-current characteristics of a photojunction cell; parameter, level of illumination.

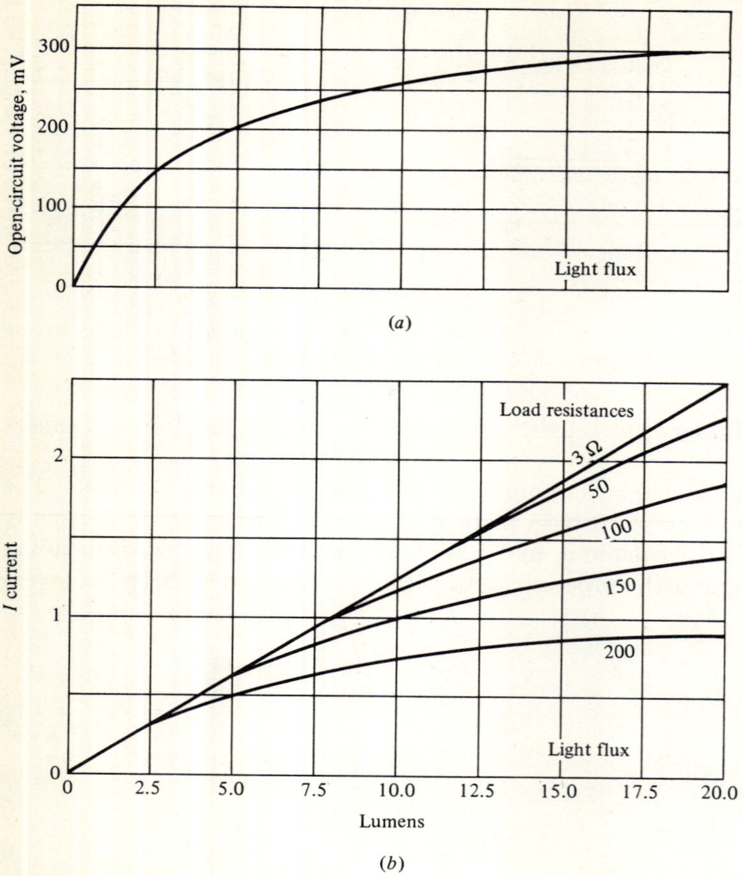

Fig. 11-25 (*a*) Open-circuit output voltage of a photovoltaic cell as a function of the incident light flux; (*b*) current output of the same cell for different load resistances.

S and *B* or *S* and *T*. A simple circuit for the operation of a photojunction cell is shown in Fig. 11-23*b*. The voltage-current characteristic of the system in the absence of illumination is nonlinear and has the character of a diode (see Fig. 11-24, curve *a*). If light falls on the cell, the characteristics change as shown in curves *b* and *c*. It should be noted that the output current $-I$ is not zero even if the applied voltage E_a is zero. These cells furnish an output current $-I$, which is approximately proportional to the incident light flow without the need of a bias voltage.

The open-circuit output voltage increases in approximately logarithmic fashion with the level of illumination (Fig. 11-25*a*) and is independent of the cell area. The short-circuit output current increases linearly with the level of illumination (Fig. 11-25*b*) and is proportional to the cell area. For a typical *n*-on-*p*-type silicon cell, the short-circuit current is of the order of 10 mA/lm.

The spectral sensitivity characteristic of the germanium *p*-*n* alloy photojunction cell is shown in Fig. 11-26*a*. The cell responds to wavelengths from the near-ultraviolet (200 nm) to the infrared region (about 2,000 nm) with a maximum sensitivity in the vicinity of 1,500 nm. Some cells are also sensitive in the x-ray region, as well as for incident α- and γ-rays and electrons. The x-ray sensitivity of the RCA-type cell SL 2205 is shown in Fig. 11-26*b* for an incident radiation of the 90-kV x-rays. As with the photoconductive cells, the spectral range in which a cell is sensitive can be extended into the infrared region by cooling. The

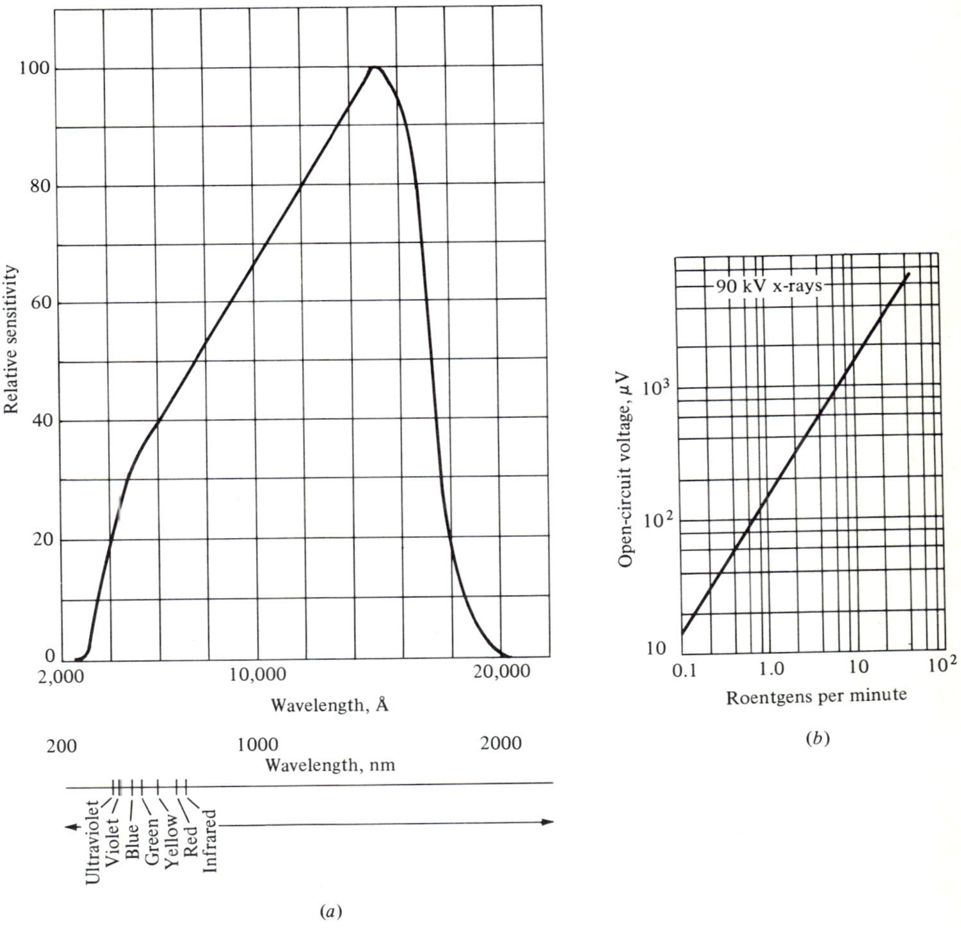

Fig. 11-26 (*a*) Spectral sensitivity characteristic of a *p*-*n* alloy photojunction cell; (*b*) x-ray sensitivity of a silicon photovoltaic cell. Incident radiation is 90-kV x-rays.

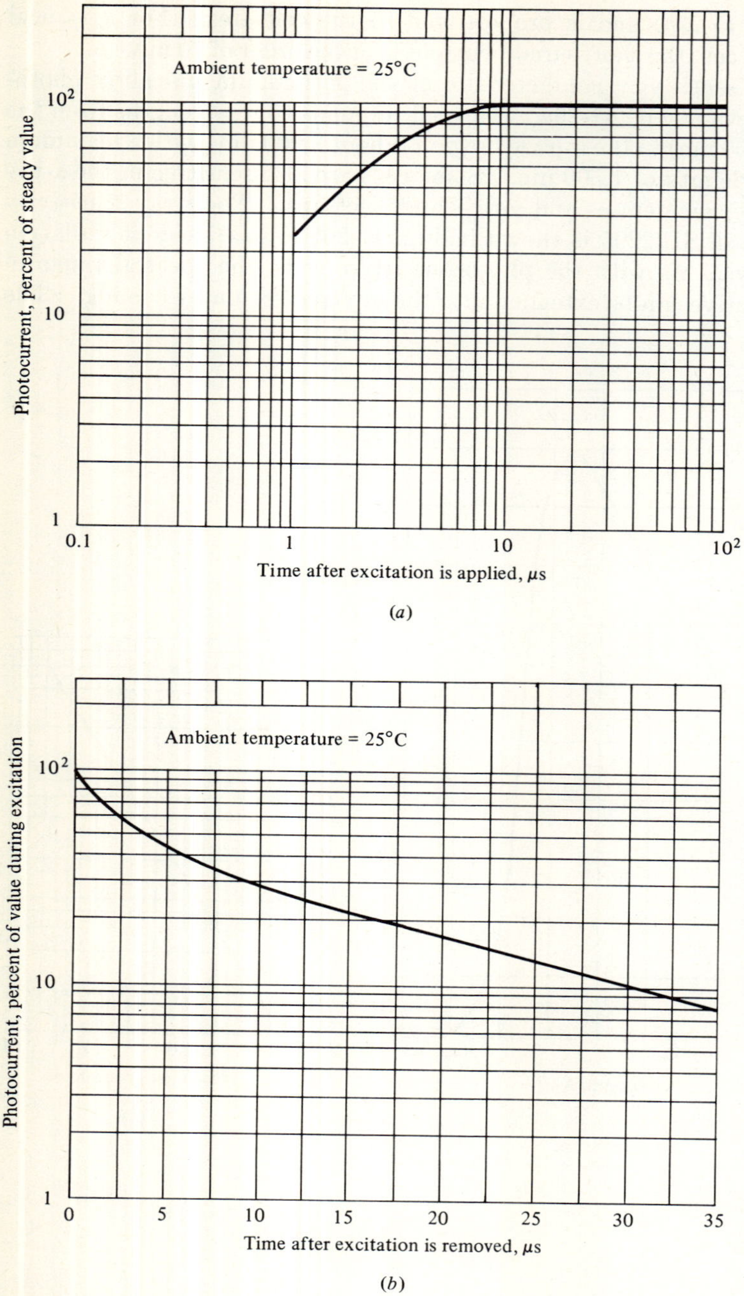

Fig. 11-27 Germanium *p-n* alloy photojunction cell. (*a*) Rise characteristic; (*b*) decay characteristic.

temperature range in which most of these cells operate satisfactorily is from -40 to $+50°C$. An increase of temperature raises the dark current. For instance, for the commercial *p-n* alloy photojunction cell, the dark current changes from 1 μA at 0°C to about 10 μA at 20°C.

N-on-*p* silicon (*photovoltaic*) cells operate satisfactorily in a temperature range from -100 to $+125°C$, i.e., far in excess of the range in which photojunction cells can operate. Temperature variation has little effect on the short-circuit current but may have a profound effect on the open-circuit voltage. The open-circuit voltage decreases rapidly when the cell temperature is raised.

The time response of the germanium *p-n* alloy photojunction cell is considerably faster than that of the photoconductive cell just described. As may be seen from Fig. 11-27*a* and *b*, the buildup time (i.e., the time delay between the application of a light excitation and the steady-state condition) and the decay time are of the order of *microseconds*. Therefore, photojunction cells of this type can be used for sound pickup from motion picture films; however, this response is still slow compared to that of photoemissive cells.

REFERENCES

"RCA Semiconductor Products Databook," RCA Corporation.

Walston, J. A., and J. R. Miller (eds.): "Transistor Circuit Design," McGraw-Hill, New York, 1963.

Cleary, J. F. (ed.): "Transistor Manual," General Electric Co., 1964.

Lion, K. S.: Semiconductor Photoelectric Devices in "Instrumentation in Scientific Research," McGraw-Hill, New York, 1959.

Semiconductor Electronics Education Committee series:

- Adler, R. B., A. C. Smith, and R. L. Lougini: "Introduction to Semiconductor Physics," vol. 1.
- Gray, P. E., D. DeWitt, A. R. Boothroyd, and J. F. Gibbons: "Physical Electronics and Circuit Models of Transistors," vol. 2.
- Searle, C. L., A. R. Boothroyd, E. J. Angelo, Jr., P. E. Gray, and D. O. Pederson: "Elementary Circuit Properties of Transistors," vol. 3.
- Thornton, R. O., D. DeWitt, P. E. Gray, and E. R. Chenette, "Characteristics and Limitations of Transistors," vol. 4.

Slater, Lloyd (ed.): "Bio-Telemetry," Symp. Pub. Div., Pergamon, Oxford, 1963.

Pohl, Robert W., "Elektrizitätslehre," Springer-Verlag, Berlin, 1957.

12
Amplifiers

BASIC AMPLIFIERS

The following description of amplifiers is presented in three parts. First, the basic amplifiers (*open-loop amplifiers*) are described to demonstrate the mechanism of amplification and introduce the principal amplifier characteristics. Second, amplifiers with feedback are described to show how feedback influences these characteristics. The third part deals with operational amplifiers, i.e., basic amplifier blocks which can be used for the implementation of almost any experimental amplifier or analog computer requirements by the application of external feedback circuits.

An amplifier, like any other instrumentation element, is determined by three groups of basic characteristics: input, transfer, and output characteristics. The input quantity applied to an amplifier is, in general, a voltage or a current, rarely a power. The output is a voltage, a current, or a power. The transfer characteristic is the gain and, depending on the input and output signals, a distinction is made among voltage, current, or power gain. When the input is expressed in voltage (E_i) and the output in current (I_o), the transfer characteristic frequently is given the name *transconductance* because the output/input ratio $I_o/E_i = G$ has the formal dimension of a conductance (see Eq. 9-6). Voltage,

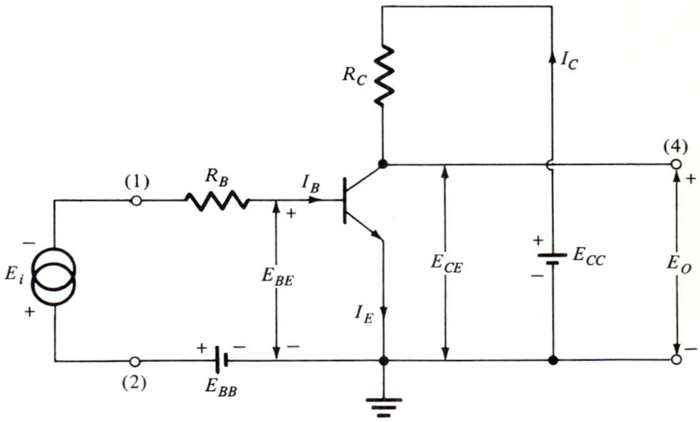

Fig. 12-1 Basic circuit diagram of a common-emitter amplifier, using an *n-p-n* transistor.

current, or power gain are by no means equivalent; an amplifier may have a voltage gain of less than one, while at the same time it may have a substantial current or power gain. Of course, the different types of gain may be calculated if, say, the voltage gain as well as the input and output impedances and the load impedance are known.

A circuit diagram for a single-stage common-emitter amplifier circuit, the most widely used transistor amplifier, is shown in Fig. 12-1 for an *n-p-n* transistor.

The signal E_i is applied to the input terminals 1 and 2; the amplified output E_o appears between terminal 4 and ground. The voltage E_{BB} provides a forward bias for the base-emitter junction, causing a current I_B in the base-emitter circuit. The voltage E_{CC} provides a reverse bias for the collector-base junction resulting in a current I_C in the collector circuit. The operating characteristics of a transistor are illustrated in Fig. 12-2*a*. Stable and safe operation of the transistor requires that the transistor be used within the "active region" of the characteristic, i.e., in the area enclosed by lines 1 to 5. Line 1 is the lower limit of the area where the transistor output curves are reasonably linear. Line 2 is the maximum allowable collector current. Curve 3 is the maximum rated value of power ($E_{CE} \times I_C$) which heats the transistor. Line 4 is the maximum allowable collector-to-emitter voltage. Line 5 is the minimum useful collector current, i.e., the current I_C for which I_B is zero.

Figure 12-2*a* is specific for a given transistor but not for a given operating condition. To find the operating condition in a circuit, it is necessary to select a collector-supply voltage E_{CC} (which must be smaller than that indicated by the limit line 4) and a load resistor R_C (which must be so large that the value E_{CC}/R_C does not exceed the limit line 2). Once these two parameters are chosen, one can construct the load line L by connecting point E_{CC} on the abscissa with point E_{CC}/R_C on the ordinate as shown in Fig. 12-2*a*. The prevailing

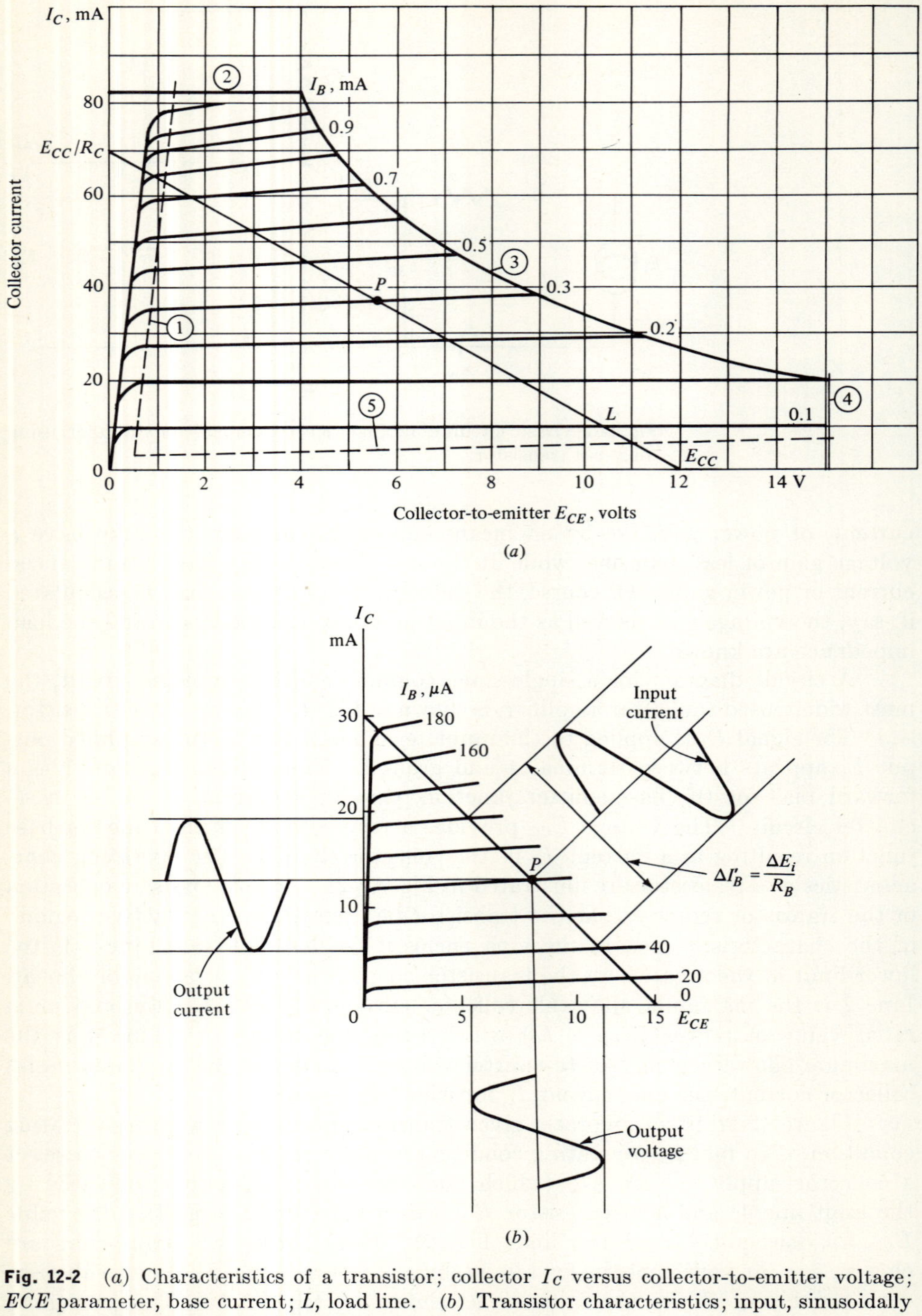

Fig. 12-2 (*a*) Characteristics of a transistor; collector I_C versus collector-to-emitter voltage; *ECE* parameter, base current; *L*, load line. (*b*) Transistor characteristics; input, sinusoidally varying base current; output current and output voltage.

operating condition in a circuit can be found from the point of intersection between the load line and the transistor characteristic for a given input (base) current I_B. The output voltage for a given input I_B can be read on the abscissa, the output current on the ordinate. Figure 12-2*b* shows the output voltage and the output current for an input signal I_B that varies sinusoidally with time. Amplifiers for this type of symmetrical signal are generally biased so that the quiescent operating point P (operating condition for an input signal equal to zero) is in the middle of the characteristic or slightly below the middle in order to reduce the average current and the demand for a high-current power supply.

Input and output impedance The input impedance for a signal applied to the input terminals is approximately R_B. The base-emitter junction resistance is of the order of a few hundred ohms or less and frequently is neglected. The output impedance is approximately R_C. The output resistance of the transistor, which is in parallel with R_C, is large (of the order of 100,000 Ω) and frequently can be neglected.

Gain The incremental voltage gain $\Delta E_o/\Delta E_i$ of a transistor amplifier may be found as follows. If the input signal is zero ($E_i = 0$), then

$$E_{BE} = E_{BB} - I_B R_B$$

or

$$I_B = \frac{E_{BB} - E_{BE}}{R_B}$$

provided $I_B > 0$. Since the transistor is operating in the active region,

$$I_C = \beta I_B \qquad \text{and} \qquad I_E = (\beta + 1) I_B$$

Thus,

$$\begin{aligned} E_o &= E_{CE} = E_{CC} - I_C R_C \\ &= E_{CC} - \beta I_B R_C \end{aligned}$$

The static parameters E_o and I_C establish a quiescent operating point for the amplifier circuit as expressed above and as indicated by the load line in Fig. 12-2*a*.

If a signal E_i is now injected, the current in the base changes to

$$I'_B = \frac{E_{BB} - E_{BE} + E_i}{R_B}$$

and E_o changes to

$$E'_o = E_{CC} - \frac{(E_i - E_{BE} + E_{BB})}{R_B} \beta R_C$$

Since E_{CC} and E_{BB} are fixed quantities and serve only to establish the operating point and since E_i and $E_i R_C/R_B$ are small compared to E_{BB} and E_{CC}, the incremental expressions become

$$\Delta I'_B = \frac{\Delta E_i}{R_B}$$

and

$$\Delta E'_o = \frac{\Delta E_i \beta R_C}{R_B}$$

Thus the incremental gain is

$$A = \frac{\Delta E_o}{\Delta E_i} = -\beta \frac{R_C}{R_B} \tag{12-1}$$

The minus sign indicates that there is 180° phase inversion between the input and output of the amplifier.

Nonlinear distortion If the signal applied to an amplifier is small, the transfer characteristic can be considered linear, and the gain is constant as expressed by Eq. (12-1). Nonlinear or amplitude distortion arises if the signal is large or if the operating point is located at a point in the I_C-E_{CE} diagram where the transistor characteristics are curved or unequally spaced. The distortion can be predicted analytically from the characteristics; it can be found experimentally by measuring the output voltages E_o for different input voltages E_i. Any deviation of the E_o-E_i function from a straight line indicates a nonlinear distortion. Figure 12-3*a* shows examples for the generation of amplitude-distorted signals. The curve f is the transfer function of a transistor amplifier. This function is linear in the center part only. A small sinusoidal signal a applied to the input in this region appears undistorted in the output as a signal a'; a large sinusoidal input signal b leads to a distorted output b'. An unsymmetrical distortion c' is observed if the input signal c is applied using an operating point close to the upper or lower curvature of the function f. Nonlinear distortion leads to the generation of harmonics if an alternating signal is applied. It can lead to intermodulation if a signal containing two or more frequencies is applied, i.e., the output contains, in addition to the applied frequencies, a frequency proportional to the sum or the difference of the applied frequencies. The amount of distortion can be measured with a harmonic analyzer (frequency analyzer).

Frequency distortion An ideal amplifier would amplify any signal with the same gain regardless of its frequency. The gain of practical amplifiers decreases with increasing signal frequency as shown in Fig. 12-3*b*. Two effects contribute to the loss of gain at higher frequency. First, the values α and β of a transistor depend on frequency because of the finite transit time of charges through the depletion layer. This effect is much more pronounced for β (therefore, for common-emitter circuits) than for α (common-base circuits). The gain of the common-emitter circuit begins to decline at frequencies above 10^4 Hz. It reaches the cutoff point $\beta_{\text{cutoff}} = 0.707 \times \beta_{\text{max}}$ between 10^4 and 10^6 Hz. It

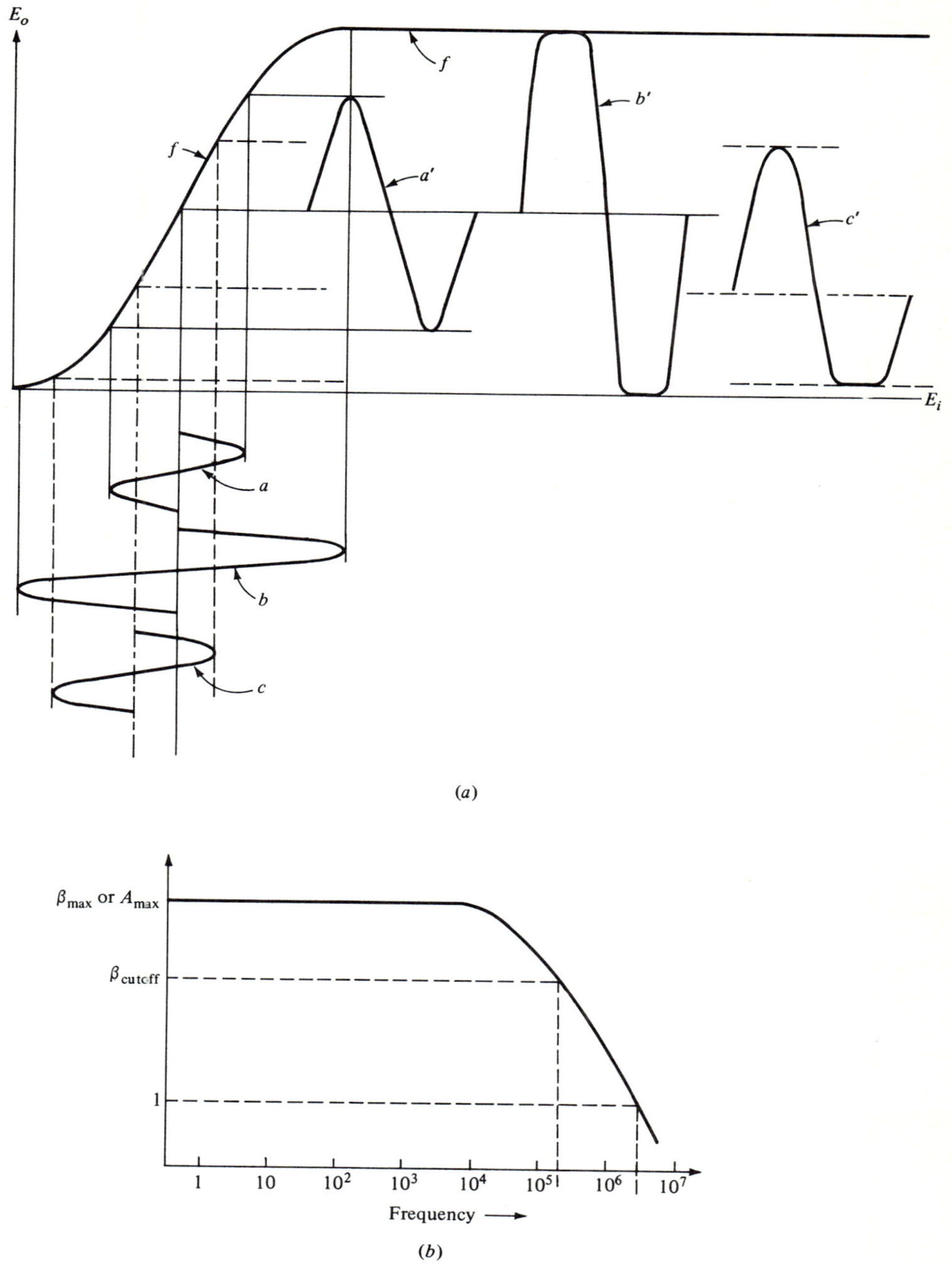

Fig. 12-3 (*a*) Nonlinear or amplitude distortion. *a*, small input signal; *a'*, undistorted output; *b*, large input signal; *b'*, symmetrically distorted output; *c*, unsymmetrical input signal; *c'*, unsymmetrically distorted output. (*b*) Frequency distortion.

decreases to unity at about 10^6 to 10^7 Hz or as high as 10^9 Hz depending on the type of transistor used. The frequency range of the common-base circuit is much larger; the cutoff point for α is of the order of 10^8 Hz. A second reason for the decline of gain at higher signal frequencies is the unavoidable stray capacitance between the components and the wiring in transistor amplifiers. This effect causes a shunt across the amplifier; it is more pronounced in multistage than in single-stage amplifiers. However, amplifiers with a gain of unity at a frequency of more than 100 MHz are commercially available.

Phase distortion Because of the presence of reactive elements (e.g., stray capacitance) in amplifiers, there is generally a phase shift between the input and the output signal of an amplifier. For many applications, this phase distortion is of no consequence; but it may have an influence on the operation of an amplifier if the output, or a part of it, is fed back to the input. In such cases phase distortion limits the frequency range over which the feedback mechanism is effective. Also, a phase shift which is different for the different components of a complex waveshape alters the waveshape. This condition is not disturbing if the amplifier is used for sound reproduction (because the acoustic response of the human ear is independent of the relative phase of harmonics) but may be disturbing for measurements.

Many circuits have been described for single-stage amplifiers, *n-p-n* or *p-n-p* transistors, low or high input and output impedances, dc or ac signals, amplifiers for operation at low frequencies (<1 MHz) and in the high-frequency range (order of 100 MHz), and for low-power or high-power amplification. The main disadvantages of a single-stage amplifier without feedback are limited gain, instability of the gain and operating point resulting in drift at the amplifier output, and low-power efficiency.

MORE COMPLEX OPEN-LOOP AMPLIFIERS

Cascade amplifiers If a gain is desired larger than a single amplifier stage can furnish, a multistage or cascaded amplifier may be used, i.e., the signal output of a first amplifier stage is again amplified by a second stage. If A_1 is the gain of the first stage and A_2 the gain of the second, then the overall gain A is equal to the product A_1A_2. Several methods of connecting stages are available: direct coupling and reactance coupling. A direct-coupled amplifier permits the amplification of dc and ac signals; a reactance-coupled amplifier permits amplification of ac signals only.

Figure 12-4 shows the equivalent circuit of a direct-coupled amplifier consisting of stages I and II. The input signal, E_i or I_i, is applied to terminals 1 and 2 and generally contains an ac and a dc component. The output signal from the first stage appears amplified at terminals 3 and 4. The amplified signal contains the same components as the input signal, but frequently also an additional fixed dc component E'_o which has its origin in the first amplifier stage.

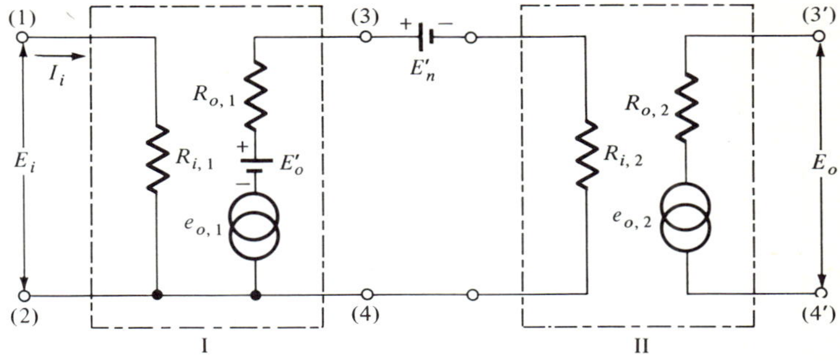

Fig. 12-4 Two-stage amplifier; dc coupled.

If the input signal is zero, the output signal is not zero but has the finite value E'_o, called *quiescent voltage*. The quiescent voltage may shift the operating point of the second stage to a level that is undesirable unless counteracted by a modified bias voltage. An equally undesirable effect may be caused by the quiescent voltage E'_o if a sensitive meter is connected to the output of an amplifier as shown in Fig. 12-5*a*. The voltage E'_o may overload the meter so that the variations of the amplified signal cannot be read. Reducing the sensitivity of the meter by a shunt or a series resistor will generally be of no help since such a procedure will reduce the sensitivity for the measurement of ΔE_o. The remedy is the insertion of a compensating voltage source E'_n equal to E'_o and of a polarity opposite to E'_o so that the last two terms in the equation of Fig. 12-5*a* cancel each other. There are several means to produce the compensating voltage E'_n: by a separate source (as shown) or by deriving it from the bias supply of the amplifier with a resistive voltage divider.

Examples of practical two-stage amplifiers are shown in Fig. 12-5*b* and *c*. The circuit of Fig. 12-5*b* uses two *n-p-n* transistors. The resistances R_3 and R_4 are dimensioned so as to produce the proper bias for the second stage, which includes compensation for the quiescent voltage E'_o at the output from the first stage. The circuit of Fig. 12-5*c* uses an *n-p-n* transistor followed by a *p-n-p* transistor, a combination which may be followed by further similar systems of *n-p-n*/*p-n-p* couples.

While this compensation of E'_o provides for the proper bias of the second stage, it does not compensate for any *variation* of E'_o or E'_n and its effect on the cascade amplifier. If the voltage E'_o is of the order of several volts and if the dc component of the signal at the output from the first stage is of the order of microvolts, then both the voltage E'_o and the compensating voltage E'_n must be constant by better than 1 part in 10^6 or else the output variation due to the signal and that due to variation of $E'_n + E_o$ are indistinguishable. That means that the lower limit for which dc amplifiers can be used is determined by the amount of drift and is, with good operational amplifiers of the order of 10 to 100 μV, within a temperature range of 30 to 40°C.

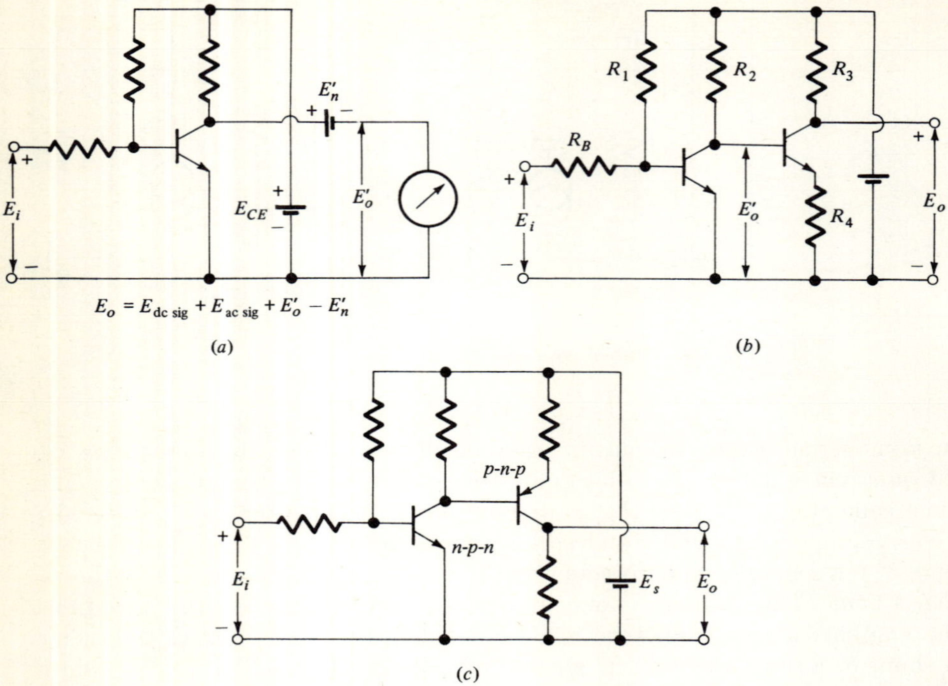

Fig. 12-5 (*a*) Amplifier with quiescent current compensation by means of a voltage source E_n'. (*b*) Two-stage transistor amplifiers, direct coupled; two *n-p-n* transistors. (*c*) Two-stage transistor amplifier, direct coupled, *n-p-n*/*p-n-p* operation.

If the signal has a pure ac character or is converted into an ac signal by chopping or modulation (see below), reactance methods can be used that are substantially free from the described difficulties. The most frequently used interstage-coupling method for ac signals is a resistance-capacitance network illustrated in Fig. 12-6*a*. The input signal E_i causes a potential variation at the output of the first amplifier stage, point P_1. The potential variation of P_1 is applied to the input of a second stage via a capacitive-resistive network $C_C R_2$. R_2 represents the input resistance of stage II. The frequency characteristic of the CR network has been treated in Chap. 5 and is redrawn in Fig. 12-6*b*, curve *a*. Any low-frequency variation of the potential at P_1 is not transferred to the second stage; therefore, any slow variation caused by drift at any part of the circuit to the left of the capacitor C_C is suppressed in the output signal. The RC-coupled amplifier also shows attenuation of the gain at higher frequencies for reasons stated in connection with Fig. 12-3 so that the overall frequency characteristic is that of Fig. 12-6*b*, curve *B*.

Another circuit with reactive coupling between amplifier stages, the transformer-coupled amplifier, is shown in Fig. 12-7. Transformers can be built with a wide variety of input/output ratios and impedances. They permit

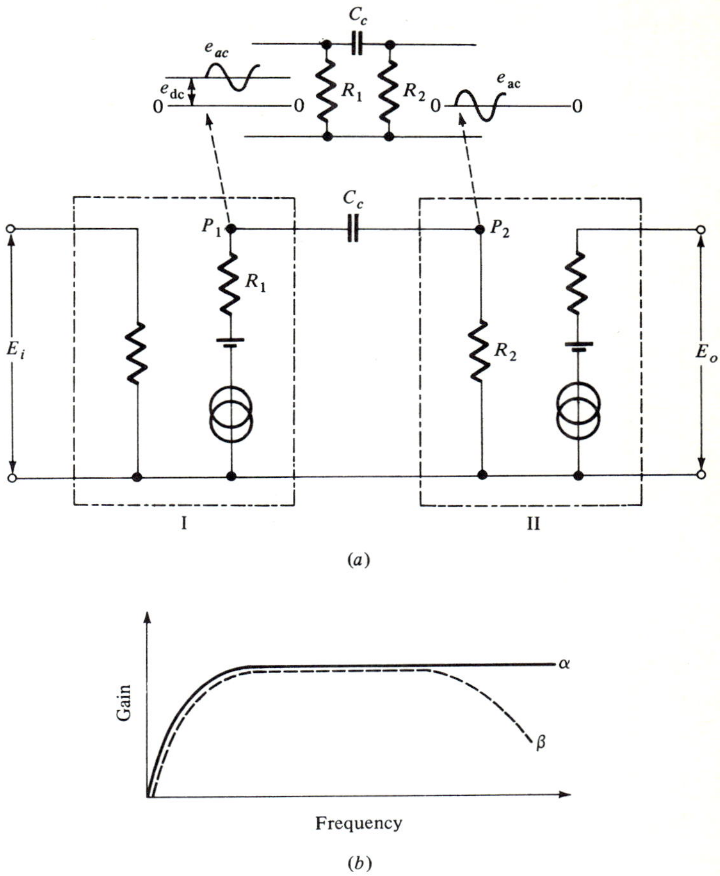

Fig. 12-6 (*a*) Two-stage *RC*-coupled amplifier (for ac operation). Insert, *RC*-coupling network. (*b*) Frequency characteristics of *RC*-coupled amplifiers, see text for explanation.

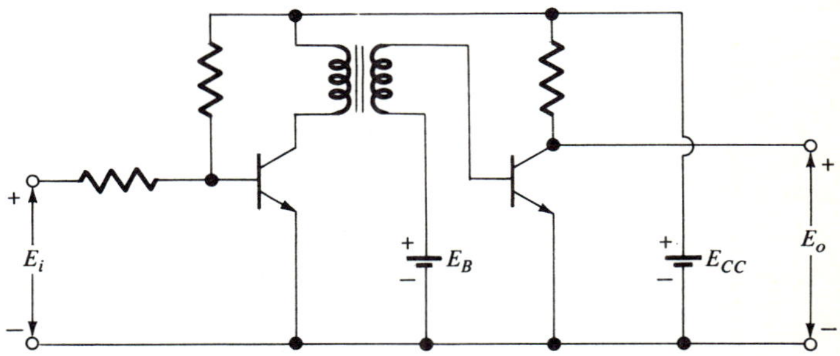

Fig. 12-7 Two-stage transformer-coupled amplifier (for ac operation).

impedance matching as well as coupling without passing direct current. However, they may cause magnetic stray fields and they are sensitive to external magnetic stray fields. Therefore, their use may lead to hum (pickup of line-frequency magnetic stray fields) and to undesirable feedback conditions and instability. Transformers frequently are used in the output of amplifiers and as coupling elements between transistors and power vacuum tubes. They are used less frequently as coupling elements between transistor stages.

There are other methods of interstage coupling, for instance, resonance circuits which may be tuned to a single frequency or those which employ band filters to operate within a limited bandwidth. Other interstage coupling systems employ nonlinear elements, diodes or glow lamps, but their use is limited to special applications.

Differential or balanced amplifiers Figure 12-8*a* shows a simplified circuit diagram of a difference or balanced amplifier. The amplifier has two transistors which should be well matched so as to have nearly equal operating characteristics and temperature coefficients. The same bias is applied to both transistors and the same collector currents pass the two equal resistors R_{C1} and R_{C2}. Any remaining small difference between both halves of the circuit can be balanced out by adjusting the potentiometer R so that for $E_i = 0$, output voltage $E_o = 0$. Because of the symmetrical construction, the compensation is nearly perfect and drift is reduced considerably. Signal voltages E_i may be applied between terminal 1 and ground, between terminal 2 and ground, or between terminals 1 and 2. The circuit frequently is used for meter amplifiers (so-called volt-ohm-meters).

If the signal is applied between the input terminals 1 and 2 with the polarity as shown, the resistances R_1 and R_2 act as a voltage divider so that half the input is applied to one transistor and raises the base potential, while the other half of the voltage is applied to the other transistor, lowering its base potential by the same amount. The difference of potential between both bases appears amplified at the output terminals 3 and 4. The output may be applied to a load, a meter or recorder, or it may be coupled directly or by means of an RC network to a further stage.

A different result is obtained if two voltages of equal magnitude and of the same polarity are applied, one between terminal 1 and ground, the other between terminal 2 and ground. The application of two voltages with equal polarity or, if alternating current, of equal magnitude and phase (*common-mode application*), causes a change of potential at either output terminal 3 or 4 by the same amount and in the same direction or polarity so that the potential *difference* in the output remains zero. Two signals applied in common mode to a difference amplifier do not produce an output signal (they are "rejected").

This *common-mode rejection* feature of the balanced amplifier is particularly useful for the elimination of spurious signals caused by electric stray fields when shielding of the experimental object is difficult, such as in electrocardiography. The equivalent circuit is illustrated in Fig. 12-8*b*. Two electrodes are attached

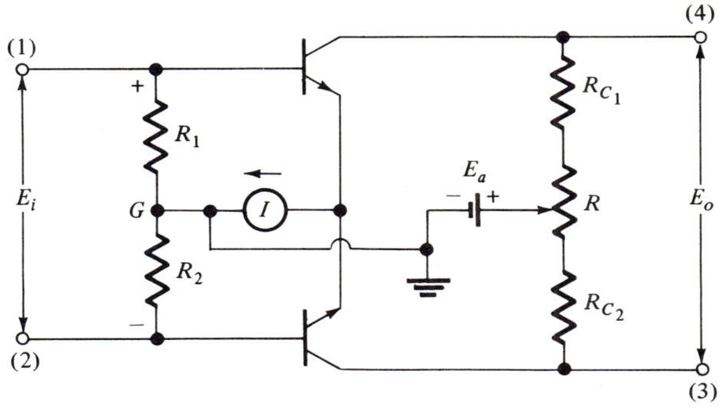

(a)

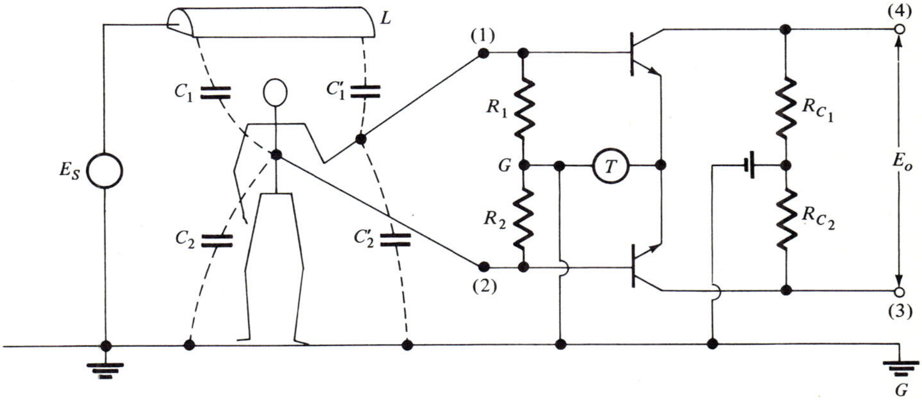

(b)

Fig. 12-8 (*a*) Circuit diagram of a differential amplifier. *I*, constant-current source. (*b*) Example of common-mode rejection by means of a differential amplifier.

to the subject: a chest and a wrist electrode. The electrocardiac potential between them is of the order of 10 to 30 mV and oscillates in a typical fashion with a frequency of the order of 1 Hz. This voltage is applied in a differential mode to the input terminals 1 and 2 of the balanced amplifier and, therefore, appears amplified in the output. The subject is surrounded by an electric stray field caused by a line-voltage source E_s, which may be grounded on one side and connected to a metallic object, such as a fluorescent light fixture L, on the other side. Such an electric stray field causes the induction of a voltage at the electrodes equivalent to the applied voltage between L and ground (the line voltage, e.g., 115 V 60 Hz) and the partial capacitances C_1, C_2 and C_1', C_2', which form a voltage-dividing network. The equivalent circuit shows that a spurious ac voltage arises between the chest electrode and ground and another one between the wrist electrode and ground. These two voltages are approximately alike and they are, of course, of the same polarity (phase). They may be of the order of 10 V, and thus they may exceed the signal (cardiac) potential by several orders of magnitude. If the center point G of the amplifier is grounded, it is seen that one of these voltages appears between terminal 1 and ground, the other one between terminal 2 and ground. Thus, they are applied to the amplifier in common mode, and ideally they will not appear (or appear greatly reduced) in the output of the amplifier. A figure of merit for the performance of a balanced amplifier exposed to a common-mode signal and a differential-mode signal is the common-mode rejection ratio which is somewhat loosely defined as the ratio of the common-mode signal at the input divided by the common-mode signal at the output of a balanced amplifier. Balanced amplifiers have been built with common-mode rejection ratios exceeding a value of 10^6.

Class A, B, C operation; push-pull amplifiers Amplifiers for the linear amplification of symmetrical signals are usually operated so that the quiescent point is in the middle of the transistor characteristic as described above and as illustrated in Figs. 12-2 and 12-9*a*, point A. This mode of operation is called *class A operation.* If the signal varies in one direction, e.g., from zero to a positive maximum, it can be more advantageous to shift the operating point to a lower I_c value as shown in Fig. 12-9*a*, point B. Such *class B operation* allows a range of input and output signals over almost twice the range available with class A operation. The class B circuit also reduces the current I_c for the no-signal condition and, therefore, makes less demand on the bias sources and power supplies.

It is also possible to employ class B operation for symmetrical signals if used in connection with a push-pull amplifier as illustrated in Fig. 12-9*b*. The input signal E_i is divided into two equal partial voltages E_i' and E_i'', either by a resistive voltage divider or by a center-tapped transformer as shown. If the input voltage swings to the positive maximum, as indicated by the polarity signs, transistor Q_1 will be conductive and, provided operation takes place in the linear region of the characteristics, a positive amplified half-wave will appear at the output. During this half-cycle the transistor Q_2 will be negatively biased and will block any flow of current. During the next half-cycle the transistor action is reversed, the transistor Q_2 produces in the output a negative half-wave. If the two half-waves are added, a complete sine wave will be produced in the output, fairly free of nonlinear distortion. Figure 12-10*a* is a graphic presentation of the operation of a push-pull amplifier.

Figure 12-10*b* shows the characteristic of a single-stage amplifier with the operating point C very negative. If the input signal is large enough to extend into the positive region (cross-hatched area), a short positive pulse will arise in the output. Such *class C amplifiers* can be used, for example, in oscillators; they permit the periodic excitation of oscillations in a tuned LC circuit.

Vibrating reeds, choppers, and chopper amplifiers It has been stated that the amplification of small dc signals causes difficulties because the inherent instability (*drift*) of a dc amplifier gives rise to an output signal which is indistinguishable

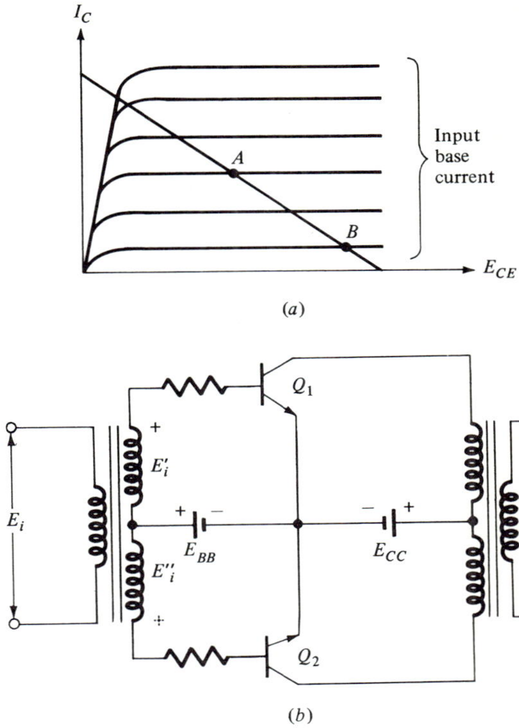

Fig. 12-9 (*a*) Transistor characteristics and load line. *A*, operating point for class A operation; *B*, for class B operation. (*b*) Circuit diagram of a push-pull amplifier with two transistors, Q_1 and Q_2.

from that caused by the small dc input signal. If the dc signal is of the same order of magnitude or smaller than that caused by drift, it is preferable to convert the dc input signal to ac, amplify with an ac amplifier, and demodulate the output into dc. The conversion of a dc signal into ac can be done with a vibrating reed or with a chopper.

The vibrating reed is a variable capacitor as shown in the schematic diagram Fig. 12-11. Its operation is similar to that of the capacitive microphone modified so that instead of the voltage E being constant and the frequency and amplitude of the capacitance variation between a membrane and a fixed electrode changing, the capacitance variation caused by a driving mechanism D proceeds at a constant rate and excursion and the voltage E_i, i.e., the dc input signal, changes. The instantaneous voltage e_i furnished from the dc input E_i via the vibrating-reed capacitor to the input of the amplifier is

$$e_i = E_i R \frac{dC}{dt}$$

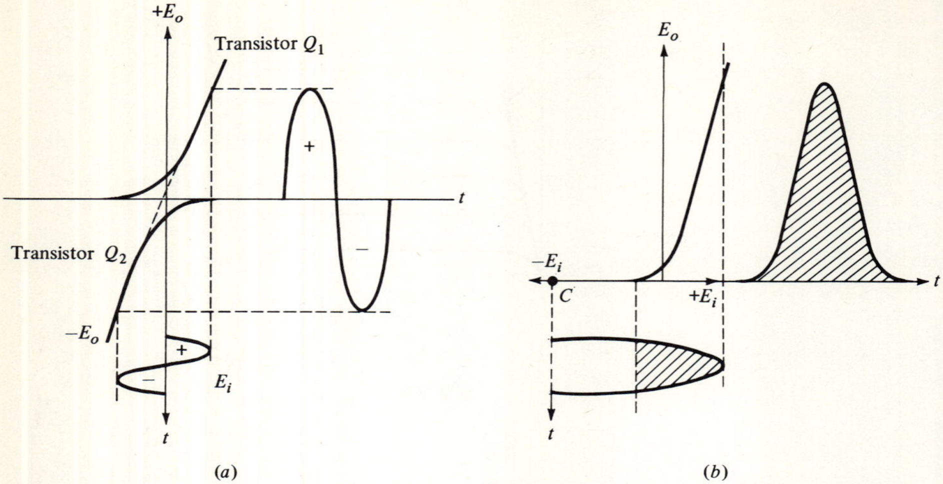

Fig. 12-10 (a) Transfer characteristics of two transistors I and II, in class B mode; (b) illustration of a single-stage-transistor amplifier in class C operation mode.

provided that the variation of E_i is slow compared with the variation of the capacitance [that is, $C(dE_i/dt) \ll E_i(dC/dt)$]. The capacitor C is charged when its capacitance increases and is discharged by an equal amount when its capacitance decreases, so that there is no loss of charge and, for a full symmetrical cycle of the capacitance variation, no loss of power from the source. The driver D furnishes the energy or power for the dc-to-ac conversion. Therefore, vibrating-reed systems are frequently called *electrometer amplifiers.*

Another device for the conversion of the dc input signal into an ac signal is the chopper, schematically illustrated in Fig. 12-12a. The chopper is a mechanical switch actuated continuously by a magnetic-driver system (not shown). It closes alternatively the contacts M and G or N and G. In position M, the input signal is grounded and thus reduced to zero. In position N, the output signal is grounded and reduced to zero. The sequence of events in this system is shown in Fig. 12-12b. The dc input signal, assumed to be +1 mV, appears between point A and ground. Through the action of the modulating contacts MG, the signal is suppressed during half of the chopper cycle; this signal appears between point B and ground. If the amplifier is not inverting and has a gain of 1,000, the amplifier output signal between point C and ground is in phase

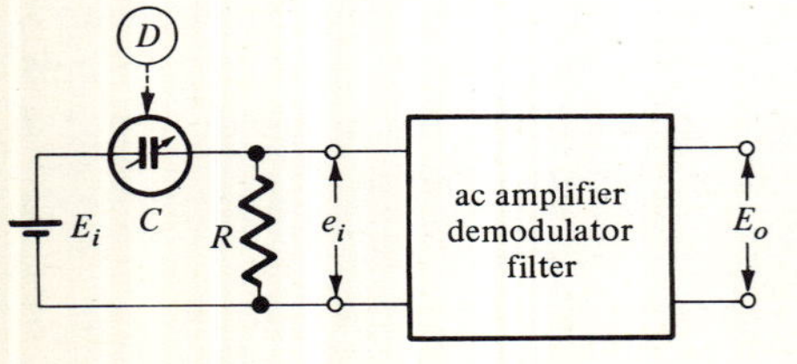

Fig. 12-11 Dc amplification with vibrating reed C and driver D. The dc input E_1 is converted to ac, amplified with an ac amplifier and reconverted to dc afterward.

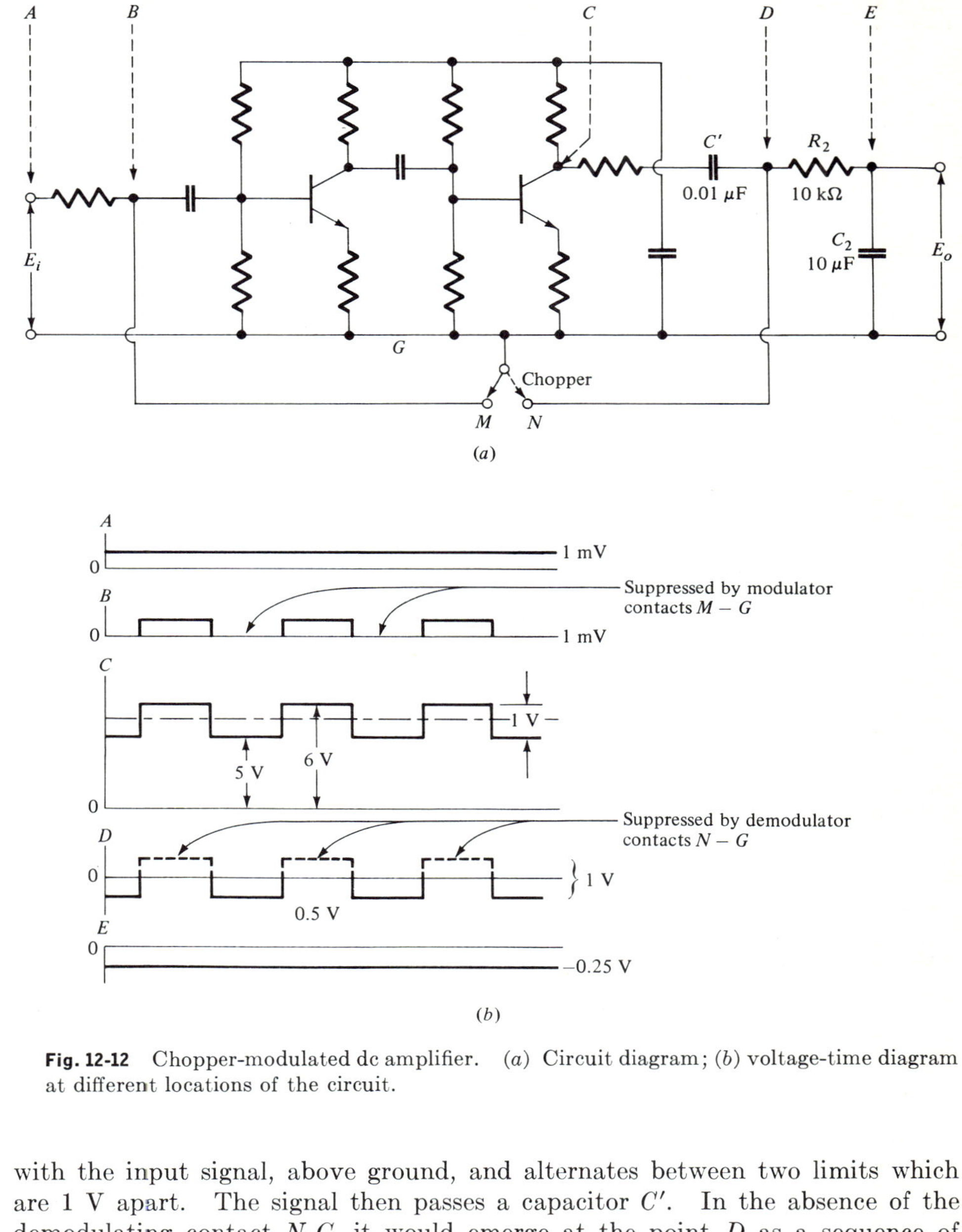

Fig. 12-12 Chopper-modulated dc amplifier. (*a*) Circuit diagram; (*b*) voltage-time diagram at different locations of the circuit.

with the input signal, above ground, and alternates between two limits which are 1 V apart. The signal then passes a capacitor C'. In the absence of the demodulating contact N-G, it would emerge at the point D as a sequence of square waves alternating between the values $+500$ and -500 mV. Through the effect of the demodulating contacts N-to-ground, the positive part of the square wave is suppressed so that the resulting square wave alternates between -500 mV and ground. The signal then passes an average forming (integrating) network consisting of a large resistor R_2 and a large capacitor C_2 and appears at the output terminals as a 250-mV dc signal.

The circuit shown in Fig. 12-12*a* can be used for the measurement of small dc voltages with a gain of 250. It can also be used as an auxiliary device for the reduction of drift effects in operational amplifiers (see below). The disadvantage of the chopper amplifier is generally its poor high-frequency response. Mechanical choppers driven at 60 or 400 Hz can be used for signals varying at a frequency from zero to several tens of hertz. Solid-state choppers (ac-modulated transistors) can be used up to frequencies in the kilohertz range. The maximum operating frequency is limited by the average forming network R_2C_2 in the amplifier output. Chopper-amplified amplifiers also are likely to introduce noise of the chopper frequency into the output. The theoretical limit of the signal frequency is half that of the chopper frequency.

FEEDBACK SYSTEMS

The characteristics of any amplifying system can be changed by feeding a part of the output signal back to the input. A mechanical example of a feedback system is shown in Fig. 12-13. The circuit contains a motor M mechanically coupled to the movable contact arm of a potentiometer P. If an input signal E_i with the indicated polarity is applied to the motor, the motor will rotate in the direction of the arrow *a*. The rotation will cause an increase of the potentiometer output E, which is applied to the amplifier input terminals 1 and 2. The output signal E_o from the amplifier is connected to a voltage divider B as well as to a meter V. The output from the voltage divider is $E_f = kE_o$, where $0 < k < 1$. For instance, k may be 0.01, i.e., E_f is 1 percent of the output signal E_o. The voltage E_f is fed back to a summing network S in a polarity such that the voltage across the motor is

$$E_M = E_i - E_f \tag{12-2}$$

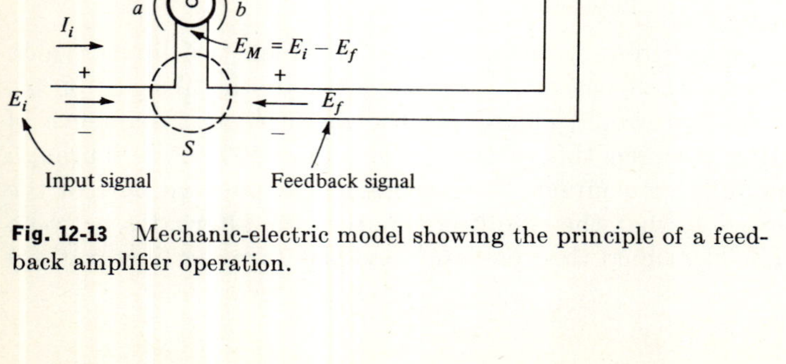

Fig. 12-13 Mechanic-electric model showing the principle of a feedback amplifier operation.

(negative, inverse, or degenerative feedback). If $E_f > E_i$, the motor will turn in the direction of the arrow b. The system comes to equilibrium and the motor stops rotating if $E_M = 0$, i.e., if $E_i = E_f = kE_o$. The gain of the amplifying *system* is then

$$A_f = \frac{E_o}{E_i} = \frac{E_o}{kE_o} = \frac{1}{k}$$

i.e., the gain is determined only by the setting of the voltage divider B. In the example mentioned above, the gain is $1/0.01 = 100$, provided the gain A of the amplifier without feedback is large enough to produce at the output the voltage E_o. The gain A_f of the feedback amplifier *system* is less than A, the gain of the "open-loop" amplifier, but it is stable and independent of any change of parameters within the amplifier, such as a variation of the power-supply voltage or the aging of the components.

The input impedance of the feedback system is E_i/I_i by definition. If the input voltage is applied directly to the motor, a rather large input current I_i will flow. However, in the illustrated system the input voltage E_i is counteracted by the feedback voltage E_f which, at balance, is equal to E_i and of the same polarity, so that I_i is zero and the system has the effect of an infinite input impedance. It follows that inverse feedback increases drastically the input impedance of an amplifying system.

The output impedance of any system can be measured by applying a load resistance to the output terminal and observing a drop of the output voltage (Fig. 2-4). If an (additional) resistance R'_L is connected in parallel with the meter V, the meter will temporarily indicate a drop of E_o. The decrease of E_o will cause a decrease of E_f and, in further sequence, a disturbance of the balance condition Eq. (12-2). The motor will move until equilibrium is reestablished, and the system produces the effect of an infinitely small output impedance. It follows that inverse voltage feedback drastically decreases the output impedance of an amplifier system.

The physical phenomena and relations described in the preceding paragraphs can be stated more generally and in quantitative terms in connection with the purely electric feedback system shown in Fig. 12-14*a* and *b*. The system contains an amplifier of gain A. An input signal E_i is applied to the system. The amplifier furnishes an output E_o of which a fraction k is fed back to the amplifier input in a negative sense, i.e., in polarity such that the difference $E_i - E_f = E$ appears at the amplifier input terminals.

The gain of the amplifier is

$$A = \frac{E_o}{E}$$

The input voltage to the system is

$$E_i = E + E_f = \frac{E_o}{A} + kE_o$$

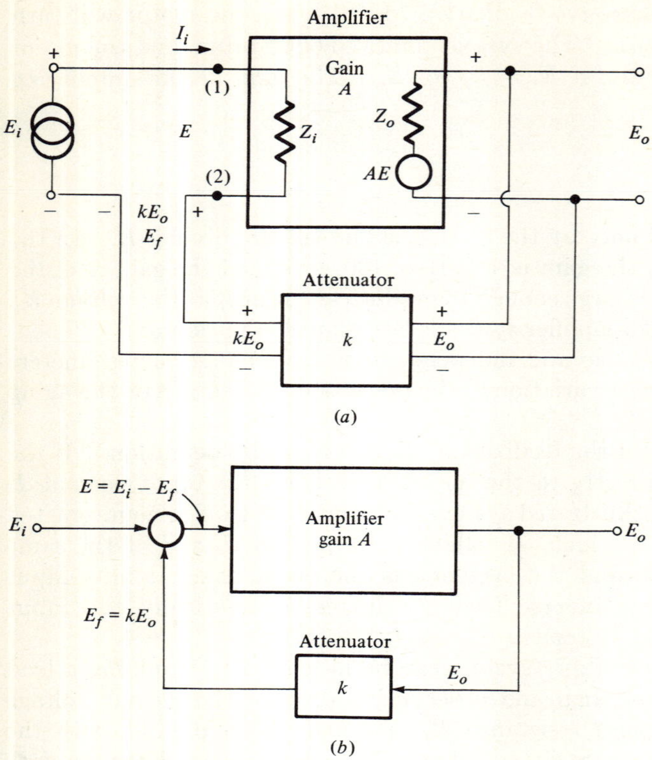

Fig. 12-14 Feedback amplifier. (*a*) Circuit diagram; (*b*) schematic diagram.

The output voltage is E_o.

The gain of the *system* (amplifier plus feedback) is

$$A_f = \frac{E_o}{E_i} = \frac{E_o}{E_o/A + kE_o} = \frac{A}{1 + kA}$$

If the gain A of the amplifier is very large, so that $kA \gg 1$, then

$$A_f = \frac{A}{kA} = \frac{1}{k}$$

i.e., the gain of the system (amplifier plus feedback) is stabilized. It is independent of the open-loop gain A of the amplifier or of any variation of A. When the gain A or the amount kA fed back is large, the signal E appearing at the amplifier input terminals is very small.

By definition the input impedance is $Z_{i,f} = E_i/I_i$. Since

$$E_i = E_o\left(\frac{1}{A} + k\right)$$

and since

$$I_i = \frac{E}{Z_i} = \frac{E_o}{AZ_i}$$

it follows that

$$Z_{i,f} = \frac{E_o(1/A + k)}{e_o/AZ_i} = Z_i(1 + kA)$$

i.e., the input impedance of an amplifier system with negative feedback is always higher by a factor of $(1 + kA)$ than the input impedance of the same amplifier without feedback.

The output impedance of the system is defined as $Z_{o,f} = E_o/I_o$. According to Fig. 12-14a,

$$E_o = AE - I_oZ_o$$

For $E_o = 0$, the signal appearing at the amplifier input is $E = kE_o$, so that

$$E_o = kE_o - I_oZ_o$$

It follows that

$$Z_{o,f} = \frac{AkE_o - I_oZ_o}{I_o} = AkZ_{o,f} - Z_o$$

hence,

$$Z_{o,f} = \frac{Z_o}{1 + kA}$$

i.e., the output impedance of an amplifier system with negative feedback is always lower by a factor of $1/(1 + kA)$ than the output impedance of the open-loop amplifier.

In general, the output of an amplifier is a linear function of the input only over a limited range. If the amplitude of the input signal exceeds the linear range, nonlinear distortion or amplitude distortion sets in, i.e., a small signal is amplified with a gain A different from that of a large signal. Since in a feedback amplifier with a loop gain $kA \gg 1$ the gain is independent of A and only depends on the constant feedback attenuation factor k, amplitude distortion is practically eliminated or highly reduced. Another way to understand this effect is the following. If a signal E_i is applied to the input of a system (amplifier plus feedback), it is instantaneously balanced by a feedback signal E_f, which is almost equal to E_i, so that the voltage E between the input terminals is small. An amplitude variation ΔE_i of considerable magnitude causes an amplitude variation ΔE at the amplifier which is well within the linear range so that amplitude distortion is practically eliminated.

The frequency response of an amplifier is also affected by negative feedback.

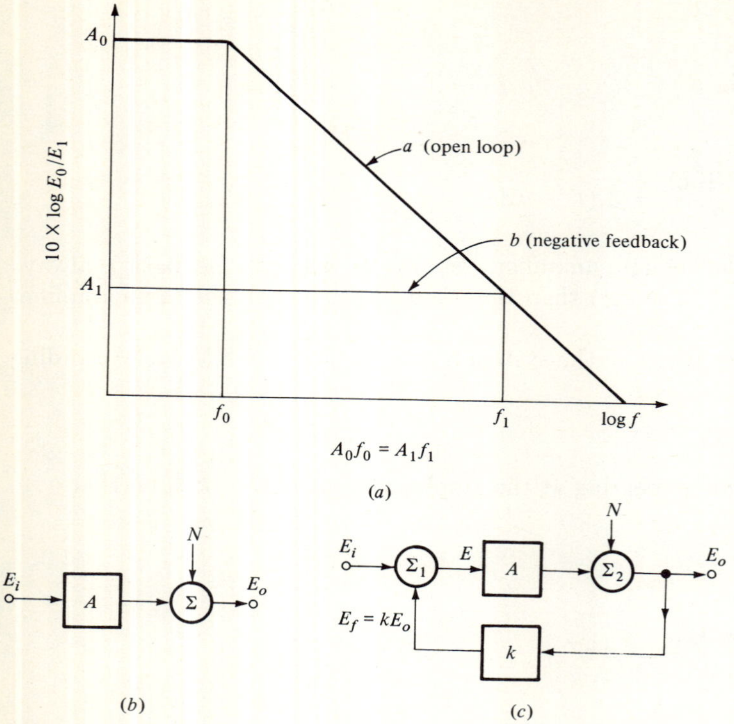

Fig. 12-15 Frequency characteristic of an amplifier illustrating the improvement of the frequency response through feedback. (*a*) Open-loop amplifier A and noise source N; Σ, summation point; (*b*) feedback amplifier and noise source; Σ_1 and Σ_2, summation points; (*c*) schematic diagram.

Figure 12-15*a* shows the frequency response curve *a* for an open-loop amplifier. If negative feedback is applied (curve *b*) the gain is reduced, but the frequency response at this lower gain extends over a considerably wider region than that obtained without feedback.

Noise (such as thermal noise, shot noise, or spurious signals generated in an amplifier) appears at the output of an open-loop amplifier as shown in Fig. 12-15*b*; the output is $E_o = AE_i + N$, where N is the noise voltage. The effect of negative feedback on the noise level or the signal-to-noise ratio is illustrated in Fig. 12-15*c*. The voltage appearing at the amplifier input terminals is

$$E = E_i - kE_o$$

The output signal is

$$\begin{aligned} E_o &= AE + N \\ &= A(E_i - kE_o) + N \\ &= \frac{1}{1 + kA}(AE_i + N) \end{aligned}$$

for $kA \gg 1$,

$$E_o = \frac{E_i}{k} \pm \frac{N}{kA}$$

For example, if $A = 100$ and $k = 0.1$, the signal component S in the output is $10E_i$ and the noise component is $N/10$; the signal-to-noise ratio S/N is 100. Without feedback, using an amplifier of a gain $A = 10$ (Fig. 12-15c), the signal component would again be $10E_i$ but the noise component would be N; the signal-to-noise ratio S/N would be only 10.

While the application of negative feedback strongly reduces any noise that originates within the amplifier, it has no effect on noise that originates in a stage preceding the amplifier, and it does not improve the signal-to-noise ratio inherent in an applied input signal.

In the foregoing examples, it was assumed that the attenuation k was constant and independent of frequency, such as obtained if the attenuating element k is a resistive voltage divider. If reactive elements or filters are used as attenuating elements in the feedback loop, the attenuation varies with the frequency of the input signal and so does the gain of the amplifier. For instance, if the element k contains a filter which only permits the passage of a narrow frequency band, then the gain of the amplifier is reduced for this band while signals of frequencies outside this band are amplified with the full gain A of the amplifier. Such devices can be useful if, in the course of a measurement, an undesirable frequency band is to be suppressed (e.g., 60 Hz). On the other hand, if the element k has a characteristic to pass signals of all frequencies with the exception of a narrowband, such as a parallel LC resonance circuit, then the amplifier will be frequency selective; it will amplify with high gain only signals of that frequency which are blocked by the filter.

An example of a practical feedback system is shown in Fig. 12-16a. Feedback is accomplished by the resistor R_B. The amplifier is inverting, so that the feedback effect of R_B is negative or degenerative. The input current is I_i; therefore, the input voltage is[1]

$$E_i = I_i R_a$$

The collector current is $I_c = \beta I_B$ and the output voltage is

$$E_o = I_f R_b = -(\beta I_B + I_f) R_L \tag{12-3}$$

The base current is the sum of the input current I_i and the feedback current I_f

$$I_B = I_i + \frac{E_o}{R_B} = \frac{E_i}{R_a} + \frac{E_o}{R_B} \tag{12-4}$$

Substituting Eq. (12-4) into Eq. (12-3) furnishes

$$E_o = -\beta\left(\frac{E_i}{R_a} + \frac{E_o}{R_B}\right) R_L - \frac{E_o}{R_B} R_L$$

[1] This and the subsequent derivations are simplified by assuming the resistance and the offset voltage in the base-emitter junction to be zero. The base-emitter junction is forward-biased and thus has a low resistance. The offset voltage is of the order of 0.6 V.

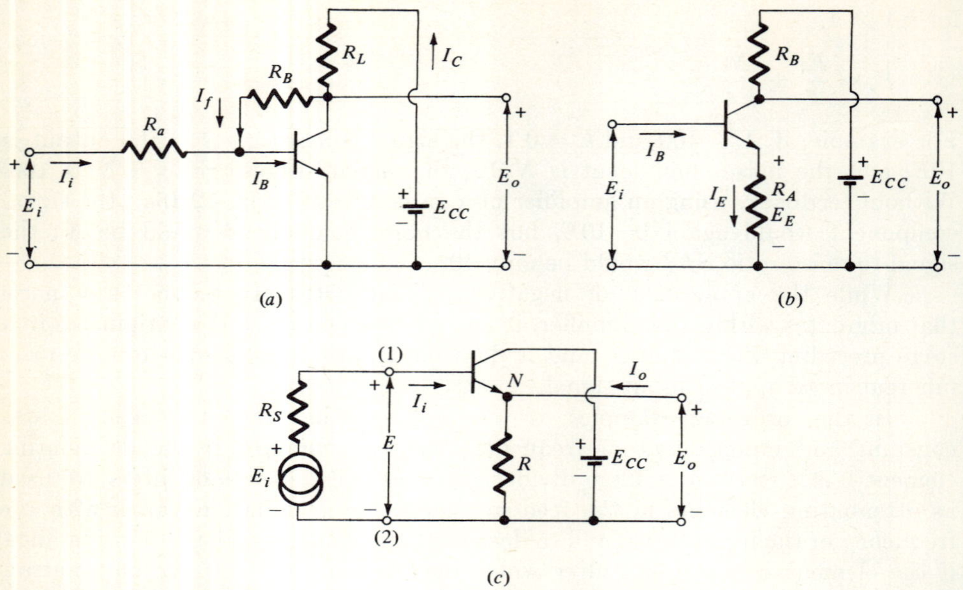

Fig. 12-16 Practical circuit diagram of a transistor single-stage feedback amplifier. (*a*) Feedback is accomplished by means of the resistor R_b; (*b*) circuit diagram of a transistor feedback amplifier with feedback resistor R_A in the common-emitter line; (*c*) circuit diagram of an emitter follower.

Solving for E_o/E_i,

$$A_f = \frac{E_o}{E_i} = -\frac{\beta R_L}{R_a}\frac{R_B}{R_B + R_L(\beta + 1)}$$

If $\beta R_L \gg R_B$, the gain approaches

$$A_f = \frac{R_B}{R_A}$$

i.e., the gain depends only on the ratio of the resistances R_B/R_A and is independent of the gain β of the transistor. The input resistance to the amplifier is R_A; the output resistance is

$$R_o = \frac{R_B R_L}{R_B + R_L(\beta + 1)}$$

If $\beta R_L \gg R_B$, then

$$R_o = \frac{R_B}{\beta}$$

A different type of feedback amplifier is shown in Fig. 12-16*b*. The feedback resistor in this circuit is R_A. The input signal is

$$E_i = E_E = I_E R_A = (\beta + 1) I_B R_A$$

The output signal is

$$E_o = R_B \beta I_B$$
$$= \beta R_B \frac{E_i}{(\beta + 1) R_A}$$

so that the gain of the amplifier system is

$$A_f = \frac{E_o}{E_i} = \frac{\beta R_B}{(\beta + 1) R_A}$$

If $\beta \gg 1$, the gain of the amplifier system is

$$A_f = \frac{R_B}{R_A}$$

i.e., the gain depends only on the ratio R_B/R_A and is independent of the gain β of the transistor. The input resistance is $(\beta + 1)R_A$; the output resistance is R_A.

A frequently used amplifier with 100 percent negative feedback ($k = 1$) is the voltage follower shown in Fig. 12-16c. Feedback is accomplished by the resistance R; there need be no resistor in the collector circuit. The circuit does not furnish a voltage gain but a considerable power gain. An input signal E_i is applied to terminals 1 and 2 of the amplifier and causes a potential difference E between base and ground. The output signal is

$$E_o = (\beta + 1) R I_i$$

The input signal is

$$E_i = (\beta + 1) R I_i + R_s I_i$$

The gain of the system is

$$A_f = \frac{E_o}{E_i} \frac{(\beta + 1) R}{(\beta + 1) R + R_s}$$

for $\beta R \gg R_s$, the gain is $A'_f = 1$.

The input resistance for an applied signal E_i is

$$R_i = \frac{E_i}{I_i} = (\beta + 1) R$$

The output resistance is obtained as follows. The input current is

$$I_i = \frac{E_i - E_o}{R_s}$$

The current sum at the node N yields

$$(\beta + 1)I_i = \frac{E_o}{R} - I_o$$

Substituting for I_i and rearranging,

$$-E_o\left(\frac{\beta + 1}{R_s} + \frac{1}{R}\right) = -I_o - E_i\frac{\beta + 1}{R_s}$$

For E_i = const., the output impedance is

$$R_o = \frac{\Delta E_o}{\Delta I_o} = \frac{1}{1/R + [(\beta + 1)/R_s]}$$

$$= \frac{RR_s}{R_s + R(\beta + 1)}$$

This is the equivalent resistance for R and $R_s/(\beta + 1)$ in parallel. If $R_s = 0$ or if β approaches ∞, the output resistance R_o approaches zero.

The emitter follower exhibits no phase inversion; the output signal follows the input with regard to amplitude and phase. Both input and output signals have a common ground so that grounding and shielding are simplified.

OPERATIONAL AMPLIFIERS

Since the characteristics of an amplifier system depend largely on that of its feedback loop, it is economical to build amplifiers in the form of universal building blocks (*operational amplifiers*) that can be adapted to a variety of tasks and with a variety of desirable physical properties by merely connecting external feedback loops to them. Not every amplifier can be used for this purpose. A conductive path from the output to the input of an amplifier might offset the operating conditions of the first stage or of the entire amplifier. However, amplifiers can be designed that are not adversely affected by the addition of a variety of external feedback loops. Historically, such operational amplifiers were developed first in connection with elements used in analog computers.

An operational amplifier is generally a high-gain dc amplifier with differential input and single-ended output as shown in Fig. 12-17*a*. The common input point 3 is usually grounded and so is the reference point 5 of the output. In most circuit diagrams these ground connections are silently assumed and the operational amplifier is drawn as in Fig. 12-17*b* or *c*, sometimes also without the grounded bar as in Fig. 12-17*d* or *e*. Ideally, the gain of the operational amplifier and its input impedance should be large; its output impedance should be close to zero; and the output voltage E_o should be zero when the input voltage E_i is zero. Some applications also require a large bandwidth, from dc to several megahertz and sometimes to several hundred megahertz. The response time of the amplifier should be practically zero. Power is usually supplied in a symmetrical fashion, i.e., the power supply is grounded at the center and the positive

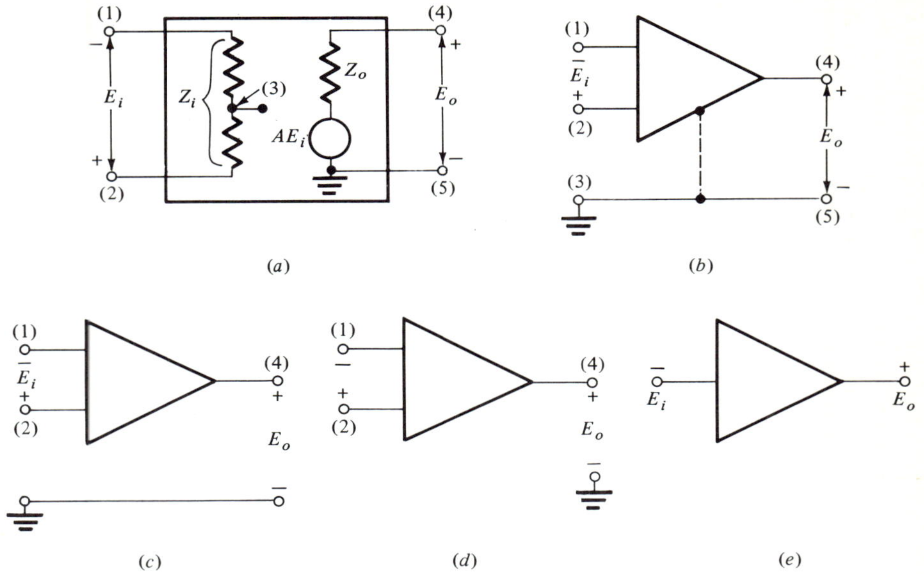

Fig. 12-17 Symbols of operational amplifiers. (*a*) Showing the input and output equivalent circuits; (*b*) same as (*a*); the ground bar 3 to 5 is shown separately and connected to the amplifier (dotted line); (*c*) same as (*b*); ground connection not shown; (*d*) same as (*b*); ground bar not shown, most frequently used symbol; (*e*) inverting amplifier with a gain of one (inverter). Polarity symbols: The output terminal 4 is positive if a negative potential versus ground is applied to terminal 1 or a positive potential to terminal 2.

and negative terminals are connected to the amplifier. The power supply connections are generally omitted from the circuit diagram.

The generalized and, in most cases, desirable frequency response of an operational amplifier is shown in Fig. 12-18. The gain (*logarithmic scale*) has a plateau and diminishes with increased frequency. The roll-off of gain with frequency is essentially uniform at a rate of 6 dB/octave (or 20 dB/decade). The gain is unity at a frequency f_H.

Figure 12-19*a* shows an operational amplifier in an open-loop circuit, i.e., without feedback. Two voltages E_1 and E_2 are applied to the input terminals 1 and 2; an output E_o will be observed at terminal 4, i.e., between terminal 4 and ground. The input-output characteristic is shown in Fig. 12-19*c*. As long as the difference between the two input voltages, $E_1 - E_2$, is large, the amplifier will be driven to saturation at either side. In the vicinity of the region $E_1 = E_2$, the output voltage changes rapidly and linearly with the applied voltage difference. The same effect will occur with the circuit in Fig. 12-19*b* where the input voltage to terminal 2 is zero and that applied to terminal 1 varies from negative values through zero to positive values. The open-loop gain is expressed by the slope of the characteristic $\Delta E_o/\Delta(E_1 - E_2)$ or $\Delta E_o/\Delta E_i$.

Commercial operational amplifiers have open-loop gains varying between

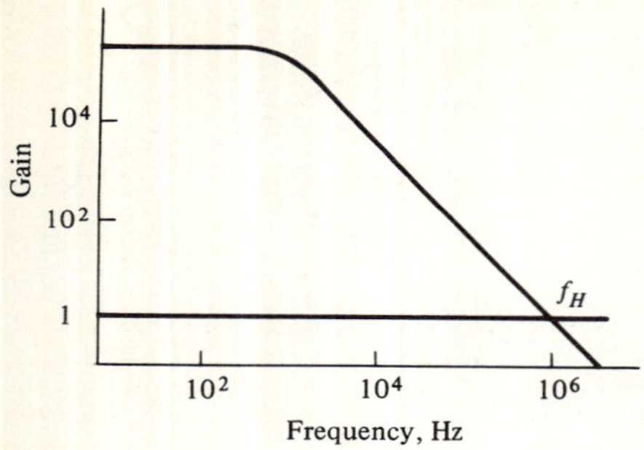

Fig. 12-18 Gain-frequency characteristic of an operational amplifier.

10^4 and 10^8. In the open-loop operation, the amplifier can be used as a voltage comparator or as a voltage-operated switch, changing the polarity of the output in response to small variations of the input voltage. The input impedance of an open-loop operational amplifier is usually about 10^4 to 10^8 Ω, although amplifiers with 10^{12}-Ω input impedances are available, so that the amplifier input currents are negligible; no current passes from the source to either input terminal 1 or 2:

$$I_1 = 0 \qquad I_2 = 0 \tag{12-5}$$

Simple operational amplifiers with feedback used for the amplification of single-input signals (in contrast to differential inputs) are illustrated in Figs. 12-20 to 12-23. Two basic circuits are available, the inverting amplifier (Figs.

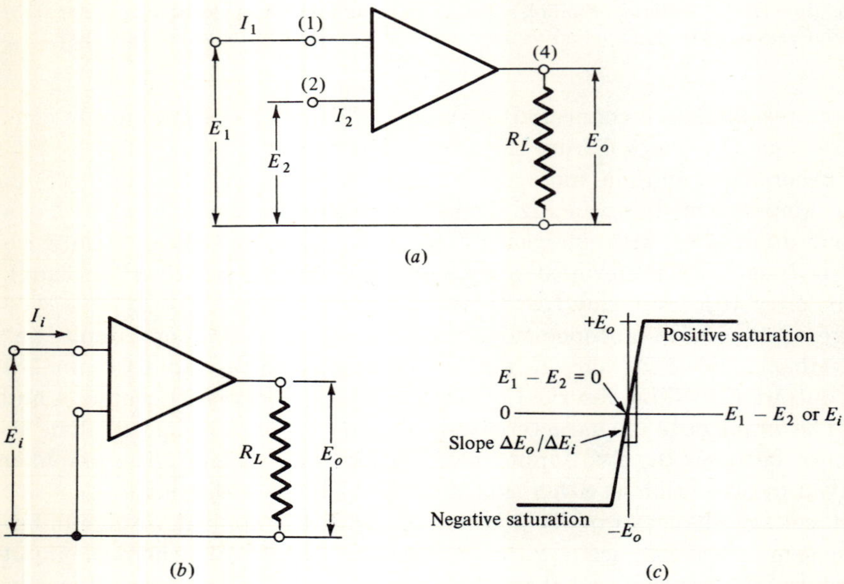

Fig. 12-19 Open-loop operation of operational amplifiers. (*a*) For differential inputs E_1, E_2; (*b*) for single input E_i; (*c*) input-output characteristic of an open-loop operational amplifier.

12-20 and 12-21) and the noninverting amplifier (Figs. 12-22 and 12-23). In the inverting amplifiers, the input signal is applied between terminal 1 and ground; terminal 2 is connected to ground. The output signal appears between terminal 4 and ground, terminal 4 being of polarity opposite that applied to terminal 1. In the noninverting amplifiers (Fig. 12-22), the input signal is applied between terminal 2 and ground; terminal 1 is connected in the feedback loop. The output signal appears again between terminal 4 and ground, terminal 4 being of the same polarity as the input signal applied to terminal 2.

Figure 12-20*a* shows the inverting operational amplifier with feedback loop. Terminal S is called the *summing point*. According to Eq. (12-5), no current flows into terminal 1; it follows that $I_i = I_f$. The input current is

$$I_i = \frac{E_i - E_s}{R_i}$$

The feedback current is

$$I_f = \frac{E_o - E_s}{R_f}$$

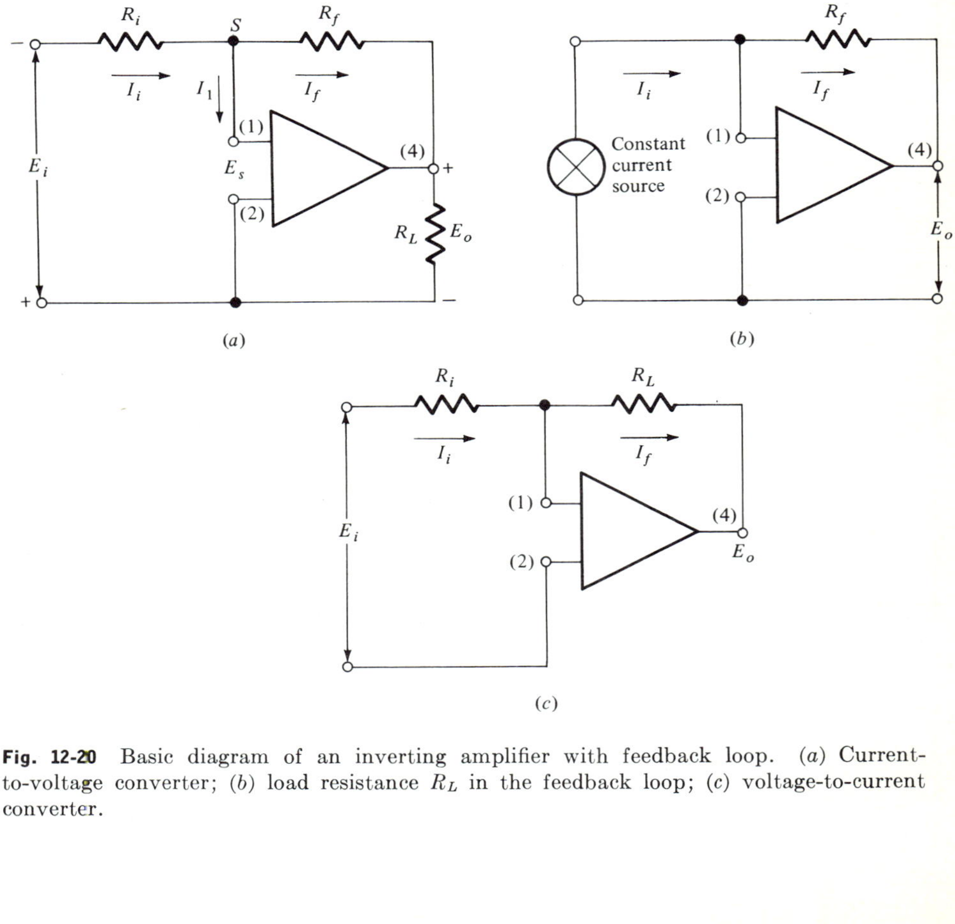

Fig. 12-20 Basic diagram of an inverting amplifier with feedback loop. (*a*) Current-to-voltage converter; (*b*) load resistance R_L in the feedback loop; (*c*) voltage-to-current converter.

The output voltage is

$$E_o = -AE_s$$

so that

$$E_s = -\frac{E_o}{A}$$

If the gain A is very large, E_s approaches zero, so that

$$\frac{E_i}{R_i} = -\frac{E_o}{R_f}$$

$$E_o = -\frac{R_f}{R_i}E_i$$

The feedback operation has the effect of reducing the voltage E_s between terminals 1 and 2 to zero:

$$E_s = 0 \tag{12-6}$$

Conditions (12-5) and (12-6) govern the behavior of any operational amplifier. It should be noted that the gain of the inverting amplifier is determined by the feedback network, i.e., by R_f and R_i. The gain is independent of the open-loop-amplifier characteristics, provided the gain A of the amplifier is large; that is, $E_s = -E_o/A$ is negligible compared to E_i or E_o. The gain of an operational amplifier can be made adjustable by replacing R_i, R_f, or both, by variable resistors.

The input impedance of the inverting amplifier is R_i and is generally much lower than that of the open-loop operational amplifier. The input current $I_i = E_i/R_i$ continues as feedback current I_f as shown in Fig. 12-20*a*.

If the input to the inverting amplifier is a current source (Fig. 12-20*b*), i.e., if the current is limited by the source rather than by a separate resistance R_i, then the circuit acts as a current-to-voltage converter. The output voltage is

$$E_o = R_f I_i$$

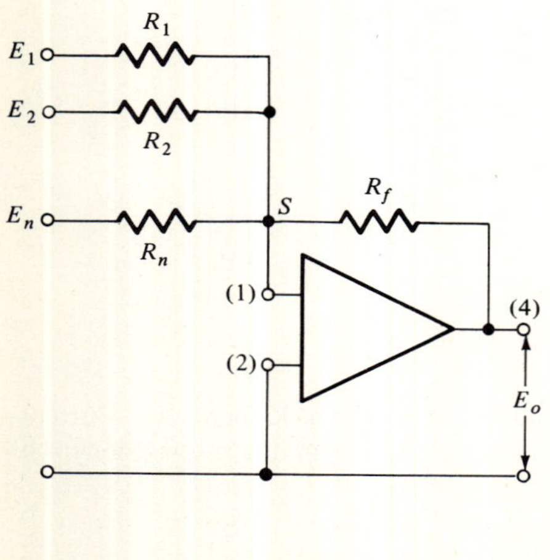

Fig. 12-21 Inverting summing or adder (subtractor) circuit; the output E_o is proportional to the sum of the input voltages $E_1, E_2, \ldots, E_n$.

If the load is inserted in the feedback loop as shown in Fig. 12-20c, the circuit can be used as a constant-current source or as a voltage-to-current converter. Terminal 1 of this source is at ground potential since terminal 2 is grounded and there is no potential difference between terminals 1 and 2. The current through the load resistance R_L is $I_f = I_i$, i.e., the current I_f is independent of the resistance R_L.

Figure 12-21 shows an inverting adder circuit with several input voltages $E_1, E_2, \ldots, E_n$ connected to the input of the operational amplifier by series resistors $R_1, R_2, \ldots, R_n$. A resistor R_f is inserted in the feedback loop. The circuit acts as an adding network. The output is

$$E_o = -\left(E_1 \frac{R_f}{R_1} + E_2 \frac{R_f}{R_2} + \cdots + E_n \frac{R_f}{R_n}\right)$$

i.e., the output equals the sum of all voltages E_i, each voltage multiplied by an adjustable weight factor

$$\frac{R_f}{R_1}, \frac{R_f}{R_2}, \ldots, \frac{R_f}{R_n}.$$

If all series resistances are alike, i.e., for $R_1 = R_2 = \cdots = R_n = R_i$, the output is

$$E_o = \frac{R_f}{R_i}(E_1 + E_2 + \cdots + E_n)$$

The circuit is identical with that shown in Chap. 2, Fig. 2-9. However, the use of the operational amplifier as an adding element has the advantage that the voltage between terminals 1 and 2 in Fig. 12-21 is reduced to zero by the feedback condition, while the corresponding voltage E_o across the load resistance R_L in Fig. 2-9 remains finite and thus gives rise to an interaction between the sources E_1, E_2, and E_n. Subtraction may be accomplished by changing the polarity of one or more of the input voltages $E_1, E_2, \ldots, E_n$.

The basic circuit for the use of a noninverting amplifier is shown in Fig. 12-22a. The input signal is applied between terminal 2 and ground. The current I_a is equal to the current I_f

$$I_a = I_f$$

The current in loop a is

$$I_a = \frac{E_i + E_s}{R_a}$$

The feedback current is

$$I_f = \frac{E_o - (E_i + E_s)}{R_f}$$

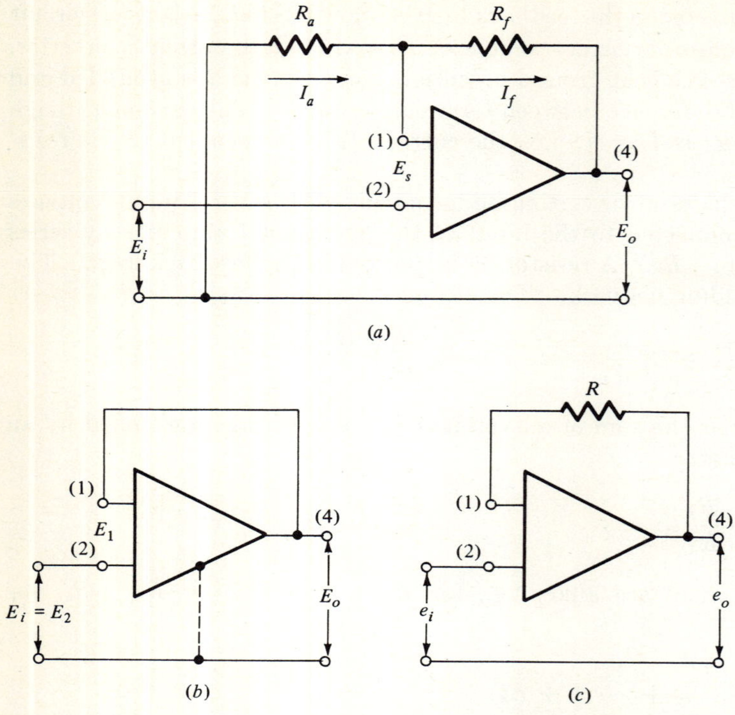

Fig. 12-22 Noninverting amplifier, basic diagram. (*a*) The output at terminal 4 is of the same polarity as the input applied to terminal 2; voltage followers. (*b*) Without resistor in the feedback loop. (*c*) With resistor R in the feedback loop (same gain).

From Eq. (12-6) it follows that $E_s = 0$, so that

$$\frac{E_i}{R_a} = \frac{E_o - E_i}{R_f}$$

solved in terms of E_o,

$$E_o = \left(1 + \frac{R_f}{R_a}\right) E_i$$

The closed-loop gain of the noninverting amplifier is $1 + (R_f/R_a)$. The input impedance is practically infinite; no current flows into terminal 2. The output impedance is practically zero.

A frequently used application of the noninverting amplifier is the voltage follower shown in Fig. 12-22*b*. The voltage between ground and terminal 2 is E_i. Because of the feedback conditions [Eqs. (12-5) and (12-6)], the voltage E_i between ground and terminal 1 is the same as E_i, and so is E_o

$$E_o = E_1 = E_2 = E_i$$

The input impedance for E_i is practically infinite and the output impedance of E_o is near zero. The circuit acts as a power amplifier for a high impedance input with a voltage gain of one. Since no current flows into terminal 1 and, therefore, through the feedback loop, any arbitrary (but finite) resistance can be inserted into the feedback loop as shown in Fig. 12-22*c*.

The voltage follower with a voltage gain of one can be converted into a follower with a gain larger than one (*electrometer amplifier*) by reducing the amount of feedback as shown in Fig. 12-23*a*. A potentiometer with the partial resistances nR and R is provided between amplifier output and ground. The output voltage is

$$E_o = (1 + n)E_i$$

The input impedance of this amplifier can be very high, provided the basic amplifier, by itself, has a high input impedance. Input currents into terminal 2 may be of the order of 10^{-11} A and lower.

The noninverting amplifier can also be used as an adder circuit. An example is shown in Fig. 12-23*b*. From the dimension of the feedback network R, R, it follows that the potential of terminal 1 is

$$E_1 = E_o \frac{R}{R + R} = \frac{E_o}{2}$$

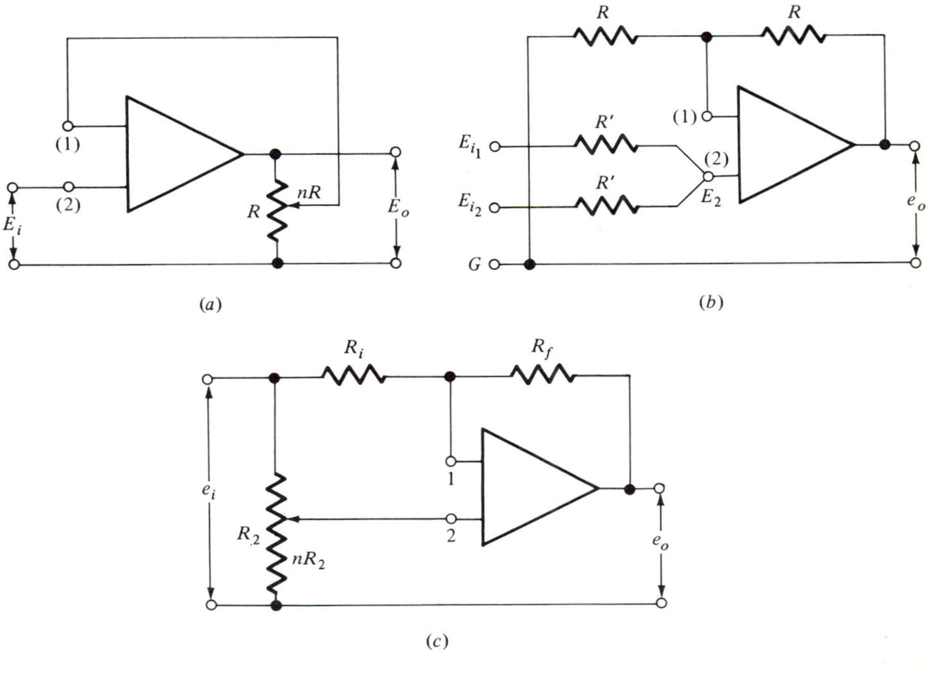

Fig. 12-23 (*a*) Electrometer amplifier for high-input impedance and adjustable gain; (*b*) noninverting adder circuit; (*c*) amplifier for continuous bipolar adjustment, see text for explanation.

The potential of the other input terminal 2 must be the same because of Eq. (12-6). Since

$$\frac{E_{i_1} - E_2}{R'} + \frac{E_{i_2} - E_2}{R'} = 0$$

$$E_{i_1} + E_{i_2} = 2E_2 \qquad \text{or} \qquad E_2 = \frac{E_{i_1} + E_{i_2}}{2}$$

it follows that

$$E_o = E_{i_1} + E_{i_2}$$

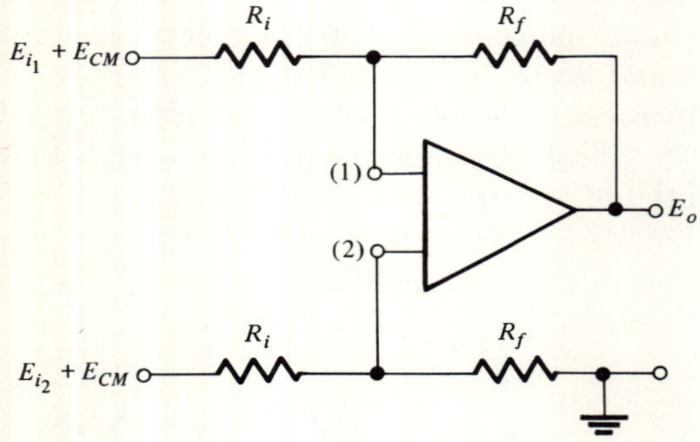

(a)

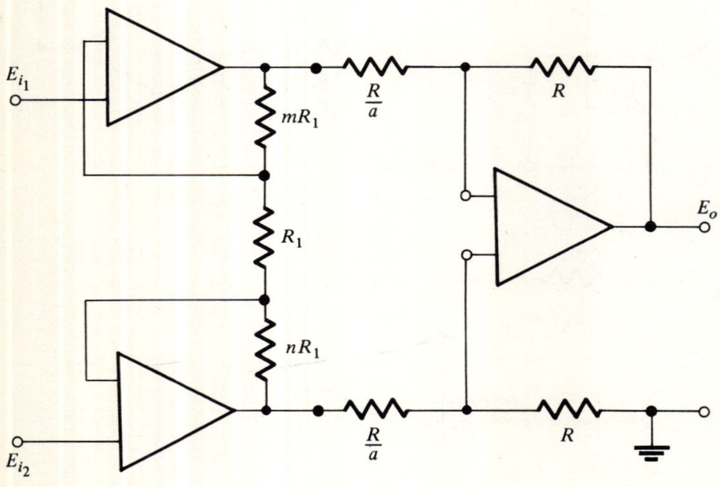

(b)

Fig. 12-24 (a) Operational amplifier for differential input-to-single-ended output; (b) amplifier with high-input impedance for differential input-to-single-ended output.

This circuit can be used for more than two inputs as well as for addition and subtraction. It may also be combined with any number of input circuits and weighting networks.

Figure 12-23c shows a very useful amplifier that provides continuous bipolar adjustment of the gain from negative values (inverting), through zero to positive (noninverting) values. In the midposition of the potentiometer R_2 ($n = 0.5$) the gain is zero. For $R_f = R_i$, the output is

$$E_o = (2n - 1)E_i$$

while for the general case R_f, the response is

$$E_o = \left(1 + \frac{R_f}{R_i}\right) n - \frac{R_f}{R_i} E_i$$

A differential input-to-single-ended-output amplifier is illustrated in Fig. 12-24a. Applied to the two inputs are the voltages E_1 and E_2 plus their common-mode components E_{CM}. The output voltage is

$$E_o = \frac{R_2}{R_1} (E_{i_1} - E_{i_2})$$

Because of the use of a symmetrical subtractor, the output voltage is proportional to the difference of both components; the common-mode term disappears.

A better solution of the common-mode-rejection differential amplifier is shown in Fig. 12-24b. It is particularly applicable where the common-mode-rejection feature has to be combined with very high input impedance. The output is

$$E_o = a(1 + m + n)(E_{i_2} - E_{i_1})$$

REACTIVE CIRCUITS

If the resistances R_i and R_f in one of the circuit diagrams described above are replaced by reactive circuit elements Z_i and Z_f, respectively, the basic equation for the inverting amplifier changes to

$$E_o = -\frac{Z_f}{Z_i} E_i$$

Circuits of this type can be used for a number of applications, such as integrators, differentiators, or time-lag elements.

A simple integrator circuit is shown in Fig. 12-25a. Since $i_i = i_f$, it follows that

$$\frac{e_i}{R} = -C \frac{de_o}{dt}$$

therefore,

$$e_o = -\frac{1}{RC} \int_0^t e_i \, dt$$

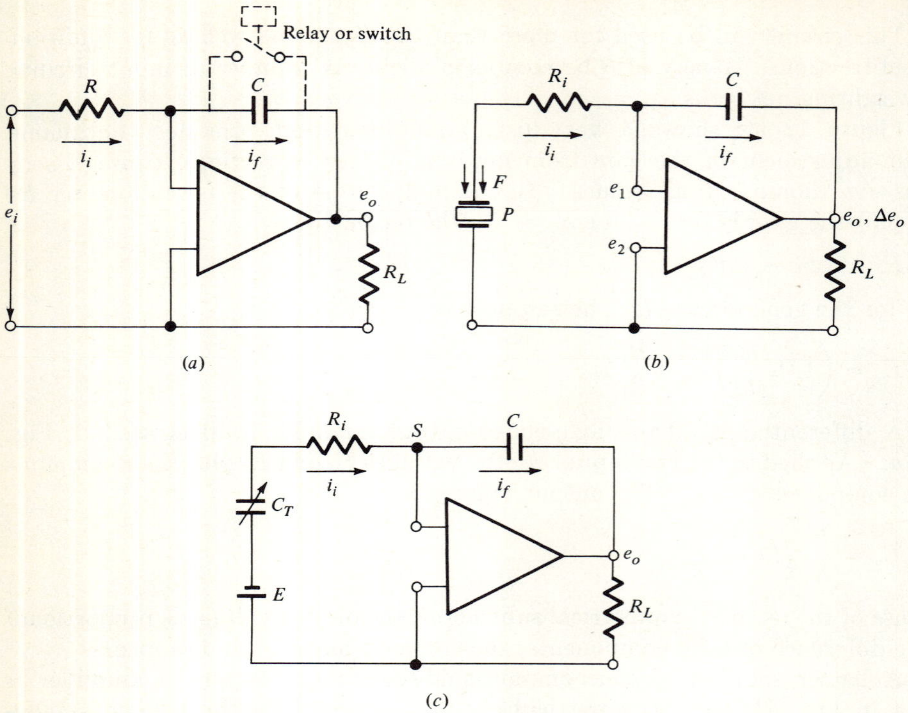

Fig. 12-25 (*a*) Circuit diagram of an integrator; (*b*) circuit diagram of a charge amplifier with a piezoelectric transducer P in the input; (*c*) with a variable capacitance C_T and a constant voltage E in the input.

The output voltage is proportional to the (negative) time integral of the instantaneous input voltage. The time constant is RC.

If e_i is zero, no current flows in the feedback loop and, theoretically, the charge across capacitor C does not change. However, any offset current in the feedback loop persisting for any length of time will charge or discharge the capacitor and may cause an error in integration. When the running time of the integrator (time during which charge accumulates) or its holding time (time during which the voltage across the capacitor is stored for later read-out) is short, a simple relay can be used as shown in Fig. 12-25*a* (dotted lines). The relay is normally closed and sets the initial voltage to zero. The relay is open during the integrating and holding period. If both of these periods are short compared to the time constant RC, a small offset current is of minor influence and the error is small. Another means of reducing the effect of an offset current consists of using a large capacitor C, but such a procedure will reduce the output and, hence, the sensitivity of the integrator and may cause leakage conductance errors. The best method of reducing the error is, of course, to keep the offset current negligibly small.

Closely related to the integrator circuit is the charge amplifier shown in Fig. 12-25*b*. A piezoelectric transducer P is connected to the input of the charge amplifier. A compressional force F acting on the transducer causes the occurrence of a charge $Q = \int i\,dt$ between the electrodes attached to P. The same current-time integral appears in the feedback loop. The output voltage is

$$E_o = \frac{1}{C}\int i\,dt = -\frac{Q}{C} + \text{const.}$$

The charge amplifier can also be used in connection with capacitive transducers converting a change of capacitance into a proportional change of voltage. An example is shown in Fig. 12-25*c*. The capacitive transducer C_T in series with the constant-voltage source E and the resistor R_i are connected to the input terminals of the charge amplifier. Again, the charge applied from the input at the summing point S must be equal to that coming from the feedback

$$Q = C_T E = -Ce_o$$

$$dQ = -C\,de_o$$

$$de_o = E\frac{dC_T}{C} + C_T\frac{dE}{C}$$

While the source E and the capacitor C_T are fixed (and after transient phenomena are eliminated), no voltage will appear at the output terminals. Any variation dC_T will cause a change de_o of the output. In most applications E is constant; therefore $dE = 0$, and the second term of the equation is reduced to zero.

The basic circuit diagram of a differentiator is shown in Fig. 12-26*a*. The input current is

$$i_i = C\frac{de_i}{dt}$$

The feedback current i_f must be equal to i_i so that

$$-e_o = i_i R = RC\frac{de_i}{dt}$$

This circuit, although operating satisfactorily at small values of de_i/dt, has some practical limitations; it is not dependably stable and its gain for high frequencies (noise) is larger than that for lower frequencies (signal). However, these errors can be corrected to some extent by inserting a small capacitor C_i in parallel with the resistor R_d and a small resistor R_i in series with the capacitor C_d as shown in Fig. 12-26*b*. The purpose of C_i and R_i is to limit the gain of the differentiator at higher frequencies and to provide better dynamic stability. C_i and R_i are frequently determined experimentally.[1]

The input resistance of a noninverting operational amplifier is usually

[1] A theoretical approach may be found in Philbrick Researches, Inc., "Applications Manual for Computing Amplifiers," p. 49, 1966.

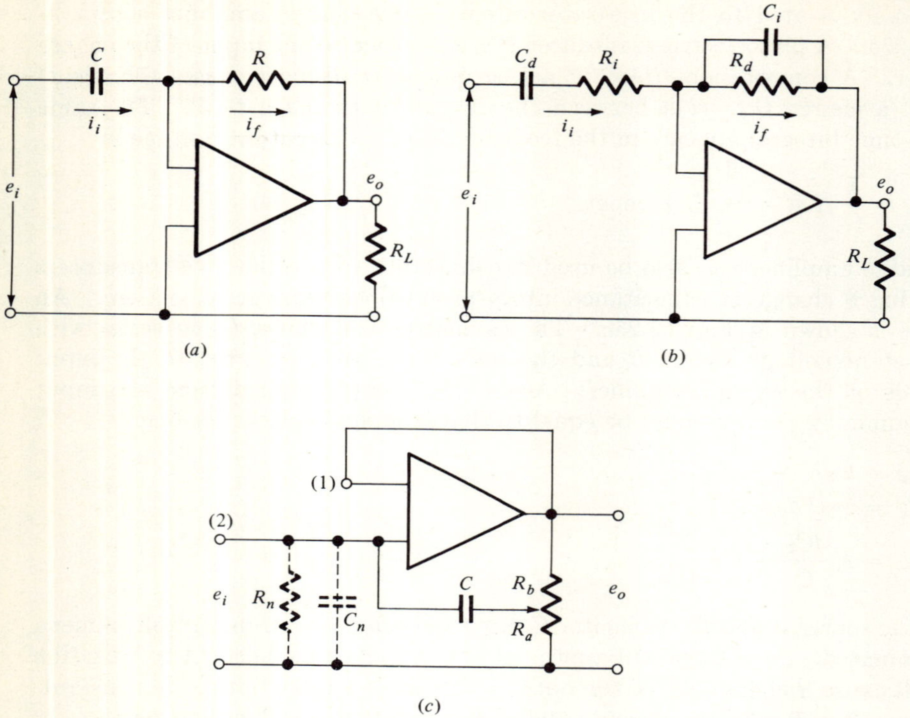

Fig. 12-26 (*a*) Circuit diagram of a simple differentiator; (*b*) practical circuit diagram of a differentiator similar to that in *a* but stabilized by the addition of a small capacitor C_i and a small resistor R_i; (*c*) circuit diagram of an amplifier with cancellation of the effect of the input capacitance C_n.

high, of the order of 10^8 to 10^{14} Ω. Such high input resistances are desirable for the measurement of voltages from high impedance sources, like ionization chambers or biological microelectrodes. However, the natural input capacitance C_n, which is always present, causes a load for such sources and raises the effective time constant R_nC_n, which in turn restricts the use of such amplifiers to very low frequencies. The circuit shown in Fig. 12-26*c* permits the compensation or cancellation of the input capacitance of an amplifier. The basic circuit is identical with that shown in Fig. 12-22*b*, but it contains a capacitive feedback loop connected to a potentiometer in the output circuit. The total effective capacitance appearing in the input circuit between terminal 2 and ground is

$$C_t = C_n - C \frac{R_a}{R_a + R_b}$$

By appropriate setting of the potentiometer, the term $CR_a/(R_a + R_b)$ can be made equal to C_n so that the entire input capacitance C_t is reduced to zero.

NONLINEAR CIRCUITS

Many semiconductor elements have nonlinear voltage-current characteristics. Such elements can be combined with operational amplifiers to produce elements with a response characteristic of the form $e_o = f(e_i)$. The function $f(e_i)$ may be an approximation, over a limited range, of a quadratic or root function, a logarithmic or exponential, sine or cosine, or any other desired function. These nonlinear elements may be used in the input circuit or in the feedback circuit of the amplifier. If used in the feedback circuit, the response is the reverse of that which ensues if the nonlinear system is in the input circuit. An example is shown in Fig. 12-27*a*. A diode D is inserted in the input circuit of an inverting amplifier with the feedback resistance R_f. The output voltage E_o versus E_i is plotted in Fig. 12-27*b*. Figure 12-27*c* shows the position of the diode D (in the same direction) and of the resistor R_i interchanged. The result appears in Fig. 12-27*d*; the characteristic is the reverse of that in Fig. 12-27*b*.

Figure 12-28*a* shows a limiter or bounding circuit. Two zener diodes in series and with opposite polarity are inserted in the feedback loop, in parallel with a resistor R_f. In the voltage range where none of the diodes is conductive,

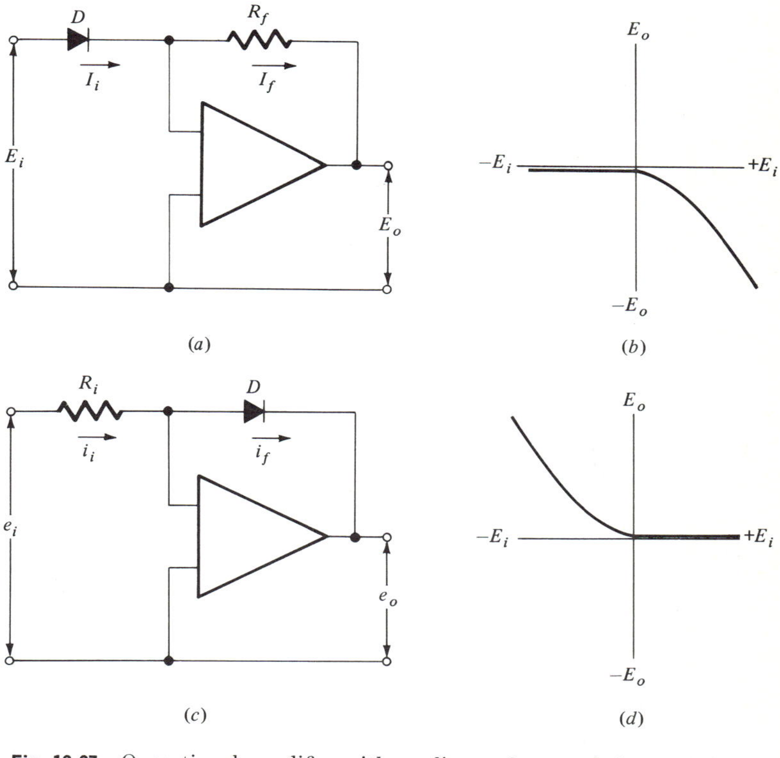

Fig. 12-27 Operational amplifier with nonlinear characteristic. (*a*) Circuit diagram with a diode in the *input* circuit; (*b*) transfer characteristic; (*c*) same as *a* but diode in the *feedback* circuit; (*d*) transfer characteristic.

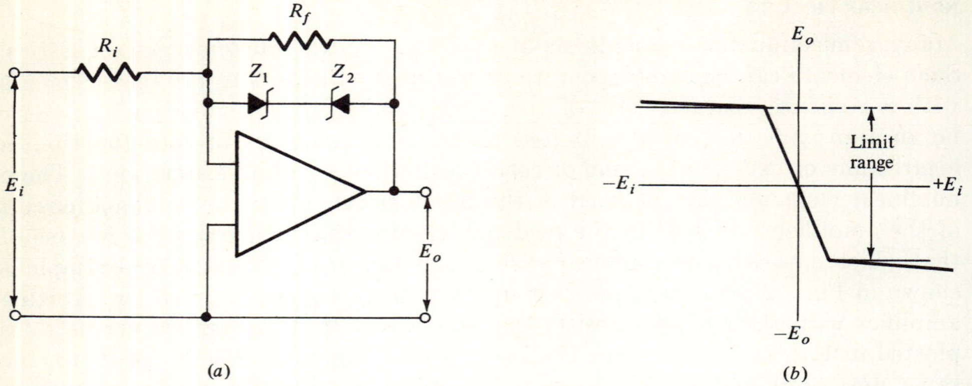

Fig. 12-28 Limiter or bounding circuit. (a) Circuit diagram; (b) transfer characteristic.

the resistor R_f is the only element to provide feedback. (In the absence of this resistor and within the range where none of the zener diodes are conductive, the circuit acts as an open-loop amplifier and the output E_o varies drastically with a variation of the input voltage E_i.) In the region where one or the other zener diode becomes conductive, the resistance in the feedback circuit is reduced to almost zero, and the output voltage does not grow any further. Circuits of this kind are useful to limit the amplifier output to a safe value. The limiting range can be varied by the selection of the zener diodes.

One of the elements for which numerous applications exist is the logarithmic amplifier. Such amplifiers can be constructed with the help of diodes or transistors which exhibit a logarithmic voltage-current characteristic. The logarithmic characteristic extends frequently over a considerable range of current, sometimes over almost 10 decades, i.e., from about 10^{-12} to 10^{-2} A. The voltage drop across the diode in this range extends from about 0.1 to 0.6 V. An example of a circuit is shown in Fig. 12-29a. The logarithmic element, either a transistor in

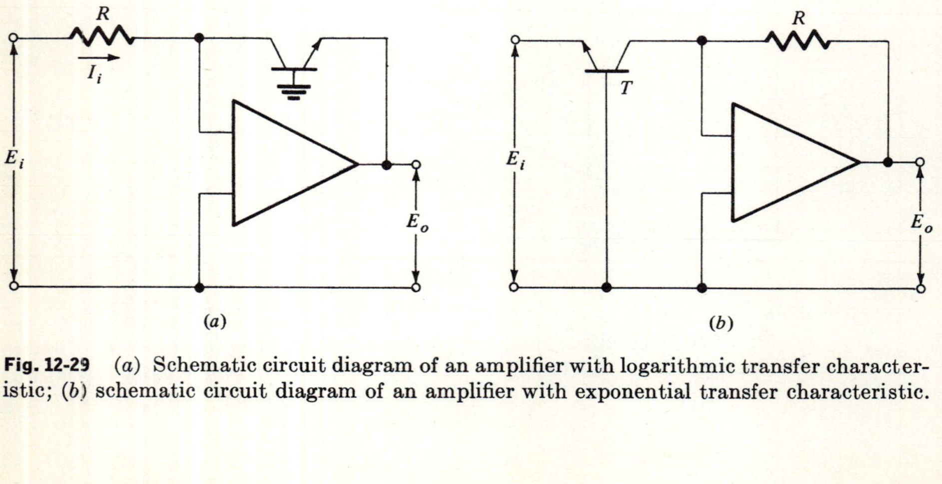

Fig. 12-29 (a) Schematic circuit diagram of an amplifier with logarithmic transfer characteristic; (b) schematic circuit diagram of an amplifier with exponential transfer characteristic.

any one of several modes of operations or a diode, is positioned in the feedback circuit of a logarithmic amplifier. The output voltage is

$$E_o = -E_o' \log \frac{E_i}{I_o'}$$

where E_o' and I_o' are constants relating to the transistors or diodes. Circuits which contain matched sets of transistors and which are temperature-compensated are commercially available (*logarithmic transconductors*). A circuit which furnishes antilogarithms is shown in Fig. 12-29*b*. The output is

$$E_o = I_o' \exp\left(\frac{-E_i}{E_o'}\right)$$

Logarithmic circuits as well as transconductors for other functional relationships can be used in many analog computation systems, such as for the determination of the logarithm of a ratio or in multiplication and division. Also, systems have been designed for clippers, multivibrators, oscillators, and many other applications.

REFERENCES

Thornton, R. D., C. L. Searle, D. O. Pedersen, R. B. Adler, and E. J. Angelo, Jr.: "Multistage Transistor Circuits," vol. 5, Semiconductor Electronics Education Committee series.

Thornton, R. D., L. G. Linvill, E. R. Chenette, H. L. Ablin, J. N. Harris, A. R. Boothroyd, and J. Willis: "Handbook of Basic Transistor Circuits and Measurements," vol. 7, Semiconductor Electronics Education Committee series.

"Handbook of Operational Amplifier Applications," Burr-Brown Research Corp., Tucson, Ariz., 1963.

"Applications Manual for Computing Amplifiers," Nexus-Philbrick, Inc., Dedham, Mass.

13
Electronic Analog Circuits

The basic electronic amplifier with its numerous variations and applications has been described in Chap. 12. There are a considerable number of other analog circuits that can be constructed with the help of electronic elements, such as diodes, transistors, and electronic tubes. The more important circuits will be described here.

RECTIFIERS AND POWER SUPPLIES

Half-wave rectifiers The basic rectifier circuit is the half-wave rectifier illustrated in Fig. 13-1*a*. The circuit contains an ac source S which is frequently a transformer operated from the ac power line. The ac source furnishes a sinusoidal voltage of the instantaneous value e_i, which is shown by a dotted line in the Et diagram, Fig. 13-1*b*. Whenever terminal N is positive with respect to G, diode D conducts and a current i flows in the circuit (Fig. 13-1*c*). When terminal N is negative, diode D acts as an open switch and no current passes the circuit. The instantaneous output voltage e_o, that is, the voltage across the load resistor R_L, is plotted by a solid line in Fig. 13-1*b*. Only one-half of each cycle is used; both output voltage and output current appear in only every other half-

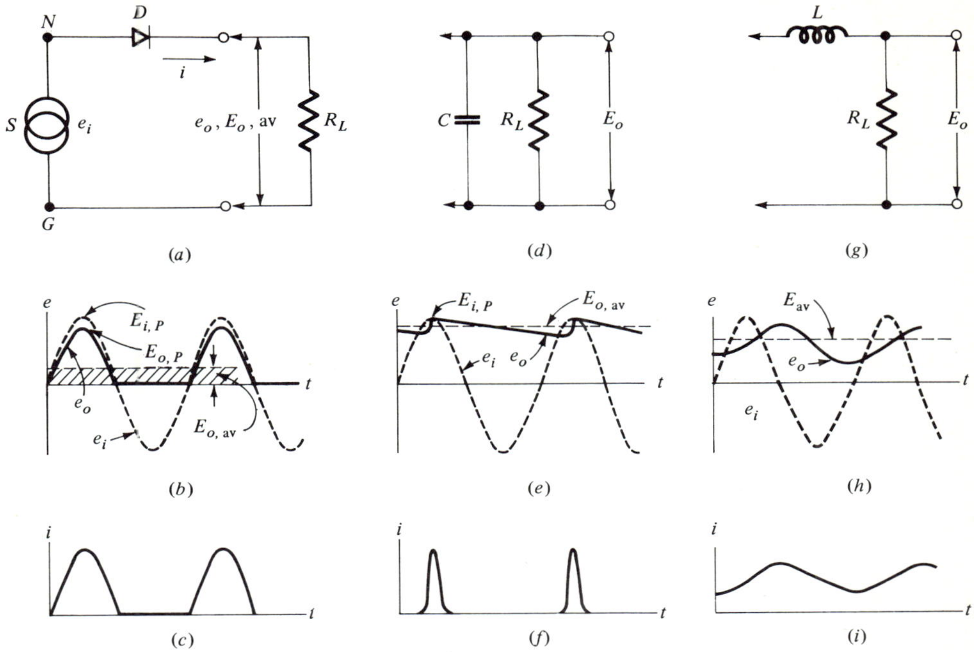

Fig. 13-1 Schematic diagram of the basic half-wave rectifier circuits. (*a*) to (*c*) Resistive load circuits; (*d*) to (*f*) capacitive-resistive load circuits; (*g*) to (*i*) inductive-resistive load circuits. (*b*), (*e*), (*h*) Voltage-time diagrams; dotted lines, input (source) voltages; solid lines, output voltages.

wave. The average output voltage is

$$E_{o,\text{av}} = \frac{1}{T}\int_0^T e_o\,dt$$

or, expressed in terms of the peak-input voltage $E_{i,p}$

$$E_{o,\text{av}} = 0.318E_{i,p}$$

This is one-half of the average value of a sinusoidal current of the peak value $E_{i,p}$ [see Eq. (7-1)].

The output voltage may be considered a dc voltage of the value E_{av} and a superimposed unsymmetrical ac voltage, the ripple voltage, which has the same fundamental frequency as the input voltage E_i. The efficiency of the half-wave rectifier circuit is low. Even if the transformer and the diode are assumed to be free of losses, the volt-ampere product at the input $E_{\text{ac,rms}} \times I_{\text{ac,rms}}$ is 3.49 times larger than the useful dc output power $E_{\text{av}} \times I_{\text{av}}$. Furthermore, since the dc current I_{av} passes through the windings of the transformer, the transformer becomes partially saturated and its efficiency decreases.

If a capacitor C is connected in parallel with the load resistor R_L, as shown

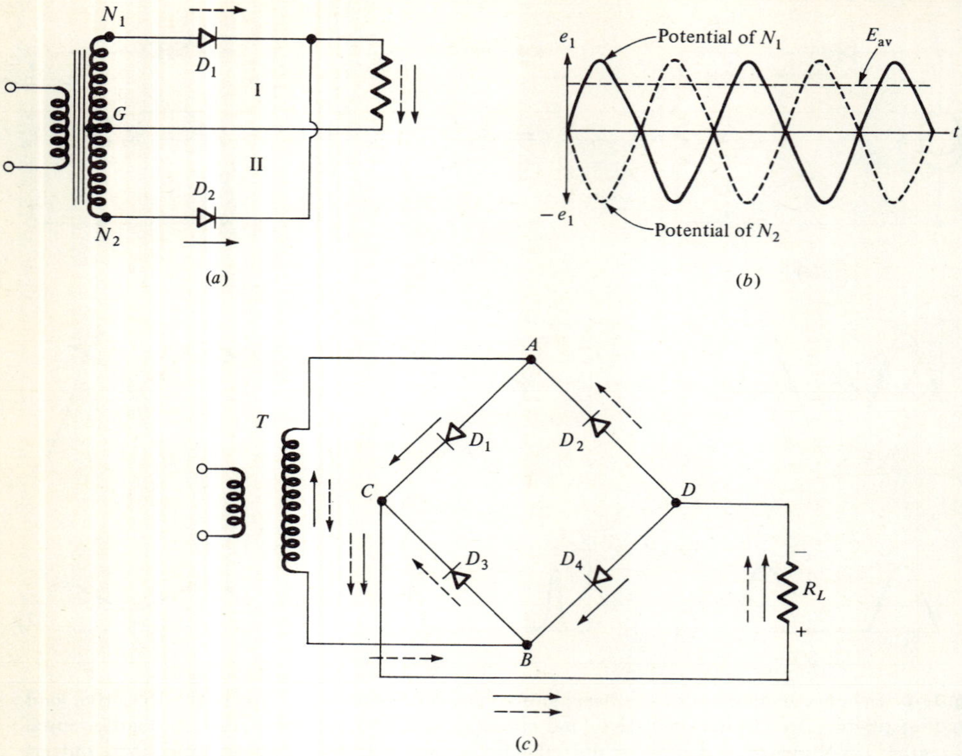

Fig. 13-2 Full-wave rectifier circuit. (*a*) Circuit diagram; (*b*) potential-time diagram; (*c*) full-wave rectifier circuit using a bridge rectifier system T, transformer; R_L, load resistor; D_1 to D_4, diodes. Solid arrows signify current flow during first half-wave, dotted arrows during second half-wave cycle.

in Fig. 13-1*d*, the capacitor will be charged during the current cycles to the peak value of the applied ac voltage $E_{i,p}$. The capacitor will be partially discharged during the time interval between the current cycles (Fig. 13-1*e*). The current which charges the capacitor flows only during the short time interval when the instantaneous output voltage e_o is smaller than the instantaneous input voltage e_i (provided the voltage drop in the source and that across diode D are negligible). The average output $E_{o,\text{av}}$ will be higher than in a circuit without the capacitor (see Fig. 13-1*a*) and can reach the value of the peak voltage $E_{i,p}$ if the time constant CR_L is large. The ripple voltage is then very small and may be negligible for many applications.

If an inductance is connected in series with the load resistance as shown in Fig. 13-1*g*, the inductance will act as a high reactance for the ripple voltage, in particular for the high-frequency components, and an *Et* characteristic as shown in Fig. 13-1*h* will result.

The most frequently used rectifier circuit is the full-wave rectifier illustrated in Fig. 13-2*a* and *c*. The circuit consists of two partial circuits I and II, joined together at the point G, which is the center tap of the transformer. Each of these partial circuits is identical with that shown in Fig. 13-1*a*. The function of the full-wave rectifier circuit is evident from the figures and the voltage-time diagram, Fig. 13-2*b*. The average (dc) voltage E_{av} at full-wave rectification is $0.637 \times E_p$, that is, twice as high as that of the half-wave rectifier circuit. The amplitude of the ripple voltage is considerably smaller than that of a half-wave rectifier. The ripple frequency is twice that of the applied ac voltage so that a capacitor connected in parallel to the load resistance or an inductor connected in series with the load will be twice as effective as in a half-wave rectifier circuit. The transformer is not saturated by dc current.

Figure 13-2*c* shows a circuit which accomplishes the same as the full-wave rectifier but uses a bridge-rectifier system and no center-tapped input transformer. The operation is as follows: In a first half-cycle the polarity of point A may be positive, that of B negative. A current from the transformer output A will flow in the direction of diode D_1 from A to C. At C, a current cannot flow in the direction against diode D_3; instead the current will flow from C in the direction of the solid arrow, through load resistor R_L and through diode D_4 to the negative pole B. During the next half-cycle, point B will be positive, point A negative. A current will pass from B via diode D_3 to point C, and in the direction of the dotted arrows through R_L to D and through diode D_2 to the negative pole A.

Both the full-wave rectifier and the bridge-rectifier circuit can be used in connection with a capacitive load (as shown for half-wave rectification in Fig. 13-1*d* or with an inductive load (Fig. 13-1*g*). They are frequently used in connection with more complex filters to eliminate or strongly reduce the ripple voltage.

When the point N in Fig. 13-1*a* is positive, the diode will act as a short circuit and the voltage drop across the diode is negligible for many applications. When the point N is negative, the diode acts as an open switch as pointed out above. The voltage across the diode is then composed of the potential e_i, developed by the source plus that across the load resistance R_L. Since in that part of the cycle no current passes the circuit, the voltage drop across R_L is zero; only the source voltage appears across the diode. This voltage appears in the inverse direction, i.e., in the direction in which the diode does not conduct. The voltage reaches a maximum value $E_{i,p}$ (*peak inverse voltage*), and the diode must be dimensioned to withstand this voltage without the danger of a breakdown.

The situation is different for the capacitive-load circuit in Fig. 13-1*d*. The inverse voltage across the diode is again composed of the potential E_i and that across the load E_o. But this load E_o incorporates a voltage across a capacitor which is charged to the peak voltage $E_{i,p}$ and, if the time constant R_LC is large, remains practically at that level to the end of the blocking half-cycle so that the total peak inverse voltage reaches a maximum of $2 \times E_{i,p}$. This potential difference can be dangerously high and may cause a breakdown of the diode. For example, for an applied sinusoidal voltage $E_{i,\text{rms}} = 1{,}000$ V, the peak inverse voltage is $2 \times \sqrt{2} \times E_{\text{rms}} = 2{,}830$ V.

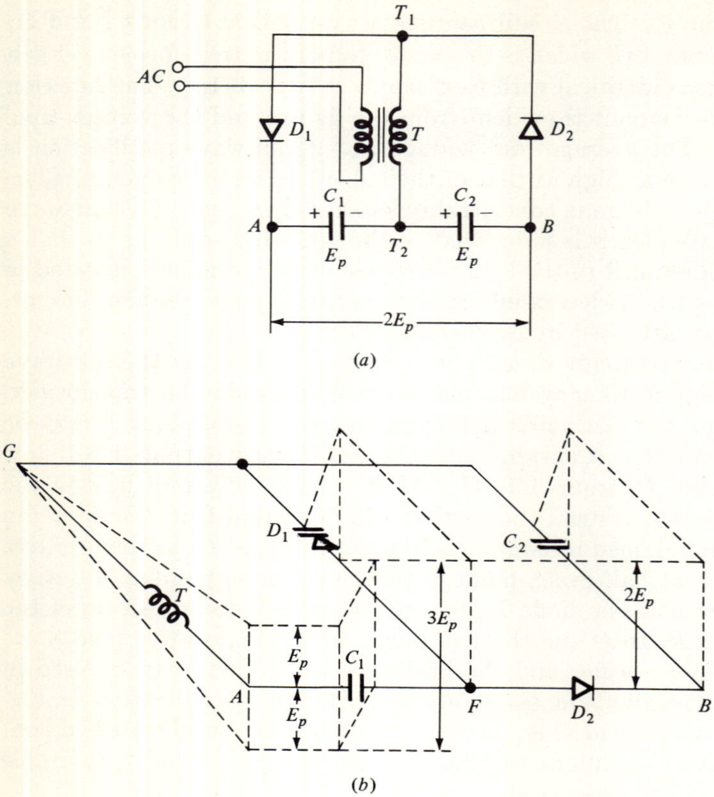

Fig. 13-3 (a) Voltage doubler. The ac voltage arising between the transformer terminals $T_1 - T_2$ is rectified in one half-cycle by the diode D, and the peak voltage E_p is stored in the capacitor C_1. In the other half-cycle the voltage is stored in C_2. The voltage between the points A and B is then twice the transformer peak voltage ($2E_p$). (b) Voltage multiplier, schematic diagram.

Circuits which develop a peak voltage higher than that applied to the circuit can be used for doubling, tripling, etc., of a voltage (*voltage multipliers*). An example is shown in Fig. 13-3a. The circuit comprises a transformer T which furnishes, during a first half-cycle, a sinusoidal voltage of a peak value E_p. This voltage is rectified by a diode D_1 and stored in a capacitor C_1. During the alternate half-cycle, a voltage E_p develops across the diode D_1 and is stored in the capacitor C_2. It can be seen that between points A and B a voltage arises which reaches the value $2E_p$. Several other circuits for voltage multiplications exist. A simple circuit suitable for voltage doubling is shown in Fig. 13-3b.

The operation of a rectifier system can lead to undesirable fluctuation of the dc output voltage if the rectifier circuit is preceded by an inductive-voltage stabilizer (*constant-voltage transformer*). Some of these stabilizers keep the ac-rms voltage constant but in doing so may cause a variation of the transformer output waveshape or peak voltage. If such a voltage stabilizer is followed by a rectifier system and if the output from the rectifier system depends on the peak voltage at its ac input, the dc output will be very poorly stabilized; in fact, it may vary more than without the ac stabilizer.

VOLTAGE AND CURRENT REGULATORS

The generalized diagram of a voltage regulator is shown in Fig. 13-4*a*. The purpose of the regulator is to keep the voltage E_o across the load resistor R_L constant, i.e., nearly independent of a variation of the input voltage E_i and also nearly independent of a variation of the load resistance R_L. A measure of the performance of a voltage regulator is the regulation constant K_E, that is, the ratio of fractional variation of the output voltage to the fractional variation of the input voltage

$$K_E = \frac{\Delta E_o/E_o}{\Delta E_i/E_i}$$

In the less frequent case where the regulator is used to reduce the influence of variations of the load resistance R_L only, a figure of merit for the performance of the regulator is the stabilization constant

$$K_{E'} = \frac{\Delta E_o/E_o}{\Delta I_o/I_o}$$

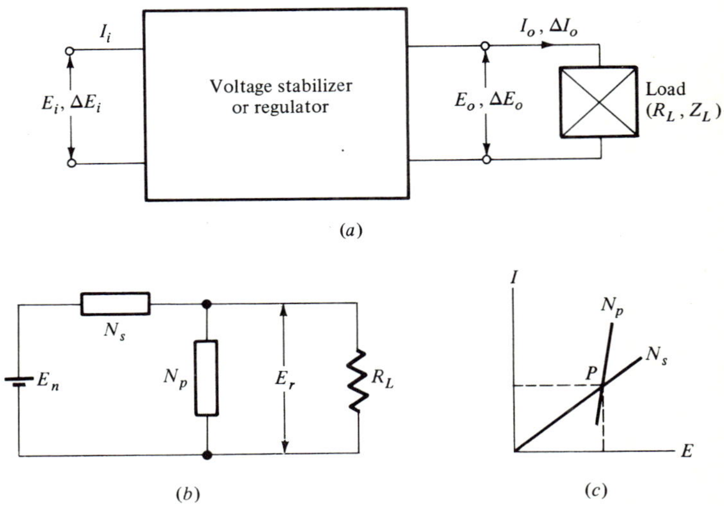

Fig. 13-4 (*a*) Schematic diagram of voltage (or current) stabilizer or regulator; (*b*) voltage regulator with nonlinear circuit elements; (*c*) *EI* characteristics of the components N_p and N_s.

Alternatively, the incremental output resistance $r_{d,o} = \Delta E_o/\Delta I_o$ characterizes the performance of a regulator for variations of the load resistance. The smaller the factors K_E, $K_{E'}$, or $r_{d,o}$ are, the better is the regulation.

Two general schemes are primarily used for the construction of voltage regulators: one involving the use of nonlinear circuit elements and the other employing degenerative (negative feedback) networks.

The general circuit diagram of a voltage regulator with *nonlinear circuit elements* is shown in Fig. 13-4*b* and *c*. The circuit is a voltage divider in which at least one component, N_s or N_p, has a nonlinear characteristic. Generalized voltage-divider circuits of this kind have been described in the introduction to Chap. 8. For the application to voltage regulators, the incremental resistance of N_p ($= r_{d,p} = \Delta E_p/\Delta I_p$) should be small; i.e., the EI characteristics of N_p at the operating point should be steep. N_s may be a linear-circuit element (resistor).

N_p may be a glow-discharge tube (VR tube, Fig. 8-9) or a zener diode (Fig. 11-3), and N_s a resistor. The zener diode with a resistor is the most frequently used voltage-regulator system; its stability is limited by the thermal stability of the zener diode. The most effective operation results from a combination of nonlinear elements whereby N_s has a large incremental resistance, such as the voltage-current characteristic of an ionization chamber (Fig. 8-3), or the plate-current, plate-voltage characteristic of a pentode, or the collector-current versus collector-emitter voltage characteristic of a transistor (Fig. 12-2), while N_p has a very small incremental resistance, such as the voltage-current characteristics of a zener diode.

A simple *degenerative* circuit for voltage regulation of the series-control type is illustrated in Fig. 13-5*a*. The system uses negative feedback but no special amplifier. The unregulated voltage E_i is applied between terminal T_1 and ground; the regulated output voltage appears between T_2 and ground. A zener diode maintains the potential of point P constant at the level E_2 so that

$$E_o + E_{be} = E_Z = \text{const.}$$

If E_o decreases by the amount Δe, then E_{be} increases by the same amount. Since the current is

$$I = KE_{be}$$

a variation of E_{be} by Δe causes a current

$$I = K(E_{be} + \Delta e)$$

The output voltage again depends primarily on the stability of the zener voltage and, to some degree, on noise and temperature stability of all other components. Some practical figures are: temperature coefficient of the zener diode, 2.2 mV/C; temperature coefficient of E_{be}, -2.5 mV/C; load stability of circuit, 0.5 to 5 percent; and noise stability, 2 to 100 mV. The zener diode stability can be improved considerably with temperature-stabilized zener diodes.

A circuit diagram of a voltage stabilizer which uses an amplifier is shown in

Fig. 13-5*b*. The unregulated voltage E_i is applied to the input terminals and gives rise to a current I which passes through a control element C_s. The element C_s causes a voltage-controlled variation of current and thus a change of the output voltage E_o. The output voltage is applied to a subsequent element or stage with a load resistance R_L. A voltage divider R_1, R_2 is connected across the output, and a voltage E_f is derived from it. E_f is proportional to E_o and is applied to one input of the differential amplifier A. The other input to A is taken from a reference voltage source E_{ref}. If $E_f = E_{\text{ref}}$, the amplifier input is zero as is the amplifier output, and the system is in equilibrium. If the output voltage E_o drops in response to a decrease of the unregulated voltage E_i or of the load resistance R_L, then the voltage E_f decreases too, and gives rise to an error signal $E_f - E_{\text{ref}}$ at the amplifier input. The error signal is amplified and appears at the amplifier output. This output signal, if applied to the control element C_s in a degenerative mode, causes an increase of the current I and of the voltages E_o and E_f until the error signal disappears and the previous level of E_o is reestablished.

Voltage regulators are commercially available in the form of *integrated circuits* of the size of a small transistor. They are usually quite complex, containing more than 20 transistors and connected circuitry. They can deliver load currents up to 600 mA. The load regulation (output impedance $r_{d,o}$) is of the order of 20 mΩ. The ripple rejection is of the order of 0.002 percent/V. These can be used for larger loads by adding a power transistor.

Generally, *current regulators* consist of a voltage-regulated source and a large resistor or impedance in series with the source and the load resistor. If the series resistor is large compared to the load resistor, the current will be determined primarily by the series resistor as explained in Chap. 2. Instead of a large linear resistor $R = E/I$, a nonlinear circuit element N can be used and operated at a point of its characteristic where the incremental resistance $r_d = \Delta E/\Delta I$ is large (i.e., where the characteristic approaches that of a parallel to the X axis). An example of such a system is shown in Fig. 13-4*b*. A transistor operated in common-base configuration exhibits very high output impedance. If a varying load is connected to the output of the transistor and the base-bias voltage is kept constant, the output current will remain constant within wide limits of load variations. A typical transistor characteristic illustrating the collector current versus collector voltage for common-base operation is shown in Fig. 13-5*c*.

A simple practical circuit for a current regulator is shown in Fig. 13-5*d*. The base current is fixed by the base bias E_{ref} and the large base resistor R_b. The fixed-base current I_b causes a constant current I_o in the collector-emitter circuit. This current is independent of changes of the input voltage E_i or the load R_L.

The problem solved by the circuit illustrated in Fig. 13-5*d* is different from that solved by the voltage-to-current converter shown in Fig. 12-20*c*. The circuit in Fig. 13-5*d* *stabilizes* the current I_o passing R_L, that is, it makes the current I_o nearly independent of any variation of R_L or E_i that may occur in time. In contrast to this, the output current I_f of the circuit in Fig. 12-20*c* is dependent

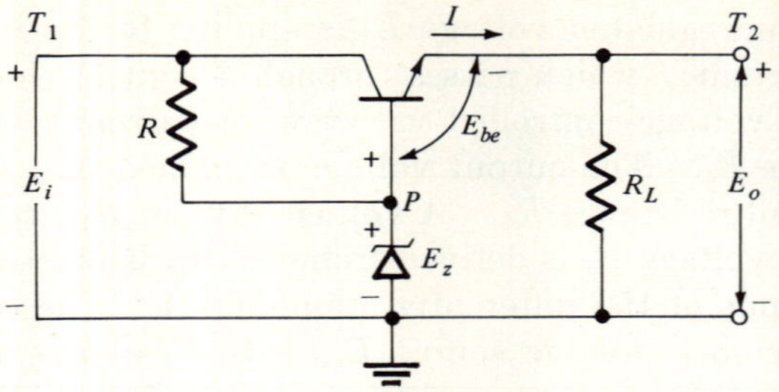

T_1
I
T_2
+
−
R
E_{be}
+
E_i
P
R_L
E_o
+
E_z
−

(*a*)

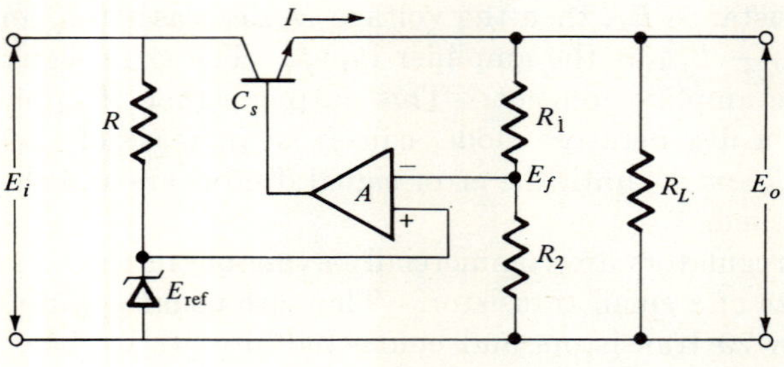

I
C_s
R
R_1
E_i
A
E_f
R_L
E_o
+
−
R_2
E_{ref}

(*b*)

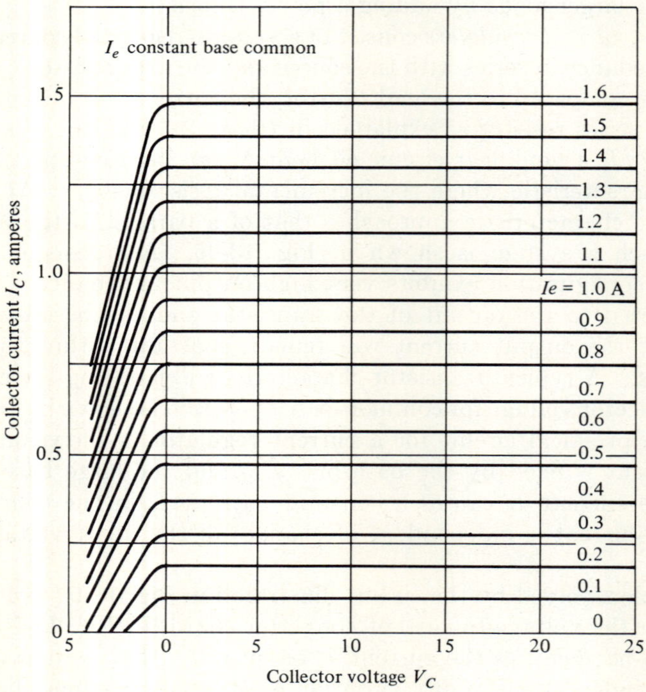

I_e constant base common
1.6
1.5
1.4
1.3
1.2
1.1
Ie = 1.0 A
0.9
0.8
0.7
0.6
0.5
0.4
0.3
0.2
0.1
0
1.5
1.0
0.5
0
5
0
5
10
15
20
25
Collector current I_C, amperes
Collector voltage V_C

(*c*)

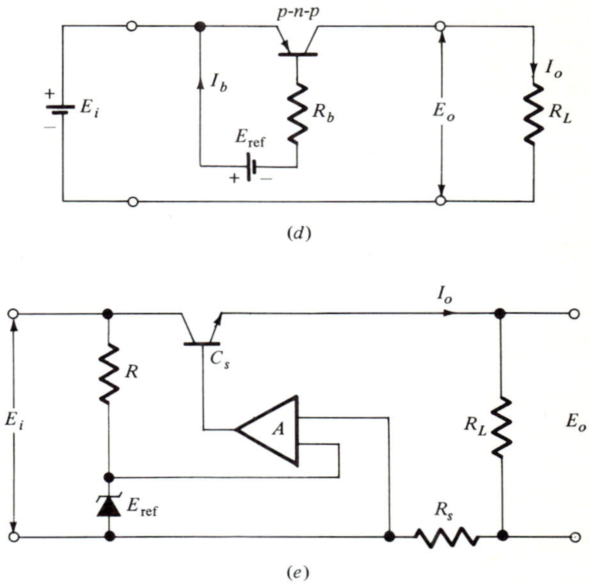

Fig. 13-5 (*a*) Voltage regulator, degenerative type. (*b*) Degenerative voltage regulator with amplifier A. (*c*) Characteristics, collector current versus collector voltage of a transistor. The characteristics form the basis of a current regulator. (*d*) Circuit diagram of a current regulator. (*e*) Circuit diagram of a voltage regulator.

on the corresponding input voltage. I_f is, in fact, a faithful replica of E_i and is simultaneously independent of variations in the load resistance R_L.

A circuit diagram of a current regulator is shown in Fig. 13-5*e*. It stabilizes the output current by comparing the voltage across R_s ($= IR_s$) with the reference voltage E_{ref}. Note the similarity between this circuit and that of Fig. 13-5*b*. Current regulators have been built with a current regulation for a 50 percent change of the load resistance of better than $1:10^5$.

OSCILLATORS

In general, an oscillator is a source. In most cases it does not transform or convert information like a transducer or a modifier, but it has the inherent capability to do so and on occasions it has been used so. For instance, an oscillator may convert dc signals into corresponding ac signals in the sense that the output level or the output frequency or phase or other physical quantity is a reproducible function of the input signal.

A frequently used oscillator type is the LC oscillator illustrated in the schematic diagram of Fig. 13-6*a*. It contains an LC circuit, also called a *tank circuit*,

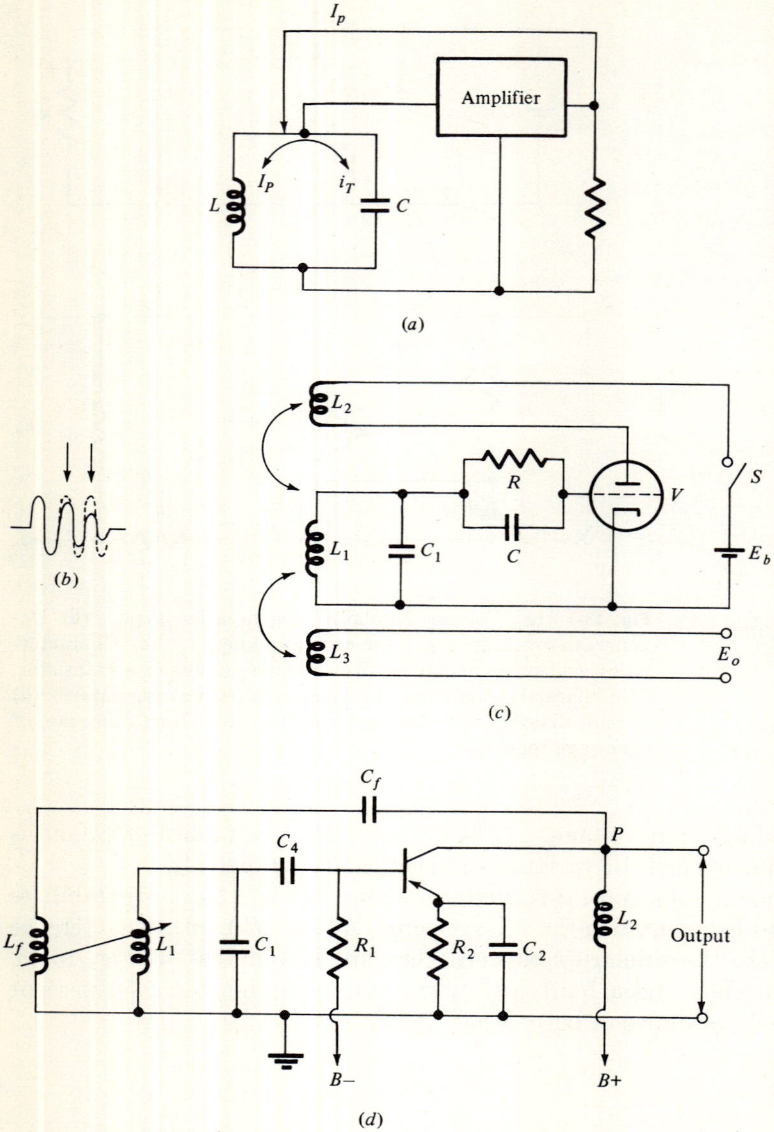

Fig. 13-6 Operation principle of an oscillator. (a) Schematic diagram; L_1C_1 tank circuit. The voltage across the circuit is applied to an amplifier. At the proper time or phase, indicated by the arrows in (b). (b) The amplifier causes a current pulse I which recharges the capacitor C so as to maintain proper oscillation amplitude. (c) Vacuum-tube oscillator. L_1C_1, tank circuit; L_2, feedback coil; L_3, coil for magnetic coupling of the output E_o. R, C, capacitive-resistive network to cause proper grid bias. (d) Transistor oscillator. L_1C_1, tank circuit; L_f, feedback coil; L_2, radio-frequency choke; R_2, C_4, biasing network.

as described in Chap. 7. If the capacitor C is charged by an outside source, it will discharge through the inductor L. The change of current in the inductor will induce a voltage in the inductor, which, in turn, will recharge the capacitor. The process repeats itself, but because of unavoidable losses in the circuit, the energy as well as the voltage across the capacitor and the current in the inductor will diminish in each cycle, causing a damped oscillation as shown in Fig. 13-6*b*. If in each cycle, at times indicated by the arrows, a current or voltage pulse is injected into the circuit, the losses can be replaced and a sinusoidal oscillation of constant amplitude will be generated. For this purpose it is necessary that (1) the injection of the current or voltage pulse occurs at the proper time or in the proper phase with respect to the natural oscillation of the LC circuit (i.e., 0° or 360°) and (2) the energy injected is high enough to cover the losses, i.e., an amplifier of sufficient gain is included in the oscillator circuit. The frequency of the output current is determined by the capacitance of C and the inductance of L, that is, $f = 1/(2\pi \sqrt{LC})$.

A considerable number of circuits can be used for the practical embodiment of the schematic circuit shown in Fig. 13-6*a*. Transistor oscillators are frequently used for small-output requirements and for class A operation (see Fig. 12-9*a*). Vacuum-tube oscillators are preferred for class C operation (see Fig. 12-10*b*) and for high-voltage oscillators which may cause voltage breakdown in transistors.

A basic oscillator with a vacuum-tube amplifier is shown in Fig. 13-6*c*. The vacuum tube exhibits a 180° phase shift between grid voltage and plate voltage. Another 180° phase shift is introduced by the inductive coupling L_2-L_1, which represents a transformer so that a total phase shift of 360° takes place.

The grid-leak bias system RC requires additional explanation. When the switch S in the plate-supply line is first closed, the capacitor C is uncharged, and the grid bias (voltage between grid and cathode) is zero. The tube conducts immediately and the system begins to produce oscillations. The grid then begins to draw a current during the short times when it is driven to a high positive value. This grid current charges the capacitor C with the indicated polarity, i.e., the grid assumes a highly negative average potential as required for class C operation. The capacitor partially discharges in every other half-cycle. The result of the RC grid-leak system is that it causes a high efficiency of the oscillator combined with a zero grid bias at the start of the operation which allows an easy start of the oscillations. The output from an oscillator is usually taken from the tank circuit by means of an inductive coupling L_1-L_3.

Figure 13-6*d* shows a transistor-operated oscillator circuit. The transistor is used in the common-emitter configuration; the LC tank circuit is applied between base and ground. The tank circuit is separated with respect to direct current from the base by a capacitor C_4 so as not to disturb the proper bias connections for the transistor. The feedback voltage is taken off the collector, point P. The dc connection is applied via a choke L_2, which constitutes a large impedance for the radio-frequency current. The feedback connection, which carries only alternating current is applied through a capacitor C_f.

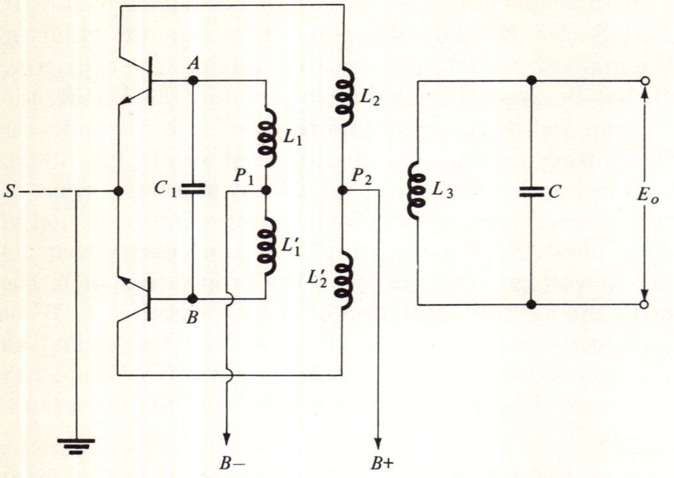

Fig. 13-7 Push-pull oscillator. C_1, L_1, L_1', tank circuit; L_2, L_2' feedback coil; L_3, output coupling. See text for explanation.

A useful modification of the basic *LC* circuit, the push-pull oscillator, is shown in Fig. 13-7. If cut along the symmetry lines, the push-pull system is identical with two *LC* oscillators of the type shown in Fig. 13-6*c*. The advantage of the symmetrical system over a single-ended circuit is that the points P_1 and P_2, which carry the bias potentials for the transistors, are at ground potential with respect to alternating current. Therefore, there is no need for the insertion of RF chokes in the supply lines. This may appear to be a minor point, but RF chokes may introduce undesirable (*parasitic*) oscillations of a frequency different from that determined by the *LC* circuit. If properly built, the oscillation frequency is determined by the inductance $L_1 + L_1'$ and the capacitance C_1, or by the stray capacitance between points *A* and *B*. The push-pull oscillator is, in general, very stable as far as the frequency and the output voltage level are concerned.

Instability of the output voltage is frequently caused by variation of the supply voltages. It can also be caused by variation of the tuning of the feedback coupling or of the temperature of the components. Variation of the frequency is caused by the same parameters, primarily by temperature variation of the tank circuit. Where extreme frequency stability is required, the crystal oscillator generally is used.

A quartz crystal cut of the proper orientation with respect to the crystal axes exhibits piezoelectric properties, i.e., a mechanical compression of the crystal will cause the appearance of a potential difference between opposite surfaces. Inversely, the application of a voltage causes a deformation (contraction or expansion) of the crystal. If an ac voltage is applied, an oscillation of the crystal will ensue. The amplitude is maximum at resonance. The resonant frequency depends on the orientation of the cut and the thickness of the crystal; the thinner the crystal, the higher the resonant frequency.

The crystal is usually mounted between two metal plates as shown in Fig. 13-8*a*. Such a configuration acts as a series or as a parallel-resonant circuit, depending on the frequency of the applied voltage. Compared to an *LC* circuit (Fig. 13-8*b*), the resonance of a quartz crystal is extremely sharp and very constant. The invariance of the quartz crystal depends somewhat on the orientation of the cut. A remaining frequency variation due to temperature variation of a crystal can be corrected by operating the crystal in an oven at constant temperature. Stabilities of one part in 10^6 to 10^8 can be obtained. The crystal can be used in place of a tank circuit or in a feedback line. There are many circuits that can be used. An example (crystal-controlled Hartley oscillator) is shown in Fig. 13-8*d*.

The oscillator circuit shown in Fig. 13-9*a* is based on an entirely different principle. The circuit contains a transistor Q_1 operated in the common-emitter mode. The collector output, point *C*, is fed back to the base input via a multiple *RC* network. The purpose of the *RC* network is to create a 180° phase shift between the output and the input, which causes and maintains an oscillation of the output at the desired frequency. Variation of the frequency over a relatively wide range can be accomplished by changing simultaneously either the values of the capacitances C_1, C_2, and C_3 in the feedback network or those of the resistances R_1, R_2, and R_3. The output is essentially a sinusoidal oscillation of good amplitude and frequency stability. The 180° phase shift can also be obtained by other *RC* networks, for instance, with a Wien bridge (Fig. 7-31).

There are a great number of other oscillating circuits. Most of them are variations of those already described such as the glow-lamp (relaxation) oscillator described in Chap. 8 (Fig. 8-11) or the negative-resistance oscillator described in Chap. 11 (Fig. 11-4).

Oscillators may be operated so that the oscillation can be suppressed

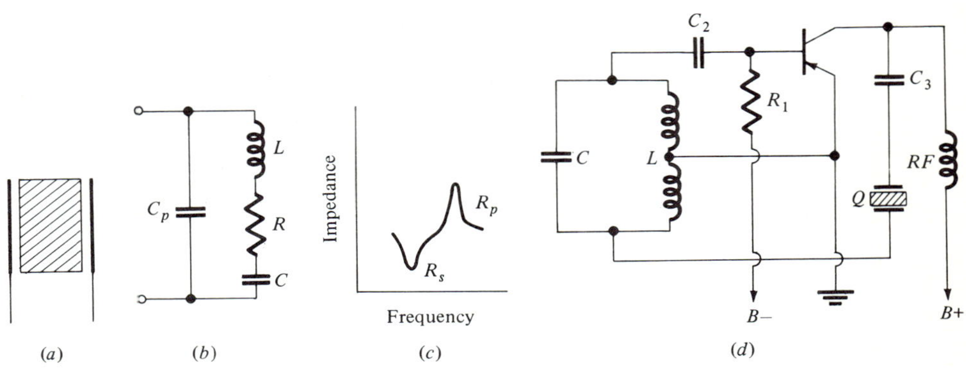

Fig. 13-8 (*a*) Piezoelectric crystal *Q*, mounted between two electrodes. (*b*) Equivalent circuit of piezoelectric crystal arrangement as shown in (*a*). (*c*) Frequency characteristic of network shown in (*a*). (*d*) Crystal-controlled oscillator; C_1L, tank circuit; *Q*, piezoelectric crystal.

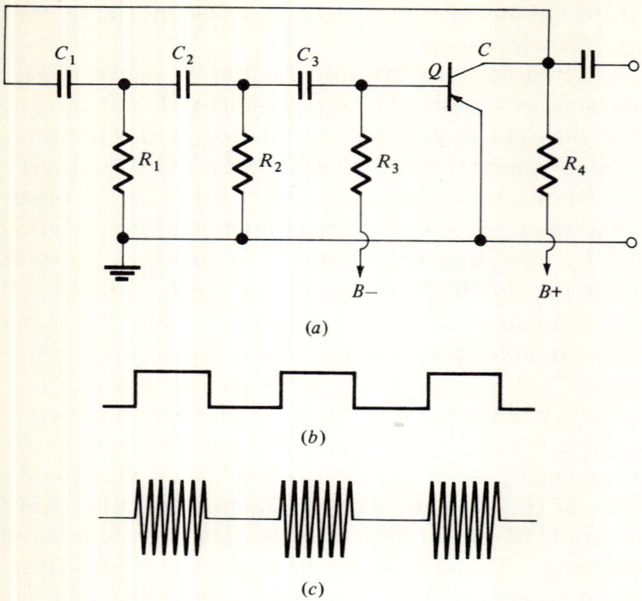

Fig. 13-9 (*a*) Phase-shift oscillator. The networks C_1R_1, C_2R_2, and C_3R_3 cause a 180° phase shift which is required to maintain oscillations; blocking oscillator diagram. (*b*) Applied modulation. (*c*) Frequency waveshape.

(blocked) by the injection of an external signal. If this quenching signal has the form of a square wave, the output will be modified in the form of a wave-train as shown in Fig. 13-9*b*. The quenching circuit can also be part of the oscillator (*blocking oscillator*) and consists, in general, of an *RC* network in the base or emitter circuit of the oscillation transistor. If this transistor is operated as a class C amplifier and if the time constant of the *RC* network is sufficiently large, the potential of the base (or that of the grid in a vacuum-tube oscillator) will become increasingly more negative so that a cutoff point is reached and the oscillation stops. The capacitor will then discharge through the resistor until oscillation conditions are reestablished.

MODULATORS

If an oscillator is operated so that its output amplitude or frequency is invariable (*constant wave, CW, operation*) it obviously cannot convey information. It can carry information only if the output can be modulated in accordance with an injected signal. The principle of the modulation process has been described in Chap. 7. The following is a description of the electronic circuits used for modulation.

The basic circuit of a modulator is shown schematically in Fig. 13-10*a*.

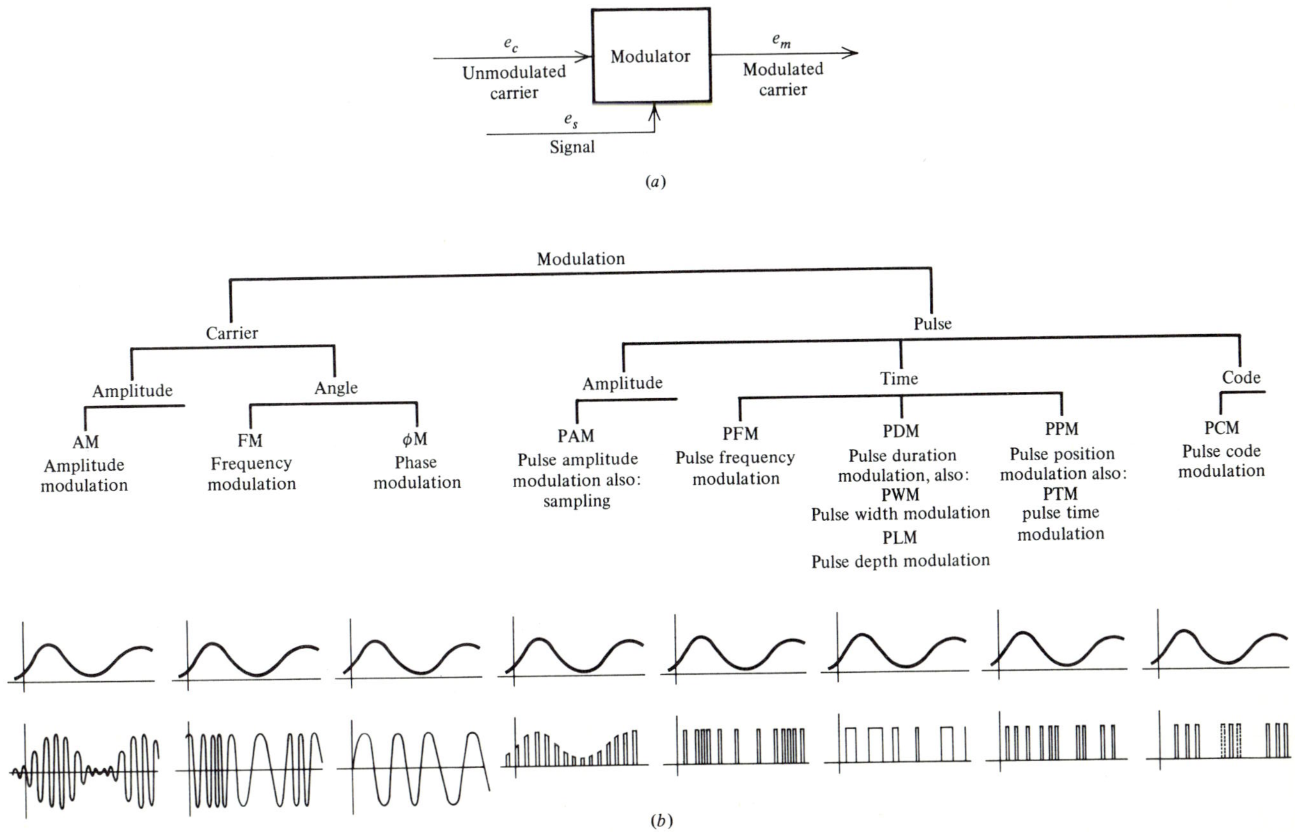

Fig. 13-10 (*a*) Schematic diagram of a generalized modulation system; (*b*) synopsis of modulation methods for continuous-carrier systems and pulses.

An unmodulated carrier e_c and an information-carrying signal e_s are fed into the modulator; a modulated carrier e_m emerges from the output terminals. The unmodulated carrier may be continuous, such as a sine wave, or it may be a sequence of pulses. The modulation consists most frequently of a variation of the carrier *amplitude* (amplitude modulation, AM, Fig. 7-23) or of the carrier *frequency* (frequency modulation, FM, Fig. 7-25) although other means for modulation are known, in particular if the carrier is a sequence of pulses. Figure 13-10*b* shows a synopsis of modulating systems.

AMPLITUDE MODULATORS

A circuit diagram of an amplitude modulator is shown in Fig. 13-11*a*. Amplitude modulators are frequently operated at high-voltage levels (up to several thousand volts) where vacuum tubes are more suitable than transistors. The high-frequency carrier voltage e_c is applied to the grid of the tube. The low-frequency signal e_s is injected into the plate circuit via the coupling transformer T. The effective plate-supply voltage (instantaneous value) $e_{b,\text{eff}} = E_b \pm e_s$ determines the gain of the tube. If the effective supply voltage is small (for $e_{b,\text{eff}} = E_b - e_s$), the gain of the tube is small, and the amplitude of the modulated carrier at the output is small. If $e_{b,\text{eff}}$ is large (for $e_{b,\text{eff}} = E_b + e_s$), the amplitude of the modulated carrier is large. The modulation is linear over a small part of the modulation characteristic only; higher harmonics generated by the nonlinearity of the modulator are practically eliminated in the output by the tank circuit CL which is tuned to resonance with the carrier frequency. The coupling transformer T avoids the flow of the dc plate current through the signal source. The modulating signal can also be applied in series with the carrier to the grid circuit of the triode.

A balanced modulator circuit is shown in Fig. 13-11*b*. This circuit pro-

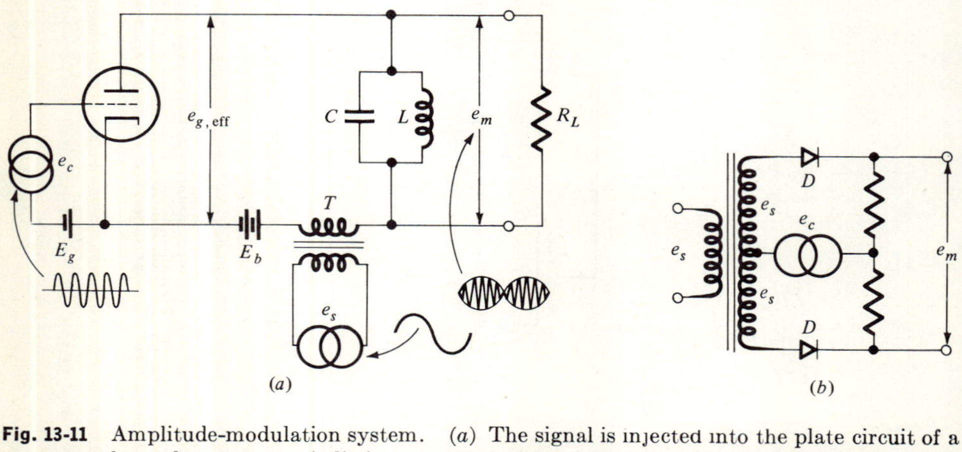

Fig. 13-11 Amplitude-modulation system. (*a*) The signal is injected into the plate circuit of a vacuum tube and causes a periodic increase of the effective plate voltage in the rhythm of the signal; (*b*) balanced amplitude-modulator circuit.

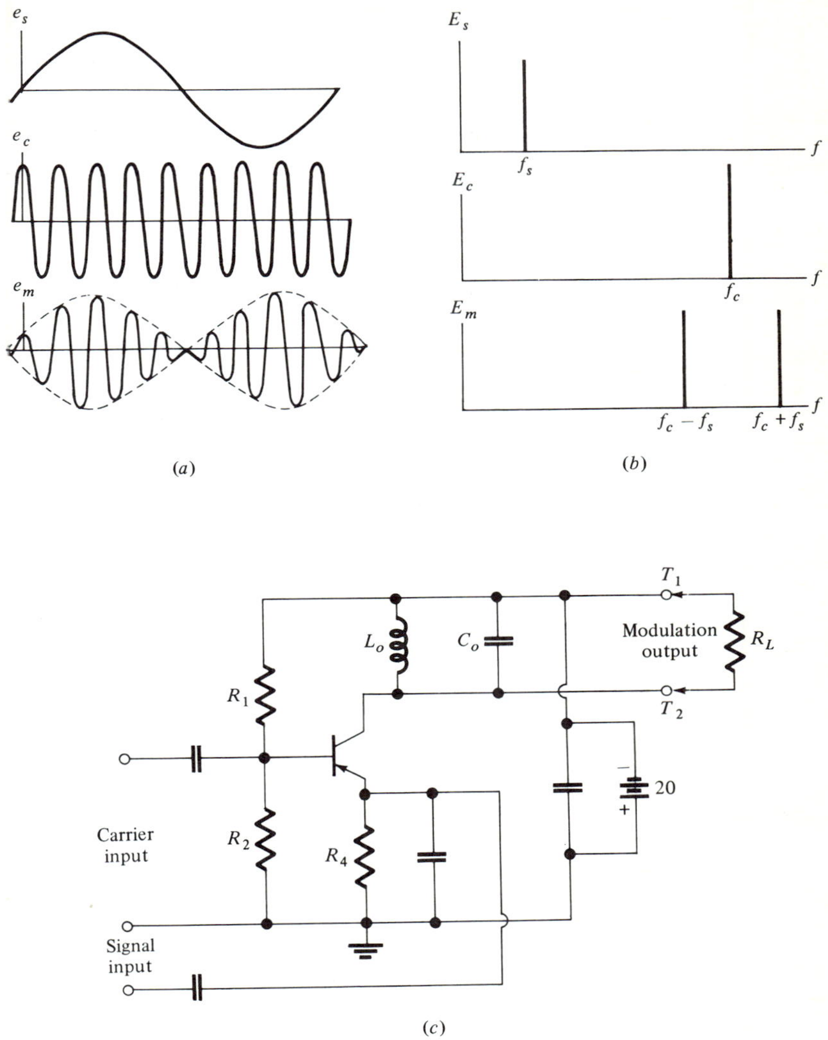

Fig. 13-12 (a) Signal e_s, carrier e_c, and modulated output e_m from a balanced modulator as shown in Fig. 13-11. (b) Frequency diagram. The modulated output does not contain the original carrier frequency f_c but only the sidebands $f_c - f_s$ and $f_c + f_s$. (c) Transistor-operated amplitude-modulator circuit.

duces a suppressed-carrier modulation wave; i.e., the output does not contain the original carrier frequency f_c, but it contains the sum of the carrier and signal frequencies $f_c + f_s$ and their difference $f_c - f_s$. These waveforms, both as a function of time and as a function of frequency, are shown in Fig. 13-12a and b. The circuit has the advantage of remarkably linear operation.

An amplitude-modulator circuit using transistors (for application at low-power levels, carrier input approximately 0.1 V rms) is shown in Fig. 13-12c. The circuit is a common-emitter amplifier with fixed bias produced by the voltage divider R_1 and R_2. The emitter current is determined by a resistor R_4. The signal is applied to the emitter circuit and changes the instantaneous emitter-base potential. The modulated output appears in parallel with the tank circuit

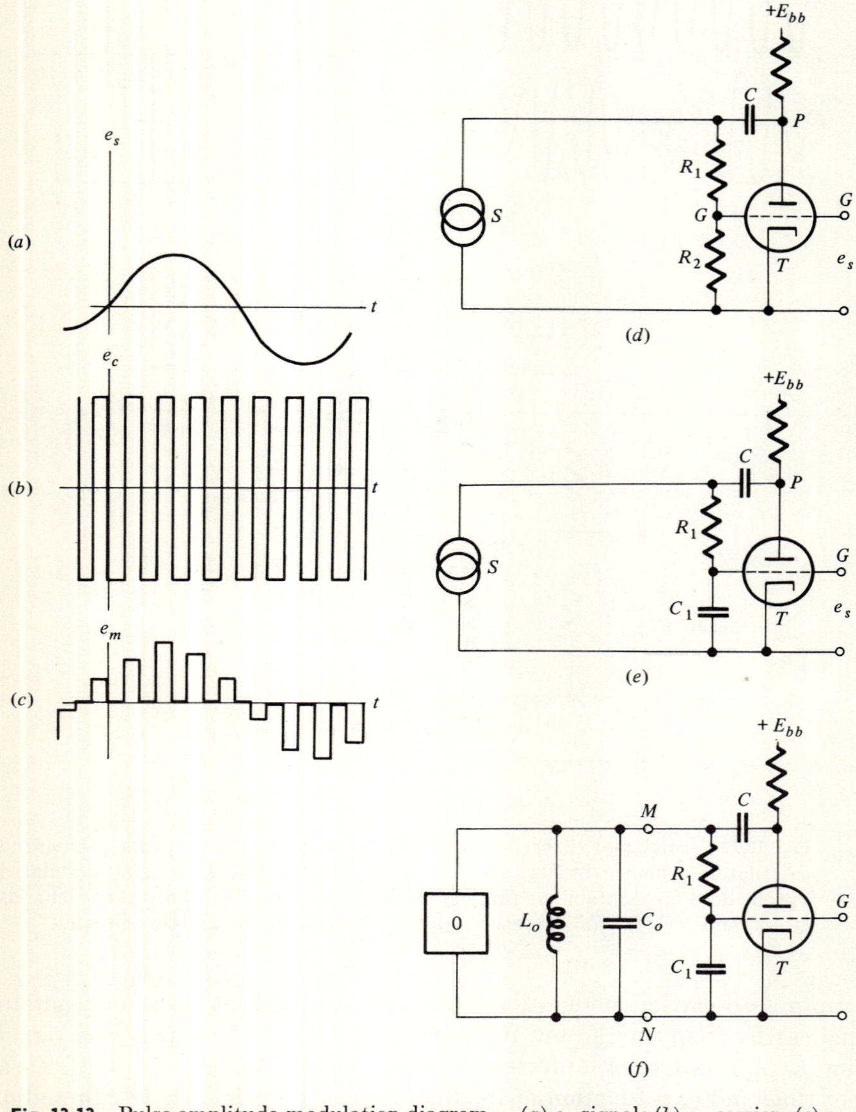

Fig. 13-13 Pulse amplitude-modulation diagram. (*a*) e_s, signal; (*b*) e_c, carrier; (*c*) e_m, modulated output; (*d*) to (*f*) frequency modulation by means of a reactance tube *G*; see text for explanation.

L_oC_o between terminals T_1 and T_2. R_L is the load resistance of the succeeding stage. A pulse amplitude-modulation diagram for a sinusoidal signal is illustrated in Fig. 13-13.

FREQUENCY MODULATION

The basic principle of frequency modulation is shown in Fig. 7-25. The object is to vary the frequency of an oscillator in accordance with a signal. In Chap. 7 this was accomplished directly by the mechanical setting of the variable capacitor in a tank circuit. Instead of a capacitor, a variable inductor could be used. Systems of this type are advantageous if the signal at the output of a transducer appears in the form of a capacitance or inductance variation. For application in frequency modulators, the microphone is directly connected to the tank circuit so that its capacitance is added to that of the tank capacitor. The sum of both capacitances determines the frequency of the oscillator.

Frequently, the signal is given in the form of a voltage-time or current-time function. The conversion of a voltage variation into a frequency variation requires a reactance-tube circuit, which is explained with the aid of Fig. 13-13*d* to *f*.

Figure 13-13*d* shows an ac source S usually operated in the radio-frequency range. The source is connected to a load circuit consisting of a triode T and a voltage divider R_1, R_2. The large capacitor C blocks the direct current in the triode circuit but offers practically no impedance to the radio frequencies. For the sake of simplicity, the grid bias and plate supply are omitted. The resistances R_1 and R_2 are large; the current flowing from S through R_1 and R_2 is negligible.

The vacuum triode T represents a resistive load. Whenever the source voltage S rises from zero, the plate potential (point P) rises too, as well as the grid voltage (point G), so that a current flows through the triode. This current can be controlled by an outside potential e_s applied between grid G and ground which effectively controls the tube transconductance.

In Fig. 13-13*e*, the voltage divider R_1, R_2 is replaced by a resistance-capacitance network R_1C_1 (whereby $R_1 \gg 1/\omega C_1$). Because of the 90° phase shift between the voltage across the resistance and that across the capacitance, the grid potential is now also 90° phase shifted, and the entire circuit acts now as an inductive reactance. The magnitude of this reactance can also be changed by application to the grid of an external dc or ac potential e_s. With direct current applied, the reactance increases by a fixed amount; with alternating current applied, the magnitude of the reactance varies sinusoidally.

In Fig. 13-13*f*, the fixed source S is replaced by an oscillator O incorporating a tank circuit C_oL_o. This circuit is in parallel with the reactance tube. The reactance tube represents a capacitance which is added to that of the tank capacitor and which changes the frequency of the oscillator in the rhythm of the applied modulation signal e_s. The modulated output from the combination oscillator plus reactance tube is taken from the tank circuit (points M and N) or is taken by inductive coupling from the tank coil.

Phase modulation can be considered a special case of frequency modulation whereby the change of the carrier frequency is less than 1 Hz.

DEMODULATORS (DETECTORS, DISCRIMINATORS)

A modulated high-frequency signal can be transmitted over wires or it can be converted by a transmitting antenna into an electromagnetic wave, transmitted as such at some distance and reconverted into a voltage-time function by a receiving antenna. The information contained in the voltage-time function can then be extracted by a demodulation process.

AMPLITUDE DEMODULATORS, DETECTORS

Detection or demodulation of amplitude-modulated signals is achieved by rectification and subsequent low-pass filtering. A simple form of an AM detector is shown in Fig. 13-14*a*. The AM input signal e_i is rectified by the diode D. The rectified current then passes the RC filter. The time constant of the RC filter must be such that the carrier frequency f_c is filtered out, but the signal frequency f_s passes the filter. Consequently, two conditions have to be met for a satisfactory operation of an amplitude demodulator: the carrier frequency f_c must be large compared to the highest signal frequency f_s', that is,

$$f_c \gg f_s'$$

and the filter must pass the (rectified) low frequency f_s' but not the high frequency f_c, that is,

$$f_c < \frac{1}{RC} < f_s'$$

The input impedance of the simple diode detector is low so that a loss of power of the input signal may occur. Where this loss of power is undesirable, amplification of the input signal or the use of an emitter-follower prior to demodulation may be needed. Alternately, the diode can be replaced by a transistor so that the transistor combines the role of the amplifier with that of the rectifier. An example is shown in Fig. 13-14*b*.

The rectifier characteristic can be modified by inserting a bias supply in the rectifying circuit, for instance, at point B in Fig. 13-14*a*. The characteristic $e_o = f(e_i)$ of such a circuit can approach that of a square function. Square-law rectification can be advantageous in detector circuits if the input signal e_s or the modulation index m is small.

FREQUENCY DEMODULATORS, DISCRIMINATORS

Demodulation of frequency-modulated signals involves essentially two operations: conversion of frequency changes into amplitude variations and rectifi-

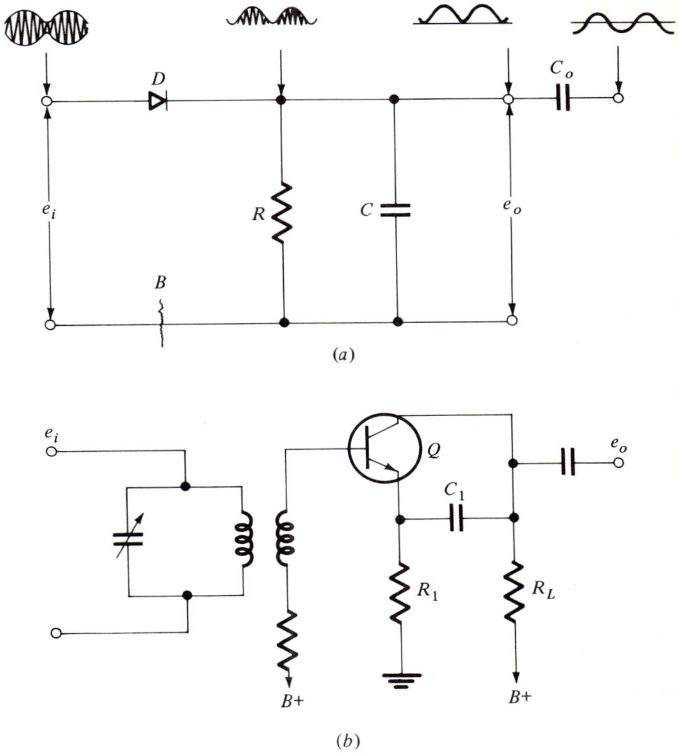

Fig. 13-14 (*a*) Diode amplitude demodulator. The incoming signal e_i is rectified by the diode D. The carrier component is filtered out by the low-pass filter RC. The dc component is eliminated by the capacitor C_o. (*b*) Transistor-operated amplitude demodulator. The transistor Q acts as a rectifier and amplifier for the incoming signal e_i.

cation and filtering of the amplitude-varying signal. These two processes are usually combined in one circuit.

Figure 13-15*a* shows a circuit diagram of a simple discriminator, the slope detector. It consists essentially of a tank circuit L_oC_o tuned to a frequency f_o which is slightly higher or lower than the carrier frequency f_s of the input signal e_i. As the frequency of the input signal e_i changes by Δf, the impedance of the tank circuit changes by ΔZ which results in a voltage change across diode D and across the output circuit consisting of the resistor R in parallel with the capacitor C. The RC filter eliminates the high-frequency components. The output voltage e_o is proportional to the change of frequency Δf. The output is a faithful reproduction of the input signal as long as the operation of the circuit is limited to the linear part of the resonance curve as illustrated in Fig. 13-15*b*. However, the discriminator has the disadvantage that it is sensitive to amplitude

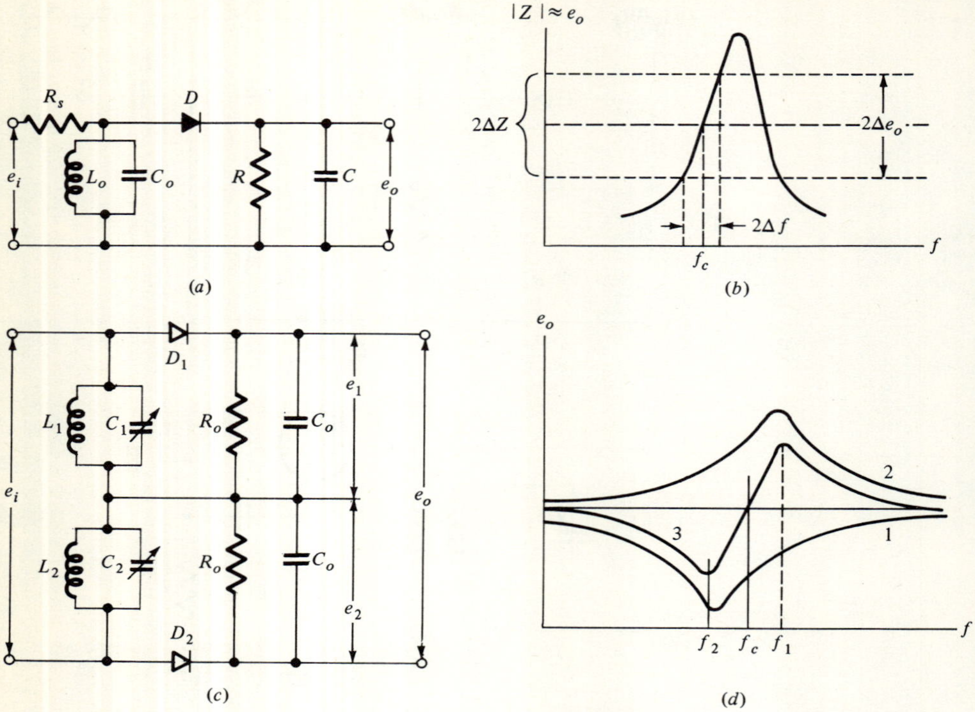

Fig. 13-15 (a) Slope detector for the demodulation of a frequency-modulated signal e_i; (b) signals; (c) circuit diagram of a balanced-slope discriminator; (d) (1) frequency characteristics of resonance circuit C_1L_1; (2) of resonance circuit C_2L_2; (3) of the combined effect of both resonance circuits.

variations of the input signal. To reduce this effect, the discriminator should be preceded by a device which limits the amplitude of the incoming signal to a constant value. Limiters are described below.

The characteristic e_o versus e_i can be made linear over a longer range with the balanced discriminator shown in Fig. 13-15c. The circuit contains two resonance circuits L_1C_1 and L_2C_2. The first resonance circuit is tuned slightly above the carrier frequency; the other one is tuned below the carrier frequency by the same amount. The output is $e_o = e_1 + e_2$. The two resonance curves and the resultant response characteristic are shown in Fig. 13-15d.

A number of similar discriminator circuits have been described that are more sensitive but also more complex than the slope detector. Of these, the Foster-Seeley[1] circuit is most widely used. The circuit is illustrated in Fig. 13-16a and operates as follows. The FM input signal e_m is first led through a limiter A and is then applied to the discriminator circuit in two ways. First, it is injected via the capacitor C_o to the center tap of the coil L_2. The potential

[1] D. E. Foster and S. W. Seeley, *Proc. IRE*, **25**:289 (1937).

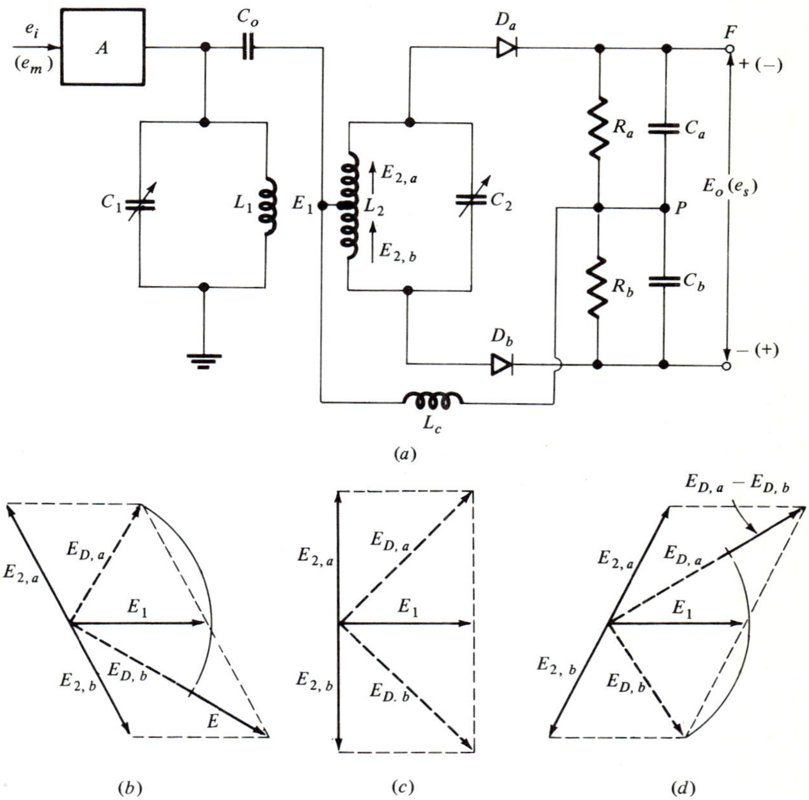

Fig. 13-16 (*a*) Circuit diagram and operational vector diagrams of the Foster-Seely FM discriminator; (*b*) for e_i at a frequency lower than f_c; (*c*) for e_i tuned for center frequency, f_c; and (*d*) for f_{ei} higher than f_c.

at the center tap is E_1. This potential is fixed, i.e., it does not change phase or amplitude if the frequency of the input signal changes. Second, the output from the limiter is applied to the tank circuit L_1C_1 and, by inductive coupling, to the tank circuit L_2C_2. Both tank circuits are tuned to resonance for the center frequency f_c. A change of frequency of the input signal changes the phase angles as shown in Fig. 7-16*d*. The tank circuit is inductive at signal frequencies f_s lower than the resonance frequency f_r; it is resistive at $f_s = f_r$; and it is capacitive at frequencies f_s higher than f_r. The voltage induced in the center-tapped coil L_2 is composed of two partial voltages E_{2a} and E_{2b}. A dc return path leads from the connecting point P via an RF choke L_c back to the center tap of L_2.

There are now the following combinations of voltages applied to the two diodes D_a and D_b: the vector sum of E_1 and E_{2a} is applied to the diode D_a; the vector sum of E_1 and E_{2b} is applied to the diode D_2. Of these voltages, E_1 is constant while the phase angles of E_{2a} and E_{2b} vary with the frequency of the input signal. The vector diagrams for three different input frequencies are shown in Fig. 13-16*b* to *d*. The dc output e_o is simply the difference of the two vectors $\mathbf{E}_{\mathrm{D_a}} - \mathbf{E}_{\mathrm{D_b}}$. For the center frequency, the dc outputs E_a and E_b across the resistances R_a and R_b are equal; the output is zero. For lower or higher frequencies the outputs are either positive or negative as is evident from Fig. 13-16*b* to *d*.

MULTIVIBRATORS

Multivibrators are switching elements, similar in operation to manually operated switches or electrically operated relays. As illustrated schematically in Fig. 13-17*a*, they are activated by the pulsed electric input signal itself, rather than by an electric or electromagnetic auxiliary signal. A multivibrator has two stable states only, i.e., if in Fig. 13-17 a condition should arise which can be expressed by an intermediate position of the contact arm N between A and B, such a condition would cause an instability in the multivibrator circuit which would correct itself immediately by establishing a new contact position $N \rightarrow A$ or $N \rightarrow B$.

A multivibrator can be constructed by combining the two inverters A and B of Fig. 13-17*b* in such a way that the output of A is connected to the input of B and the output of B is connected to the input of A. A circuit diagram of a simple multivibrator is shown in Fig. 13-18*a*. The circuit contains two transistors Q_1 and Q_2. The voltage across Q_1 is subdivided by voltage-dividing resistors R_1 and R_4; the voltage across R_4 is applied to the base of the transistor Q_2. The voltage across Q_2 is divided by resistors R_2 and R_3; the voltage across R_3 is applied to the base of the transistor Q_1. The circuit has two stable states (*bistable* or *flip-flop operation*): either Q_1 is conductive (is on) while Q_2 is off, or Q_2 is on and Q_1 is off. The following consideration will show that either of these conditions constitutes a stable state.

When the circuit is first energized, a transient condition may prevail such that both transistors may carry some current. However, since the characteristics of the two transistors will never be completely alike or since the circuit may never be completely symmetrical, one transistor will always draw more

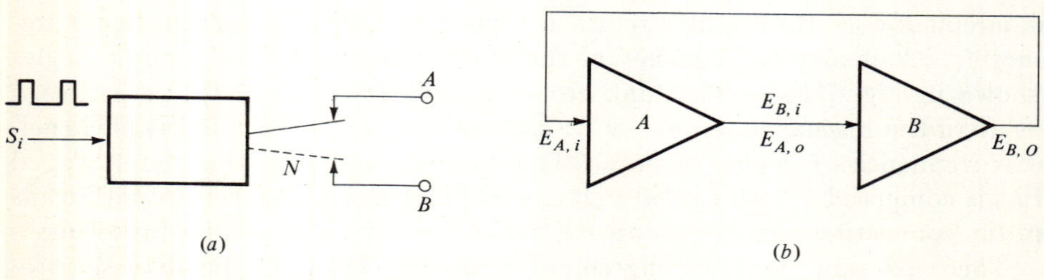

Fig. 13-17 (*a*) Schematic diagram of a multivibrator; (*b*) the output from an inverter A applied to the input of inverter B and the output from B applied to the input of A.

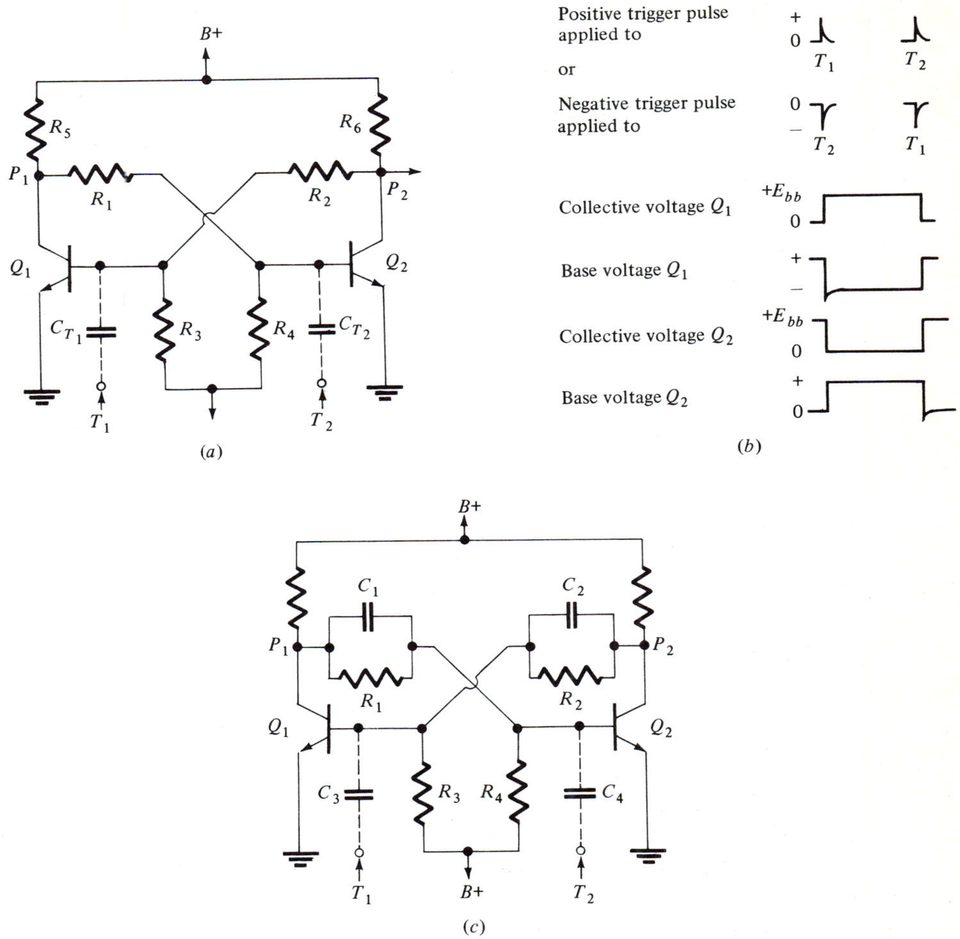

Fig. 13-18 (*a*) Circuit diagram of a bistable multivibrator; (*b*) synopsis of voltage variations in different parts of a bistable multivibrator; (*c*) bistable multivibrator, circuit diagram incorporating the commutating capacitors C_1 and C_2.

current than the other. If the transistor Q_1 draws a current that is slightly higher than that of Q_2, then its collector voltage (the potential of P_1) will be lowered more than that of Q_2. This change of potential is transferred via the resistor R_1 to the base of Q_2 and causes a reduction of the current of Q_2. This reduction has the effect of increasing the collector potential of Q_2. The increase of the potential of P_2 is transferred via R_2 to the base of Q_1, increasing the collector current of Q_1. These processes will continue until the transistor Q_2 is turned off and Q_1 is turned on to saturation. The circuit is then in one of its two stable states. The bistable multivibrator will remain in this state until a trigger pulse applied to the circuit changes the operation to that of the other stable state.

The trigger pulse can be applied in two different ways. A negative pulse can be applied to the base of transistor Q_1 (for an *n-p-n* transistor as illustrated in Fig. 13-18*a*). Such a pulse has the effect of temporarily cutting off the collector current of Q_1 and turning on that of Q_2. Alternatively, a positive pulse can be applied to the base of Q_2, causing an instantaneous on condition of Q_2 and an off condition of Q_1. The output from the multivibrator is taken either between point P_1 and ground or between point P_2 and ground. A synopsis of voltage variations in different parts of a bistable multivibrator is shown in Fig. 13-18*b*.

The operation of a bistable multivibrator can be improved, and the transition from one stable state to the other can be accelerated by replacing the coupling resistors R_1 and R_2 by capacitive-resistive networks R_1C_1 and R_2C_2 as shown in Fig. 13-18*c*. The main purpose of capacitors C_1 and C_2 (*commutating capacitors*) is to improve the high-frequency response of the trigger action. Their capacitance value is not critical.

If sequential operation of the type *A-B-A-B*, etc., is required, the bistable multivibrator can be triggered either by pulses of the same polarity applied to alternate inputs or by pulses of alternate polarities applied to the same input. It is also possible to apply pulses of one polarity to the circuit through a steering system which automatically applies the triggering pulse to the proper transistor base. A circuit diagram of a multivibrator (using *n-p-n* transistors) is shown in Fig. 13-19*a*. In such a multivibrator the base of the on transistor is always positive, that of the off transistor is negative. One possible mode of operation of the steering circuit is then to direct a positive trigger pulse to the (negative) base of the off transistor. This can be accomplished by connecting the anodes of the two diodes D_1 and D_2 to a common point M of a voltage divider (resistors R_a and R_b) that is sufficiently positive so that a trigger pulse can flow only from the positive point M to the negative base, but not from the positive point M to the positive base.

Figure 13-19*b* shows the circuit diagram of an astable or free-running multivibrator. It differs from the bistable multivibrator by the capacitive coupling between the two transistor stages (in contrast to the resistive coupling of the bistable multivibrator). The stability considerations for the time immediately after energizing are the same as those described before. A slight difference between the two transistors Q_1 and Q_2 causes a transient condition which leads to the lapse of the circuit into a stable state. Assuming as before that Q_1 is then on and Q_2 is off, the capacitor C_2 is charged to almost the value of $B+$, the negative potential being applied to the base of transistor Q_2 and keeping it in a cutoff state. The capacitor will then discharge through resistor R_1 with a time constant of about C_2R_1. As soon as the negative base potential is sufficiently reduced, Q_2 will begin to conduct until Q_2 is completely on and Q_1 is off. The cycle proceeds then in the reverse direction. The astable multivibrator furnishes a sequence of pulses between P_1 and ground (or P_2 and ground). If $C_1R_2 = C_2R_1$, the square wave will be symmetrical, i.e., the duration of a duty

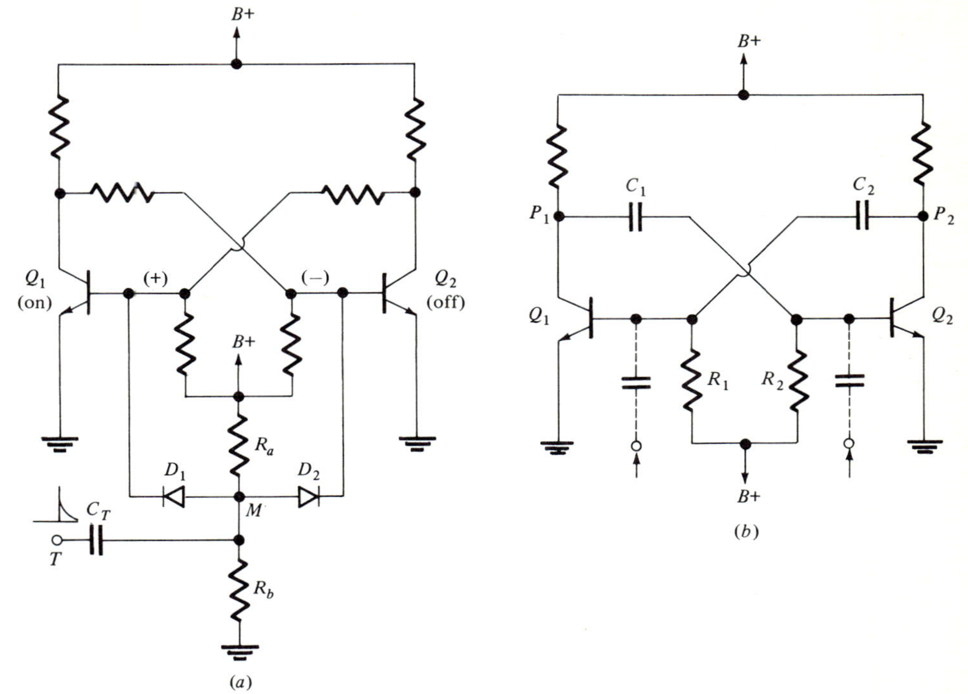

Fig. 13-19 (*a*) Bistable multivibrator with steering system consisting of the diodes D_1 and D_2. A trigger pulse applied to T causes the transition from one stable state into another; a further trigger pulse causes the reconversion into the first stable state. (*b*) Astable or free-running multivibrator, producing continuously a sequence of pulses between P_1 and ground or P_2 and ground.

cycle will be the same as that of the pause between two duty cycles. A pulse diagram is shown in Fig. 13-20*a*.

The free-running multivibrator can be synchronized with another oscillator by applying a sequence of trigger pulses into one of the transistor bases.

Besides the astable and the bistable operation, a multivibrator can be built with one stable state (monostable, one-shot, or flyback circuit). An example is shown in Fig. 13-20*b*. The circuit has one resistive coupling between stages and one capacitive coupling. If a negative trigger input is applied to transistor Q_1, it will cause Q_1 to change from an on to an off state. The multivibrator remains in this state for a time determined by the time constant CR of the circuit and returns to the stable state Q_1 on and Q_2 off. Monostable multivibrators can be used to convert trigger signals of variable lengths or heights into those of constant lengths and heights.

Multivibrators can be used in many instrumentation systems, such as counters, timers, oscillators, switching devices, dc-level detectors, time-delay devices, relays, pulse-shaping devices, and many more.

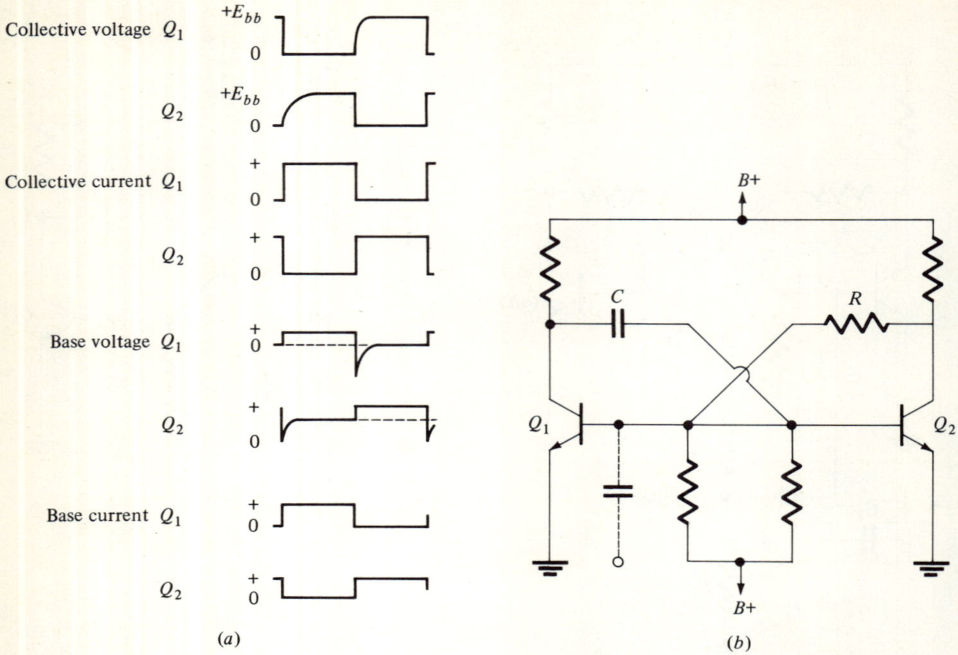

Fig. 13-20 (*a*) Synopsis of voltages and currents in different parts of an astable multivibrator; (*b*) circuit diagram of a monostable or flyback multivibrator.

DIODE-OPERATED ANALOG CIRCUITS

These are diode-operated circuits or networks that change their *EI* characteristics in response to the level of the applied signal. A diode can be considered a voltage-controlled switch and can be used as such in analog circuits. A number of such circuits are described below. Other circuits in which the diode can be used as a switching device for digital applications are described in Chap. 14.

Figure 13-21*a* and *b* shows the symbol and the *EI* characteristics of a diode when it is considered a voltage-controlled switch. Details about the voltage-current characteristics such as a curvature in the vicinity of the zero point can be neglected. When the anode is positive with respect to the cathode, the diode conducts with negligible forward resistance; when the anode is negative with respect to the cathode, the diode acts as an insulator or as an open switch.

An example of the application of a diode switch is shown in Fig. 13-21*c* and *d*. When the input voltage E_i is negative, the diode is forward-biased and behaves as a short circuit; therefore, there will be no voltage drop across it, that is, $E_o = 0$. When the input voltage is positive, the diode behaves as an open circuit, and the total input voltage appears at the output, that is, $E_o = E_i$. The resistance R is inserted to limit the current through the diode when the input voltage is negative.

The point P at which the slope changes can be varied by the insertion of a

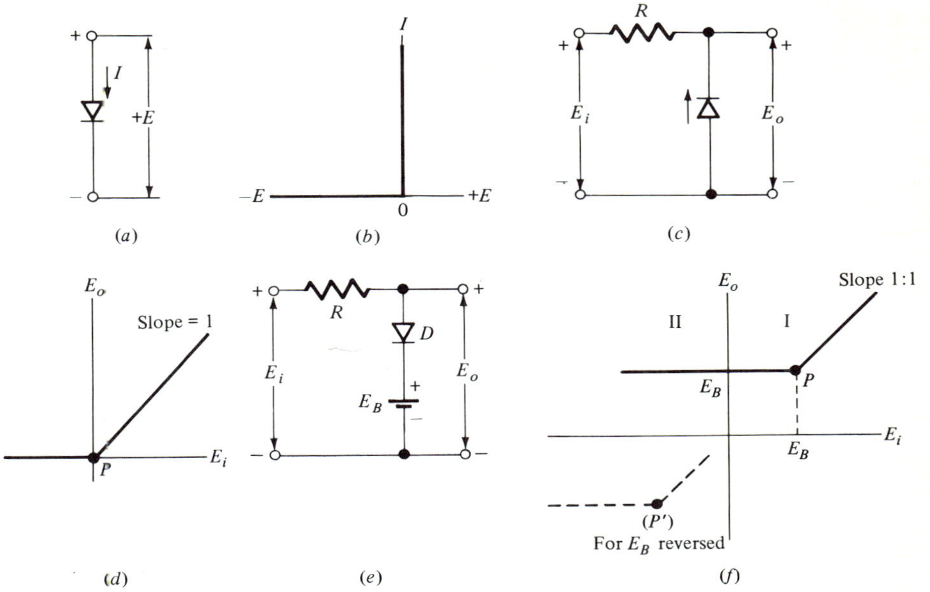

Fig. 13-21 Diode switch operation. (*a*) Symbol, polarity, and current direction of a semiconductor diode or generalized diode; (*b*) idealized *EI* characteristic; (*c*) circuit diagram of a diode switch with resistor R in the input; (*d*) transfer characteristic of the circuit; (*e*) same circuit but diode D biased by a source E_B; (*f*) transfer characteristics, solid line for polarity of E_B as shown; dotted line for polarity of E_B reversed.

bias voltage E_B in series with the diode as illustrated in Fig. 13-21*e* and *f*. The voltage applied to the diode D is composed of E_B and the applied input voltage E_i. A current I will pass through the diode as long as E_i, in the polarity shown, is smaller than E_B (or is negative). As long as a current passes through the diode, it can be considered a closed switch so that the output voltage is $E_o = E_B$. If the input voltage E_i exceeds the bias voltage E_B, the diode switch opens, and the output voltage is $E_o = E_i$. Reversal of the polarity of E_B has the effect that the point P comes to lie in the third quadrant as indicated by point P' and the dotted lines.

The slope of that part of the transfer characteristic for which the diode switch is closed can be varied by the insertion of a resistor R_D in series with the diode. Figure 13-22*a* shows a circuit of this type. As long as the input voltage E_i is smaller than the bias voltage E_B, the diode acts as an open circuit, and the output voltage E_o is equal to the input voltage E_i. If the input voltage, with the polarity as shown in Fig. 13-22*a*, exceeds the bias voltage E_B, the circuit acts as the voltage divider illustrated in Fig. 13-22*b*. The transfer function has a breakpoint P at $E_i = E$. The slope of the characteristic for $E_i < E_B$ is one, and for $E_i > E_B$, the slope is $R_D/(R + R_D)$ as shown in Fig. 13-22*c*.

The fact that the slope $R_D/(R + R_D)$ of the characteristic of a circuit like

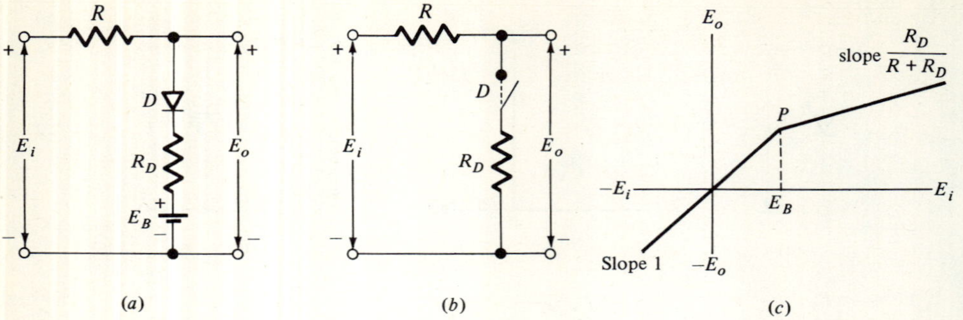

Fig. 13-22 (*a*) Circuit diagram of biased diode combined with resistive voltage divider RR_D; (*b*) equivalent circuit, diode D and bias E_B replaced by a switch S; (*c*) transfer characteristic.

the one shown in Fig. 13-22*c* can be varied over a wide range by varying the resistors R and R_D and that the position of the breakpoint P can be varied by the variation of the polarity and the magnitude of the bias voltage E_B offers the possibility of synthesizing networks that approximate, by linear segments, almost any desired nonlinear EI characteristic such as a quadratic or a logarithmic function. An example is shown in Fig. 13-23*a* and *b*.

This circuit is used to approximate a quadratic curve. The input to the circuit is E_i, the output is the current I, and the relationship which is to be approximated is $I = kE_i^2$ for the current range $0 < I < 9$ mA.

When E_i is negative (polarity shown in Fig. 13-23*a*), there will be no current in the circuit as all three diodes are open. When E_i becomes positive, diode D_o closes and a current $I = E_i/R_o$ will flow. The slope of the IE curve, therefore, becomes $G_o = 1/R_o$. When $E_1 < E_i < E_2$, both D_o and D_1 are conducting, and

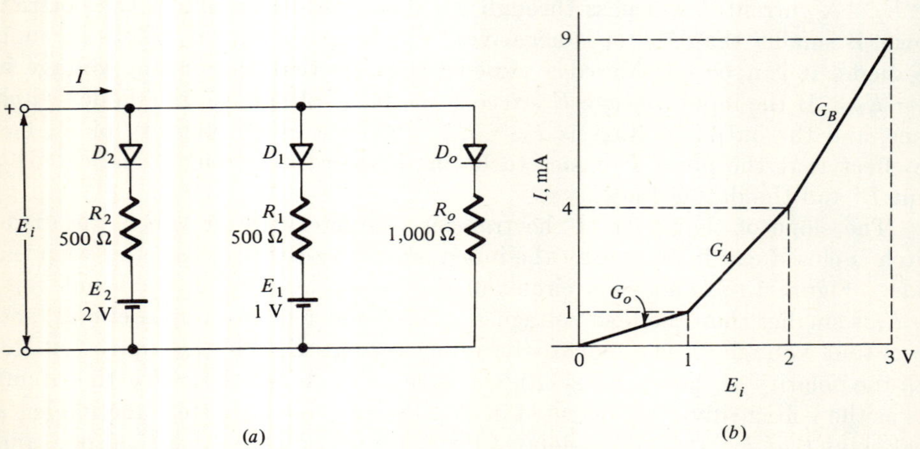

Fig. 13-23 (*a*) Diagram of a circuit which causes curved EI characteristic by synthesis of linear segments; (*b*) approximately quadratic characteristic formed from three segments.

the slope of the curve becomes $G_A = G_o + 1/R_1$. When $E_i > E_2$, all three diodes are conducting; all three resistors are connected in parallel; and the resulting slope of the IE curve is the sum of the three conductances, $G = G_o + G_A + G_B$.

Clipping circuits The diode switch circuit shown in Fig. 13-24*a* contains two basic diode switch circuits of the type shown in Fig. 13-22; the two diodes and their bias sources are in opposite directions. If an input voltage E_i is applied with the polarity as shown in Fig. 13-24 and if $E_i < E_1$, both diodes are open, and the output voltage E_o is equal to the input voltage E_i. For $E_i > E_1$, the

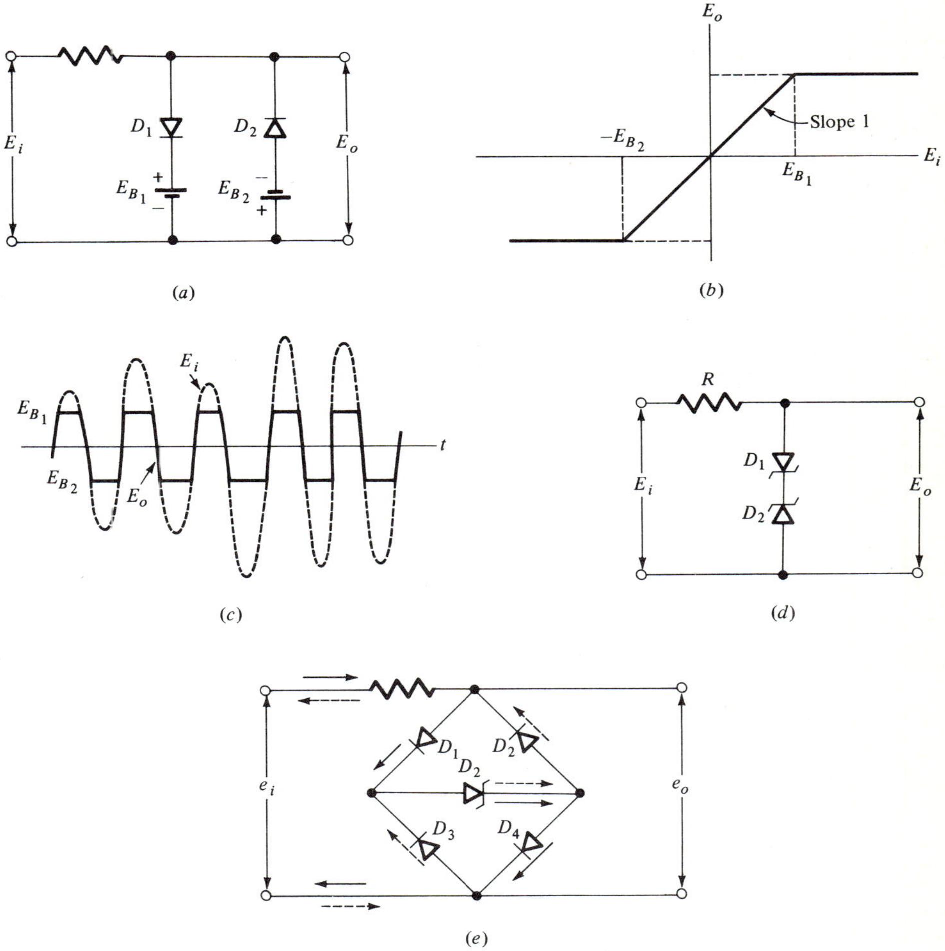

Fig. 13-24 Clipper or slicer. (*a*) Circuit diagram for a symmetrical clipper; (*b*) *EI* characteristic; (*c*) clipper action on a sinusoidal input E_i causing an approximately square-wave output E_o; (*d*) the same clipping effect as shown in (*c*) can be obtained with two zener diodes connected back to back; (*e*) circuit diagram for symmetrical clipping or slicing.

diode D_1 becomes a short circuit, and the output voltage E_o remains at the value of the bias voltage E_1. For an applied input voltage of opposite polarity, an analogous process will occur at the diode D_2. If E_1 and E_2 are alike, a transfer characteristic will result like that shown in Fig. 13-24*b*. If the applied voltage E_i alternates periodically (Fig. 13-24*c*) and varies with respect to amplitude and frequency, then the output voltage closely resembles a square wave of constant amplitude E_1 and of the same fundamental frequency as the input voltage, provided that the excursion of the input voltage is large compared to the bias voltage E_1. This clipper circuit can be used to separate any information contained in the frequency from that contained in the amplitude of the applied voltage.

The same result can be obtained by a circuit like that shown in Fig. 13-24*d*. The circuit contains two zener diodes connected back to back and a series resistor R. If the input e_i is positive, diode D_2 is conducting, but D_1 is open until the zener voltage breakpoint is reached. At that point, both diodes are conducting and the output voltage e_o can rise no farther. The same condition exists for the opposite polarity, so that the output e_o approaches a square-wave function like that shown in Fig. 13-24*c*.

If symmetry of the positive and of the negative half-waves is essential, and if a matched pair of zener diodes is not available, the modified bridge circuit shown in Fig. 13-24*e* may offer some advantages. As is evident from the arrows signifying the flow of current in the positive (solid arrows) and in the negative (dotted arrows) half-cycles, the current flows through the zener diode D_z in each half-cycle in the same direction. Therefore, the clipping points in both half-cycles are identical, provided that the characteristics of the diodes D_1 to D_4 are reasonably alike.

CLAMPING CIRCUITS, DC RESTORERS

Figure 13-25*a* shows an RC network frequently used as a coupling element between stages in multistage amplifiers. If a repetitive, bipolar, symmetrical signal e_i is applied to the input as shown in Fig. 13-25*b*, the output waveform will be bipolar and the average output $E_{o,\text{av}}$ will be zero. If the signal is unipolar (Fig. 13-25*c*), the output e_o will again be bipolar, and the average output $E_{o,\text{av}}$ will be zero. It is not evident from the output signal what the average (dc) value of the input signal is. If the dc value carries information, such information will be lost.

However, the dc value in the output can be restored with the clamping circuit shown in Fig. 13-25*d*. If the resistance of R is large compared with the forward resistance of the diode and if the time constant CR is large compared with the duration of one cycle of the repetitive input signal e_i, then the output waveform consists of the input waveform e_i pushed up by the dc voltage E_c developed across the capacitor C. This dc voltage E_c is approximately equal to the peak value E_i of the input voltage. The negative peak of the output waveform is then said to be held or clamped at the reference level (*zero potential*). Reversing the direction of the diode D reverses the potential of the charge of the capacitor C. The output waveform is then the input waveform depressed

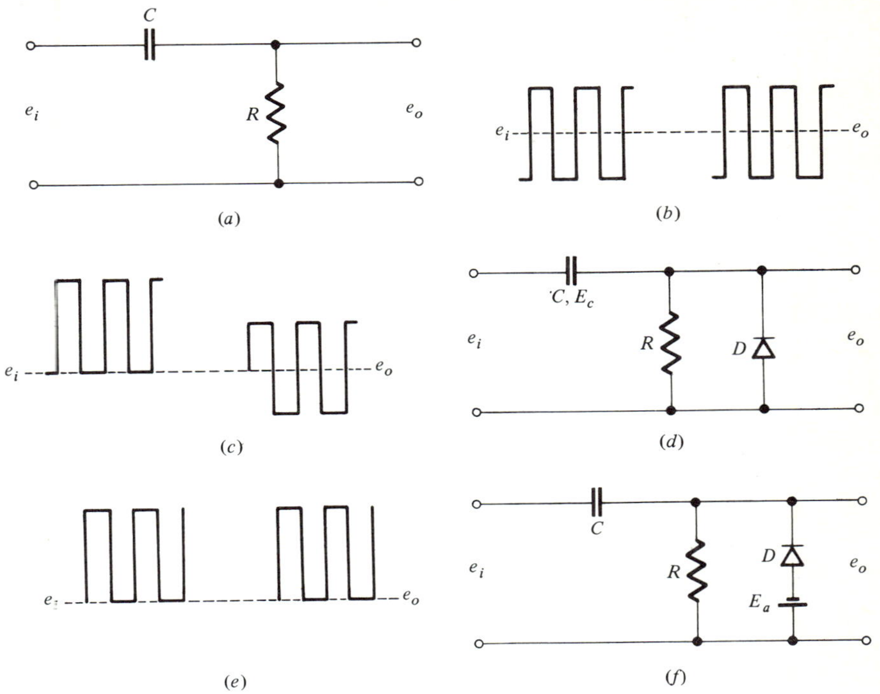

Fig. 13-25 Clamping circuit; see text for explanation.

by the voltage across the capacitor; the positive peak of the ac output signal is then clamped to the reference potential. The reference voltage may be increased or decreased by adding a dc voltage source in series with the diode of whatever magnitude or polarity is desired.

REFERENCES

Landee, R. W., D. C. Davis, and A. P. Albrecht: "Electronic Designers Handbook," McGraw-Hill, New York, 1957.

Mileaf, Harry (ed.): "Electronics Five, Six and Seven," Hayden, New York, 1967.

Zimmermann, H. J., and S. J. Mason: "Electronic Circuit Theory," Wiley, New York, 1959.

Mason, S. J., and Henry J. Zimmermann: "Electronic Circuits, Signals and Systems," Wiley, New York, 1960.

14
Digital Logic Elements and Circuits

INTRODUCTION

In an analog system the information is usually represented by the magnitude of the signal. For instance, in a display system like that illustrated in Fig. 14-1*a*, the signal is analogous to the displacement d of a pointer P along a calibrated scale. In such an analog system the signal and the information represented by the signal vary a continuous, stepless, fashion. The output signal from an analog transducer, be it voltage, temperature, frequency, phase, time, reactance, or any other physical quantity, is generally a continuous function of the transducer input signal.

The situation is different for the digital-display system illustrated in Fig. 14-1*b*. The signal which carries the information is generally presented by the on or off markings of discrete levels in a counting system. For instance, in Fig. 14-1*b* the signal consists of three separate figures or characters: 2, 6, and 7. The display of these characters is discontinuous. They are arranged in three vertical columns and in each column within discrete boxes according to their magnitudes (0, 1, 2, . . . , 9). They can change by nothing smaller than one unit of the least significant digit (see below). To find the total value of the display, multiply each figure with its column number and add the obtained values. In the

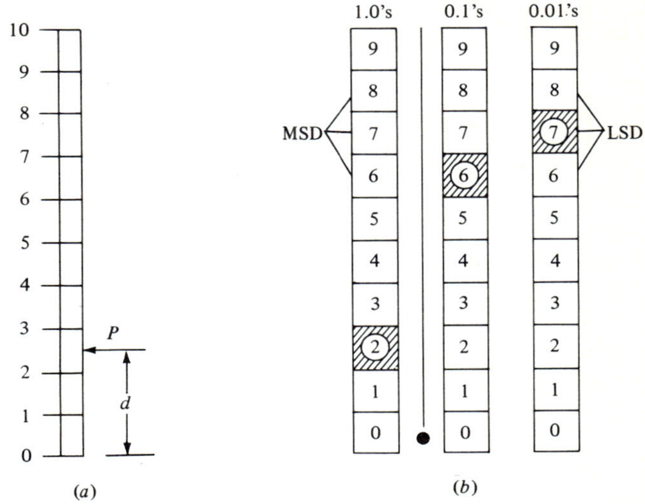

Fig. 14-1 Display faces of instruments. (*a*) For analog signals; (*b*) for digital signals.

example of Fig. 14-1*b*, the value is

$$A = (2 \times 1.0) + (6 \times 0.1) + (7 \times 0.01) = 2.67$$

A change within the left digit (1.0) causes a considerably larger change in the sum than a change of position in the right digit (0.01). Therefore, the left digit is called *the most significant digit* (MSD), and the digit at the right is *the least significant digit* (LSD).

There is no variety of direct-acting transducers similar to those in analog systems but furnishing a digital output. Analog transducers combined with analog-to-digital converters take the place of digital output transducers. Only the shaft-position encoder furnishes a true digital output. Of the different existing technical variations, a relatively simple encoder is shown in Fig. 14-2. It

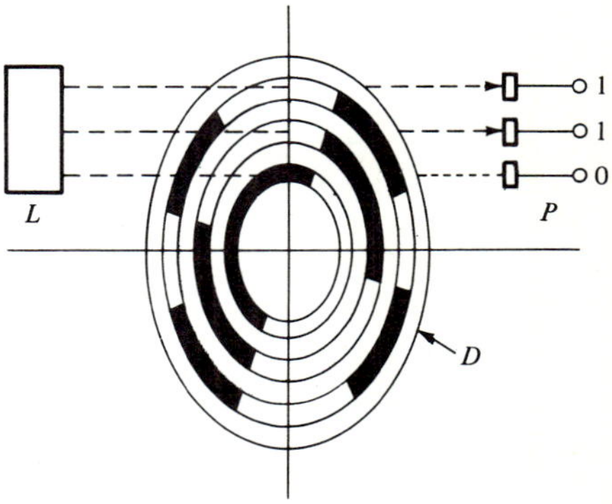

Fig. 14-2 Schematic diagram of a shaft-position encoder.

consists of a transparent disk carrying an opaque pattern in concentric circular bands. The light passing through the transparent portions is received by three photosensitive transducers. In the position shown in Fig. 14-2, the outputs from the three photoelectric cells are 0, 1, 1, respectively. If the disk is rotated 45° counterclockwise, the output from the same photoelectric transducer is 1, 0, 0. The intensity of the light falling on the photosensitive cells and the magnitude of the photoelectric output signal do not enter into the result as long as the photoelectric output signal is above the noise level. The subsequent digital equipment recovers the signal by reconstructing the position of the encoder disk at any time. The system does so by the selective passing of input signals into output channels through electronic switches (called *gates*) that are set by the incoming signals. Each binary digital circuit element can exist in either of two states: 0 or 1, on or off, true or false. There are several gate types used in digital circuits; they are illustrated in Fig. 14-3.

AND gates The commonly used symbol of an AND gate with two inputs A and B is shown in Fig. 14-3*a*, and that of an AND gate with three inputs in Fig. 14-3*b*. Figure 14-4*a* shows the circuit of an AND gate for two input signals E_A and E_B. The AND gate provides an output signal E_o only if inputs E_A and E_B are present (are 1). In the absence of input signals from the sources S_A and S_B, both diodes D_A and D_B conduct, and the output E_o is zero. If an input signal

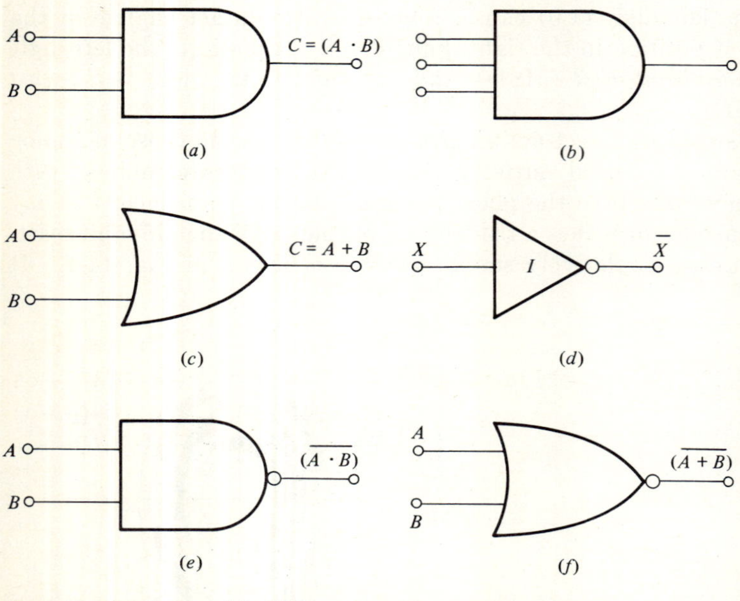

Fig. 14-3 Symbols. (*a*) AND gate with two inputs, A and B; (*b*) AND gate with three inputs; (*c*) OR gate with two inputs A and B; (*d*) inverter; (*e*) NAND gate; (*f*) NOR gate.

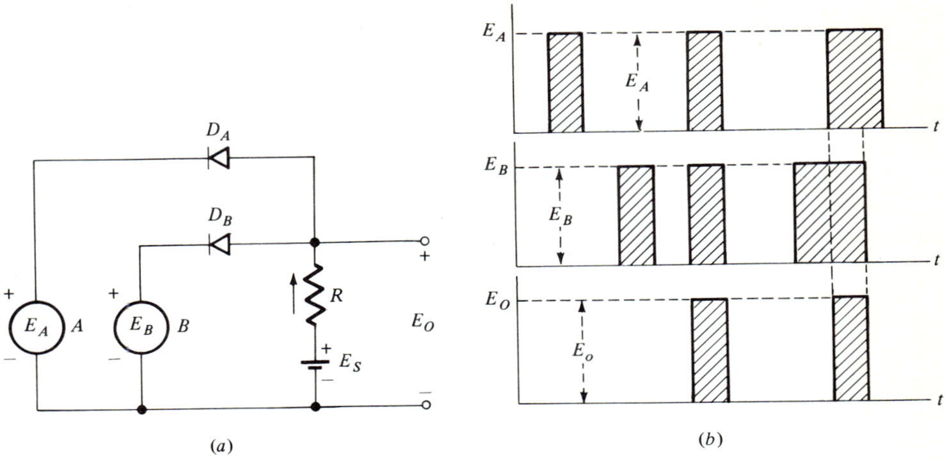

Fig. 14-4 (*a*) Circuit diagram of an AND gate for two input signals E_A and E_B; (*b*) *Et* diagram of an AND gate.

E_A is applied to the diode D_A and if $E_A > E_B$, the diode D_A will stop conducting. However, the diode D_B still conducts, and the output remains zero. If two signals of a magnitude E_B or larger appear at the same time at the sources S_A and S_B, then both diodes will be open, and an output $E_o = E_B$ will result as shown in Fig. 14-4*b*. Table 14-1 shows the output signal for all possible combinations of input signals (*truth table*). It is evident that an output signal occurs only when both inputs (or generally all inputs) are on. The output C in terms of the input signals A and B is expressed by the symbolic logic equation

$$C = A \cdot B \qquad \text{(alternatively } C = AB\text{)}$$

(read: C equals A *and* B).

The circuit diagram of Fig. 14-4*a* is, of course, only one example of an AND gate. Gates can be built in a variety of ways using diodes, zener diodes, transistors, relays, etc. Most gates are built in the form of integrated circuits and are sometimes considerably complex.

If one of the input signals is a pulse train with an amplitude greater than or equal to E_E and the other is a time-varying signal with an amplitude less than E_B,

Table 14-1

A	B	$C\ (= A \cdot B)$
0	0	0
0	1	0
1	0	0
1	1	1

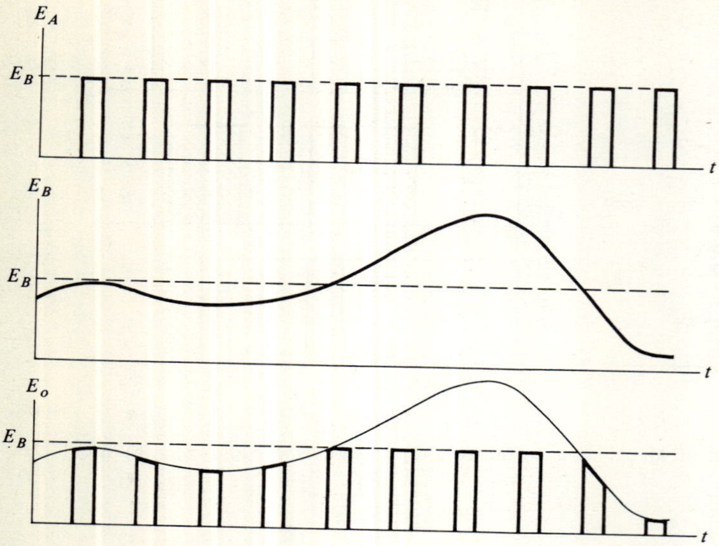

Fig. 14-5 Sampling or pulse modulation; see text for explanation.

the circuit in Fig. 14-4*a* can be used as a pulse-amplitude modulator or sampler as illustrated in Fig. 14-5. If the amplitude of the modulating signal is greater than E_B, then the output will not rise further than E_B; the modulation is *clipped.*

The AND gate can also be used for the detection of a coincidence of pulses, a problem frequently encountered in nuclear instrumentation or in certain statistical problems. For this purpose, two nuclear detectors (e.g., scintillation counters S_A and S_B) are provided as illustrated in Fig. 14-6. The arrival of

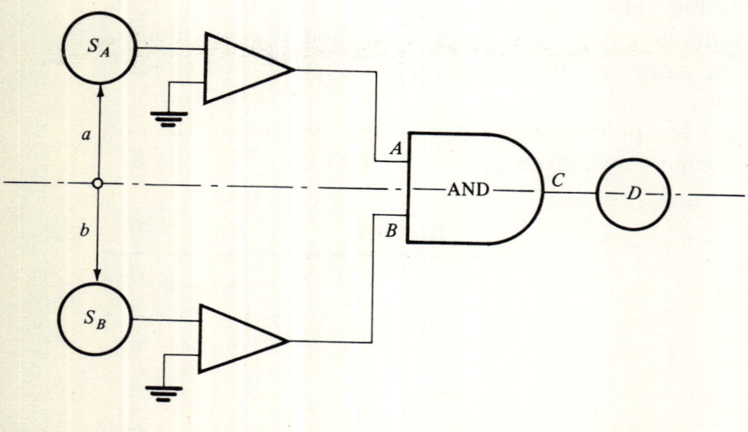

Fig. 14-6 Schematic diagram of coincidence circuit. S_A and S_B, scintillation counters. The outputs are amplified and applied to an AND gate. An output C appears only if the signals A and B coincide.

nuclear particles a and b at the detectors causes the formation of current pulses in the scintillation counters. The pulses are amplified to the proper levels and applied to the input terminals of an AND gate. If the pulses arrive at different times at terminals A and B, they will be blocked by the AND gate and will cause no output signal. Only current pulses arriving simultaneously at A and B will cause an output signal at C and will be recorded by the counter D. Coincidence circuits are not limited to two inputs but can be used for three or more coinciding signals by using AND gates with three or more inputs. Gates that indicate the anticoincidence of two events, i.e., that appear only when the two events do *not* occur simultaneously also exist (NAND gates) and are described below.

OR gates Figure 14-7*a* shows a circuit of an OR gate for two input signals E_A and E_B. The OR gate provides an output (that is, E_o is 1) if either the input E_A *or* the input E_B (*or* both inputs E_A and E_B) are 1. An output 0 appears only if all inputs are 0. With no signals applied to the input terminals of sources S_A and S_B, diodes D_A and D_B are reverse-biased by the bias source E_B, and the output E_o is 0. If a voltage source E_i which exceeds the bias voltage E_s is applied to D_1, diode D_A is forward-biased, and a signal E_o appears at the output. The same is true if a signal exceeding E_s is applied to source terminal S_A. If two signals E_A and E_B are applied and if both are alike, the output will be the same as if only one of the input signals were applied. But if, say, E_A is larger than E_B, then the output E_o will be that imposed by the larger input as illustrated in Fig. 14-7*b*. The two circuits S_A-D_A-P and S_B-D_B-P in Fig. 14-7*a* act as a buffer. They permit a separate control of the potential of P by the two sources S_A and S_B, but they block the flow of any signal from S_A to S_B or reverse.

The commonly used symbol of an OR gate with two inputs A and B is

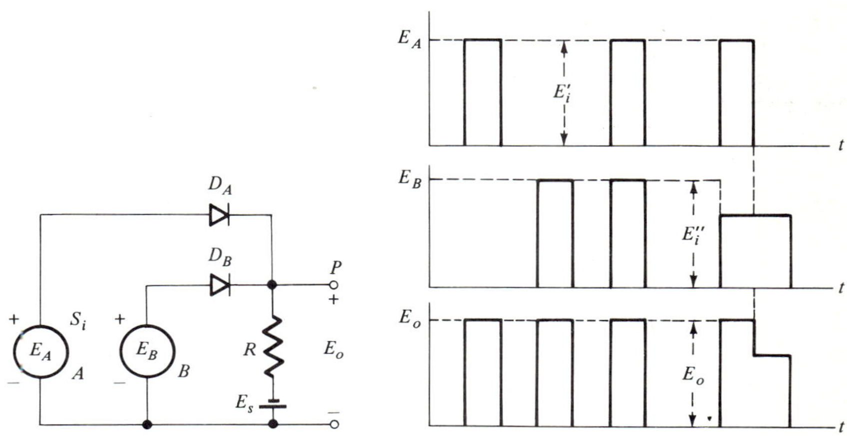

Fig. 14-7 OR gate. (*a*) Circuit diagram; (*b*) *Et* diagram of an OR gate.

Table 14-2

A	B	$C = A + B$
0	0	0
0	1	1
1	0	1
1	1	1

shown in Fig. 14-3*c*. The output C in terms of the input signals A and B is expressed by the symbolic logic equation

$$C = A + B$$

(read: C equals A *or* B). A truth table for the OR gate with two inputs is shown in Table 14-2.

Inverter gates The inverter gate can be described as an inverting amplifier with a voltage gain of one. If the input to an inverter is a logical 1, the inverter output is 0; if the input is 0, the output is 1. The relationship is shown in the truth table, Table 14-3. In general terms, if the input to the inverter is A, the output is $\bar{A}$ (read: NOT A). The inverter gate symbol is shown in Fig. 14-3*d*.

NAND gates The symbol of the NOT-AND gate or NAND gate for two inputs A and B is shown in Fig. 14-3*e*. The NAND gate consists of an AND gate followed by an inverter. If the inputs to a NAND gate are A and B, the output is $(\overline{A \cdot B})$. If the inputs are $\bar{A}$ and $\bar{B}$, the output is $(\overline{\bar{A} \cdot \bar{B}})$. By the application of De Morgan's theorem (see below), the output $(\overline{\bar{A} \cdot \bar{B}})$ can be converted into the form $(A + B)$.

NOR gates The NOT-OR gate or NOR gate is an OR gate followed by an inverter. The customary symbol of a NOR gate with two inputs is shown in Fig. 14-3*f*. If the inputs to the NOR gate are A and B, the output from the NOR gate is $(\overline{A + B})$. If the inputs are $\bar{A}$ and $\bar{B}$, the NOR gate output is $(\overline{\bar{A} + \bar{B}})$ which, as will be shown below, can be converted into $(A \cdot B)$.

Logic circuits An example for the application of an AND gate in digital logic systems is the equation

$$D = A \cdot \bar{B} \cdot \bar{C}$$

Table 14-3

A	$\bar{A}$
0	1
1	0

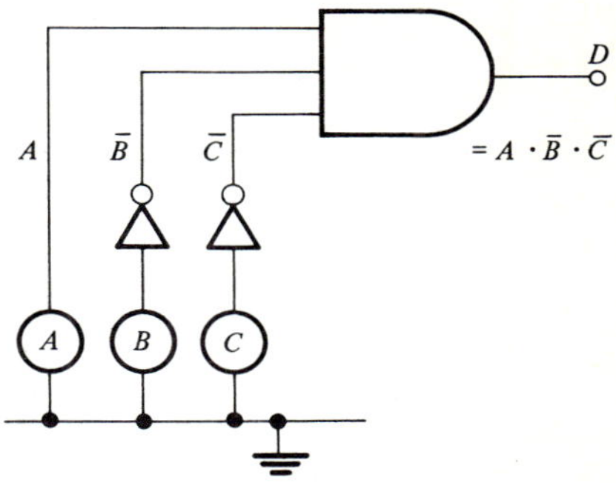

Fig. 14-8 Logic circuit for the equation $D = A \cdot \bar{B} \cdot \bar{C}$.

The logic circuit for this equation is shown in Fig. 14-8. Each input A, B, or C can be either 0 or 1. The truth table for this circuit is shown in Table 14-4 and indicates that of all (eight) possible input conditions, only one causes an output of 1, namely, $A = 1$, $B = 0$, and $C = 0$.

The circuit of Fig. 14-8 can be used for the identification of chemical or biological constituents in complex systems. If the blood of a donor is mixed with that of a recipient, one of two effects will result. If the two blood samples are of the same blood group (are compatible), the blood cells, after mixing, remain smoothly suspended in solution. If they are not compatible, the sample will agglutinate, i.e., form clumps. Such an agglutination may be fatal if it occurs within the human body. Therefore, testing the blood prior to transfusion is required. This is done by mixing small samples of blood with reference substances (for instance, unknown serum with accurately typed reference blood cells). A simplified table of the expected mixing reaction is shown in Table 14-5. The logic circuit is shown in Fig. 14-9. The inputs (A_1), (A_2), and (B)* are either 1 if agglutination occurs, or 0 if the two mixed constituents are compatible. The outputs D(O), D(A_1), and D(B) are signals which indicate the unknown blood group.

* Letters in parentheses in the text as well as in Fig. 14-11 and Table 14-5 refer to the customary symbols for blood groups.

Table 14-4

A	B	C	$D = A\bar{B}\bar{C}$
0	0	0	0
0	0	1	0
0	1	0	0
0	1	1	0
1	0	0	1
1	0	1	0
1	1	0	0
1	1	1	0

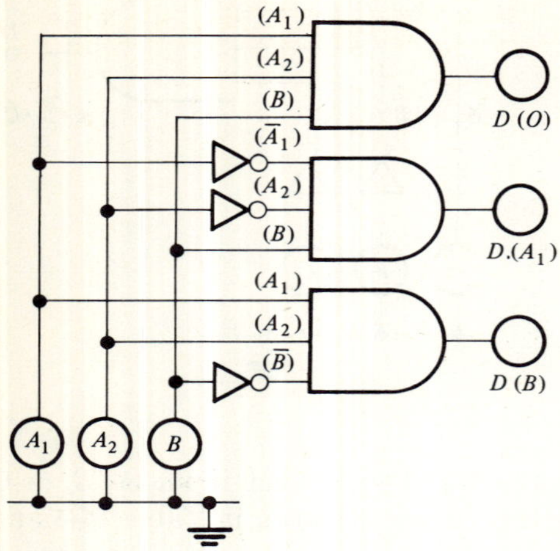

Fig. 14-9 Logic circuit for a blood group testing system; see text for explanation.

Most logic systems are combinations of different gate types, i.e., AND, OR, NAND, and NOR gates. For example, the logic equation

$$D = A \cdot B + C$$

may be implemented by the circuit in Fig. 14-10. The signals A and B are applied to the AND gate A_1. The output signal from A_1 as well as the signal C are applied to the OR gate B_1. The result, signal D, appears at the output of the OR gate B_1. A truth table for this circuit is shown in Table 14-6.

Boolean[1] logic equations The design of a logic circuit can lead to considerable complications. It is advisable, therefore, to simplify the logic equation before

[1] George Boole (1815–1864), English mathematician and logician.

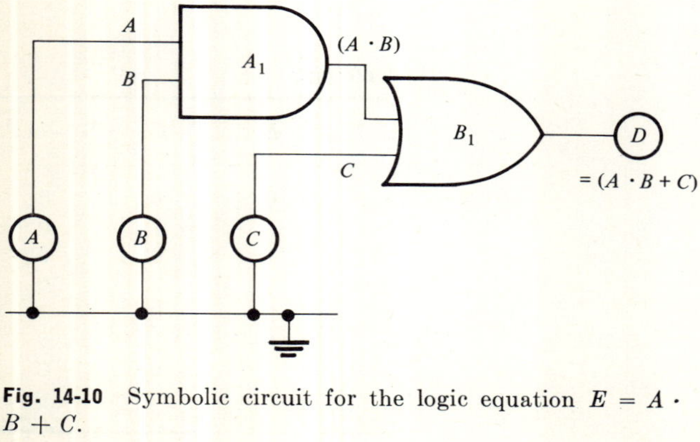

Fig. 14-10 Symbolic circuit for the logic equation $E = A \cdot B + C$.

Table 14-5

Unknown serum	*Reference cells*		
Blood group	(A_1)	(A_2)	(B)
(O)	1	1	1
(A_1)(A_2)	0	0	1
(B)	1	1	0

Letters in parentheses signify the common symbols of blood grouping.

implementing it. Logic equations can be manipulated in much the same way as standard algebraic equations. Some of the rules for manipulating logic expressions are (Fig. 14-11 illustrates logic expressions 1 to 8):

1. $X + 1 = 1$ — whether X is 1 or 0, the output is 1
2. $X + 0 = X$ — when $X = 1$, the output is 1; when $X = 0$, the output is 0
3. $X \cdot 0 = 0$ — whether $X = 1$ or 0, the output is 0
4. $X \cdot 1 = X$ — when $X = 1$, the output is 1; when $X = 0$, the output is 0
5. $X + X = X$ — when $X = 1$, the output is 1; when $X = 0$, the output is 0
6. $X \cdot X = X$ — when $X = 1$, the output is 1; when $X = 0$, the output is 0
7. $X \cdot \bar{X} = 0$ — when $X = 1$, $\bar{X} = 0$, and the output is 0; when $X = 0$, $\bar{X} = 1$, and the output is 0
8. $X + \bar{X} = 1$ — when $X = 1$, $\bar{X} = 0$, and the output is 1; when $X = 0$, $\bar{X} = 1$, and the output is 1
9. $X + Y = Y + X$, $X \cdot Y = Y \cdot X$ — Commutative law
10. $(X + Y) + Z = X + (Y + Z)$, $(X \cdot Y) \cdot Z = X \cdot (Y \cdot Z)$ — Associative law
11. $XY + XZ = X \cdot (Y + Z)$ — Distributive law
12. $X + YZ = (X + Y)(X + Z)$

An example which demonstrates how a logic equation can be simplified with the help of the rules discussed above is the following. Assume an equation is given in the form

$$D = \bar{A}(BA + C) \tag{14-1}$$

Table 14-6 Truth table for $D = AB + C$

A	B	C	$D = AB + C$
0	0	0	0
0	0	1	1
0	1	0	0
0	1	1	1
1	0	0	0
1	0	1	1
1	1	0	1
1	1	1	1

With the application of the distributive law, it can be transformed to

$$D = \bar{A}BA + \bar{A}C$$

since $\bar{A} \cdot A = 0$ (rule 7)

$$D = (0 + \bar{A}C) = \bar{A}C \tag{14-2}$$

A circuit to solve Eq. (14-1) requires two AND gates, one OR gate, and one inverter. Equation (14-2) requires only one AND gate and one inverter.

De Morgan's[1] theorem offers additional methods for the transformation and

[1] Augustus De Morgan (1806–1871), English mathematician and logician.

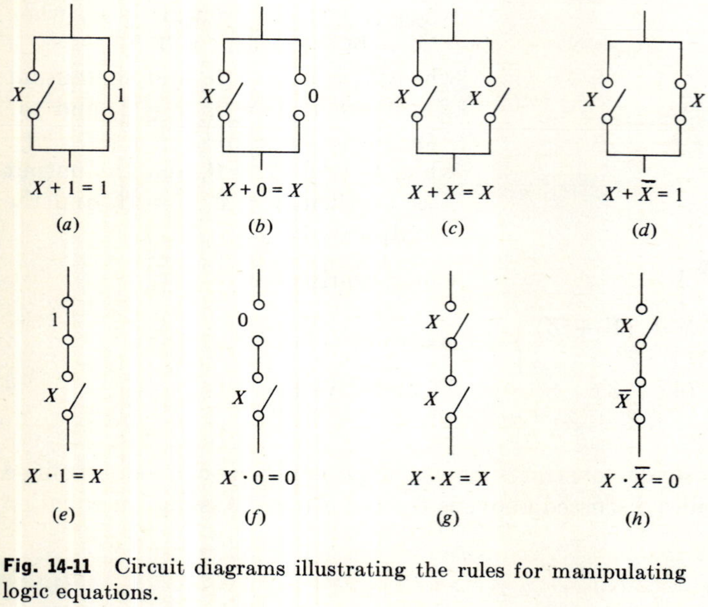

Fig. 14-11 Circuit diagrams illustrating the rules for manipulating logic equations.

simplification of logic expressions. The theorem states that

$$\overline{(X + Y)} = \bar{X} \cdot \bar{Y}$$

or $$\overline{X \cdot Y} = \bar{X} + \bar{Y}$$

In other words, inversion of an expression of logical operation results (1) in an inversion of each variable and (2) in a change of each operation. De Morgan's theorem is particularly useful for the conversion of logical operations which can be implemented by NAND gates or NOR gates since these gate types are frequently preferred. The following example illustrates a logic equation which is implemented for NAND gate realization.

The logic expression

$$W = \overline{\overline{[X\bar{Y}Z + XY\bar{Z} + \bar{X}YZ]}}$$

can be converted to

$$W = \overline{[\overline{(X\bar{Y}Z)} \cdot \overline{(XY\bar{Z})} \cdot \overline{(\bar{X}YZ)}]}$$

This expression is well suited for implementation with NAND gates. A circuit for it is shown in Fig. 14-12.

The following is a logic expression which is converted so that it is suitable for implementation with NOR gates:

$$\begin{aligned} W &= (X + Y) \cdot Z \\ &= \overline{[\overline{(X + Y) \cdot Z}]} \\ &= \overline{[\overline{(X + Y)} + \bar{Z}]} \qquad \text{(De Morgan)} \end{aligned}$$

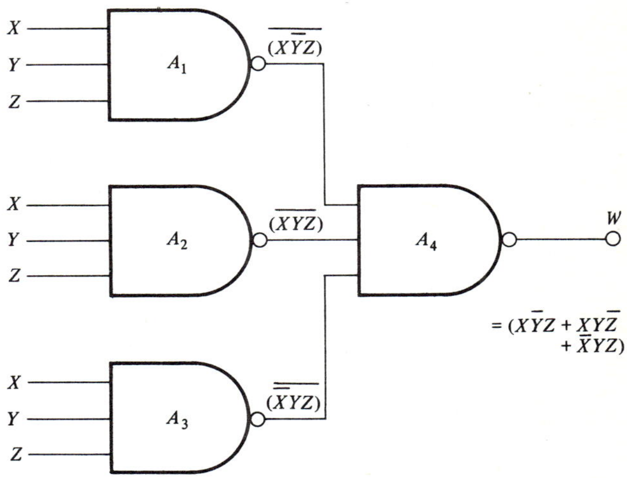

Fig. 14-12 Illustration of the implementation of the logic equation $W = X\bar{Y}Z + XY\bar{Z} + \bar{X}YZ$.

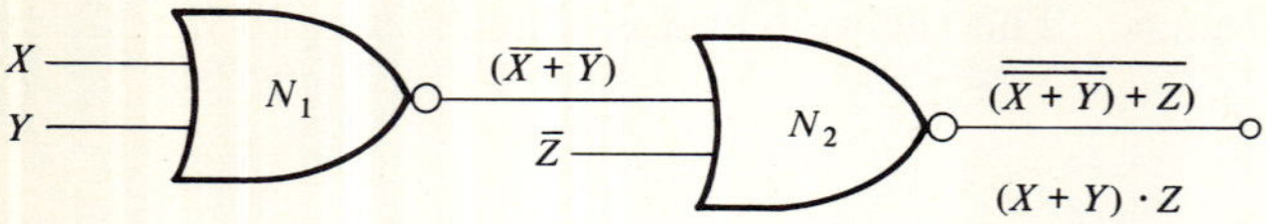

Fig. 14-13 Implementation with NOR gates of the logic equation $(X + Y) \cdot Z$.

In this form the expression can be implemented with NOR gates as shown in the logic circuit of Fig. 14-13.

Flip-flop circuits An instrumentation element frequently used in digital systems is the flip-flop or bistable multivibrator. The flip-flop has been described with other forms of multivibrators in Chap. 13, p. 294, and illustrated in Figs. 13-17 to 13-20. It has been shown that a flip-flop can be considered an aggregate of two inverters A and B, whereby the output signal of A is fed into the input of B, and the output of B into the input of A. The arrangement is redrawn in Fig. 14-14 using NOR gates N instead of inverters.

The flip-flop has two stable states which can be set or reset by applying a signal to either terminal S or R. Depending on the state of the flip-flop, the output terminal Q will be 1 or 0, while terminal $\bar{Q}$ will be 0 or 1, respectively. Once a flip-flop has been set to one of the two stable states, it will remain in that state indefinitely even if the applied signal is removed, unless it is reset by a new trigger pulse or the power is turned off. In other words, the flip-flop will "store" or "memorize" a binary digit (a *bit*), and the system can be used as a memory device.

In Fig. 13-19*a* a steered flip-flop was shown which could be triggered in either direction (*complemented*) by the application of pulses of the same direction. Complementing a flip-flop means changing the state of the flip-flop from whatever state it is into the opposite state. A flip-flop combining the properties of the R, S flip-flop with that of a steering flip-flop is frequently used (J-K flip-flop). The complementing pulse is applied to the terminal T. The customary symbol of a J-K flip-flop is shown in Fig. 14-17.

Besides its use as a memory system, a flip-flop can be used as a counter. The display face of a decimal digital counter is shown in Fig. 14-1*b* and that of a binary digital counter in Fig. 14-15. The binary digital counter operates in the following manner. Before the application of a signal, all indicators read 0. The application of a pulse to the first stage changes the indicator of the first stage to 1. The next pulse has the effect of resetting (complementing) the first stage back to

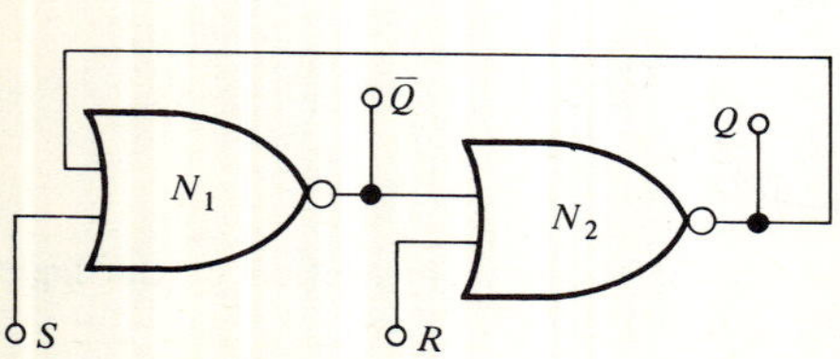

Fig. 14-14 Flip-flop arrangement composed from NOR gates N_1 and N_2; S_1 set; R_1 reset; Q, $\bar{Q}$, output terminals.

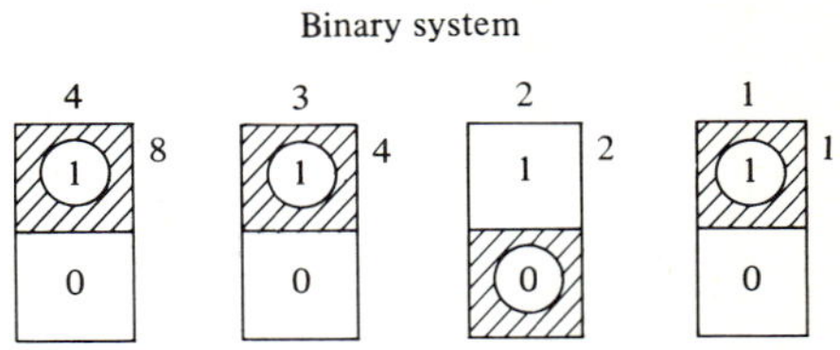

Fig. 14-15 Face diagram of a binary display system.

Indicated number in binary system notation: 1101; in decimal system notation: 8 + 4 + 0 + 1 = 13 .5

0 and advances the second stage from 0 to 1. A third input pulse changes stage 1 from 0 back to 1; stage 2 remains at 1. If the counting is interrupted at this point, the result would be:

$$\begin{aligned} &\text{Reading of the first stage } (2^0 = 1) \text{ is 1; value:} && 1 \times 1 = 1 \\ &\text{Reading of the second stage } (2^1 = 2) \text{ is 1; value:} && 1 \times 2 = 2 \\ &\text{Total value, sum:} && \underline{\underline{3}} \end{aligned}$$

If the counting is then continued, it will follow Table 14-7 up to $(2^3 - 1) = 7$ (the maximum for three stages). It will then return to 0 and repeat as before. If four stages are provided, the counter will advance to 15 and then return to 0.

Table 14-7 State table for a four-stage binary counter

Number of pulses in	
Decimal notation	*Binary notation*
	LSD
0	0000
1	0001
2	0010
3	0011
4	0100
5	0101
6	0110
7	0111
8	1000
9	1001
10	1010
11	1011
12	1100
13	1101
14	1110
15	1111
Stage:	4321
	MSD

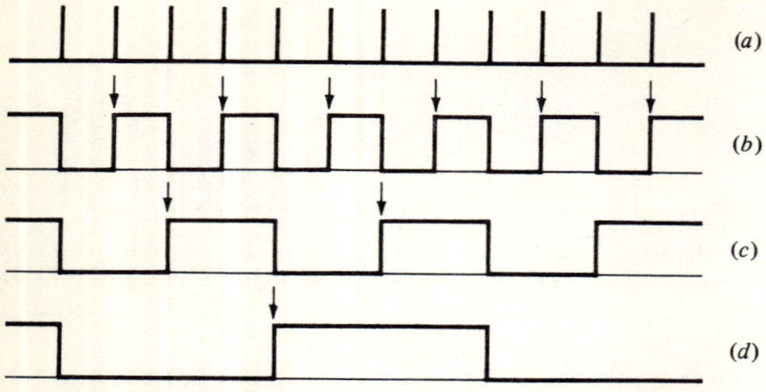

Fig. 14-16 State diagram of a three-stage counter. (*a*) Input signal (*monopolar pulse train*); (*b*) on and off states of stage 1; (*c*) on and off states of stage 2; (*d*) on and off states of stage 3.

Figure 14-16 shows the output of the same three-stage counter described above. The first row (*a*) illustrates the input signal, a monopolar pulse train. The changes of state of the individual counting stages 1, 2, and 3 are illustrated in rows (*b*), (*c*), and (*d*). If signal lights, e.g., glow lamps, are connected with the individual flip-flops, they will indicate the state of each individual counter. The transfer of the triggering action from one row to the next is done by a connection from the Q terminal of each stage to the next T terminal. This transfer takes place when the output from the preceding stage changes from 1 to 0 and is indicated in Fig. 14-16 by small arrows. Three binary stages cover a counting range from 0 to $(2^3 - 1) = 7$; n stages go from 0 to $(2^n - 1)$ and then start over again. A state table for a four-stage binary counter will be found in Table 14-7; a circuit diagram of a three-stage binary counter is shown in Fig. 14-17.

The counter circuit illustrated in Fig. 14-17 counts up from 0 to 111. However, if all connections between stages are made from $\bar{Q}$ to T rather than from Q to T, then the counter counts down. In the example of Fig. 14-17, it would count from 111 to 000. Such a counter can then be used for subtraction. It is possible to design counters that count both up and down.

The binary counting method is simple and performs adequately for many purposes. However, there is frequently a need for counters that count in the

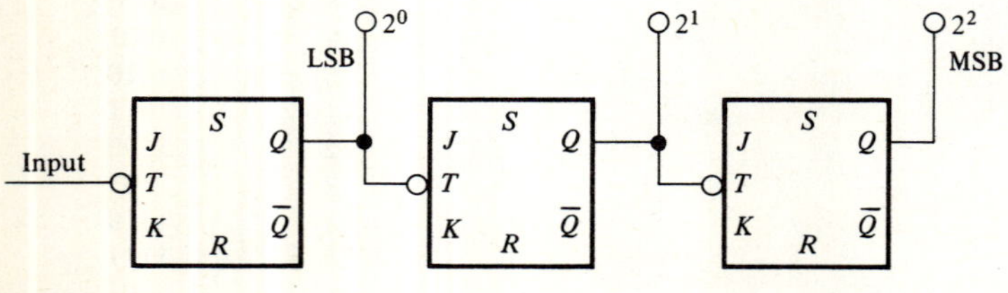

Fig. 14-17 Circuit diagram of a three-stage binary counter. LSB, least significant bit; MSB, most significant bit.

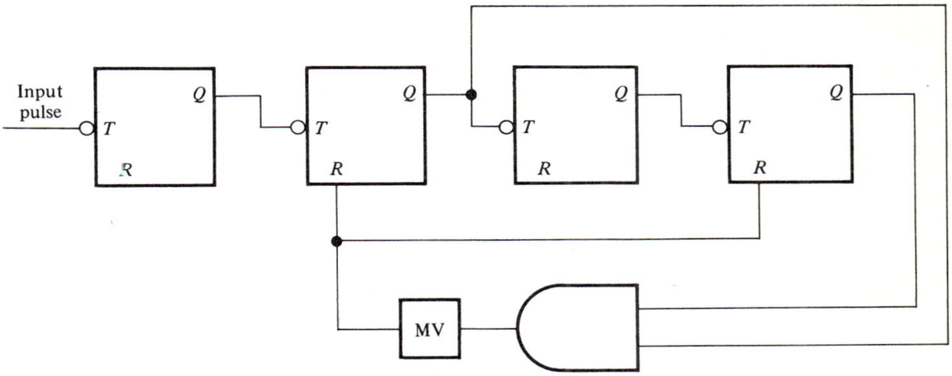

Fig. 14-18 Four-stage decade counter. At count 10 the multivibrator output causes the second and fourth stages to reset.

familiar decimal system. Decimal counters can be built in several different ways. A relatively simple circuit diagram of a decade counter is shown in Fig. 14-18 and operates as follows. One binary counter is provided that can count from 0 to 15. [The range of the counter (15) must always be as large or larger than the base of the counting system (10).] The counters are connected in a manner analogous to those in Fig. 14-17. When input signals are applied to the T input of the first counting stage, the counter will advance. At the tenth count, the counter reaches a value of 10, as determined by the second and fourth counting stages. Two signals derived from the counting stages 2 and 4 are now fed into an AND gate and cause the multivibrator MV to inject a reset signal into the counting stages. Instead of a tenth count, the counter reads 0, therefore, and counting by ten starts over again.

The multivibrator MV shown in Fig. 14-18 is a monostable time-controlled unit as described previously in Fig. 13-20, which when triggered, furnishes a square-wave pulse of sufficient length (time) and power to reset all counters required for the circuit in Fig. 14-18.

BCD code The binary system is widely used in computer applications. Another system, the *binary-coded decimal system* (BCD), is used primarily in instrumentation applications. In the BCD code, each decimal digit is expressed by a 4-bit sequence of binary code. For instance, the decimal number 9617 will be represented in BCD as

1001 0110 0001 0111

(for comparison, binary 9617 is 10010110010001). The BCD system is obviously more complex than the straight binary system, but it allows an easier reading particularly if the given decimal number has many digits. A comparison of decimal and BCD values will be found in Table 14-8.

Table 14-8 BCD/decimal comparison

Decimal	*BCD*
0	0
1	1
2	10
3	11
4	100
5	101
6	110
7	111
8	1000
9	1001
10	10000
11	10001
12	10010
20	100000

Digital-to-analog converters Frequently the problem arises of how to convert the digital output from a computer or from an instrument into an analog signal, generally a dc voltage or a current. The general scheme of digital-to-analog (D/A) converter is illustrated in Fig. 14-19, and the operation is as follows. The information may be stored in an array of flip-flops Q_0 to Q_3 (such as a register or a counter). The flip-flops are arranged so that in the on condition each provides the same output voltage. Each flip-flop is connected with a resistor and the resistances are chosen in the ratio 1:2:4:8. If the current from the

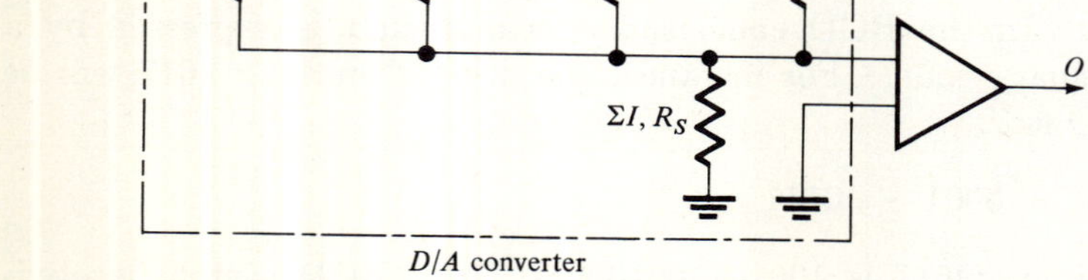

Fig. 14-19 Circuit diagram of a digital-to-analog converter connected to a set of flip-flops Q_0 to Q_3.

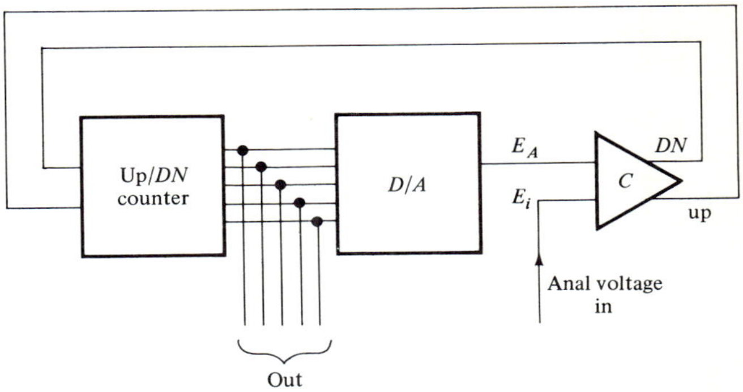

Fig. 14-20 Circuit diagram of an analog-to-digital converter D/A, digital-to-analog converter; C, comparator with two inputs and two outputs; up/down counter feeding into D/A converter.

flip-flop Q_3 is I, then that from Q_2 is $I/2$, that from Q_1 is $I/4$, and that from Q_0 is $I/8$. The resistors are connected in a current-summing network so that total current is

$$I_s = I_o + I_1 + I_2 + I_3$$

The voltage drop of the current I_s across R_s is measured with a high-impedance voltmeter, and this voltage is the digital-to-analog output from the flip-flops Q_0 to Q_3.

Analog-to-digital (A/D) conversion Conversion from analog to digital systems is more complicated than digital-to-analog conversion. The general scheme of a simple A/D converter is shown in Fig. 14-20. The A/D converter has a D/A converter as shown in Fig. 14-19. The output E_A from the D/A converter is fed into a comparator C. Also fed into the same comparator is the analog input voltage E_i which is to be given a digital representation. The comparator has two digital outputs: one indicating that $E_A > E_i$, the other indicating $E_A < E_i$. If $E_A > E_i$, one of the comparator outputs will be on, the other will be off. The presence of this signal causes the counter to decrement. Consequently, E_A, the output of the D/A converter, decreases until $E_A = E_i$. The output of the counter is then the digital equivalent of the analog input signal E_i. If $E_A > E_i$, the reverse action occurs.

Another method of A/D conversion utilizes the integration of equal pulses. The counting of the number of pulses required to raise the integral to the potential of the analog input provides the digital output.

Digital voltmeters (DVM) The operation generally resembles that of the A/D converters.

REFERENCES

Malmstadt, H. V., and C. G. Enke: "Digital Electronics for Scientists," Benjamin, New York, 1969.

McCluskey, E. J.: "Introduction to the Theory of Switching Circuits," McGraw-Hill, New York, 1965.

Hawkins, Joseph K.: "Circuit Design of Digital Computers," Wiley, New York, 1968.

Dakin, C. J., and C. E. G. Cooke: "Circuits for Digital Equipment," London Iliffe Books, Ltd., London, 1967.

appendix
Laboratory Exercises

For the student the purpose of an introductory instrumentation course is:

1. To acquire knowledge of the methods and instruments commonly used in research. Knowledge of the methods and physical mechanisms underlying the function of instruments is more important than knowledge of technical details. Knowing the underlying physical mechanisms allows their adaptation to special problems and is a permanent possession, even when new techniques are developed.
2. To become familiar with the concepts of physical quantities and magnitudes so that these concepts become manageable in a creative manner.
3. To learn general and some special experimental techniques and the application of instrumental methods to problems.

As a result of an instrumentation course, the student should have a broad knowledge and understanding of the basic instrumentation elements and principles, their applications, and their limitations. The student should be in a position to offer a choice of solutions for the more common instrumentation problems and to design or propose new methods and instruments for some special problems.

Part of this course is given in a student laboratory. Here the student should acquire practical experience by working with various methods and instruments. A better understanding of the basic physical quantities will be obtained by learning how to measure them—as distinct from defining or mathematically manipulating them. The student should also acquire the ability to think and experiment systematically.

Instruments and methods are the subject of a laboratory exercise (*the problem*) when the knowledge of their inherent physical mechanisms is required for the proper use of the instrument or the method. For instance, the response characteristic of an amplifier limits the application of the amplifier and can substantially influence the results obtained with the amplifier; therefore, amplifiers are treated in an instrumentation student laboratory. By contrast, a cathode-ray oscilloscope can be used if nothing but technical instruction is provided. Cathode-ray oscilloscopes may be used in an instrumentation course, and what little has to be learned about them can be acquired by using them, but there is no need to make them the object of study in this course.

It is more important for the student to learn how to *use* an instrument intelligently than how to *build* one. Years ago, research workers used to build galvanometers or amplifiers; nowadays, most instruments are purchased (e.g., operational amplifiers) and modified or adapted for the intended purpose. Nevertheless, having the student build an instrument by assembling it from laboratory parts can be an effective method of teaching its structure and its functioning.

The experiments cover not only the important elements, such as transducers and modifiers, but also the essential principles, such as response characteristics, feedback responses, noise and environmental influences, and some mathematical methods. The methods used in the experiments are practical and conventional, not "educational."

Experiments in an introductory course must be intrinsically simple and unsophisticated. It is more important to impress the student with the intrinsic simplicity of a natural process than with the sophistication of the equipment or the method. It requires courage to set up a simple experiment, but such an experiment is worthwhile if the student derives from it a basic concept of a physical quantity, a method, or the functioning of an instrument. It is the author's experience that a student eagerly takes up the opportunity to solve even difficult problems if he understands the basic concept of the experiment.

A student who carries out a laboratory exercise merely by following directions will gain little or no benefit from the laboratory. The student must be induced to think. The usual way to produce thought is to give the student a problem and motivate him to solve it.

Good work has been done with simple or makeshift equipment, but it does not follow that primitive equipment is needed to do good work. Time in research is frequently more costly than equipment; therefore, many research laboratories are well equipped with instruments.

Many instruments are new to the students and require adequate labeling, circuit diagrams, explanations, and technical instructions. Wherever possible we supply such technical instructions or circuit diagrams (sometimes simplified diagrams). We have developed some *ad hoc* instruments and boards with the circuit diagrams painted on the surface and the different parts of the diagram accessible through terminals and jacks. In several experiments we have given preference to commercial equipment and training sets.[1]

Rarely will an experiment work satisfactorily right after it has been assembled. If the student is aware that this may happen and if he can be reasonably certain that the commercial equipment functions satisfactorily, then such a negative outcome is not objectionable. In general, we let the student wrestle with such problems for a reasonable length of time. Frequently such things as wrong connections, wrong polarities, or unsatisfactory grounding conditions, etc., will be found. If the student does not succeed in getting the experiment to work, we expect the reasons for the failure to be found and confirmed so that the student "fails intelligently."

EXPERIMENT 1: ELECTRIC CIRCUIT DESIGN

The purpose of this experiment is to become familiar with the design of electric circuits. A number of problems are given below which require the design of specific circuits. To solve these problems it is helpful to draw a proposed circuit on paper first. The circuit should then be wired so that switches and bulbs are used to simulate other elements called for by the problem. Finally, the circuit should be tested and revised or improved if necessary.

No measurements are required in these experiments. In writing the report, it is sufficient to draw the circuit diagrams. It is essential to start from the beginning to draw circuit diagrams logically and clearly.

PROBLEMS

In all problems, where possible, one switch serves as the main switch and one light as the pilot light.

1-1. In an experimental setup a tank is equipped with a pressure switch and a temperature switch. A warning light should go on if either the pressure *or* the temperature reaches a critical value (OR circuit).

1-2. In an experimental setup a tank is equipped with a pressure switch and a temperature switch. A warning light should go on only if both the pressure *and* the temperature have reached a critical value (AND circuit).

1-3. Two animal cages are connected by a door that opens in both directions. Design a one-switch system which energizes one light if the door opens to one side and another light if the door opens to the other side.

1-4. Design a circuit for a hall light that permits turning the light on or off from a downstairs switch or from an upstairs switch, so that if either switch is turned, the light will go on if it was off and will go off if it was on.

1-5. Design a circuit similar to that in problem 1-4 but with three switches.

1-6. Three television cameras are in service in an experimental setup. A cen-

[1] The use of commercial equipment in our laboratory or the description of it in this book does not constitute an endorsement of a certain instrument or of manufacturer.

tral operator must be able to choose which of the three pictures is to be recorded. If the camera chosen is turned on, a light by the selector switch lights up and a light at the camera glows to indicate to the camera operator that his picture is being recorded (omit pilot lights).

1-7. Design a circuit with a three-position selector switch so that in each position of the switch two bulbs are energized, one with full power and one with adjustable (resistance-controlled) power (omit pilot lights).

1-8. A person building an acoustic system wants to connect two speakers with a variable resistor so that as the loudness of one speaker increases, the loudness of the other decreases.

1-9. One speaker is controlled by a rheostat and a switch is provided so that another speaker can be added and regulated by the same rheostat.

1-10. Same as 1-9, but the addition of the second speaker should not affect the loudness of the first.

1-11. In some psychological experiments rat cages are used that have a metal grid of, say, six bars isolated from each other. To expose a rat to an electric shock, a voltage is applied to two groups of bars. The rat soon learns to avoid the shock by standing on bars of the same polarity. Design a "scrambler," i.e., a motor-driven switching arrangement which changes the polarities of the bars so that the rat will be exposed to shock, during a full switch rotation, no matter what bars it touches. (Furnish design diagram only.)

Available equipment

1 panel No. 1
6 bulbs and sockets
3 switches, single-pole, single-throw (SPST)
2 switches, single-pole, double-throw (SPDT)
2 switches, double-pole, double-throw (DPDT)
1 selector switch, single-pole, five positions
1 selector switch, double-pole, five positions
1 variable resistor, 50 Ω, 3 W
1 transformer, 115/6.3 V
(Test equipment is available from the instructor.)

EXPERIMENT 2A: ELECTRIC CIRCUIT ANALYSIS

BACKGROUND INFORMATION

Experiment Fig. 2A-1 shows a schematic diagram of a constant-voltage source E with two terminals T_1 and T_2. The voltage or *potential difference between the terminals* is E_o, which is equal to E. The voltage E_o can be measured with any instrument (voltmeter) which does not cause a change in the circuit (e.g., does not cause polarization of the source E). Experiment Fig. 2A-2 is another schematic diagram of the circuit in Experiment Fig. 2A-1 but uses a different symbolic presentation: the terminal T_2 of the source E is grounded, i.e., its absolute

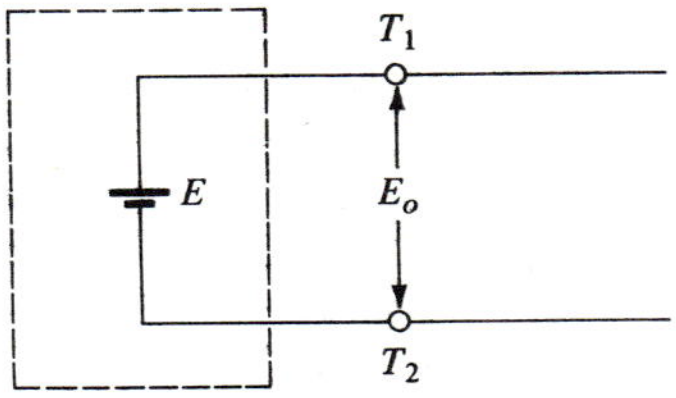

Experiment Fig. 2A-1

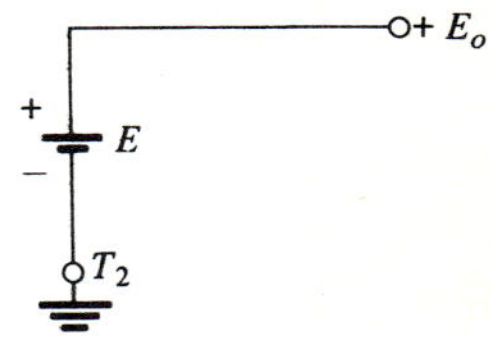

Experiment Fig. 2A-2

potential is assumed to be zero. The (absolute) potential of T_1 is then $E_o = E$, that is, the potential difference between T_1 and ground is equal to E_o.

In practice, every source has an internal resistance R_i (sometimes called R_o, output resistance or output impedance) as indicated in Experiment Fig. 2A-3. As long as no load resistance is connected to terminals T_1 and T_2, no current flows in this circuit; the terminal voltage is $E_o = E$. The voltage E_o can be measured with any voltmeter that does not cause a flow of current, i.e., that has practically an infinite resistance. Electrometers, potentiometric instruments, and certain types of electronic voltmeters fulfill this condition.

If a load resistance R_L is connected to the output terminals as shown in Experiment Fig. 2A-4, a current I will flow in the circuit. The voltage which causes the flow of current is E. The resistances in the circuit are R_i and R_L in series; therefore, the current in the circuit is

$$I = \frac{E}{R_i + R_L} \tag{2A-1}$$

(provided the meter resistance R_M is large, $R_M \gg R_i$).

The current I causes a voltage drop across R_i which is $IR_i = E_i$ and another voltage drop across R_L which is $IR_L = E_o$. The sum of voltages in a closed loop is zero (Kirchhoff's voltage law)

$$E - E_{R_i} - E_{R_L} = 0$$
$$E_{R_L} = E - E_{R_i}$$
$$E_o = E - IR_i$$

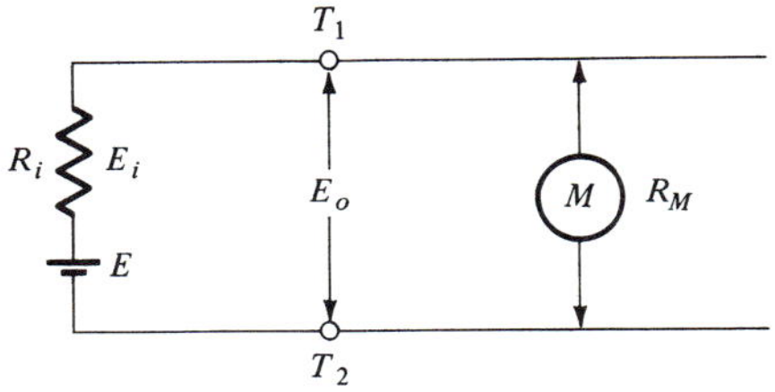

Experiment Fig. 2A-3

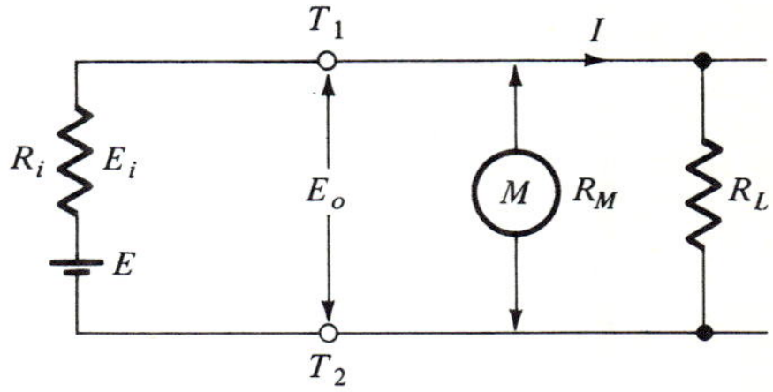

Experiment Fig. 2A-4

The output voltage E_o from a source with an internal resistance R_i is *less* than E by the amount of the voltage drop caused by the internal resistance.

If only meter M is connected to terminals T_1 and T_2 and if the meter resistance is large ($R_M \gg R_i$), then the current I is small and the voltage drop E_i caused by the voltmeter current is negligible. For this reason voltmeters generally have large resistances. If only the meter is connected to terminals T_1 and T_2 and if the meter resistance is *not* large compared to R_i, the meter indicates correctly the voltage between terminals, but this voltage is not the same as that existing in the circuit in the absence of the meter.

If both a meter of low resistance R_M and a load with the resistance R_L in parallel are applied to the terminals, then Eq. (2A-1) changes to

$$I = \frac{E}{R_i + R_c}$$

where R_c is the equivalent resistance of R_M and R_L in parallel combination

$$R_c = \frac{R_M R_L}{R_M + R_L}$$

PROBLEMS AND RECOMMENDED PROCEDURES

Given a dc source with different internal resistances and several voltmeters, measure the output voltages of the source with each voltmeter and for each internal resistance. Determine the resistances of the voltmeter from these measurements.

The voltmeter resistance R_M can be found as follows. If only the meter is connected to the terminals, the output voltage between the terminals is

$$E_o = IR_M$$

The voltage drop across the internal resistance is

$$E_i = IR_i$$

Therefore,

$$\frac{E_o}{E_i} = \frac{IR_M}{IR_i} = \frac{R_M}{R_i}$$

$$R_M = R_i \frac{E_o}{E_i} = R_i \frac{E_o}{E - E_o}$$

In this equation, R_i is known (is inscribed on the source), E_o is known (is measured with the voltmeter), and E_i is known ($E_i = E - E_o$, where E is the open-circuit voltage, measured with a voltmeter that has a very high resistance so that $R_M \gg R_i$).

The digital voltmeter (Honeywell model 333) operated at the 10-V scale has a resistance of 100 mΩ (10^8 Ω) and draws a current which can be neglected in these experiments. The circuit of Experiment Fig. 2A-5 is recommended with

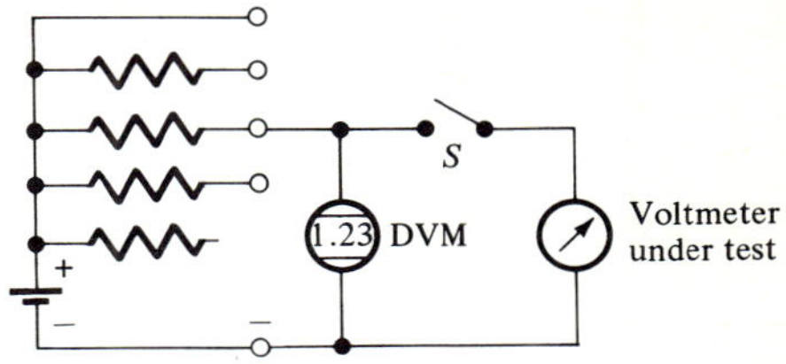

Experiment Fig. 2A-5

the digital voltmeter permanently connected in parallel to the voltmeter under observation.

Available equipment

1 Honeywell digital voltmeter, model 333
1 Weston voltmeter, model 1, ranges 3, 15, and 150 V. Use 15-V terminal and +.
1 Weston voltmeter, model 430, ranges 5, 20, and 100 V. Use 20-V terminal and −.
1 RCA vacuum-tube voltmeter Senior Volt-Ohmyst. Set to 15 V and +dc; adjust zero after warm-up.
1 source

EXPERIMENT 2B: POTENTIOMETRIC MEASUREMENTS

A simple potentiometer is shown in Experiment Fig. 2B-1. It consists of a resistance R connected to an auxiliary source E_a. Point A of the potentiometer resistance is positive with respect to B. The potential difference between A and B is E_a volts; a current $I_p = E_a/R$ flows in the circuit. A voltage e exists between the point B and the variable contact point C, point C being positive with respect to B. If point C coincides with point B, the voltage e is zero. If point C coincides with point A, then $e = E_a$. By moving the contact point from B to A, any voltage between zero and E_a can be branched off between C and B. The voltage e is proportional to the partial resistance r

$$\frac{e}{E_a} = \frac{r}{R} \qquad e = \frac{r}{R} E_a$$

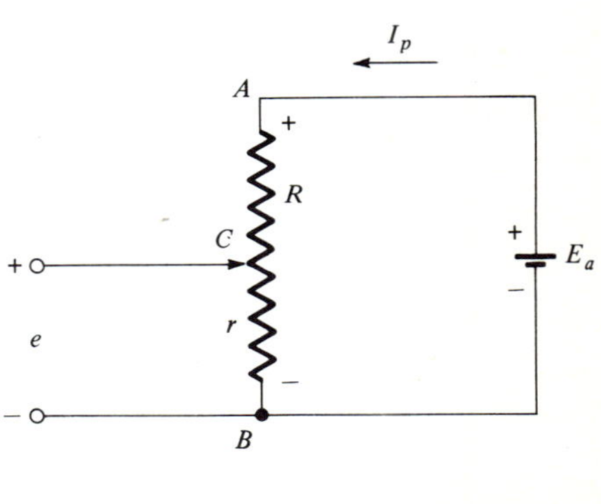

Experiment Fig. 2B-1

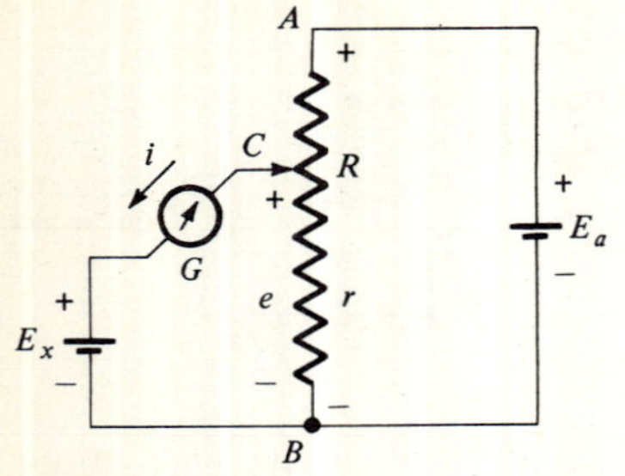

Experiment Fig. 2B-2

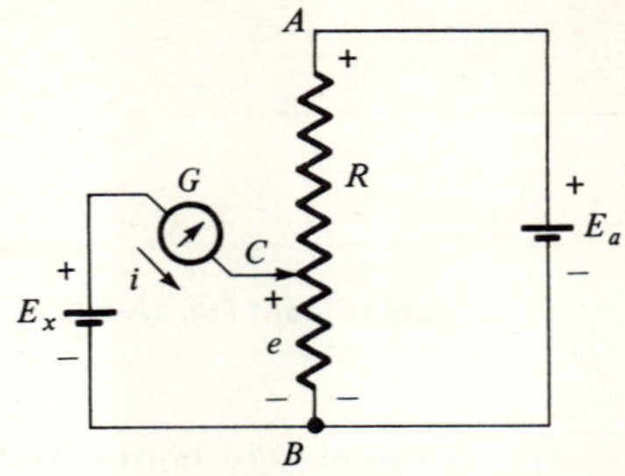

Experiment Fig. 2B-3

Experiment Figs. 2B-2 to 2B-4 show the use of the potentiometer for the measurement of an unknown voltage E_x. The source E_x is connected between point B of the potentiometer and the movable contact C through a sensitive galvanometer G. In Experiment Fig. 2B-2, the potentiometer voltage e is larger than E_x; the galvanometer indicates this inequality by a negative deflection. In Experiment Fig. 2B-3, $e < E_x$; the galvanometer shows a positive current. In Experiment Fig. 2B-4, the current in the galvanometer is zero, therefore,

$$E_x = e = \frac{r}{R} E_a$$

Knowledge of the voltage E_a is not required if the following two-step method is used. Connect a standard cell (voltage E_s) to terminals B and C with the potentiometer as shown in Experiment Fig. 2B-5 and balance the circuit:

$$E_s = \frac{r_s}{R} E_a$$

Then replace the standard cell by the unknown voltage E_x and balance again:

$$E_x = \frac{r_x}{R} E_a$$

Since

$$\frac{r_x}{r_s} = \frac{E_x}{E_s}$$

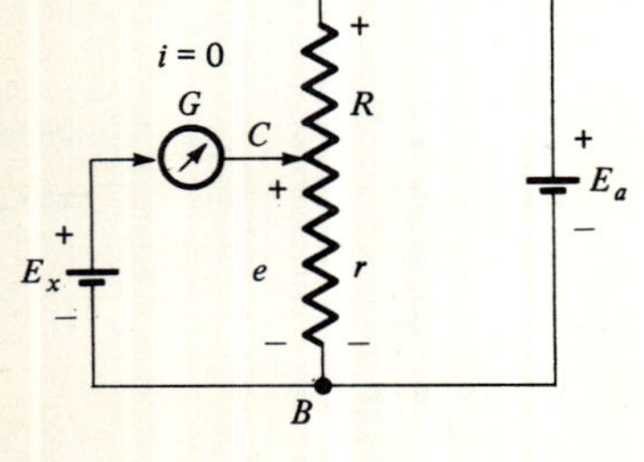

Experiment Fig. 2B-4

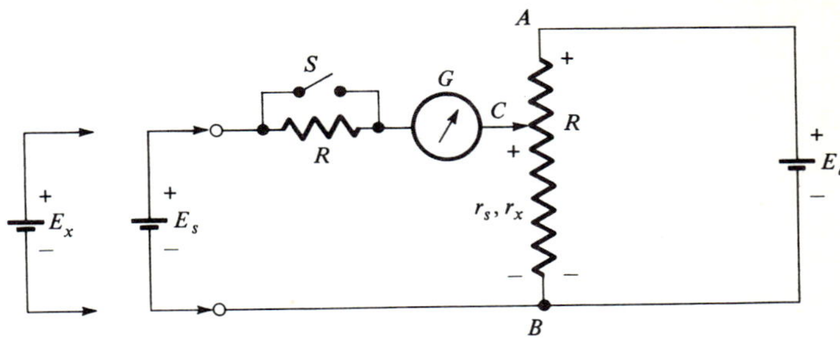

Experiment Fig. 2B-5

the unknown voltage can be found from

$$E_x = E_s \frac{r_x}{r_s}$$

The unknown voltage can be found from knowledge of the voltage of the standard cell (can be calibrated by the National Bureau of Standards) and the ratio of two resistances, i.e., the reading of the position of C on a scale. No current flows in the circuit containing the source E_x, therefore, the internal resistance R_i does not cause any error.

PROBLEM AND RECOMMENDED PROCEDURE

Construct a potentiometer using the circuit diagram of Experiment Fig. 2B-5 and measure the voltage of the given source E_x. Observe polarities. (Calibrate a thermoelement.) (Calibrate a glass electrode pH meter.) The use of the digital voltmeter (DVM) is recommended for the initial balance of the potentiometer. After initial balance, the DVM should be replaced by a galvanometer. Have the circuit checked by the instructor before connecting it to voltage sources.

Available equipment

Same equipment used in Experiment 2A plus
1 potentiometer, 10,000 Ω
2 dry batteries, 6 V each (to be used in series for E_a)
2 galvanometers

EXPERIMENT 3A: CAPACITANCE CHARGE AND DISCHARGE CHARACTERISTICS

BACKGROUND INFORMATION

Capacitance charge In Experiment Fig. 3A-1, a voltage source E is connected to a capacitor C and a resistor R in series. The voltage across the capacitor is e_c, that across the resistor is $e_r = Ri$, so that

$$Ri + e_c = E$$

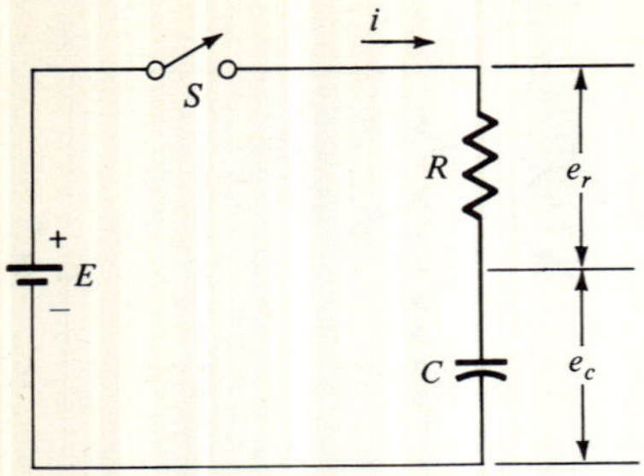

Experiment Fig. 3A-1

The current is $i = dQ/dt$. Since $Q = e_c C$,

$$i = C\frac{de_c}{dt}$$

so that

$$RC\frac{de_c}{dt} + e_c = E$$

The solution of this differential equation is

$$e_c = E(1 - \varepsilon^{-t/RC})$$

the current in the circuit is

$$i = \frac{E}{R}\,\varepsilon^{-t/RC}$$

Initially the circuit is open, i.e., for $t = 0$, the voltage across the capacitor is zero. At $t = 0$ the switch is closed, and the voltage e_c rises exponentially with time and reaches asymptotically the value of the applied voltage E. The current has a maximum value of E/R at $t = 0$ and decays exponentially to zero.

Capacitance discharge In Experiment Fig. 3A-2, the capacitor is charged to the value E and discharges through the resistance R as soon as the switch is closed,

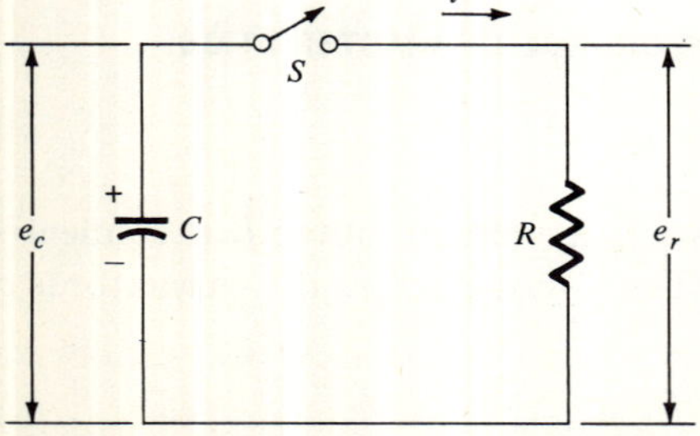

Experiment Fig. 3A-2

so that

$$RC \frac{de_c}{dt} + e_c = 0$$

The solution of this equation is

$$e_c = E\varepsilon^{-t/RC}$$

The current is

$$i = \frac{E}{R} \varepsilon^{-t/RC}$$

Both the voltage and the current decay exponentially from an initial maximum value to zero.

The product RC is the time constant T of the RC network, i.e., the time during which the capacitor voltage decays to the value $1/\varepsilon = 36.8$ percent of its initial value.

PROBLEM

Determine experimentally the charge and discharge characteristics of several resistance-capacitance combinations and compare the experimental results with those obtained from computation.

RECOMMENDED PROCEDURE

A strip-chart recorder is used for the observation of the charge and discharge characteristics. The recorder has an input resistance of the order of 10^5 Ω. It is recommended to calibrate the recorder first. Set the recorder gain control fully counterclockwise; set the recorder centering control so that the recording trace is in the center of the strip chart. Next, use a low-frequency function generator and, before connecting it to the recorder, set the output control (amplitude) to its lowest position. Connect the generator; set the frequency to 1 Hz, set to "square wave" and gradually raise the output until you get a satisfactory trace. Determine the scale of the time axis. Next, disconnect the function generator. Set the recorder gain control counterclockwise to its lowest sensitivity position; set the centering control so that the trace coincides with the right-hand margin of the strip chart. Then apply a 6-V dc source (polarity!) to the recorder and adjust the gain so that the trace occupies almost the entire horizontal span. Repeat zero- and maximum-deflection readings. Next, connect the circuit as shown in Experiment Fig. 3A-3. For charge characteristic, open switch $S2$ and close switch $S1$. For discharge characteristic, open switch $S1$ and close switch $S2$. It is recommended to start the experiment with a capacitor C of 150 μF and a resistor R of 10 kΩ. Repeat the experiment with different values of R and C. The time constant $T_1 = RC$ should be small compared to that formed by C and R_R (100,000 Ω).

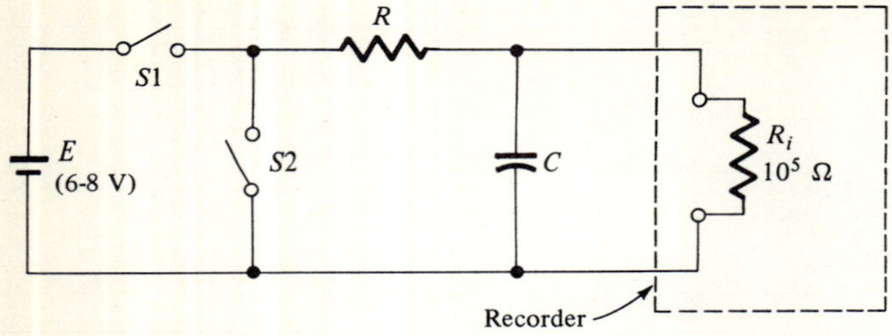

Experiment Fig. 3A-3

Available equipment

1 recorder
1 HP 202A low-frequency function generator
1 dc power supply, 0 to 10 V, battery-operated
1 connecting box
Capacitors, resistors, cables

EXPERIMENT 3B: ELECTRICAL DIFFERENTIATION AND INTEGRATION

DIFFERENTIATORS

Experiment Fig. 3B-1 shows a schematic diagram of an electrical differentiator. An input signal e_i is applied which varies with time. The output signal is proportional to the time derivative of the input signal, i.e., proportional to the rate of change or to the slope of the input function. The time derivative has a physical meaning. For instance, if the input function represents a mechanical displacement of a body, then the derivative is the velocity of the body.

If a variable voltage e_i is applied to the circuit of Experiment Fig. 3B-2, the resulting current is

$$i = C\frac{de_i}{dt}$$

i.e., the current is proportional to the time derivative of the input voltage. Usually a voltage rather than a current output is desired. If a resistor R is inserted

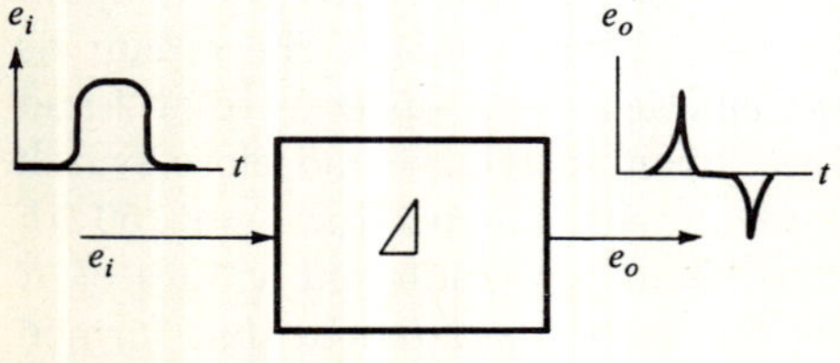

Experiment Fig. 3B-1

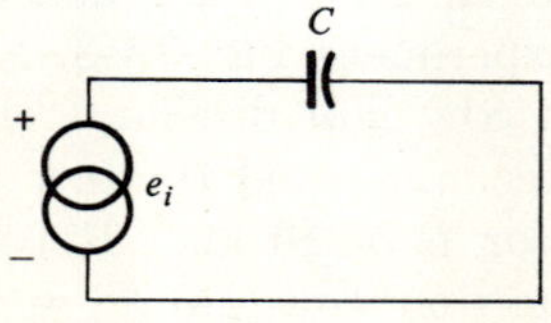

Experiment Fig. 3B-2

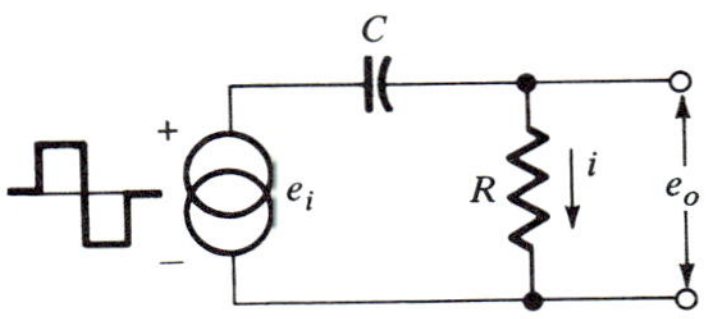

Experiment Fig. 3B-3

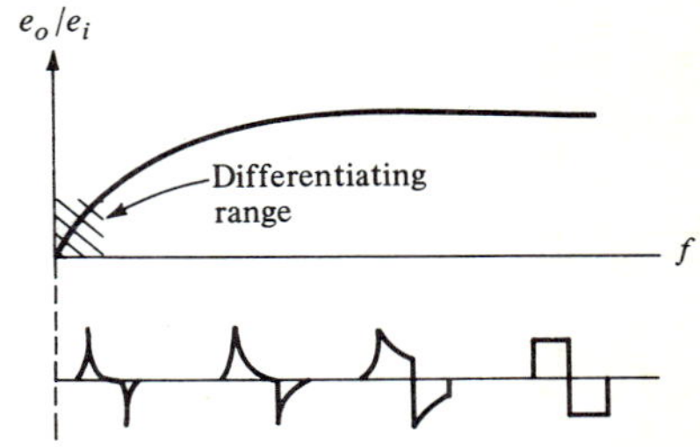

Experiment Fig. 3B-4

(Experiment Fig. 3B-3) which is so small that the current essentially is not changed by its insertion, then the voltage drop across this resistor is

$$e_o = Ri = RC\frac{de_i}{dt}$$

This equation is correct only if R is small compared to the impedance of the capacitor C. The impedance of C at a sinusoidal current of frequency f is $Z = 1/2\pi fC$, so that the condition for correct differentiation is

$$R \ll \frac{1}{2\pi fC} \qquad \text{or} \qquad RC \ll \frac{1^*}{f} \qquad \text{or } RC \ll T$$

where T is the time for one cycle of the highest frequency sinusoidal component of the input voltage e_i.

Experiment Fig. 3B-4 shows the shape of the output voltage of the differentiator as a function of frequency for a square-wave input signal. For low frequencies the network is a good differentiator, but as the frequency goes up, the voltage across the resistor becomes identical to the input voltage.

INTEGRATION

Experiment Fig. 3B-5 shows a schematic diagram of an electrical integrator. An input signal e_i is applied which may or may not vary with time. The output signal is proportional to the time integral of the input signal, i.e., proportional to the area under the input function. Also, the time integral has a physical meaning. For instance, if the input signal represents the velocity of a body, then the time integral represents the distance traveled by the body during the integration period.

* The factor 2π has been neglected.

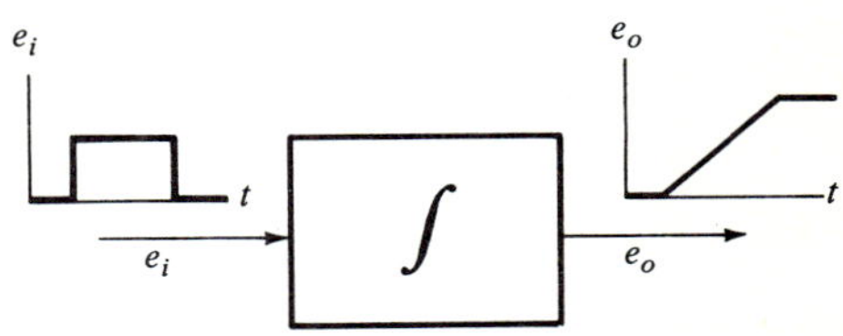

Experiment Fig. 3B-5

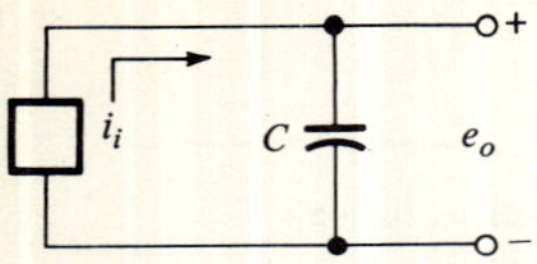

Experiment Fig. 3B-6

If a current source i_i is applied to the input terminals of Experiment Fig. 3B-6, a charge Q will accumulate in the capacitor C, and the voltage e_o between its terminals will rise:

$$Q = e_o C = i_i \, dt \qquad e_o = \frac{1}{C} i_i \, dt$$

i.e., the output voltage is proportional to the time integral of the charging current. Usually a voltage rather than a current input is used. If a large resistance R is added as in Experiment Fig. 3B-7, the current i_i in the circuit is determined by e_i/R, hence,

$$e_o = \frac{1}{C}\frac{e_i}{R} dt \qquad e_o = \frac{1}{RC} e_i \, dt$$

This equation is correct only for the case where R is large compared to the impedance of the capacitor C. The impedance of C at a sinusoidal current of frequency f is $Z = 1/2\pi f C$ so that the condition for correct integration is

$$R \gg \frac{1}{2\pi f C} \qquad \text{or} \qquad RC \gg \frac{1^*}{f} \qquad \text{or} \qquad RC \gg T$$

where T is the highest frequency sinusoidal component of the input voltage e_i.

As seen in Experiment Fig. 3B-8, the range of operation for the integrator is restricted to high frequencies. When f is substantially lower than $1/RC$, the impedance of the capacitor increases to the point where $E_c = E_i$, and the integrating action of the circuit is lost.

* The factor 2π has been neglected.

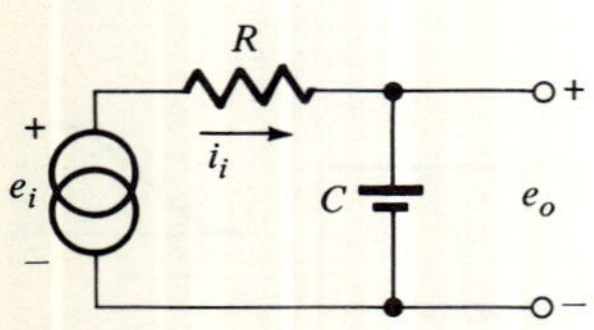

Experiment Fig. 3B-7

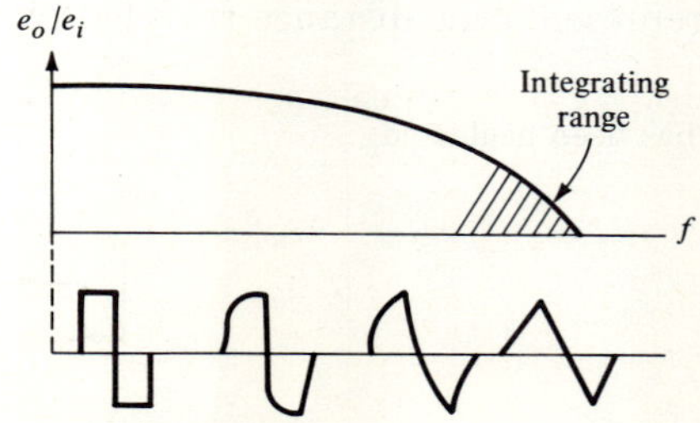

Experiment Fig. 3B-8

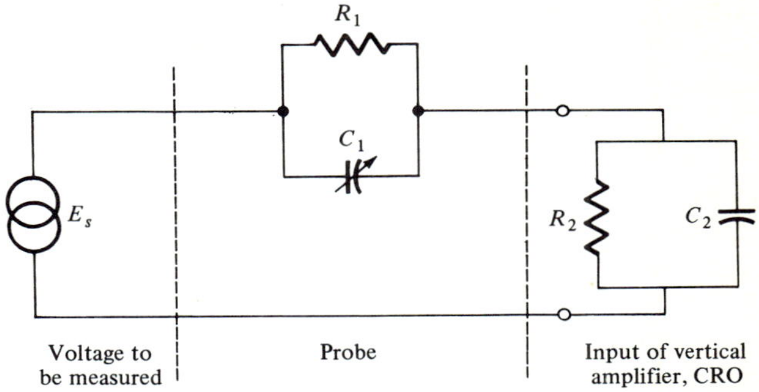

Experiment Fig. 3B-9

COMPENSATION AND DAMPING

The equivalent input circuit of an oscilloscope is a large resistance in parallel with a small capacitance. When the frequency of the input signal goes up, the capacitor's resistance decreases and shunts the signal to ground. In order to solve this problem it is necessary to compensate the input to the oscilloscope so that its frequency response is constant.

The arrangement shown in Experiment Fig. 3B-9 is most often used to compensate the input to an oscilloscope. Resistor R_1 and variable capacitor C_1 are enclosed in an oscilloscope probe so that C_1 may be set to correct the frequency response. The voltage transfer function of the complete input circuit is

$$\frac{e_o}{e_s} = \frac{R_2(1 + R_1C_1s)}{R_1(1 + R_2C_2s) + R_2(1 + R_1C_1s)}$$

If the equality $C_1R_1 = C_2R_2$ is satisfied, the voltage transfer function reduces to

$$\frac{e_o}{e_i} = \frac{R_2}{R_1 + R_2} = \text{const.}$$

The magnitude of the output trace is then related to the signal input by a constant.

In practice, R_1 is fixed to provide the desired input attenuation. C_1 is set so that when e_s is a square wave, e_o is also a square wave. If C_1 is too large, the input circuit begins to act like a differentiator and the edges of the square wave will *overshoot*. If C_1 is too small, the input circuit begins to act like an integrator and the response is *overdamped*.

PROBLEM

3B-1. Design and construct a differentiator and an integrator. Using a square-wave generator, determine the frequency range for correct operation of each circuit and compare it to its RC time constant. Also check the operation of each circuit for a triangular-wave input.

3B-2. Compensate your oscilloscope probe by using the square-wave calibrator in the oscilloscope as a signal source. Note the distortion produced by too much or too little capacitance in the probe.

RECOMMENDED PROCEDURE

Designing a differentiator or integrator requires first the determination of the frequency range over which the circuit must operate. The condition for correct differentiation is $RC \ll 1/f$. If the frequency f is about 500 Hz, then

$$RC \ll \frac{1}{500} \qquad RC \ll 0.002$$

Select values of R and C so that $RC = 10^{-5}$ for correct differentiation; for instance, $R = 10{,}000\ \Omega\ (10^4)$, then $C = 10^{-5}/10^4 = 10^{-9}\text{ F} = 0.001\ \mu\text{F}$.

For correct integration at 500 Hz, $RC \gg 0.002$. Select R and C so that the product $RC = 1$. If $R = 10{,}000\ \Omega$, then $C = 1/10^4 = 10^{-4} = 100\ \mu\text{F}$.

Available equipment

1 HP low-frequency function generator
1 oscilloscope Tektronix model 503
1 connection box
Resistors
Capacitors

EXPERIMENT 3C: DYNAMIC CHARACTERISTICS OF METERS AND RECORDERS

BACKGROUND INFORMATION

Almost all meters have a system which moves under the influence of an electric input signal until the mechanical force created by the input is equal to the restoring force in the instrument. In most cases the restoring force is exercised by a built-in spring.

The speed with which such a system moves and, therefore, the highest frequency to which the system can follow an applied sinusoidal input signal, depends in the first approximation only on the ratio of the restoring force to the inertia of the system. For the rotational movement, the speed is proportional to D/J, where D is the restoring moment and J is the moment of inertia. The damping of the system has little influence on the speed of movement and the frequency range up to which the system can be used, unless the damping is excessive.

Such systems exhibit a dynamic error when a transient input signal, such as a sudden change in level, is applied to the input, or when an ac signal of constant amplitude but variable frequency is applied.

Experiment Fig. 3C-1 shows the response of a meter system to a step function (curve S). An ideal meter would follow such a function without distortion or time lag. Practical meter systems follow three different patterns. If the system has little damping (is underdamped), it will move from the initial to the

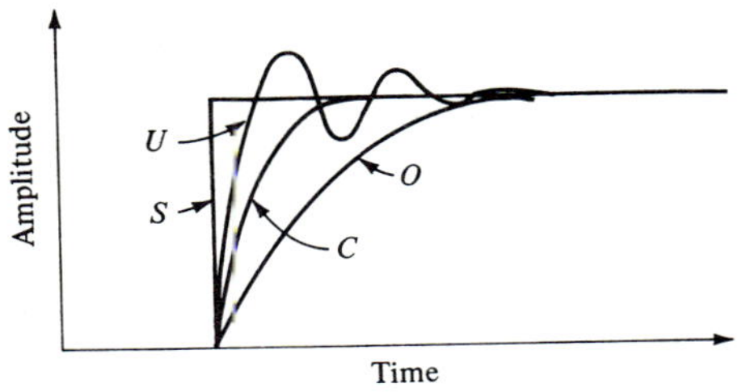

Experiment Fig. 3C-1

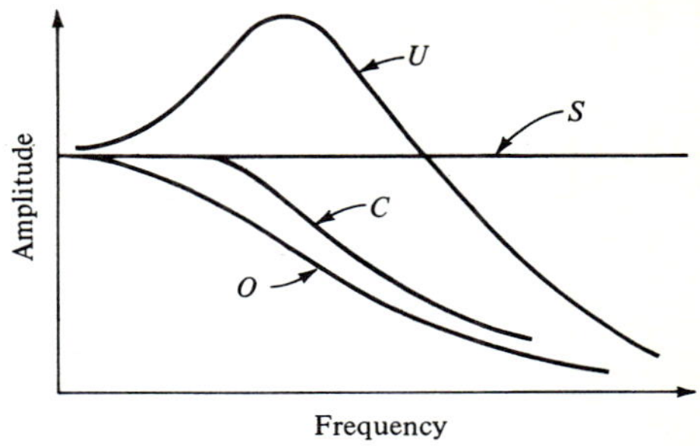

Experiment Fig. 3C-2

new equilibrium position and will oscillate around the new position as illustrated by curve *U*. If the system has an intermediate damping, it will move to the new position without oscillation and will come to rest in the shortest possible time as illustrated in curve *C* (near-critical damping, aperiodic movement). If the system is excessively damped (overdamped, curve *O*), the movement to the new equilibrium position will be slow.

The response of the same meter system to a sinusoidal input signal is shown in Experiment Fig. 3C-2. An ideal meter would follow the straight line *S*, i.e., its deflection for a given input signal level would be independent of the frequency of this signal. An underdamped system shows resonance (curve *U*), the height of the resonance peak depending on the damping. The response of an overdamped system regresses continuously as the frequency rises (curve *O*). The deflection of a near-critically damped meter system (curve *C*) remains independent of frequency up to about 60 percent of the resonance frequency (of the undamped system).

PROBLEMS

3C-1. Given four panel-type meters: one with a time constant of 0.15 s, one with 0.8 s (normal commercial type), and two with 1.5 s. The 0.15-s and 0.8-s meters are near-critically damped. One of the 1.5-s meters is overdamped, the other one is underdamped. Determine the critical frequency of the 1.5-s underdamped system. Determine the frequency ranges for which the dynamic errors of the meters are within a 10 percent limit. Apply sine waves, triangular waves, and square waves, and observe the output.

3C-2. Determine the frequency characteristic of a recorder, i.e., for each of three selected damping positions measure the deflection as a function of the applied frequency (keeping the input amplitude constant). Observe the recording traces of a sine wave, a triangular wave, and a square wave.

3C-3. Observe the frequency response of a tape recorder.

RECOMMENDED PROCEDURE

3C-1. The critical or resonance frequency is that frequency for which the amplitude of the meter is a maximum. The circuit shown in Experiment Fig. 3C-3 is

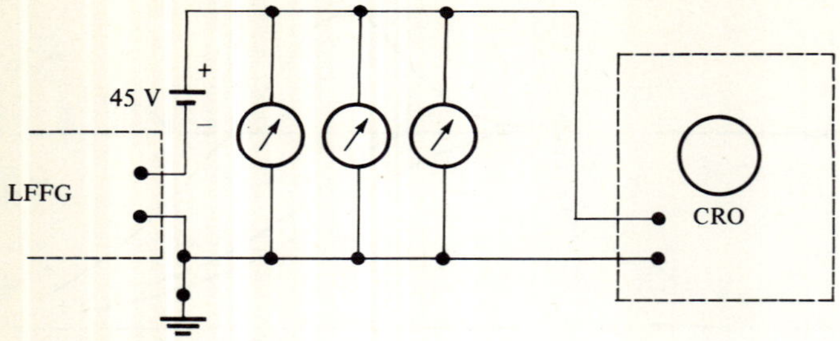

Experiment Fig. 3C-3

recommended. The frequency should never exceed 5 Hz to avoid damage to the 0.15-s meter. The output from the low-frequency function generator varies from zero to positive and negative values. Since the meters only show positive deflections, a 45-V battery is used in series with the generator.

3C-2. The applied signal frequency should never exceed 5 Hz to avoid damage of the 0.15-s meter at its resonance.

3C-3. Using the 0- to 70,000- Hz generator, apply a signal at constant amplitude to the tape-recorder input terminals. Monitor this signal with a cathode-ray oscilloscope (CRO). After recording, rewind the tape and play back into the CRO. (A low-grade tape recorder has been selected for this experiment to demonstrate the errors likely to occur with such a recorder.)

Available equipment

1 low-frequency function generator, HP model 202A
1 sine-wave generator, 5 to 600 kHz, HP model 20? CD
1 recorder MFE
4 meters, Weston, 0 to 100 V
1 tape recorder

EXPERIMENT 4A: ELASTORESISTIVE TRANSDUCERS (RESISTANCE-WIRE STRAIN GAUGES)

BACKGROUND INFORMATION

The strain gauge consists essentially of a fine wire; when exposed to strain (i.e., an elongation under the influence of stress), two physical mechanisms will cause a change of its electric resistance. First, its geometrical form will vary, i.e., the wire will show an increase of length and a decrease of its cross-sectional area. Second, its electric sensitivity will change because of solid-state variations in the metal. The result may be an increase or a decrease of the strain-gauge resistance, depending on the material of the wire. Strain gauges may be either unbonded or bonded to a paper support.

The most commonly used bonded wire strain gauge is shown in Experiment Fig. 4A-1 and consists of a grid of fine resistance wire of about 0.001-in diameter cemented to a paper support about 0.003 in thick. Two connecting leads are

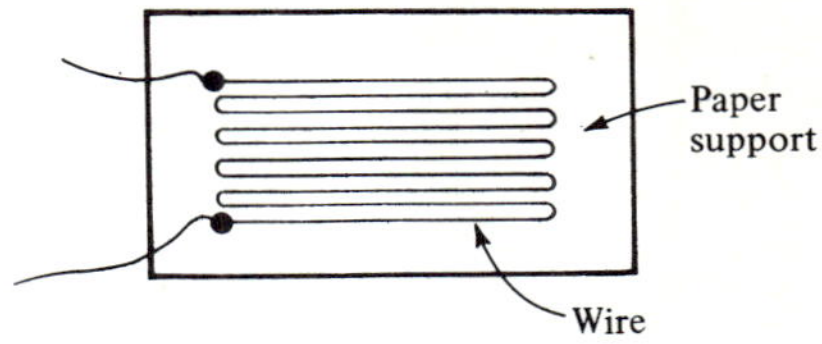

Experiment Fig. 4A-1

either soldered or welded to the fine wire. When the gauge is exposed to a strain ε, the fractional resistance variation can be expressed as

$$\frac{\Delta R}{R} = G\varepsilon$$

where G is the gauge factor. It is determined from measurements under standard conditions.

The maximum range of displacement is usually of the order of 3×10^{-3} in (70 μ); the forces needed to produce a maximum output are in the order of 10 to 100 oz.

The gauge resistance R is generally between fifty and several thousand ohms. Usually a resistance variation of the order of 1 percent for a maximum range of displacement can be obtained. (For further information on strain gauges see K. S. Lion, Electrical Input Transducers, "Instrumentation in Scientific Research," McGraw-Hill, New York, 1959, pp. 31–37.)

WHEATSTONE BRIDGE

The most common element for the conversion of a resistance variation into a corresponding voltage variation is the Wheatstone bridge (Experiment Fig. 4A-2). The galvanometer G indicates the potential difference between A and B. If the galvanometer shows no deflection, the potential of A is equal to that of B, that is, $E_x = E_s$, or $I_1R_x = I_2R_s$ and $E_a = E_b$, $I_1R_a = I_2R_b$. Therefore, the condition for balance is $R_x/R_s = R_a/R_b$ or

$$R_x = R_s\frac{R_a}{R_b}$$

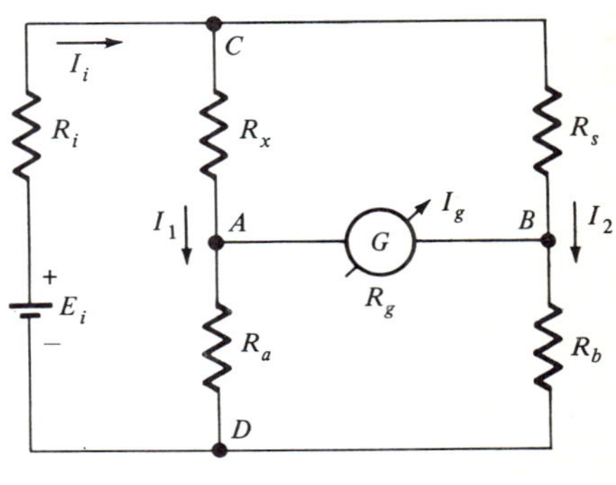

Experiment Fig. 4A-2

Two modes of operation can be used.

The balanced bridge The bridge is initially balanced. A variation ΔR_x of R_x causes an unbalance and, therefore, an output voltage between A and B. The balance is restored (manually or by means of a servosystem) by the readjustment of R_s or R_a/R_b. The resistance variation is found from $\Delta R_x = \Delta R_s R_a/R_b$ or from $\Delta R_x = R_s \Delta(R_a/R_b)$.

The unbalanced bridge The bridge is initially balanced. A variation ΔR_x of R_x causes an unbalance and an output voltage or current which is used as a measure of ΔR_x.

The output is proportional to the applied input voltage E_i and is a function of the six resistances R_i, R_x, R_s, R_a, R_b, and R_g. The sensitivity of the bridge is defined by the change of the output current or the output voltage from zero, when R_x is balanced, to the value ΔI_g or ΔE_g, when R_x is changed by the value ΔR_x, i.e.,

$$\frac{\delta E_g}{\delta R_x} = S_e \qquad \text{voltage sensitivity}$$

$$\frac{\delta I_g}{\delta R} = S_i \qquad \text{current sensitivity}$$

The values for S_e or S_i are obtained by partial differentiation with respect to R_x of the bridge equations. The computations are elementary but lengthy. The optimum condition, i.e., the condition of maximum output for a given change ΔR_x of the resistance R_x depends on a number of circumstances, such as whether the input voltage E_i is kept constant or the input current I_i, or the input power P, or the current I_1 through the resistor R_x. The optimum conditions for the selection of the bridge resistance are indicated in Experiment Table 4A-1.

Experiment Table 4A-1 Optimum bridge conditions

Bridge output	*Bridge input*	*Optimum for*
Maximum current sensitivity	E_i = const.	$R_a \ll R_b$
	I_i = const.	$R_s \gg R_b$
	I_1 = const.	$R_a \ll R_b \ll R_s$
Maximum voltage sensitivity	E_i = const.	$R_b = R_s$
	I_i = const.	$R_a \gg R_b$ and $R_s \gg R_b$
	I_1 = const.	$R_s \gg R_b$

PROBLEMS

4A-1. Design a system that measures strain in a given mechanical structure.
4A-2. Calibrate this system under stationary conditions (i.e., time invariant load).
4A-3. Find the strain in the structure when excited by a known force.
4A-4. Determine the resonant frequency of the structure, if there is any in the measurable range.

RECOMMENDED PROCEDURE

The circuit shown in Experiment Fig. 4A-3 is recommended. An amplifier C may be used for measurements at high sensitivity, but usually is not required for large loads. If the amplifier is used, it may be desirable to suppress any spurious 60-cycle noise. For this purpose a 60-cycle filter (Twin-T filter) is provided. In order to reduce the effect of 60-Hz pickup, one end of the strain gauge should be connected to ground.

A large bridge current, or a large current through the resistance R_x (the transducer), is desirable for high sensitivity, but too high a current may heat the transducer and lead to erroneous results.

The resonant frequency of the structure can be measured by using the calibrated time axis of the oscilloscope. The period T of the signal can then be read on the screen; the frequency is $f = 1/T$.

The following sequence of measurements is suggested:

1. Measure the resistance of the strain gauge unloaded.
2. Construct and test the Wheatstone bridge arrangement and determine experimentally its sensitivity.
3. Calibrate the overall system by relating the load on the beam (and the strain at the place of the strain gauge) to the output voltage.
4. Resonate the cantilever and measure its resonant frequency using the oscilloscope.

Available equipment

Strain gauges mounted on spring-aluminum cantilever (gauge factor $G = 2.04$)
Weights
Variable-gain dc amplifier

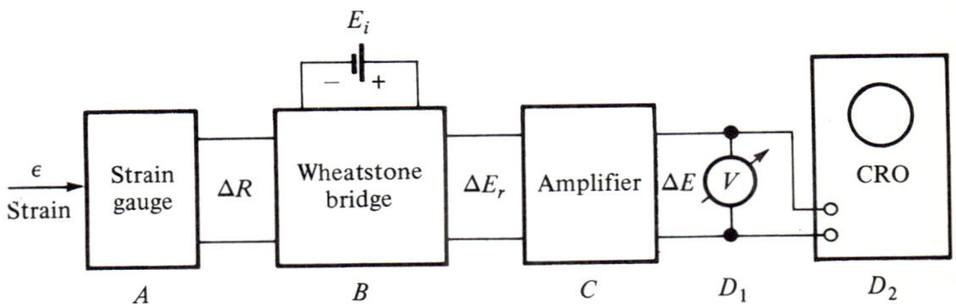

Experiment Fig. 4A-3

Vacuum-tube voltmeter
1 oscilloscope Tektronix, model 503
60-Hz filter
galvanometer
6-V battery
1,000 Ω, 10-turn potentiometer
Fixed resistors

EXPERIMENT 4B: CAPACITIVE TRANSDUCERS

BACKGROUND INFORMATION

A schematic diagram of a capacitive transducer system is shown in Experiment Fig. 4B-1. A physical quantity P (such as a force, a mechanical displacement, or a temperature variation) causes a change of the capacitance C by the amount ΔC. The transducer is connected to a modifier or signal-processing element M, which converts the capacitance or capacitance variation into an output signal (voltage or current). The output signal can be applied to an indicator or recorder or can be stored or used for the operation of a control or servosystem.

Basically, a capacitive transducer consists of two metallic or metallized plates with an overlapping area a; the plates are separated by distance d. The capacitance in picofarads can be computed from

$$C = 0.0855\, \varepsilon \frac{a}{d} \qquad \text{(4B-1)}$$

where a is expressed in square centimeters and d in centimeters. The quantity ε is the dielectric constant of the medium between the plates; its value is unity for vacuum and very close to unity for air ($\varepsilon_{air} = 1.0005$ at 0°C). It follows from Eq. (4B-1) that three basic methods exist for causing a change of capacitance in response to a mechanical movement: (1) by changing the overlapping area a (i.e., by sideways shifting of one of the plates), (2) by changing the distance d between the plates, (3) by changing the dielectric between the plates.

The capacitive transducer offers a combination of desirable physical characteristics not easily found in other transducers:

1. Its force requirements are very small. In the differential capacitor, the forces required to overcome the electrostatic attraction between the plates are of the order of several dynes or less.

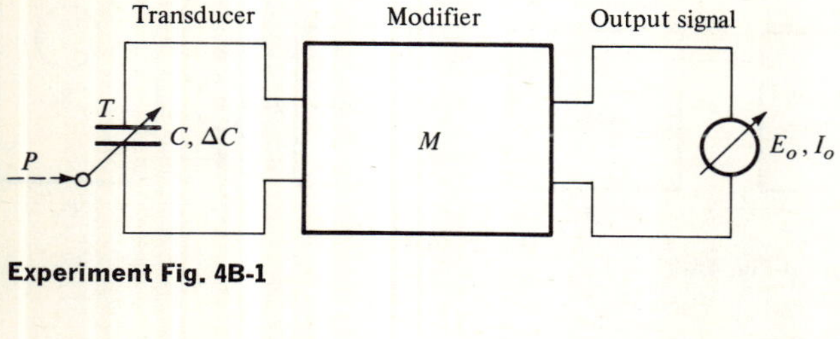

Experiment Fig. 4B-1

2. The plates themselves may have very little mass, permitting the design of transducers with highly desirable dynamic response characteristics.
3. For some applications, such as the measurement of displacement or vibration, there is no need to connect any mechanical parts to the moving member. A plate brought into the vicinity of the moving member is all that is required to measure the displacement accurately and with great sensitivity.
4. The dimensional stability can be made very high by using constructional materials such as quartz and Invar.
5. The capacitance does not depend on the conductivity of the metal electrodes or its variation with temperature.
6. The temperature coefficient of the dielectric constant of air or of most gases used as a dielectric is several orders of magnitude smaller than the temperature coefficient of materials used in resistive transducers or that of the magnetic permeability of ferrous materials used in the construction of inductive transducers.
7. The capacitive transducer can be shielded easily against the effect of external electric stray fields. It is not affected by magnetic stray fields nor does it cause an objectionable magnetic or electric stray field.

Experiment Fig. 4B-2 shows the diagram of an electric circuit which permits the conversion of a capacitance variation into a voltage or a current (T-2 bridge). The operation of this circuit is as follows: S is a radio-frequency source which furnishes symmetrical waves or sine waves of an amplitude E_1. During the positive half-cycle of the source, diode D_1 is conductive so that capacitor C_1 is charged. In the subsequent negative half-cycle, diode D_1 is nonconductive and capacitor C_1 is discharged through resistance R_1 (current i_1) and load R_L (meter, recorder, etc.) and through resistance R_2 and diode D_2. During this same half-cycle, diode D_2 is conductive, so that capacitor C_2 is charged. During the following half-cycle, C_2 is discharged through resistances R_2 and R_1, (current i_2) and resistance R_1 and diode D_1. If diodes D_1 and D_2 have the same characteristics, and if $C_1 = C_2$ and $R_1 = R_2$, then current i_1 is equal to i_2 and is of opposite polarity, i.e., the average current through R_L is zero. A meter connected in parallel to R_L or in place of R_L will show no deflection. Any variation

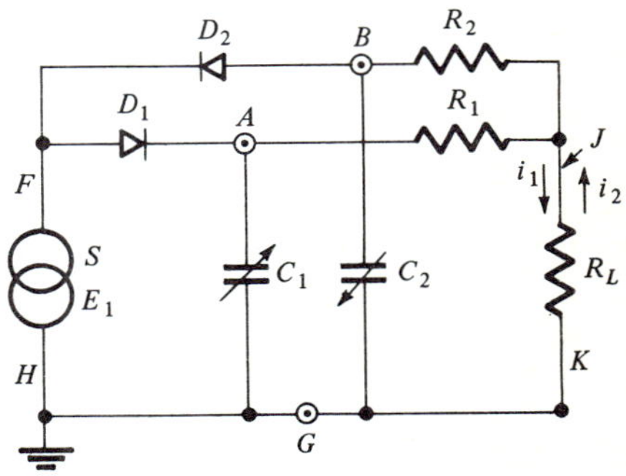

Experiment Fig. 4B-2

of C_1 or C_2 causes a deflection of the meter. For further information about this circuit see: K. S. Lion, *Rev. Sci. Instr.* **35**:353–356 (1964).

PROBLEM AND RECOMMENDED PROCEDURE

Measure the sensitivity and the linearity of a capacitive micrometer. Connect the capacitive micrometer to the T-2 unit. Determine the output voltage for different positions of the capacitive micrometer. From the values obtained, determine the sensitivity of the system (in volts per millimeter) and its linearity (deviation from a straight line (in percent output).

Available equipment

T-2 unit
Capacitive micrometer
Different output meters
1 dc amplifier

EXPERIMENT 4C: INDUCTIVE TRANSDUCER (VARIABLE DIFFERENTIAL TRANSFORMER)

BACKGROUND INFORMATION

Experiment Fig. 4C-1 shows a schematic diagram of the linear variable differential transformer (LVDT). It consists of three coils, P, S_1, and S_2, and a core C of magnetic material. An alternating current of a frequency varying between 60 Hz and 20 kHz is fed into coil P. The alternating magnetic field induces voltages E_1 and E_2 in the nearly identical coils S_1 and S_2. The output signal is $E_o = E_1 - E_2$; for complete symmetry, the output signal is zero. A displacement of either one of the coils or of the core position causes magnetic asymmetry and an output signal.

The core-displacement transducer, as illustrated in Experiment Fig. 4C-2, is most frequently used. Usually the core consists of a nickel-iron alloy and is slotted longitudinally to reduce eddy currents. The transfer characteristic (output voltage versus core displacement D) is shown in Experiment Fig. 4C-3. It is

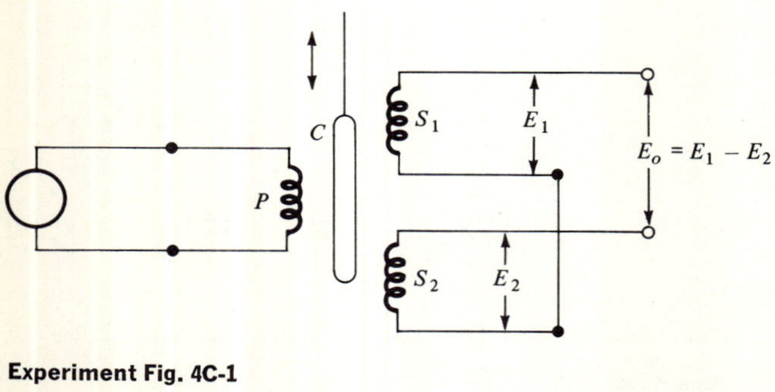

Experiment Fig. 4C-1

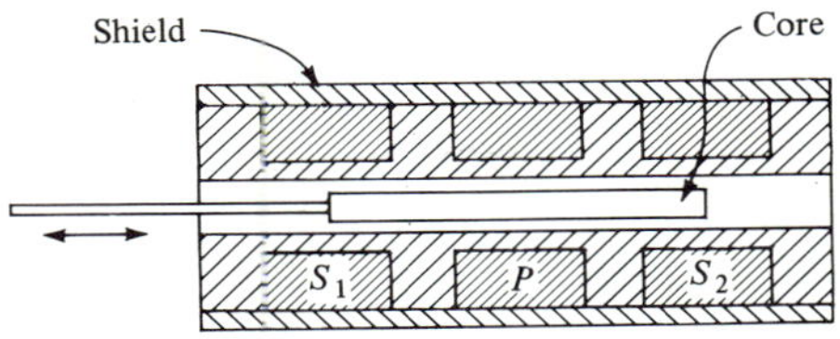

Experiment Fig. 4C-2

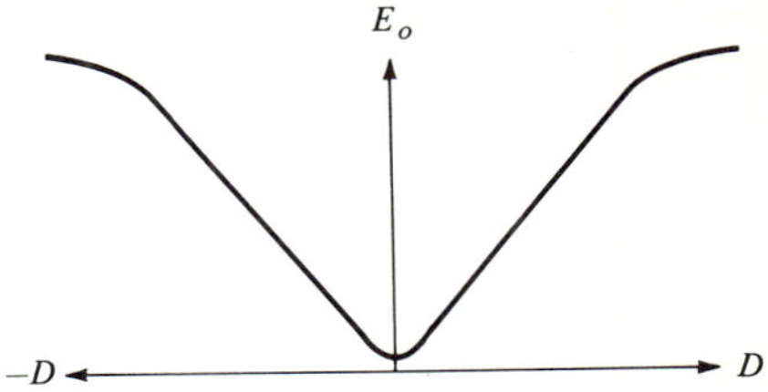

Experiment Fig. 4C-3

linear over a considerable range and flattens out on both ends. The output voltage changes its phase by 180° as the core is moved through the zero position.

The sensitivity of commercial transducers in the linear part of the transfer characteristic is of the order of 0.5 to 3 mV for a displacement of 0.001 in/V applied to the primary coil; under standard operating conditions (6.3 V at 60 Hz) the sensitivity is between 3 and 20 mV/0.001 in. The sensitivity and the output voltage can be increased by increasing the primary voltage. Such a procedure will lead to an increase of the heating of the primary coil and, thus, to a variation of the resistance and, in further sequence, of the output level and the sensitivity.

A residual voltage generally remains when the core is at the zero position. This zero voltage, which is normally less than 1 percent of the maximum output voltage in the linear range, has its origin in an incomplete magnetic or electric balance; it may consist of the fundamental frequency or of higher harmonics. Frequently it can be reduced by low-pass filters in the input circuit, by balancing the individual output circuits with resistive and capacitive shunts, or by adequate grounding.

The mechanical force requirements at the input depend on the operating conditions. The axial force which the transducer exercises on the core is zero if the core is in the center position and increases linearly with the core displacement, tending to move the core toward the axial center. The force is proportional to the square of the current in the primary coil. For a commercial transducer driven at 60 Hz and for full displacement, the force is of the order of 0.3 g (300 dyn); if operated at 400 Hz, it diminishes to 0.075 g (75 dyn). The coil system also exercises a radial force on the core which increases with the square of the primary current. This force is zero when the core is in the radial center, but in this position the core is in an unstable equilibrium since the radial force increases with the radial displacement and tends to move the core toward the periphery of the bore. A restraint of this sideways movement is required.

PROBLEM

Plot the output voltage for different displacements of the LVDT core and determine the sensitivity and the linear range (i.e., the range where the output varies less than 5 percent from a straight line).

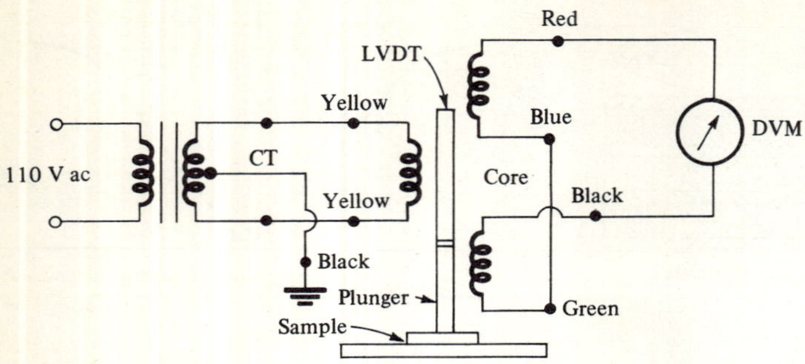

Experiment Fig. 4C-4

RECOMMENDED PROCEDURE

The circuit of Experiment Fig. 4C-4 is suggested. The LVDT is excited by 6.3 V ac taken from a transformer. The output is applied to a digital voltmeter which measures the root-mean-square value directly. A disadvantage of this setup is its low sensitivity in the vicinity of the zero position. This difficulty can be overcome by operating with a mechanical zero position different from the electrical zero position, i.e., adjusting the core position so that the output is about 0.1 V when the thickness of the test object is zero.

Available equipment

1 LVDT
1 transformer 115/6.3 V
1 digital voltmeter HP model 333

EXPERIMENT 5: IMPEDANCE BRIDGES

BACKGROUND INFORMATION

1. If in a resistive (Wheatstone) bridge (Experiment Fig. 5-1) the dc source is replaced by an ac generator E_i and the galvanometer by an ac detector, the balance condition will not be changed; i.e., the output between points A and B

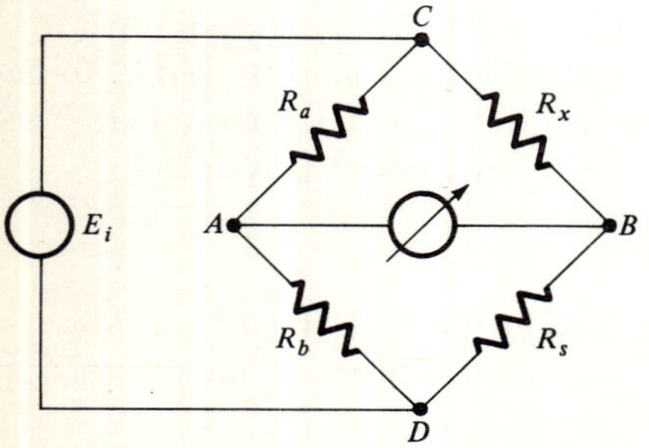

Experiment Fig. 5-1

at balance will be zero for $R_x/R_s = R_a/R_b$. The potentials of points A and B will alternate with the applied voltage, but the potentials will be equal at all times so that their potential difference is always zero.

If R_x and R_s are replaced by pure reactances X_x and Y_s, the potential difference between A and B will again be zero for $X_x/X_s = R_a/R_b$.

2. If one of the bridge arms contains a reactance X_x or an impedance Z_x, and the other bridge arms contain pure resistances (Experiment Fig. 5-2), a bridge balance cannot be obtained. Even if the absolute values of X_x (or Z_x) and the resistances are balanced, so that the magnitudes (*peak values*) of the potentials DA and DB are equal, there will be no zero output because the *instantaneous* voltages are not equal, i.e., the potentials e_A and e_B are not in phase (see Experiment Fig. 5-3). Therefore, there are two conditions to obtain a balance in impedance bridges: the impedances must be balanced (a) with respect to magnitude and (b) with respect to phase.

Condition (a): $$\frac{X_x}{X_s} = \frac{R_a}{R_b}$$

and

Condition (b): $$\frac{R_x}{R_s} = \frac{R_b}{R_a}$$

Condition (b) can be replaced by the more general expression

$$\frac{D_x}{D_s} = \left(\frac{R_b}{R_a}\right)^2 \quad \text{or} \quad \frac{Q_x}{Q_s} = \left(\frac{R_a}{R_b}\right)^2$$

where D, the dissipation factor of an impedance, is the ratio R/X and Q, the storage factor of an impedance, is the ratio X/R.

There are numerous ways in which such a double balance can be obtained; some combinations are unsuitable because they require resistance, capacitance, or inductance values that are impractical to build or to calibrate. Some bridge systems are more suitable for large, others for small values of C, L, D, Q, and R.

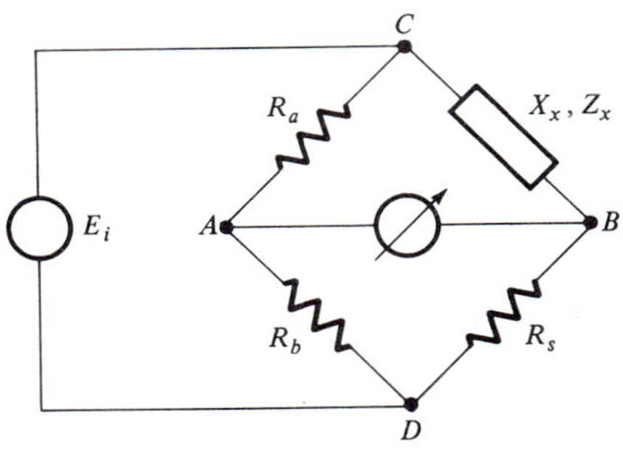

Experiment Fig. 5-2

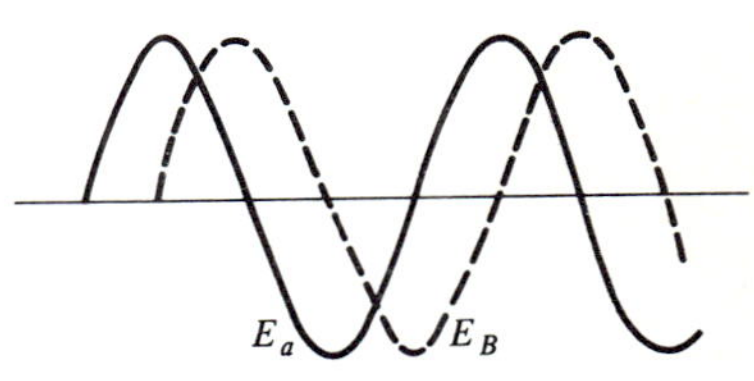

Experiment Fig. 5-3

Some permit direct readings of the unknown quantities; others are preferable because they do not require pure sinusoidal input voltages.

Impedance bridges offer the advantage that they furnish simultaneous information of the magnitude and the phase angle, or of the resistive component (i.e., charge mobility) and the reactive component (degree of polarization), or of any information contained in these quantities.

3. One of the simpler types of impedance bridges is the series-*RC* bridge shown in Experiment Fig. 5-4. The unknown impedance Z_x contains the capacitance C_x. Some power is dissipated in the impedance (for instance, in the resistance of the connecting leads or through losses in the dielectric of the capacitor), this is expressed by the equivalent series resistance R_s. The dissipation factor D_s in such a series combination is $R_s\omega C_s$. The bridge measures only the equivalent quantity C_x and D_x and not the detailed structure of the impedance which may be more complicated.

The bridge is balanced by successive adjustment of C_s and R_s. At balance

$$C_x = C_s \frac{R_b}{R_a} \cdots \tag{5-1}$$

and

$$D_x = D_s \tag{5-2a}$$

or, since $D_x = R_x/X_x$ and $D_s = R_s/X_s$,

$$R_x = R_s \frac{R_b}{R_a} \tag{5-2b}$$

If D_x and D_s are not balanced, Eq. (5-1) is still correct, but the balance (*zero reading*) is not clearly defined and only a flat minimum reading will result.

4. Since only one terminal of the bridge may be grounded, both sides of either the source or detector must be "floating." Normally, one side of the detector circuit will be grounded. The other side may, if necessary, be shielded. The source terminals of the bridge, which must be ungrounded, may then be supplied from a shielded transformer to avoid pickup. A special transformer

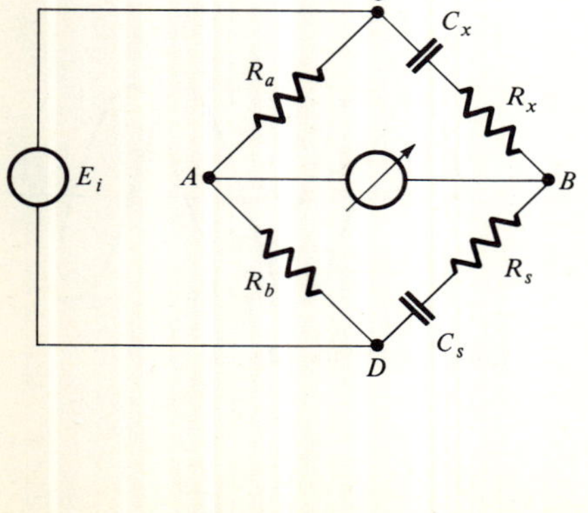

Experiment Fig. 5-4

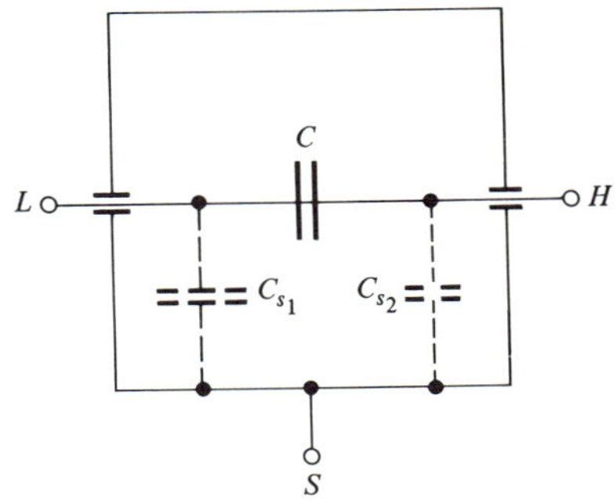

Experiment Fig. 5-5

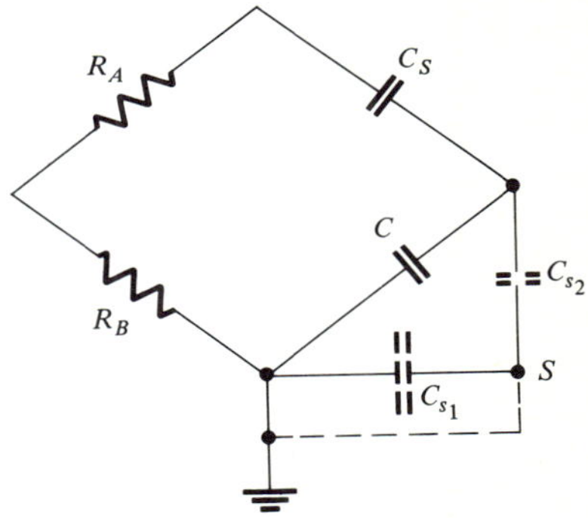

Experiment Fig. 5-6

(such as General Radio Type 578) with windings arranged in separate shielded compartments, both enclosed by another shield which may be grounded, can be used for this purpose (see Experiment Fig. 5-7).

5. A capacitance exists not only between the two terminals of a capacitor or resistor in a shielded container, but also between terminals and shield (three-terminal capacitance, Experiment Fig. 5-5). The stray capacitances C_{S1} and C_{S2} are, in general, not equal. The terminal with larger capacitance to the shield is called *low*, the other *high*. The influence of the stray capacitance is, of course, smaller if the low terminal is connected to ground. Through appropriate connection of the terminals L, H, and S in a bridge circuit, the influence of the stray capacitance can be reduced or the magnitude of C_{S1} and C_{S2} separately can be determined. (E.g., with S connected to ground, Experiment Fig. 5-6, the capacitance C_{S2} is short-circuited, and C_{S1} is connected in parallel to C_X, resulting in a reading error of C_X.)

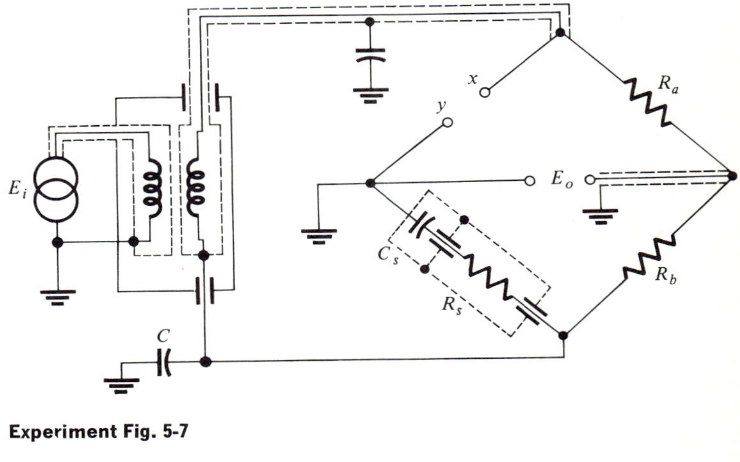

Experiment Fig. 5-7

PROBLEMS

5-1. Design, construct, and use an impedance bridge system for the measurement of the capacitance and the dissipation factor of two given capacitors.
5-2. Determine the sensitivity and the accuracy of the bridge system, as well as the influence of stray capacitance; devise means to minimize the effect of stray capacitances on the bridge measurements.
5-3. Determine the dielectric constant and the conductivity of a given material.

RECOMMENDED PROCEDURE

5-1. Assemble the bridge with due regard for shielding, pickup, etc.; the circuit of Experiment Fig. 5-7 is recommended.
5-2. Determine the capacitance and dissipation factors of the two unknown capacitors.
5-3. Determine, if possible, the three terminal capacitances of the unknown capacitors.
5-4. It is suggested that the bridge be tested with components of approximately known value before determinations are made on the unknown.
5-5. The ac source voltage may be obtained in any convenient manner (60-cycle, audio generator, etc.). Normally, a frequency of 1,000 Hz will be used which will be subject to a minimum of interfering from stray pickup.
5-6. The detector may be any device (headphones, CRO, VTVM, etc.), which will indicate the presence of a low-level ac voltage. If an ac-operated amplifier is used as one element of the detector, or if shielding of the bridge is inadequate, a 60-cycle filter may be used to remove the unwanted 60-cycle signal.

Available equipment

1 audio oscillator
1 transformer, G.R. 578 B
2 unknown capacitors, mounted
2 resistors, 10 kΩ G.R., 500 –
1 ac amplifier
1 CRO
1 60-cycle filter
1 standard resistor
1 standard capacitor

EXPERIMENT 6: FREQUENCY RESPONSE OF IMPEDANCE NETWORKS

BACKGROUND INFORMATION

In the following equations the dimensions are:

E, volts
I, amperes
L, inductance, henrys
C, capacitance, farads $= 10^6\ \mu F = 10^{12}$ pF
R, resistance, ohms

X, reactance (inductive or capacitive), ohms
Z, impedance, ohms
f, frequency, hertz
$\omega = 2\pi f$, radians per second
q, phase angle, degrees

1. If an impedance Z is connected in a circuit as shown in Experiment Fig. 6-1, its value can be found from the measurement of E and I; $Z = E/I$.

(*a*) If Z is a pure resistance,

$$Z = R = E/I_R$$

i.e., the impedance is independent of the frequency.

(*b*) If Z is a pure inductance, the magnitude of Z is

$$Z = X_L = \omega L = E/I_L$$

i.e., the impedance increases with the frequency. If the voltage is kept constant and the frequency of the source increases, the current will decrease.

(*c*) If Z is a pure capacitance, the magnitude of Z is

$$Z = X_C = \frac{1}{\omega C} = E/I_C$$

i.e., the impedance decreases with increased frequency. If the voltage is kept constant and the frequency increased, the current will increase.

2. If two resistances R_1 and R_2 are connected in series to an ac source E_i (as shown in Experiment Fig. 6-2), a current I will flow in the circuit, which is in phase with the applied voltage. The partial voltages E_{R_1} and E_{R_2} are in phase too with the current, and the sum of E_{R_1} and E_{R_2} is equal to the total voltage.

3. If a resistance R and a pure inductance L are connected in series to an ac source E_i, as shown in Experiment Fig. 6-3*a* and *b*, the partial voltages E_R and E_L are 90° out of phase, with the inductive voltage drop leading in time. The total voltage E is equal to the vector sum of E_R and E_L. (Analytically: $\sqrt{E = E_R^2 + E_L^2}$.) The current I is always in phase with the resistive voltage

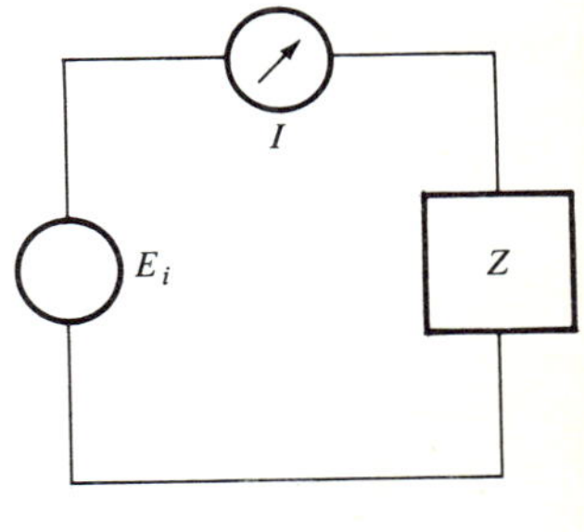

Experiment Fig. 6-1

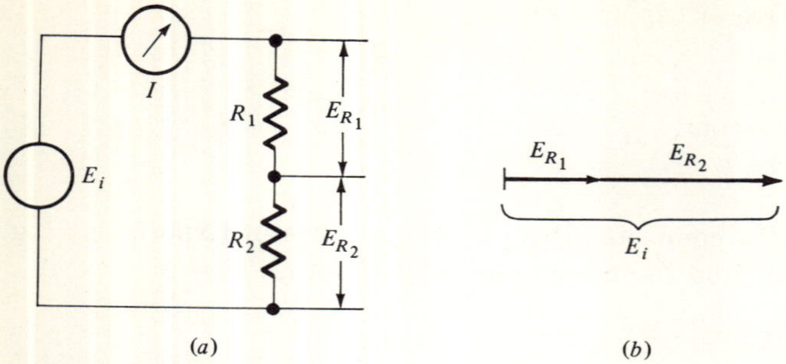

Experiment Fig. 6-2

drop and its magnitude is again determined by $I = E/Z$, where $Z = \sqrt{R^2 + \omega^2L^2}$. The phase angle between the applied voltage and the resulting current in the series circuit can be found from $\tan q = \omega L/R$ by solving for $q = \arctan (\omega L/R)$.

If the frequency of the input voltage E_i is raised from zero to increasingly higher values, the voltage drop across the inductance rises from zero and asymptotically approaches the value E_i. The voltage drop across the resistance at the frequency zero (dc) equals E_i. With increasing frequency the voltage across the resistor diminishes as shown in Experiment Fig. 6-3*c*.

4. If a resistance R and a capacitance C are connected in series to the ac source E, as shown in Experiment Fig. 6-4*a* and *b*, the partial voltages E_R and E_C are 90° out of phase, with the capacitive voltage lagging in time. The total voltage E is equal to the vector sum of E_R and E_C. The current I is always in phase with the voltage drop across the resistance, and its magnitude is $I = E/Z$, where $Z = R^2 + (1/\omega C)^2$. The phase angle between the applied voltage and the resulting current is $q = \arctan 1/\omega RC$.

If the frequency of the input voltage E_i is raised from zero to increasingly higher values, then the voltage drop across the capacitor diminishes from E_i asymptotically to zero. The voltage drop across the resistance is zero at the

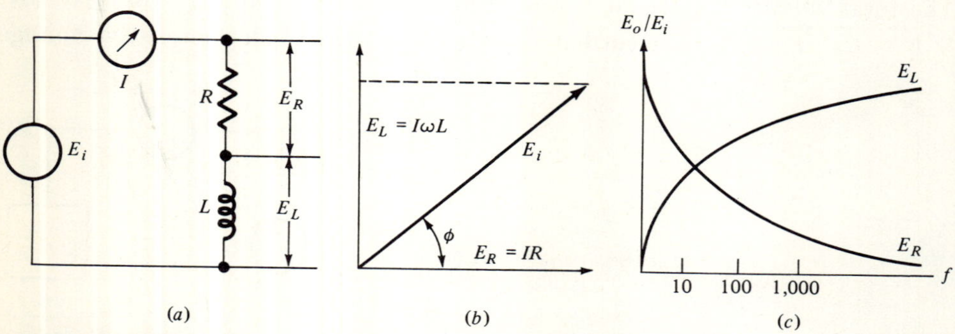

Experiment Fig. 6-3

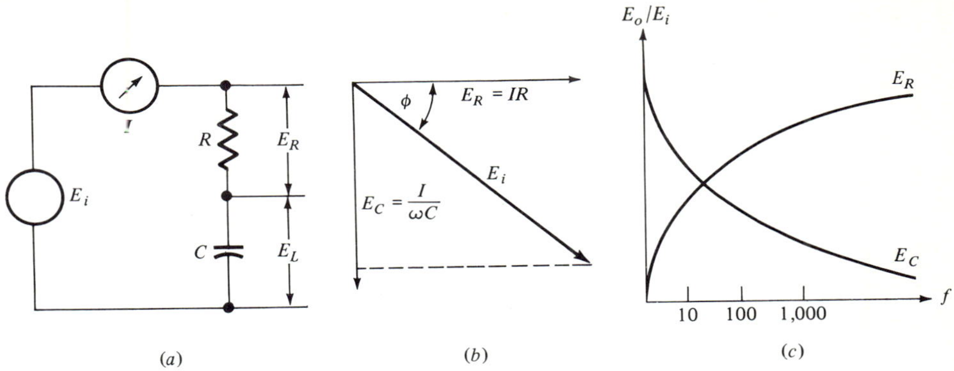

Experiment Fig. 6-4

frequency zero and rises asymptotically to the value E_i as shown in Experiment Fig. 6-4*c*. Therefore, *RC* networks can be used as low-pass filters by connecting an ac voltage to the input and using the voltage across the capacitance as the output. They can be used as high-pass filters by using the voltage across the resistor as the output.

5. If a resistance R, an inductance L, and a capacitance C are connected in series to an ac source (Experiment Fig. 6-5*a*), then the total impedance is $Z = \sqrt{R^2 + X^2}$, where $X = X_L - X_c$. The vector diagram for this circuit is shown in Experiment Fig. 6-5*b*. If a constant voltage E_i is applied and if the frequency of E_i is raised from a low value, the inductive reactance X_L will increase while the capacitive reactance will decrease. At one frequency (resonance

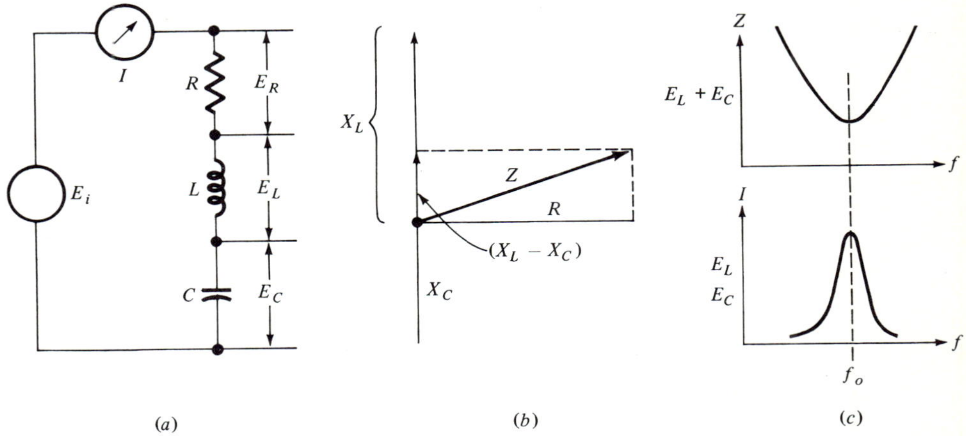

Experiment Fig. 6-5

frequency) both reactances will be alike:

$$X_L = X_C \qquad \omega L = \frac{1}{\omega C}$$

$$(2\pi f_0)^2 = \omega_0{}^2 = \frac{1}{CL} \qquad f_0 = \frac{1}{2\pi/CL}$$

At resonance, the following phenomena will be observed in an *RLC* series circuit: the reactances cancel each other; the total impedance in the circuit reaches a minimum value R. The current reaches a maximum E_i/R. The voltage across the inductive reactance reaches a maximum and so does the voltage across the capacitance reactance, but these voltages are of opposite phase. The circuit acts as a capacitive-resistive impedance at frequencies below resonance and as an inductive-resistive impedance at frequencies above resonance. At resonance, the circuit behaves like a pure resistance ($q = 0$).

PROBLEMS

6-1. Study the frequency response of simple impedance networks, i.e., of an *RC* low-pass filter, a *CR* high-pass filter, and a series resonance circuit.

6-2. Four "black boxes" are given, each containing a four-terminal network of the type shown in Experiment Fig. 6-6. The impedances Z_1, Z_2, and Z_3 consist of only passive, linear electric components, i.e., resistors, capacitors, or inductors. Some impedances may be short-circuited ($Z = 0$). Determine impedances of the network at different frequencies and find the structure of the network in each box.

RECOMMENDED PROCEDURE

6-1. The frequency-response curve is obtained by applying an ac voltage of constant amplitude and variable frequency to the input of the filter and observing the output. The circuit of Experiment Fig. 6-7 is recommended. The same voltmeter is used to monitor the input with switch S in position A and observe the output with S in position B. Plot the output versus frequency on logarithmic paper. Observation of the resonance circuit requires measurement of the current. Insert a resistance which is small compared to the reactances in the circuit (1, 10, or 100 Ω) and measure the voltage drop across this resistor.

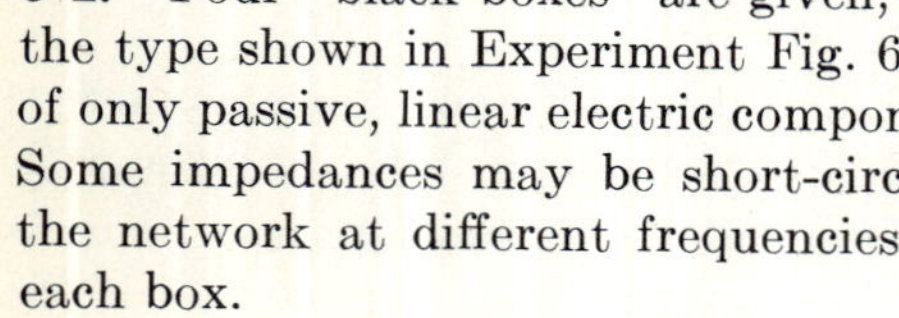

Experiment Fig. 6-6

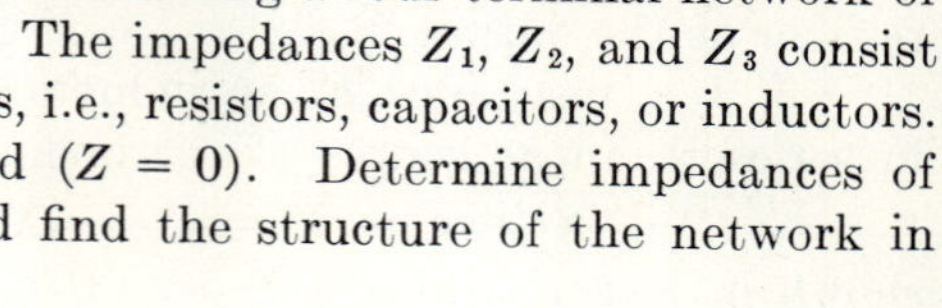

Experiment Fig. 6-7

Measure E_i and keep constant for different frequencies; measure E_L, E_c, and E_{L-C} through resonance.

Available equipment

1 audio oscillator (hp 200 D)
1 cathode-ray oscilloscope (CRO)
1 vacuum-tube voltmeter (VTVM) (hp 400 A)
1 low-pass filter
1 high-pass filter
1 series resonance circuit
4 "black boxes" (unknown impedance networks)
1 VTVM
1-, 10-, and 100-Ω precision resistors

EXPERIMENT 7: NUCLEAR RADIATION INSTRUMENTATION

BACKGROUND INFORMATION

A radioactive isotope of an element behaves chemically like a stable isotope of the same element, but it emits a radiation which permits measurement of its concentration. Since the emitted radiation of a radioactive isotope is fairly specific, it is also possible to determine the species of each isotope in a given sample material.

The most common radiation types emitted by a radioactive isotope are:

1. *Alpha particles*, i.e., helium nuclei (two protons and two neutrons). The α particles have a large size and, thus, a high probability of interaction with the material through which they are traveling. Therefore, their range is short compared with beta or gamma radiation.
2. *Beta radiation*, i.e., high-speed electrons which, because of their small size, react very little with the material through which they are traveling.
3. *Gamma radiation*, i.e., electromagnetic radiation, like light (photons) or x-rays, but of high energy, in the keV or MeV region.[1] The radiation is emitted in discrete energy levels, as shown in Experiment Fig. 7-1. The

[1] An electron of an energy of 1 MeV has the charge e and is accelerated by a voltage of 1 million volts. If such an electron falls on a target, it causes the excitation of an electromagnetic radiation whereby the entire energy of the electron is converted into a photon. The photon then has the energy of 1 MeV.

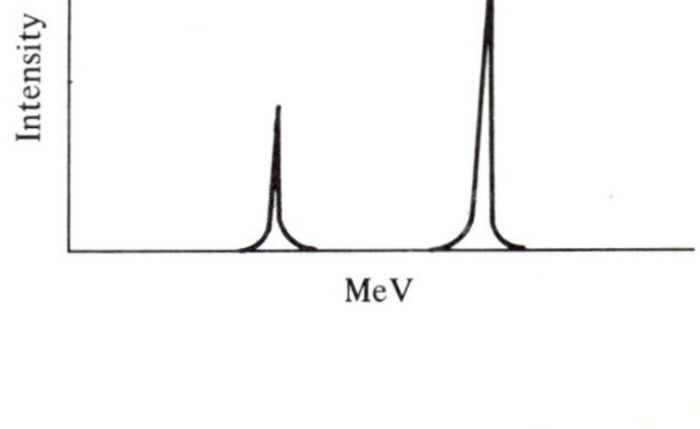

Experiment Fig. 7-1

emitted energy spectrum is characteristic for any isotope. The range of γ-rays traveling through any medium is, in general, much larger than that of α- or β-rays.

Several devices exist for the detection of nuclear radiation, such as photographic emulsions, ionization chambers, spark counters, cloud chambers, or crystal counters. The most frequently used methods of measurement and detection are the Geiger counter and the scintillation counter.

GEIGER-MUELLER TUBE

A schematic diagram of a Geiger-Mueller counter is shown in Experiment Fig. 7-2. The anode is usually a thin platinum or tungsten wire held by insulators in the center of a metal cylinder which acts as the cathode. The cylinder is thin enough, or has a thin window, to allow ionizing radiation to enter the chamber. The chamber is usually filled with an inert gas (e.g., argon) to which a small amount of a polyatomic gas (e.g., alcohol) has been added.

The property most characteristic of a tube operated in the Geiger region is the production of pulses of uniform height regardless of the number of ion pairs produced by the ionizing radiation. The mechanism of this phenomenon is as follows. The incident radiation produces a free electron, either in the gas or, more frequently, on the cathode surface. The electron is accelerated toward the anode and initiates collision ionization and an avalanche breakdown. Photons emitted from the initial avalanche region cause more electron-positive ion pairs to be produced by photoionization. These secondary electrons initiate fresh avalanche discharges and so the breakdown spreads along the anode. The region around the anode is now positively charged since the positive ions are much less mobile than the electrons. This space charge, or virtual anode, reduces the electric field strength to the point where the photon-photoelectron-avalanche reaction cannot continue. The quench gas absorbs photons and prevents the positive ions from acquiring sufficient energy to liberate electrons at the cathode, thus reinitiating the discharge phenomenon. The counter is unable to respond to two events occurring in a very short time interval. The time necessary for sweeping out the positive ion cloud around the anode and thus

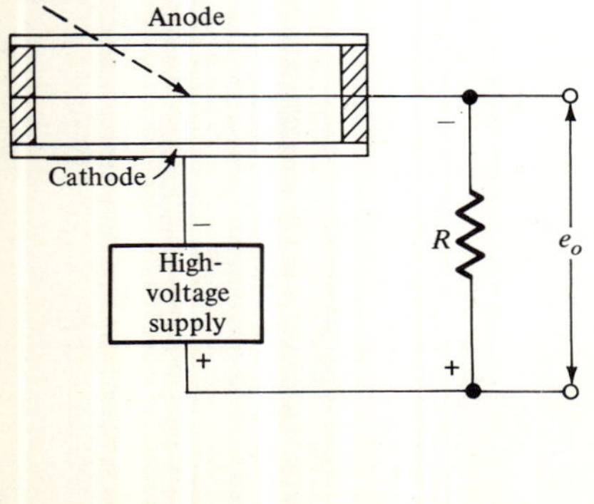

Experiment Fig. 7-2

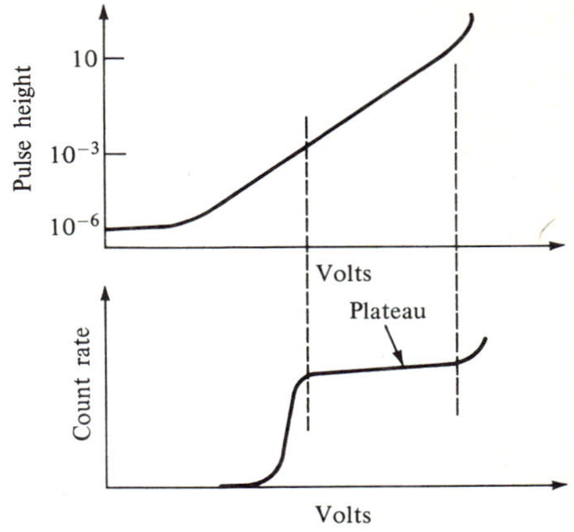

Experiment Fig. 7-3

resetting the counter varies from 30 to 300 μs for most counters. The maximum counting rate is from 3×10^3 to 3×10^4 Hz.

As the voltage across the Geiger counter is raised, the pulses observed under the influence of an incident nuclear radiation grow larger in *size*. The *number* of pulses recorded under the influence of a constant incident radiation depends, at low voltages, on the initial ionization (i.e., whether one or many electrons are produced initially). At higher voltages, the number of recorded pulses increases and becomes independent of the initial ionization and of the applied voltage. This is the plateau region, where the counter is most frequently used. Above the plateau region, pulses are observed regardless of the presence or absence of ionizing radiation. A plot of the Geiger counter characteristic is shown in Experiment Fig. 7-3.

SCINTILLATION DETECTOR

A schematic diagram of the scintillation detector is shown in Experiment Fig. 7-4. Nuclear radiation falls on a scintillator, i.e., a luminescent solid, liquid, or gaseous material which absorbs nuclear radiation and emits a light pulse. The light pulse, or as large a fraction of it as is technically possible, is directed on the cathode of a photomultiplier which converts the light pulse into an electric pulse and amplifies it. The scintillation counter has a number of advantages over

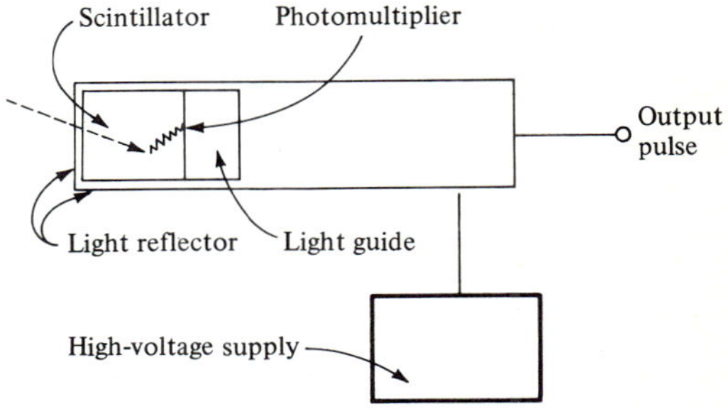

Experiment Fig. 7-4

the Geiger counter. The density of the luminescent material, in particular, the solid scintillator (such as NaI), causes a more efficient absorption of the nuclear radiation and provides a more efficient detector than the gaseous atmosphere in the Geiger counter. The amount of light produced in the scintillator is approximately proportional to the amount of energy lost by the radiation in the crystal. Therefore, the scintillation counter furnishes information of the radiation energy (pulse height) *and* radiation intensity (number of pulses per second). Also, the entire process occurs very rapidly and does not require a reset period. The maximum counting rate of the scintillation counter is essentially determined by the electronic equipment following the photomultiplier.

PULSE–HEIGHT DISCRIMINATION AND ANALYZER

Since the scintillation detector produces pulses of an amplitude proportional to the energy of the ionizing event, it is possible to increase the signal-to-noise ratio by counting only pulses of the amplitude expected from the radioisotope of interest. Gamma radiation is emitted at specific energy levels. Thus, one can determine the concentration of two or more gamma emitters if the emitted γ-rays are of sufficiently different energy levels. Experiment Fig. 7-5 is a diagram of a composite spectrum containing two levels of gamma radiation. As is apparent, the peaks are of appreciable width. This is caused by interactions of the monochromatic γ-rays with matter, by incomplete loss of energy in the crystal, and by variations of light output for a given absorbed energy for the different possible mechanisms of absorption. The signal-to-noise ratio, i.e., counts from the peak between E_1 and E_2 divided by background between E_1 and E_2, is greater than the signal-to-noise ratio would be if all counts from zero to E_2 were measured. In a like manner, by measuring from E_1 to E_2 and from E_3 to E_4 the relative concentrations of the radioisotopes producing the separate peaks can be determined. A spectrum such as that of Experiment Fig. 7-5 may be constructed by sweeping a "window" of width $\Delta E(E_2 - E_1)$ across the energy range of interest. A similar spectrum could be constructed by using a discriminator (circuit which causes only pulses larger than a selected amplitude to be measured) and making multiple measurements. Experiment Fig. 7-6 is a diagram of the type of curve which could be constructed using a discriminator. The derivative of the curve of Experiment Fig. 7-6 is the negative of the spectrum of Experiment Fig. 7-5. The

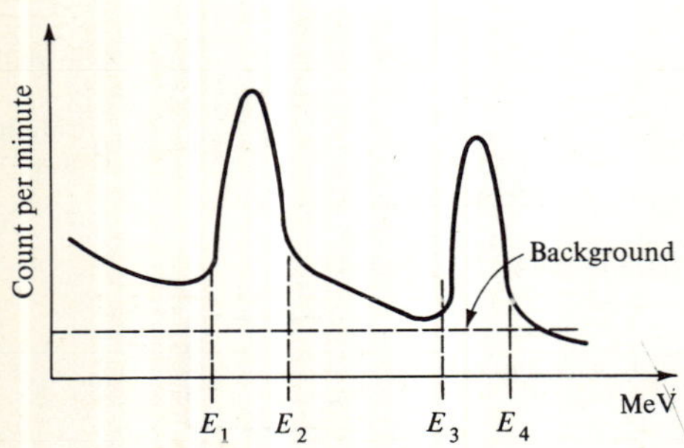

Experiment Fig. 7-5

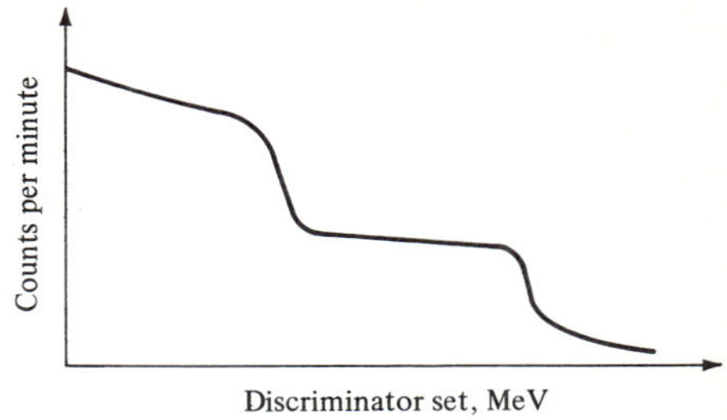

Experiment Fig. 7-6

information which can be extracted from the curve of Experiment Fig. 7-6 is similar to that which may be obtained using the pulse-height analyzer (Experiment Fig. 7-5).

PROBLEMS AND RECOMMENDED PROCEDURE

7-1. Determine and plot the operating curve for a Geiger-Mueller counter. Use a ^{137}Cs source. Set scaler discriminator to 250 mV. Set high voltage supply to its minimum value and allow equipment to warm up for at least 2 min before turning on HV supply. Advance high voltage rapidly until scaler indicates counting has begun; reduce HV and advance by 25-V steps counting for at least 1 min at each voltage. When counting rate begins to increase rapidly after plateau region, terminate the experiment. Plot data (counts per minute versus voltage).

7-2. Determine and plot the operating curve for a scintillation detector. Use a ^{137}Cs source. Set HV supply to its minimum value and allow equipment to warm up for at least 2 min before turning on HV supply. Advance HV rapidly until scaler indicates counting has begun; then, reduce HV and advance by 25-V steps counting for at least 1 min at each voltage. Terminate experiment when counting rate becomes excessive or at an applied voltage 200 V above that at which counting was first observed, whichever comes sooner. Plot data.

7-3. Using the pulse-height analyzer and scintillation detector, determine and plot the spectrum of protoactinium 234 (^{234}Pa). Calibrate the pulse-height analyzer for 1 MeV full scale with RELATIVE GAIN setting of "4" in the manner described in the section on technicalities. This will result in calibration of five ranges from 250 keV to 4 MeV full scale selectable by switching RELATIVE GAIN control. Set window width to 5 percent for measurements below 500 keV, at 2 percent for measurements from 500 keV to 1 MeV, and at 1 percent for energies greater than 1 MeV. Notice that data must be normalized for a standard window width in terms of energy before plotting data. Normalize counting rate to 20-keV window width. Plot spectrum from 250 keV to 1.5 MeV.

Available equipment

scaler
pulse-height analyzer
scintillation detector

Geiger-Mueller tube
sample changer
^{137}Cs source
^{234}Pa source

TECHNICALITIES, PICKER x-RAY CORPORATION RADIOISOTOPE TRAINING EQUIPMENT

The equipment used for this experiment is a commercial training set. Five pieces of equipment will be used: scaler, pulse-height analyzer, scintillation detector, Geiger-Mueller tube, and a sample changer.

Scaler This instrument is used for counting pulses from a source such as a Geiger-Mueller tube or a scintillation detector. It uses 5-decade counting tubes. Resolving time is 30 μs giving a maximum counting rate of 3×10^5 cycles/min. A timer which is turned on and off with the counters is built into the scaler; resetting of the timer is done manually. A switch on the back of the counter sets it to count all pulses over 25 or 250 MV. A high-voltage regulated power supply is built into the scaler to supply the correct voltages to the radiation-detecting device. The power supply controls are calibrated directly in volts. The scaler is interconnected with the remaining apparatus by means of cables on the rear of the chassis. When the scaler is used directly with either the Geiger-Mueller tube or scintillation detector, only one connection is required. This connection is a cable from the connector labeled DETECTOR INPUT to the detector. When the counter is used with the pulse-height analyzer, a cable is connected between the connector labeled PULSE INPUT on the scaler and PULSE OUTPUT on the analyzer, and another is connected between the connectors labeled H.V. OUTPUT on the scaler and H.V. INPUT on the analyzer. When the analyzer is used, the radiation detector (Geiger-Mueller tube or scintillation detector) is connected to the DETECTOR INPUT connection on the analyzer. Notice that two types of connectors are used on the rear of the chassis of both the scaler and analyzer and that the cables are not interchangeable. When used with the pulse-height analyzer, the discriminator switch on the rear of the scaler chassis should be set at 250 MV.

Pulse-height analyzer The connections to the rear of the chassis of the pulse-height analyzer have been explained above. There are three front panel controls labeled LOWER LEVEL, WINDOW WIDTH, and RELATIVE GAIN. The lower level control is a ten-turn potentiometer which sets the lower energy level of the window (E_1 in Experiment Fig. 7-5). The window width control adjusts the span of the energy window (i.e., controls the distance between E_1 and E_2 of Experiment Fig. 7-5). This span may be 1, 2, 5, or 10 percent of the full scale setting of the lower level control. That is, if the instrument is calibrated so that a setting of 1,000 (full scale) on the lower level control corresponds to 1 MeV, a window width setting of 5 percent signifies that the window width (in terms of energy) is $0.05 \times 10^6 = 50$ keV. Thus, if calibrated for 1 MeV full scale, the analyzer will produce a pulse for each particle of energy between the lower level setting and the lower level setting plus 50 keV. That is, if the lower level control is set at 500, only pulses with amplitudes corresponding to particles of energy greater than 500,000 eV and smaller than 550,000 eV will be counted. The window width position marked α (integral) indicates that there is no upper limit on the window. Thus all pulses corresponding to particles of energy greater than the lower level setting will be counted. The relative gain control changes the gain of the pulse amplifier by factors of 2. Thus, if the pulse-height analyzer is calibrated for 1 MeV full scale when the relative gain control is on 8, changing the relative gain to 4 will result in full scale of the lower level corresponding to 2 MeV. Thus, after initial calibration, the relative gain control may be used as a range control.

CALIBRATION OF PULSE-HEIGHT ANALYZER

For initial calibration of the pulse-height analyzer a γ-ray source with a well-defined peak in the 1-MeV range is desirable. ^{137}Cs is such a source with a sharp peak centered at 0.663 MeV.

To calibrate for 1 MeV, follow the following procedure:

1. Place ^{137}Cs on the scintillation detector.
2. Set the WINDOW WIDTH control on 2 percent.
3. Set the RELATIVE GAIN control on 8.
4. Set the LOWER LEVEL control on 653 (since the window width is 20 keV, a window from 653 to 673 keV will be centered on the 663-keV peak).
5. Starting with the minimum setting of the HV power supply (on scaler) increase power-supply voltage slowly until a maximum in the counting rate is observed. Make a plot of counts per minute versus voltage for this region. Set the voltage at the point where the counts per minute curve is a maximum.
6. The analyzer is now calibrated for 1 MeV full scale. Changing the relative gain control to 4 will result in 2-MeV full-scale calibration; changing to 16 will change full-scale calibration to 500 keV, etc.

EXPERIMENT 8: OPERATION OF NONLINEAR ELEMENTS

BACKGROUND INFORMATION

Linear elements Experiment Fig. 8-1 shows the generalized diagram of a four-terminal circuit element S. A quantity Q_i is applied to the input; a quantity Q_o,

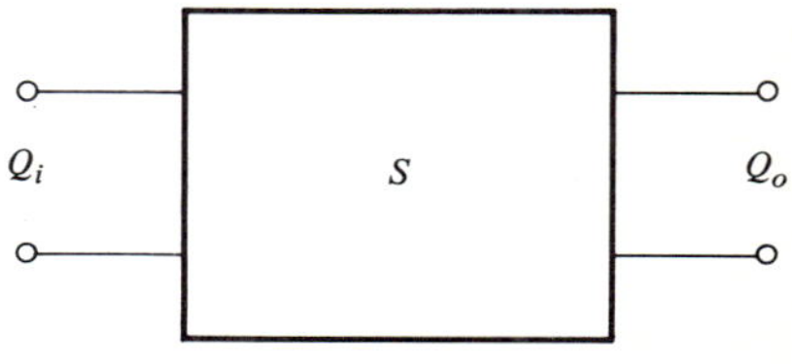

Experiment Fig. 8-1

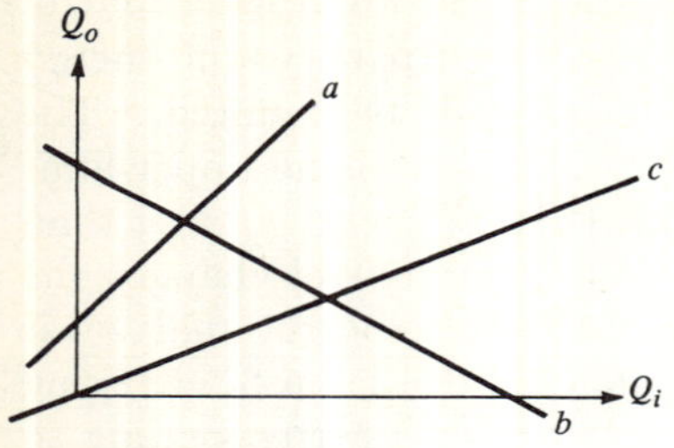

Experiment Fig. 8-2

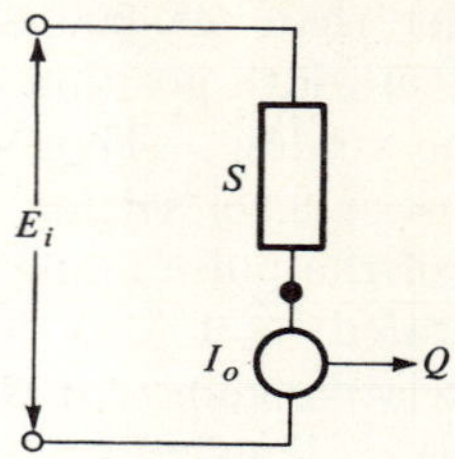

Experiment Fig. 8-3

not necessarily of the same dimension as Q_i, appears at the output. If the transfer function follows the equation

$$Q_o = mQ_i + b$$

the system S is said to be a linear network (Experiment Fig. 8-2 curves a to c). Linear networks, particularly of the kind with the characteristic of Experiment Fig. 8-2 curve c, are frequently desirable in instrumentation since they allow a simple analysis. Impedance networks containing resistive, inductive, and capacitive components are linear.

Experiment Fig. 8-3 shows a linear two-terminal network S. The current I_o is a linear function of the applied voltage E_i.

$$I_o = nE_i = \frac{1}{R} E_i$$

A two-terminal network can be converted into a four-terminal network, as shown in Experiment Fig. 8-4, by inserting in series with the network S a resistor r which is so small that the circuit parameters are practically not changed ($r \ll S$). The current I_o causes a voltage drop $E_r = I_o r$ which represents the output signal Q_o. If S is a pure resistance, then the voltage-current characteristic is a straight line as shown in Experiment Fig. 8-5. The magnitude of the resistance is expressed by the slope α of the linear characteristic.

$$R = \frac{E_i}{I} = \frac{1}{\tan \alpha}$$

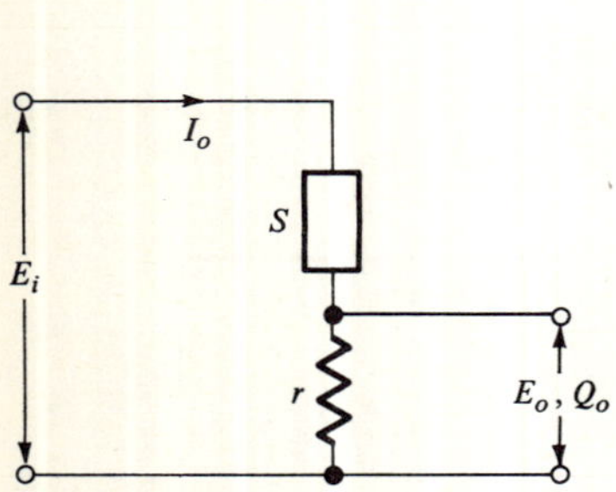

Experiment Fig. 8-4

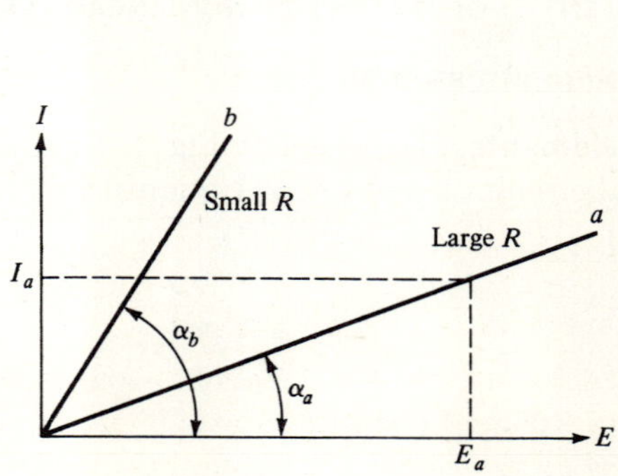

Experiment Fig. 8-5

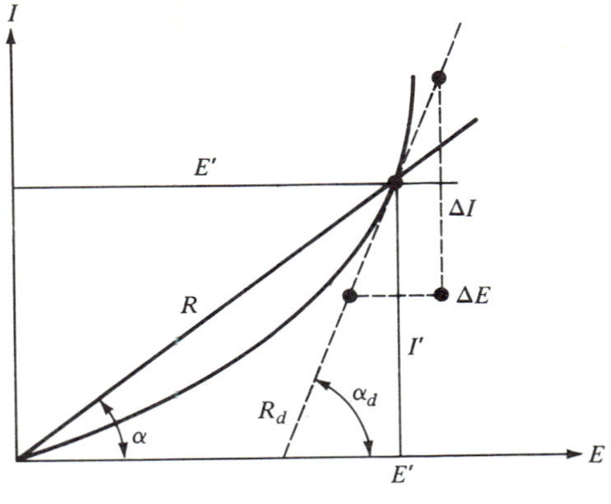

Experiment Fig. 8-6

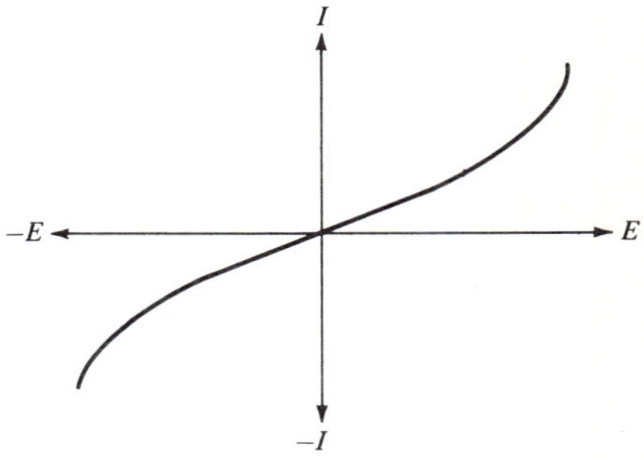

Experiment Fig. 8-7

Nonlinear elements In some elements the applied voltage causes a physical change, e.g., an increase of temperature with a resulting change of conductivity, a polarization, or a depletion of charge carriers. Such elements exhibit nonlinear characteristics; an example is shown in Experiment Fig. 8-6. For any operating point E', I', the resistance is $R = E'/I'$. However, for a small variation ΔE the current varies by ΔI. The ratio

$$\frac{\Delta E}{\Delta I} = R_d = \frac{1}{\tan \alpha_d}$$

has a physical significance. R_d, the differential resistance (sometimes called *dynamic resistance*), is the apparent resistance for a small voltage and current variation in the vicinity of the operating point. Its value is expressed by the tangent of the curve at the point E', I'. R_d varies with the location of the operating point; it can be smaller or it can be larger than R.

VARISTORS

Varistors are semiconductors usually made from silicon carbide particles in a ceramic binder. Their EI characteristics are symmetrical, i.e., they do not rectify. The origin of the nonlinear characteristic is probably in the silicon carbide–metal electrode boundary. Varistors can be used in networks to emphasize or deemphasize the effect of voltage variations. Their time constant is very short, of the order of 10^{-9} s. Experiment Fig. 8-7 shows a typical varistor EI characteristic.

THERMISTORS

Thermistors are semiconductors consisting mainly of the oxides of Mn, Ni, Co, U, or other elements and exhibiting large resistance-temperature coefficients. If operated at low current values or at constant temperature, their IE characteristics are linear. A deviation from linearity occurs if the current causes a significant

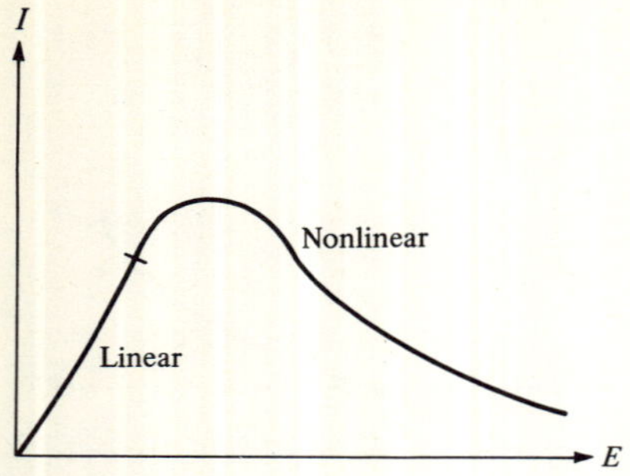

Experiment Fig. 8-8

increase of temperature. The deviation (*temperature coefficient*) may be positive or negative. The time constant of a thermistor varies between about 0.1 and 100 s. Experiment Fig. 8-8 shows the voltage-current characteristic of a typical thermistor with negative temperature coefficient.

DIODES

Junction diodes (Experiment Fig. 8-9) consist primarily of silicon or germanium. By admixture of impurities, one side is made to contain an excess of free electrons (*n-type semiconductor*), the other side an excess of holes (*p-type semiconductor*). Initial diffusion of electrons and holes creates a potential difference between the regions which prevents further flow of carriers across the junction.

If a voltage is applied in the positive direction as shown in Experiment Fig. 8-9*a*, it effectively reduces the junction potential, and a current flows as shown in Experiment Fig. 8-9*b*; i.e., with increased voltage the current increases. If a voltage is applied in the negative direction, the junction potential difference is increased, and no current flows (except a small leakage current). If the voltage is further increased, a point is reached ($-E_Z$, zener voltage) where the (negative) current begins to increase strongly (Experiment Fig. 8-9 curve *c*). (The zener voltage of ordinary diodes is beyond the range realizable with the given equipment.)

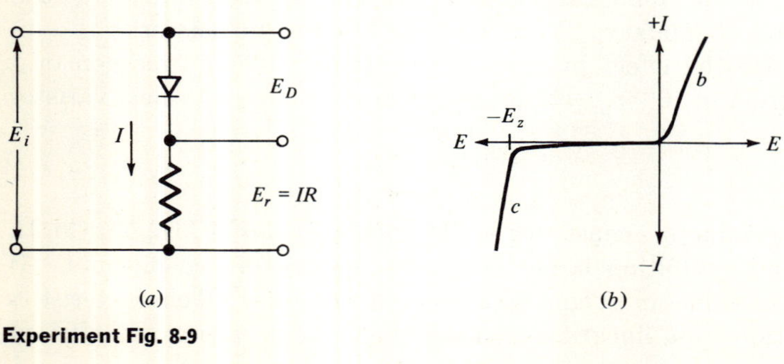

Experiment Fig. 8-9

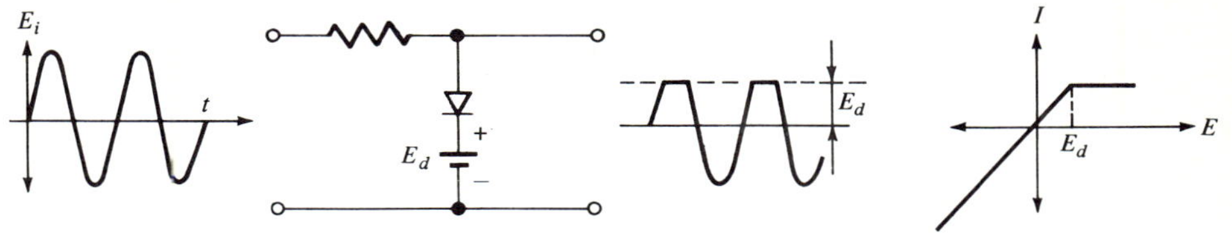

Experiment Fig. 8-10

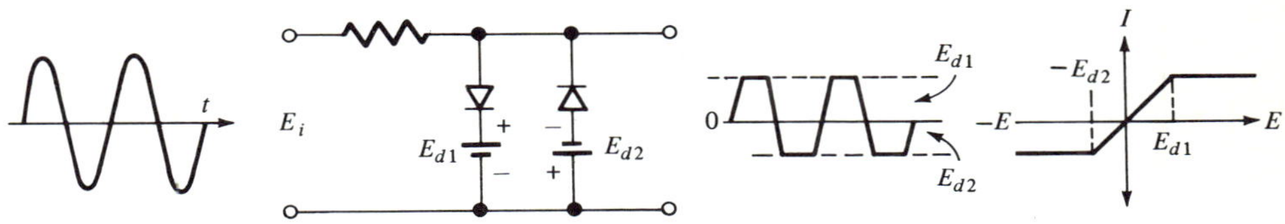

Experiment Fig. 8-11

CLIPPING CIRCUITS

If a diode is biased by a dc voltage E_d, as shown in Experiment Fig. 8-10, it will act as a clipper, i.e., the network will pass only the portion of a signal *below* the level E_d and it will cut off any signal above this value. A symmetrical clipper of the same type is shown in Experiment Fig. 8-11. The circuit of Experiment Fig. 8-12 will pass only that part of a signal which is *above* the value E_d.

CLAMPING CIRCUITS

These are also called *dc restorers*. These circuits can shift the average dc level of a given waveform signal by an amount equal to the positive or negative peak value of the waveform. The polarity depends on the direction of the diode. Experiment Fig. 8-13 shows a capacitive-resistive network without the clamping diode. If a positive square wave is applied to the input, the output is a symmetrical square wave. If a diode is added to the network as shown in Experiment Fig. 8-14, the capacitor is charged to the peak value, and the output is a negative square wave.

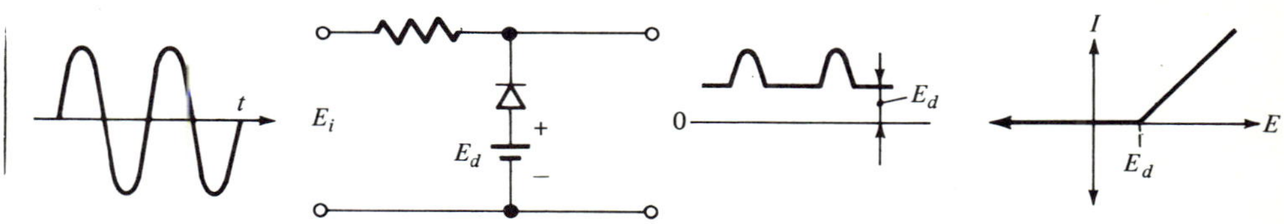

Experiment Fig. 8-12

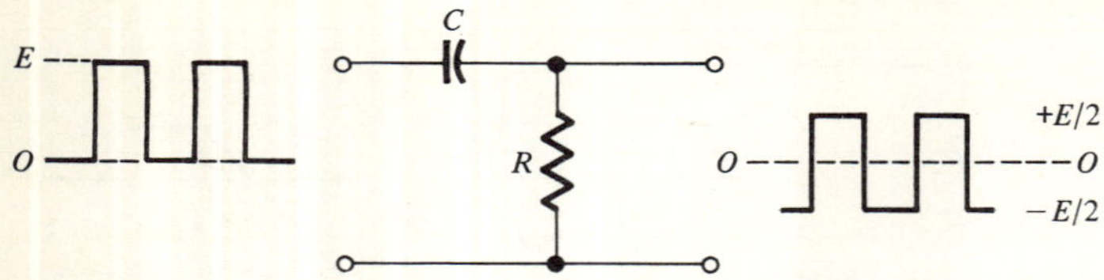

Experiment Fig. 8-13

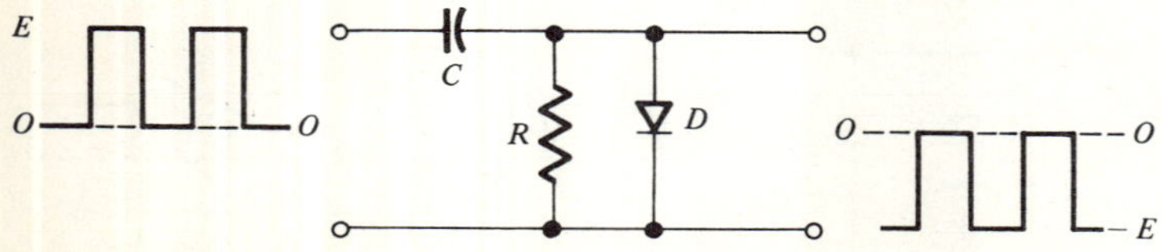

Experiment Fig. 8-14

FUNCTION SYNTHESIZERS

Almost any monotonic function can be synthesized from straight-line segments by a number of networks consisting of a diode, a resistive network, and an auxiliary voltage source. An example is shown in Experiment Fig. 8-15. A voltage E_i rising linearly from zero is applied to the network $R_1 - R_2 - D - E_d$. The diode is biased by the auxiliary source E. For $0 < E_i < E_d$, the diode acts as a closed switch and the output voltage rises with a slope determined by R_1 and R_2, that is,

$$E_o = E_i \frac{R_2}{R_1 + R_2}$$

As soon as $E_i \geqq E_d$, the diode is reverse-biased and will act as an open switch. Thus, E_o will be equal to E_i. The output voltage as a function of the input voltage is shown in Experiment Fig. 8-15*b*. By cascading several of such

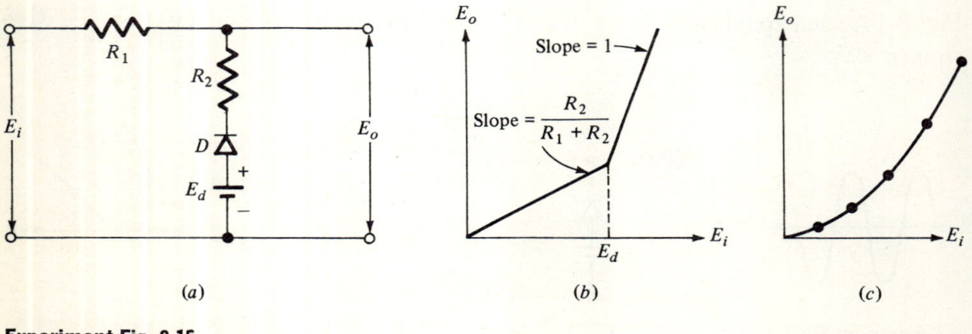

Experiment Fig. 8-15

networks, any desired monotonic function may be synthesized (Example Fig. 8-15*c*).

PROBLEMS

8-1. Investigate the characteristics of several given elements, such as resistors, diodes, zener diodes, varistors, or capacitors.

8-2. Build from laboratory equipment a function synthesizer element and investigate its characteristics with different directions of the diode and of the bias voltage.

RECOMMENDED PROCEDURE

Three different methods can be used to determine the characteristics of linear or nonlinear circuit elements: (1) the voltage and current values can be read on meters and plotted, point-by-point; (2) higher speed in the plotting of data can be obtained with the x-y recorder. The instrument furnishes a permanent record; (3) where a permanent record is not needed, the characteristic can be made to appear on the screen of a cathode-ray oscilloscope. For these experiments it is recommended that one or two experiments be carried out with the x-y recorder and all other experiments with the CRO.

Experiment Fig. 8-16 shows the general circuit diagram for measurements with the x-y recorder and with the CRO. The unknown circuit element X is connected in series with an adjustable (small) resistor R. The common connector point P is connected to the common ground connection of the recorder or CRO. For the application of the x-y recorder, a dc voltage is applied to the input terminals and is slowly raised. For the application of the CRO, an ac voltage, variable from 0 to 30 V, is applied and "sweeps" the characteristics.

The x-y recorder has a maximum sensitivity of 10 mV for full-scale deflection. Prior to applying any voltage to each channel, the sensitivity should be set to the lowest value. Apply the independently variable voltage E_i directly to the horizontally moving channel (X axis). The voltage across the resistance r should be applied to the vertically moving channel. Before bringing the pen in contact with the paper, make sure that the maximum values of E_i and E_o do not exceed the range of the recorder. Empirical calibration is usually required. Have circuit checked before applying voltages.

The output from the 30-V ac power supply sometimes contains higher harmonics which may disturb the smooth curve trace on the cathode-ray oscillo-

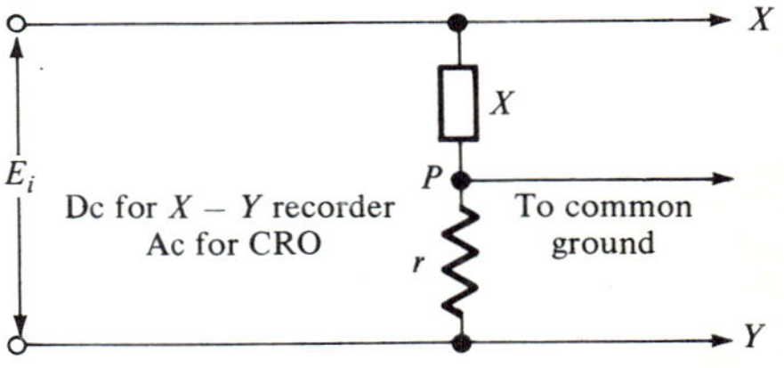

Experiment Fig. 8-16

scope. If this occurs, insert a 60-Hz filter between the output of the 30-V ac supply and the test circuit.

Available equipment

Cathode-ray oscillograph Tektronix model 503
x-y recorder, Houston Instr. Corp. HR-92
ac power source, 0 to 30 V
dc power source, 0 to 10 V
2 Eico resistance boxes
1 potentiometer 1,000 Ω
1 60-Hz low-pass filter
Capacitors, resistors, diodes, zener diodes
Varistors
Other equipment on request

EXPERIMENT 9: TRANSISTOR SINGLE-STAGE AMPLIFIER

BACKGROUND INFORMATION

Three basic circuits exist for the operation of a transistor amplifier: the common-base, common-emitter, and common-collector circuit. Of these, the common-collector circuit is most frequently used.

The basic common-emitter amplifier circuit is shown in Experiment Fig. 9-1. A bias voltage E_{cc} supplies the power for the operation. An input signal E_i is applied between the base and the ground (emitter) terminals; the output signal $E_c = E_o$ is generated between the collector and ground.

The collector characteristics of the common-emitter circuit (collector current I_c versus emitter-to-collector voltage E_{CE} for different values of the base current I_b) are shown in Experiment Fig. 9-2. The choice of the operating point (i.e., the operating conditions with zero *signal* input) determines the linearity of the amplifier, the noise figure, power consumption, maximum positive and negative output signals, input and output impedance levels, and stability of the

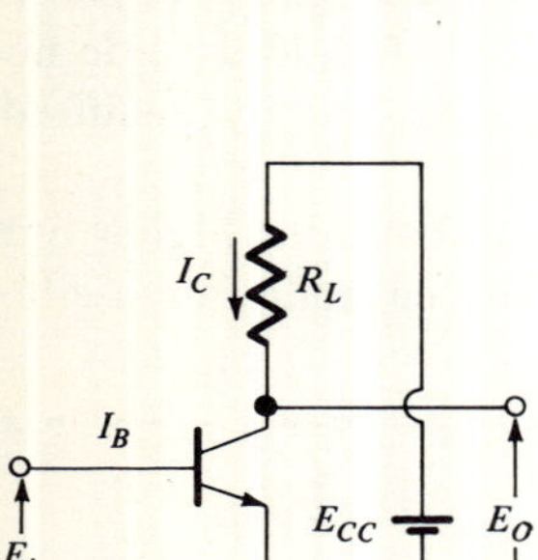

Experiment Fig. 9-1

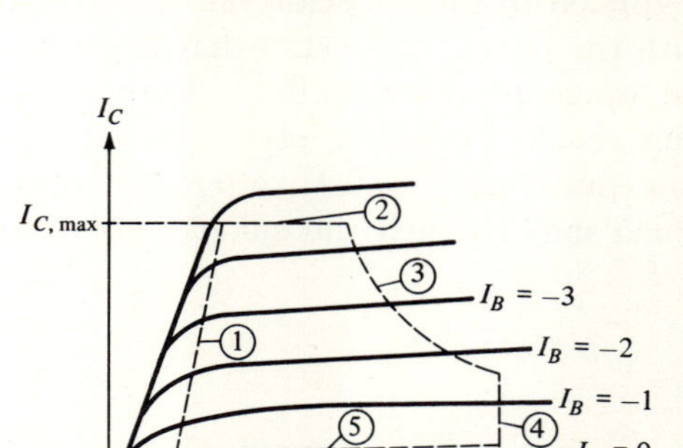

Experiment Fig. 9-2

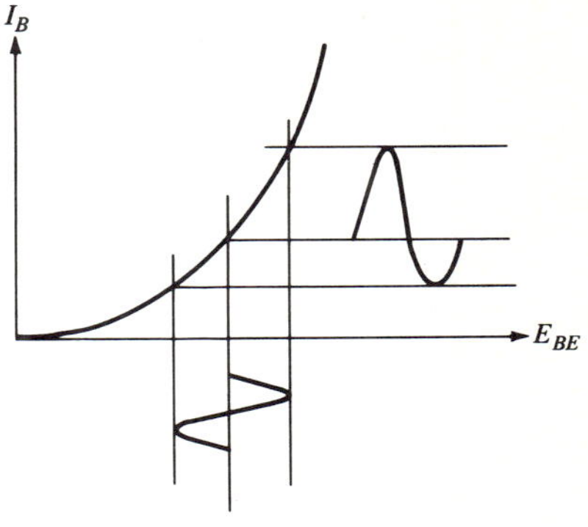

Experiment Fig. 9-3

amplifier. The operating point is ordinarily chosen within the area enclosed by lines 1 to 5. Line 1 is the lower limit of the area in which the transistor output curves are reasonably linear. Line 2 marks the maximum allowable collector current. Curve 3 is the maximum allowable power level for the transistor ($E_{CE} \times I_C$). Line 4 is the maximum allowable collector-to-emitter voltage. Line 5 is the minimum collector current, i.e., the collector current for which the base current I_b is zero. The choice of the operating point will be a compromise determined by the relative weights assigned to the above characteristics.

The characteristic describing the base current I_b versus the base-to-emitter voltage is shown in Experiment Fig. 9-3. The characteristic is nonlinear. A sinusoidal *voltage* applied to the input will cause a distorted output. The distortion is much less if a sinusoidal current is fed into the base circuit or if the sinusoidal input voltage E_b is converted into a sinusoidal base *current* signal I_b. A sinusoidal base current may be produced from a sinusoidal voltage signal if an impedance (resistance) much larger (>10 times) than the input impedance of the transistor is inserted in the base lead of the transistor (Experiment Fig. 9-4*a*). In this case, the combination of the sinusoidal voltage signal and the resistor R_i act as a current generator to produce a sinusoidal current variation. The output impedance of the voltage signal source will often be great enough to perform the function of the resistor R_i of Experiment Fig. 9-4. The sinusoidal output

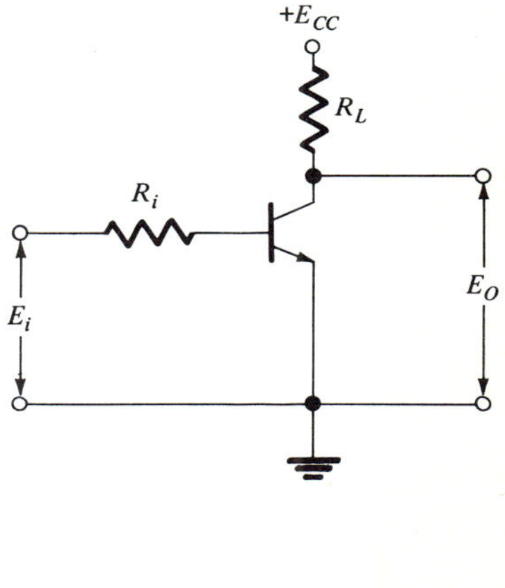

Experiment Fig. 9-4

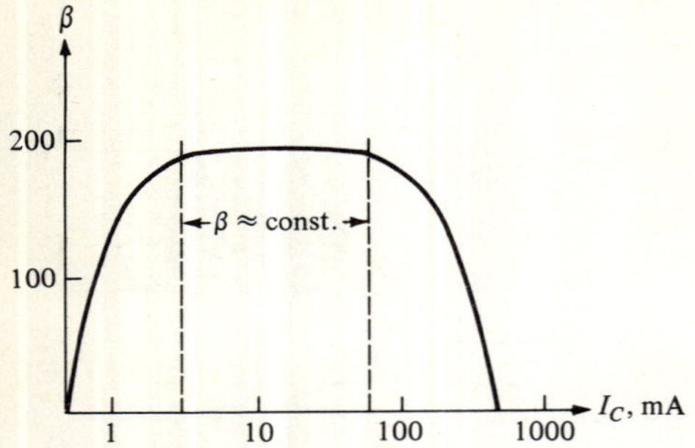

Experiment Fig. 9-5

voltage developed across the load resistance R_L by the sinusoidal collector current I_c will be a linear function of the input current if the transistor is operated at a collector-current level such that the current gain β of the transistor is approximately constant. The current gain is

$$\beta = \frac{\Delta I_c}{\Delta I_b} \qquad \text{for } E_{CE} = \text{const.}$$

The current gain β may range from about 10 to 200, depending on the transistor type and the conditions of operation as shown in Experiment Fig. 9-5. The voltage gain A_E of the amplifier shown in Experiment Fig. 9-4 is

$$A_E = \frac{\Delta E_c}{\Delta E_i} = \frac{-\Delta I_c R_L}{\Delta E_i} = \frac{-\beta \Delta E_i R_L}{\Delta E_i} = \frac{-\beta R_L}{R_i}$$

A graphical method of determining the current gain of a common-emitter amplifier is shown in Experiment Fig. 9-6. The load line L represents the conductance of the load resistor R_L. The end points of the line on the X axis are E_{cc} (the voltage across R_L when $I_c = 0$) and on the Y axis $I_c = E_{cc}/R_L$ (the current through R_L when $E_{ce} = 0$). When the input current at the base varies, e.g., from i_{b1} to i_{b2}, around the operating point P, the corresponding change of the output collector current I_c and output voltage E_o can be read on the I_c and

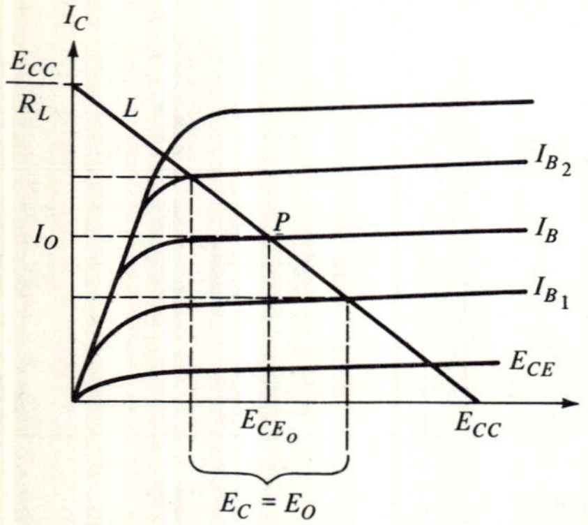

Experiment Fig. 9-6

E_{CE} axes. The position of the operating point is determined by the intersection of the load line with the quiescent base current, I_B curve. I_B is determined by the biasing network; I_C and E_{CE} are the quiescent collector current and voltage. Since in the great majority of cases the input signal will be a voltage rather than a current, the circuit of Experiment Fig. 9-7 for producing a sinusoidal base current from a sinusoidal voltage input is recommended. The sinusoidal voltage input signal may be transformed into a sinusoidal base current and, consequently, into a sinusoidal output voltage by inserting a resistor R_E in series with the emitter terminal of the transistor as shown in Experiment Fig. 9-7. As the base voltage increases, base current increases. The base-current change is multiplied by β to give collector-current variation. The emitter current is approximately equal to the collector current. Thus, the voltage across the resistor R_E will vary by an amount approximately given by $\beta \Delta i_b R_e = \Delta E_{R_E}$.

The voltage E_i is equal to the sum of the voltages across R_E and the base-emitter leads of the transistor,

$$E_{in} = E_{R_E} + E_{BE}$$

Since the voltage across the base emitter changes only slightly for large changes in base current, the quantity E_{BE} may be approximated by a constant: $E_{BE} = 0.6$ V. The transistor should be biased so that E_{BE} is on the steep part of the curve of Experiment Fig. 9-3. The voltage input is then

$$E_i = E_{R_E} + E_{Be} = E_{R_e} + 0.6$$

and

$$\Delta E_i = \Delta E_{R_E}$$

The incremental change in emitter current for a signal ΔE_i is then

$$\Delta I_e = \frac{\Delta E_{R_E}}{R_E} = \frac{\Delta E_i}{R_E}$$

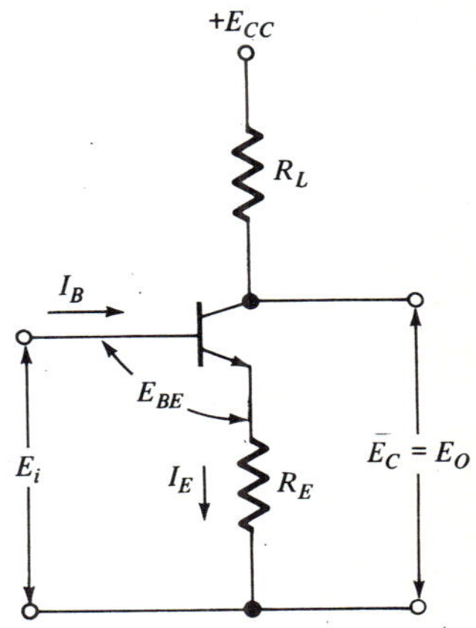

Experiment Fig. 9-7

Since emitter and collector currents are practically equal, the incremental output voltage is approximately

$$\Delta E_o = \Delta I_C R_L = -\Delta I_E R_L = \frac{-\Delta E_i R_L}{R_E}$$

For $R_L \gg R_e$, the incremental voltage gain A_v is then

$$A_v = \frac{\Delta E_o}{\Delta E_i} = \left(\frac{-R_L \Delta E_i}{R_E}\right)\left(\frac{1}{\Delta E_i}\right) = \frac{-R_L}{R_E}$$

Notice that β is not in this equation. The assumption that collector and emitter currents are approximately equal implies, however, that β is reasonably large (>20).

BIASING

The operating point of the transistor is ordinarily set by using a circuit of the type of Experiment Fig. 9-8*a*. The resistors R_1 and R_2 act as a voltage divider applying a dc voltage to the base of

$$E_B = \frac{R_2}{R_1 + R_2} E_{cc}$$

(only approximately since the currents through R_1 and R_2 are not identical but differ by the base current). The approximation assumes that

$$\beta R_e \gg R_1/(R_1 + R_2)$$

The circuit of Experiment Fig. 9-8*a* for direct current is then the same as Experiment Fig. 9-7 with E_B substituted for E_i. Thus, the quiescent emitter current (or collector current) is approximately

$$I_E \text{ or } I_C = \frac{E_{R_E}}{R_E} = \frac{[(R_2 E_{CC})/(R_1 + R_2)]}{R_E}$$

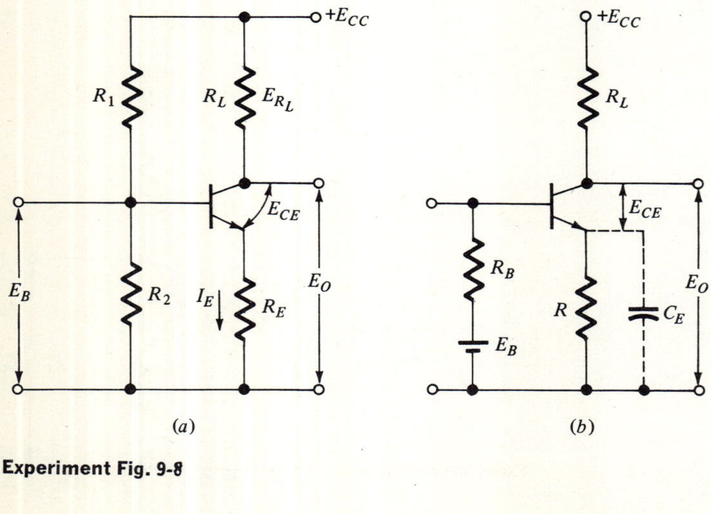

Experiment Fig. 9-8

still assuming that $\beta R_E \gg (R_1R_2)/(R_1 + R_2)$. The quiescent collector-to-emitter voltage is

$$E_{CE} = E_{CC} - (E_{R_L} + E_{R_E}) = E_{CC} - (I_C R_L + I_E R_E)$$

since $I_E = I_C$,

$$E_{CE} = E_{CC}\left(1 - \frac{R_2 R_L}{(R_1 + R_2)R_E}\right) \qquad \text{if } R_L \gg R_E$$

If the transistor is to be used as an amplifying device, E_{CE} at quiescence must be at least 1 V plus one-half of the peak-to-peak output voltage swing desired.

The voltage-divider arrangement of Experiment Fig. 9-8*a* is identical to the arrangement with a battery and resistor in series as shown in Experiment Fig. 9-8*b*. Since the stability of the dc circuits of Experiment Fig. 9-8*a* and *b* decreases with increasing voltage drop across R_B due to changing base current and decreases as the magnitude of the current through R_E increases, it is apparent that R_B should be small and R_E large for high stability. It is to be emphasized that the biasing circuits of Experiment Fig. 9-8*a* and *b* determine only the dc characteristics. Bypassing the resistor R_E with a circuit element which has low impedance for alternating current effectively removes R_E for ac signals. Inserting an element of low dc resistance and high ac resistance (an inductor) in the collector lead increases the ac impedance of the load without altering the dc characteristics.

INPUT IMPEDANCE

The input impedance of the circuit of Experiment Fig. 9-7 is

$$Z_i = \beta R_E$$

The input impedance of the circuit of Experiment Fig. 9-8 is different from Experiment Fig. 9-7 only in that the divider R_1R_2 with equivalent resistance $R_1R_2/(R_1 - R_2)$ is connected to the base lead. Therefore, the low-frequency input impedance is

$$Z_i' = Z_i \frac{R_1R_2}{R_1 + R_2}$$

where Z_i is the input impedance of the circuit of Experiment Fig. 9-7.

OUTPUT IMPEDANCE

For low source resistance ($<1{,}000\ \Omega$) the output impedance of the circuit of Experiment Fig. 9-5 is approximately

$$Z_o = \frac{r_C}{\beta}$$

where r_C is the resistance of the back-biased collector junction.

EMITTER–FOLLOWER CIRCUIT

The basic circuit of the emitter-follower (common-collector) amplifier is shown in Experiment Fig. 9-9. The explanation of the action of the load R_L of Experiment Fig. 9-9 is identical to that for the resistor R_E of Experiment Fig. 9-7. As E_i changes by an amount ΔE_i, base current I_B changes by an amount ΔI_B. This change is amplified by β to give a change of emitter current of approximately $\beta\Delta I_B$. The change in voltage across R_L is then

$$\Delta E_{R_L} = R_L\beta\Delta I_B = \Delta E_o$$

The relationship between the input and output voltages is

$$\Delta E_i = \Delta E_o + \Delta E_{BE}$$

where ΔE_{BE} is the change in base-to-emitter voltage with the change in base current ΔI_B. From Experiment Fig. 9-3 it is apparent that a very small change in E_{BE} will produce a large change in I_B, and hence (after multiplication by βR_L) a large change in output voltage. Thus,

$$\Delta E_i < \Delta E_o$$

and the voltage gain A_E is

$$A_E = \frac{\Delta E_o}{\Delta E_i} < 1$$

Since the voltage E_{BE} is on the order of 0.6 V, the output voltage is offset from the input voltage by approximately 0.6 V.

Experiment Fig. 9-10 shows the normal method of biasing the emitter-follower circuit. Biasing considerations are the same as for the circuit of Experiment Fig. 9-8*a* and *b*. Input impedance is calculated in a manner exactly

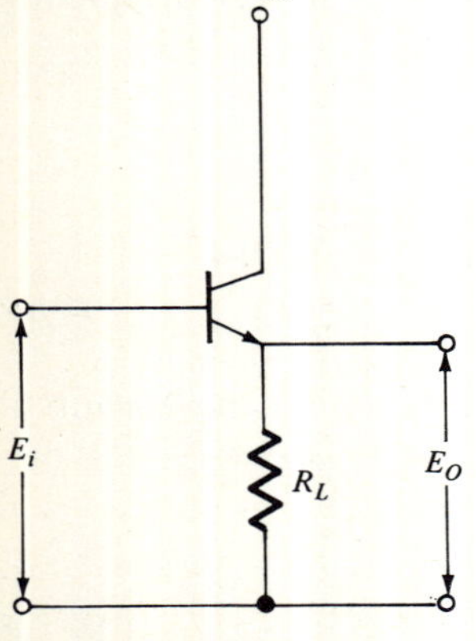

Experiment Fig. 9-9

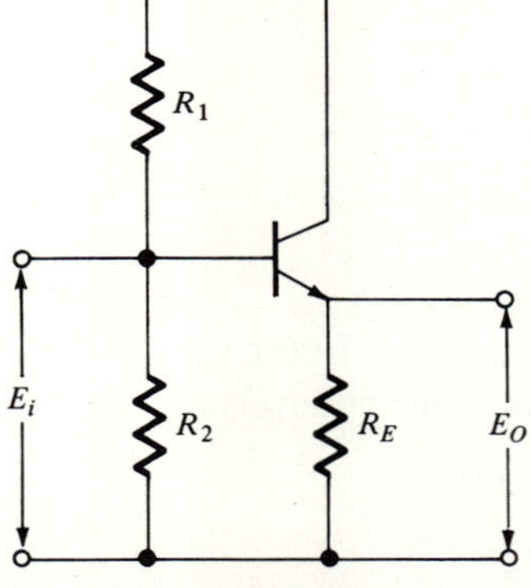

Experiment Fig. 9-10

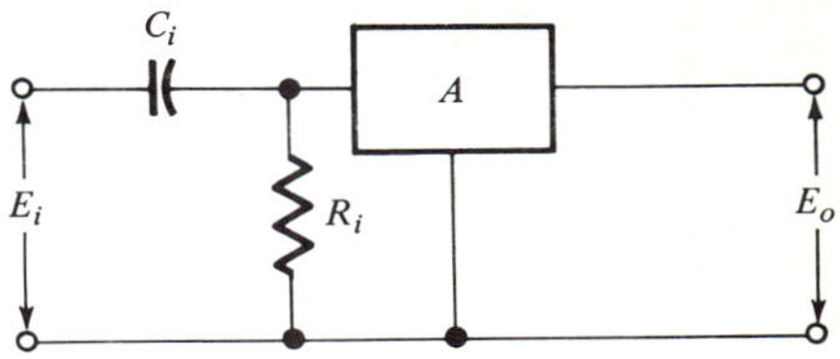

Experiment Fig. 9-11

similar to that of Experiment Fig. 9-8*a* and *b*, that is,

$$Z_i = \beta R_E \frac{R_1R_2}{R_1 + R_2}$$

For low source resistance ($<1{,}000\ \Omega$), the output impedance is approximately

$$Z_o = \frac{R_B}{\beta}$$

where R_B is again the equivalent base resistance.

LOW-FREQUENCY CUTOFF

If the signal is ac coupled into the amplifier, the magnitude of the capacitor used and the input resistance of the amplifier will determine the minimum frequency which the amplifier can amplify without undue attenuation. The low-frequency cutoff f_{cutoff} is defined as that frequency at which the voltage gain of the amplifier is reduced to 67 percent of its dc gain. Experiment Fig. 9-11 is a schematic diagram of an ac-coupled amplifier. The high-pass *RC* network formed by C_i and R_i determines the cutoff frequency. It is

$$f_{\text{cutoff}} = \frac{1}{2\pi R_i C_i}$$

PROBLEMS AND RECOMMENDED PROCEDURES

9-1. (*a*) Construct a common-emitter amplifier with a gain of 5. Characteristics should be as follows:

$E_{CC} = +10$ V
Quiescent collector current = 1 mA
Input impedance = 10K
Capable of handling an output = 3 V peak-to-peak without clipping

(*b*) Measure dc gain, output impedance, and maximum swing of output voltage signal about the quiescent level.

(*c*) Attach a capacitor C_E in parallel with R_E such that the time constant of the pair is 0.001 s. Measure ac gain from 20 Hz to 20 kHz. Apply the input signal through a capacitor. The time constant of the coupling circuit (Experiment Fig. 9-11) should be appreciably greater than that of the emitter resistor and capacitor. Measure input impedance at 2 kHz. Plot gain versus frequency on log/log paper.

9-2. (*a*) Construct an emitter-follower amplifier with the following characteristics:

Quiescent collector current = 10 mA
Input impedance = 25K
Peak-to-peak output signal at least = 3 V before clipping is observed
$E_{CC} = +8$ V

(*b*) Measure the dc output impedance of the amplifier.

(*c*) Select a capacitor C_i such that the low-frequency cutoff is $\approx$1,000 Hz. Measure ac gain from 20 Hz to 20 kHz. Measure input impedance at 2 kHz. Plot gain versus frequency measurements on log/log paper.

Available equipment

Single-stage transistor amplifier box
2 N1308 transistor, mounted on amplifier box
Ac signal generator
Vacuum-tube voltmeter (VTVM)
Voltage supply 0 to 10 V dc
Assorted resistors
Assorted capacitors
Cathode-ray oscilloscope (CRO)
HP Model 400-ac voltmeter

EXPERIMENT 10: OPERATIONAL AMPLIFIERS

BACKGROUND INFORMATION

An operational amplifier is, in general, a high-gain direct-coupled amplifier with a differential input and a single-ended output as shown schematically in Experiment Fig. 10-1. Its input impedance Z_1 and Z_2 is very high, typically 10^6 to 10^{12} Ω, so that the amplifier input current is negligible; i.e., no current passes from a source into either input terminal 1 or 2.

$$i_1 = 0 \qquad i_2 = 0 \tag{10-1}$$

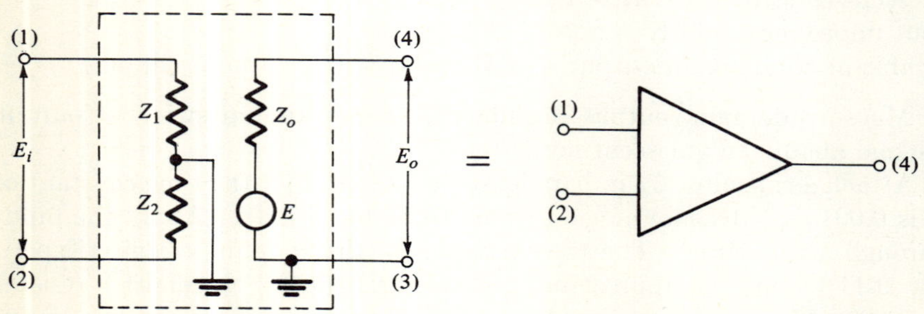

Experiment Fig. 10-1

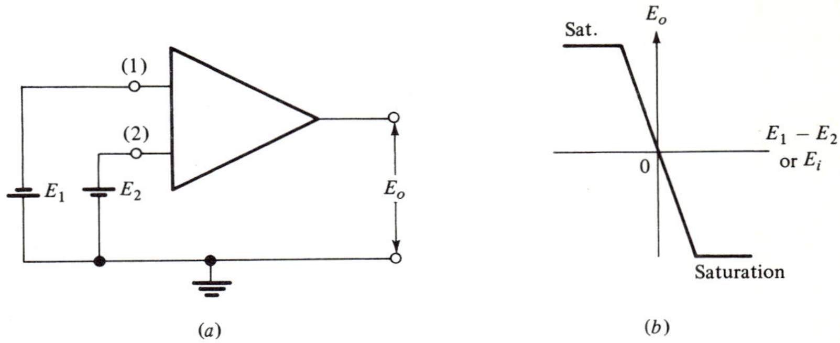

Experiment Fig. 10-2

Experiment Fig. 10-2a shows a schematic diagram of an operational amplifier in an open-loop circuit (without feedback). Two voltages E_1 and E_2 are applied to the input terminals 1 and 2 (or an input signal $E_i = E_1 - E_2$ is applied). An output E_o arises between the output terminal 4 and ground. The transfer characteristic $E_o = f(E_i)$ is shown in Experiment Fig. 10-2b. When the difference between the two input voltages $E_1 - E_o$ or $E_2 - E_1$ is large (or when the input voltage E_i is large), the amplifier is driven to saturation. When the input signal $E_i = E_1 - E_2$ is small, then the output changes linearly with the applied voltage difference. Typical open-loop gains $\Delta E_o/\Delta E_i$ in the linear range are from 10^4 to 10^8.

Operational amplifiers are rarely used in the open-loop configuration. It is the advantage of operational amplifiers that they can be modified for any particular application by the use of external feedback loops. Two basic circuits can be used: those of the inverting and the noninverting amplifier.

A circuit diagram of an inverting feedback amplifier is shown in Experiment Fig. 10-3. An input signal is applied to terminal a (between a and ground); the output appears at terminal 4. Terminal 1 is the negative terminal; i.e., the output signal at 4 is positive if terminals 1 and a are negative with respect to ground. S is the summing point. A current I flows from the source through the input resistor R_i to S. Because of the large input resistance of the amplifier, and in accordance with Eq. (10-1), no current flows from S to 1. Therefore,

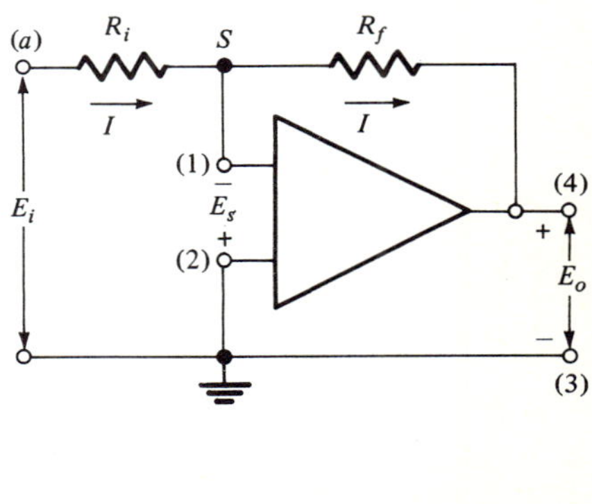

Experiment Fig. 10-3

the current I flows from S through R_f to the output terminal 4, and $I_f = I_i$. Since

$$I_i = \frac{E_i - E_s}{R_i}$$

$$I_f = \frac{E_o - E_s}{R_f}$$

where $E_o = AE_s$ or $E_s = \frac{E_o}{A}$

The gain A of the open-loop amplifier is very large, so that E_s approaches zero.

$$E_s \to 0 \tag{10-2}$$

Consequently,

$$E_o = -\frac{R_f}{R_i} E_i$$

The gain of the feedback amplifier shown in Experiment Fig. 10-3 is

$$A_f = -\frac{R_f}{R_i}$$

The minus sign indicates the inversion of the polarity of the signal.

The input resistance or impedance is R_i; the input current is $I_i = E_i/R_i$. The input resistance R_i is, in general, much smaller than that of normally used amplifiers; in particular, R_i is very much smaller than the input resistance of the open-loop operational amplifier.

Experiment Fig. 10-4 shows the circuit diagram of a noninverting amplifier. The signal is applied to terminal 2 (between 2 and ground). Resistor R_a and feedback resistor R_f are inserted. No current flows from the summing point S to terminal 1 [Eq. (10-1)]. The current I_a is

$$I_a = \frac{E_i + E_s}{R_a}$$

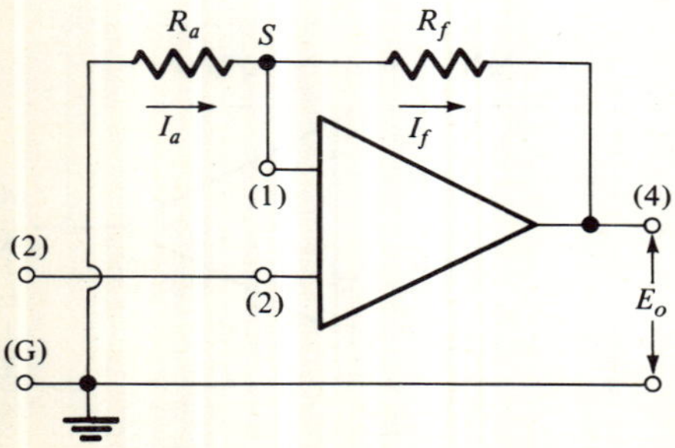

Experiment Fig. 10-4

The feedback current is

$$I_f = \frac{E_o - (E_i + E_s)}{R_f}$$

Since $I_a = I_f$ and since $E_s = 0$, [Eq. (10-2)],

$$\frac{E_i}{R_a} = \frac{E_o - E_i}{R_f}$$

so that

$$\frac{E_i}{R_a} = \frac{E_o - E_i}{R_f}$$

$$E_o = \left(1 + \frac{R_f}{R_a}\right) E_i$$

The gain of the amplifier is $1 + R_f/R_a$. The input impedance is very high; the output impedance is practically zero.

A noninverting amplifier with a gain of 1 is shown in Experiment Fig. 10-5. The feedback loop extends from the output terminal 4 to the terminal 1. The input is applied to terminal 2 and ground. The output signal is

$$E_o = E_i + E_s$$

and since E_s is practically zero, [Eq. (10-2)],

$$E_o = E_i$$

The voltage gain is 1, but the current and power gain can be considerable. The input impedance is very large; the output impedance is practically zero.

An operational amplifier-integrator is shown in Experiment Fig. 10-6. The input voltage is applied to the negative terminal through the resistance R; a

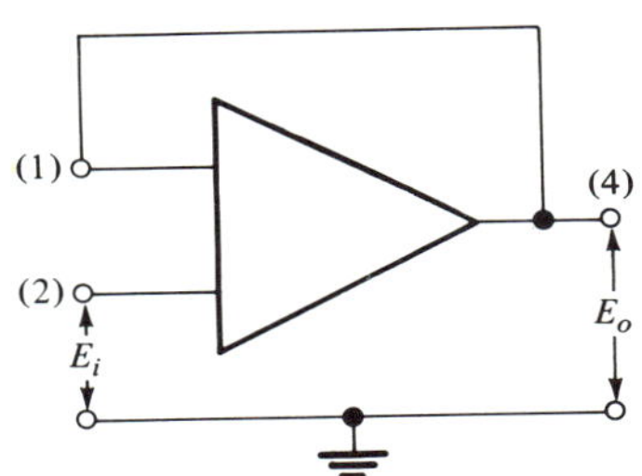

Experiment Fig. 10-5

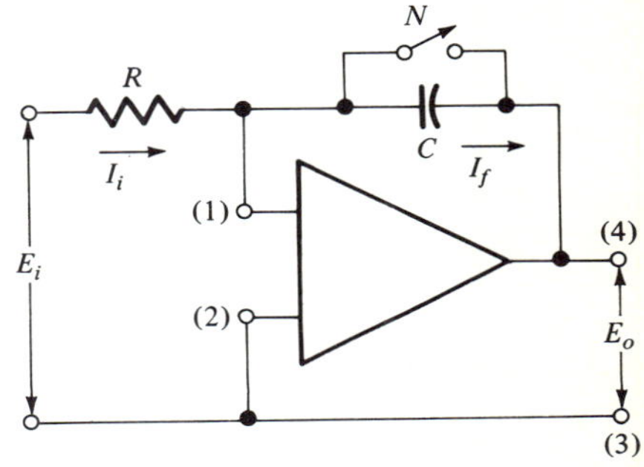

Experiment Fig. 10-6

capacitor C is placed in the feedback loop. Since $I_i = I_f$,

$$\frac{E_i}{R} = -C\frac{dE_o}{dt}$$

or

$$E_o = -\frac{1}{RC}E_i\,dt$$

i.e., the output voltage is the (negative) time integral of the input voltage. A practical difficulty encountered with the integrator is that the input offset voltage (the small bias voltage caused within the amplifier and resulting from imperfect zero adjustment) causes an input current which is superimposed on the measured current, and if integrated over a long time, leads to an erroneous output. Most modern operational amplifiers can be adjusted so that the offset voltage is close to zero.

At the end of the integration period the capacitor must be discharged (switch N in Experiment Fig. 10-6).

The basic circuit of a differentiator is shown in Experiment Fig. 10-7. The input current is

$$I_i = C\frac{dE_i}{dt}$$

which is equal to the feedback current

$$I_i = I_f = \frac{E_o}{R}$$

$$E_o = -RC\frac{dE_i}{dt}$$

There are some practical difficulties that may arise with the circuit in Experiment Fig. 10-7. The operational amplifier is subject to high-frequency instabilities which may sometimes mask the output signal. One way of bypassing this problem is to limit the response of the amplifier by placing a small capacitor C_i in parallel with the resistor R and a small resistor R_i in series with the capacitor C as shown in Experiment Fig. 10-8. This procedure limits the gain of the operational amplifier at higher frequencies and provides a better dynamic stability.

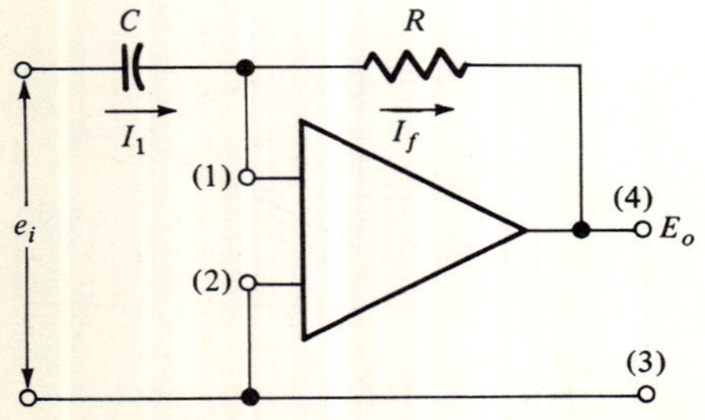

Experiment Fig. 10-7

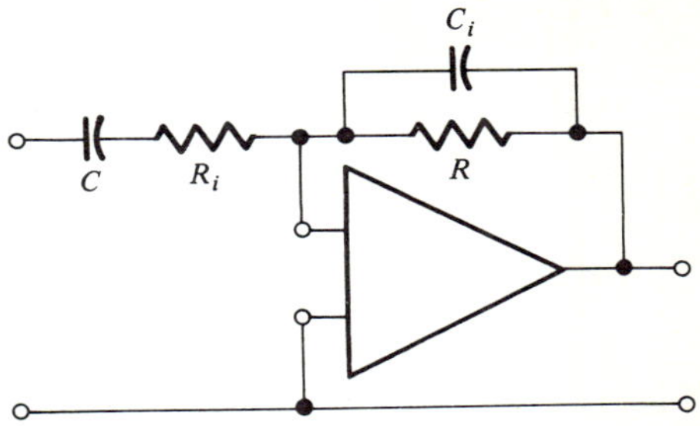

Experiment Fig. 10-8

PROBLEMS AND RECOMMENDED PROCEDURE

10-1. Using the Philbrick Operational Amplifier Manifold, construct an inverting amplifier with a gain of 1, 10, and 100. For each configuration measure the gain $A = E_o/E_i$, the input impedance $R_i = E_i/I_i$, and the output impedance $R_o = E_o/I_o$ for different loads. Use a variable dc voltage source and measure E and I with voltmeter and milliammeter. Compare results with an ac generator and CRO. It is recommended to check the offset voltage and adjust it to zero (screwdriver adjustments on the front of each of the four operational amplifiers). Input voltage E_i should be set to zero and the output measured with a sensitive milliammeter. The offset voltage will drift with time and temperature, so it is sometimes necessary to recheck the balance adjustment. The Philbrick amplifier shows saturation at an output voltage of about ± 7 V.

10-2. Construct a noninverting amplifier of the type shown in Experiment Fig. 10-4 with a gain of 100, 10, and 1. Measure the gain, the input impedance, and the output impedance.

Measurement of the input impedance is sometimes difficult as very sensitive current meters are not always available. The following method is sometimes helpful. Connect the circuit as shown in Experiment Fig. 10-9. Apply a fixed dc input voltage E_i and provide a large resistance of known value R_s, $10^6\ \Omega$ or higher, bypassed by a switch N. With the switch N closed, measure the gain $A = \Delta E_o/\Delta E_i$. Then open switch N. The voltage applied to the input is then

$$E_i' = E_i - I_i R_s \tag{10-3}$$

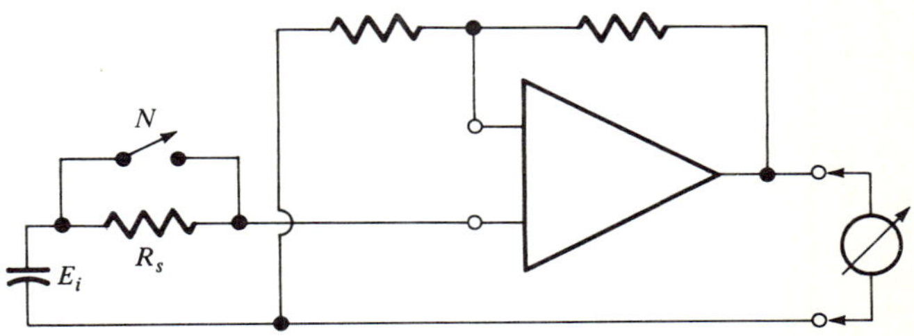

Experiment Fig. 10-9

which causes an output $E'_o \cdot E'_i$ can be found since

$$E'_i = E'_o/A$$

E_i and R_s are known and I_i can be calculated from Eq. (10-3). The input impedance is

$$R_i = E_i/I_i$$

10-3. Construct an integrator as shown in Experiment Fig. 10-6. Calibrate with a dc source and a stop watch, and use it with a square-wave source and CRO.

10-4. Construct and investigate a differentiator circuit like that shown in Experiment Fig. 10-7. Observe the output voltage for E_i = const. (dc), E_i slowly varying, and for a sharp step increase of E_i. Also use a square-wave generator for E_i and observe the output with a CRO. If high-frequency instabilities are encountered, limit the response of the amplifier as shown in Experiment Fig. 10-8.

Available equipment

Philbrick Operational Amplifier Manifold
Assorted resistors and capacitors
Sine- and square-wave generator
Cathode-ray oscilloscope
Additional equipment on request

EXPERIMENT 11: DIGITAL CIRCUIT ELEMENTS

BACKGROUND INFORMATION

Physical information, such as a temperature or a pressure, can be expressed in two different ways: analog form or digital form. In the analog form, the carrier of information, such as a voltage or the position of a pointer on a meter scale, is made analogous to the physical information (Experiment Fig. 11-1*a*). In the

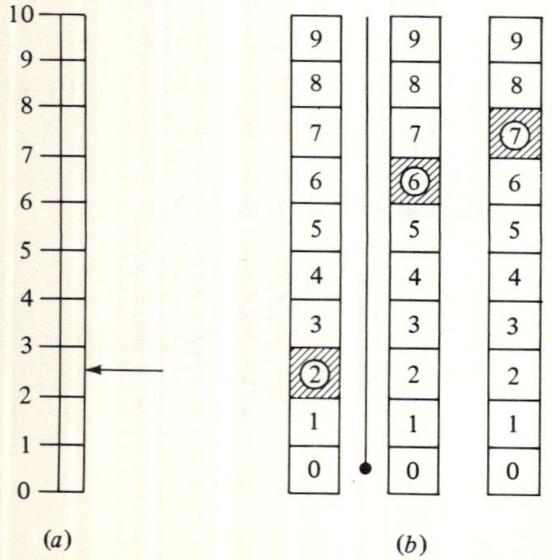

Experiment Fig. 11-1

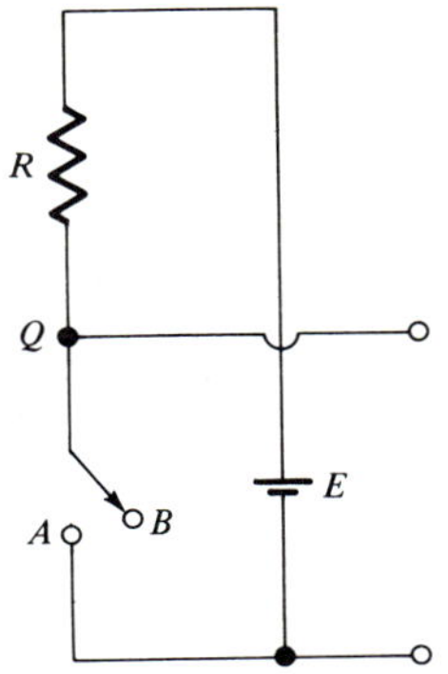

Experiment Fig. 11-2

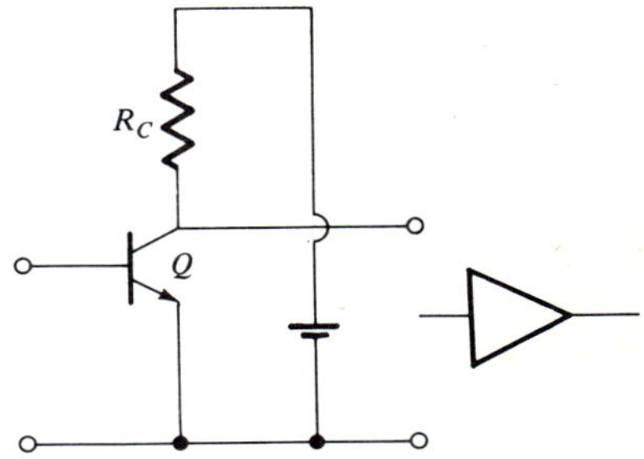

Experiment Fig. 11-3

digital form, the information is presented by the on or off markings of discrete levels in a counting system (Experiment Fig. 11-1*b*). Digital circuit elements have a number of useful functions. They can serve as electrically actuated switches, counters, pulse-forming elements, oscillators, memory elements, and others.

Basically, a digital circuit element is a switch with two states: on or off or 0 and 1 as shown in Experiment Fig. 11-2. The circuit shows a voltage source E, a resistor R, and a mechanically movable switch S. The point Q represents the output. If the switch is in the position A, the output Q is zero; if the switch is in the position B, the output Q is $+E$.

Fig. Experiment 11-3 shows a transistor circuit acting as a switch that is actuated electronically (by a voltage or a current) rather than mechanically. If the transistor is open, i.e., if no base current is applied, then there is no collector current and, consequently, no voltage drop across the collector resistance R_C; the potential of the point Q (the output potential) is $+E$ volts. If the base current is applied, a current will flow in the collector circuit; point Q will be tied to ground and its potential will be zero. The transistor acts as an inverter. If the input current (base current) is turned on (1), the output, i.e., the potential, of Q will be 0. If the input is off (0), the output potential will be 1. This condition can be presented in the form of the truth table in Experiment Table 11-1. The symbol of an inverter is shown on the right side of Experiment Fig. 11-3.

Experiment Fig. 11-4 shows two inverters coupled so that the output of each inverter is applied to the input of the other inverter. Assuming that the

Experiment Table 11-1

Input	*Output*
0	1
1	0

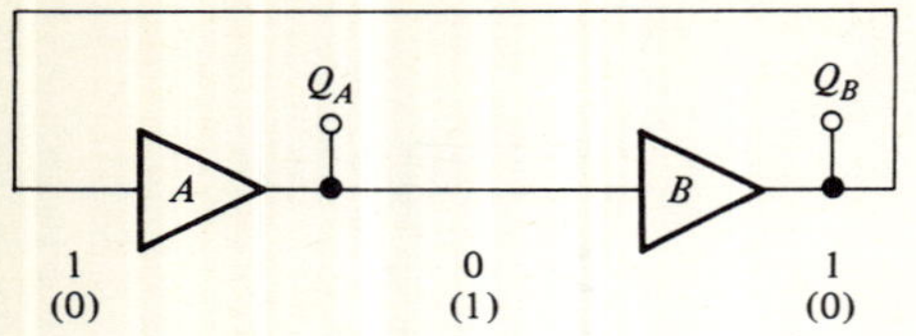

Experiment Fig. 11-4

input of the inverter A is 1, then its output will be 0. An input of 0 to the inverter B causes its output to become 1 which is the input of A and which will cause a stable configuration. Whether *this* stable configuration will result when the two inverters are connected, or whether the opposite outputs will appear (signals in parentheses in Experiment Fig. 11-4) is a matter of chance, provided the two inverters are built alike. The system tends to come to a stable equilibrium. The state of the output of each inverter can be checked at the test points Q_A and Q_B.

Experiment Fig. 11-5 shows a modification of the basic two-inverter system. A second input is provided with each inverter which permits a change of the input and output of one inverter (and, consequently, also of the other inverter). In the on-off position shown in Experiment Fig. 11-5, the inverter B is the only

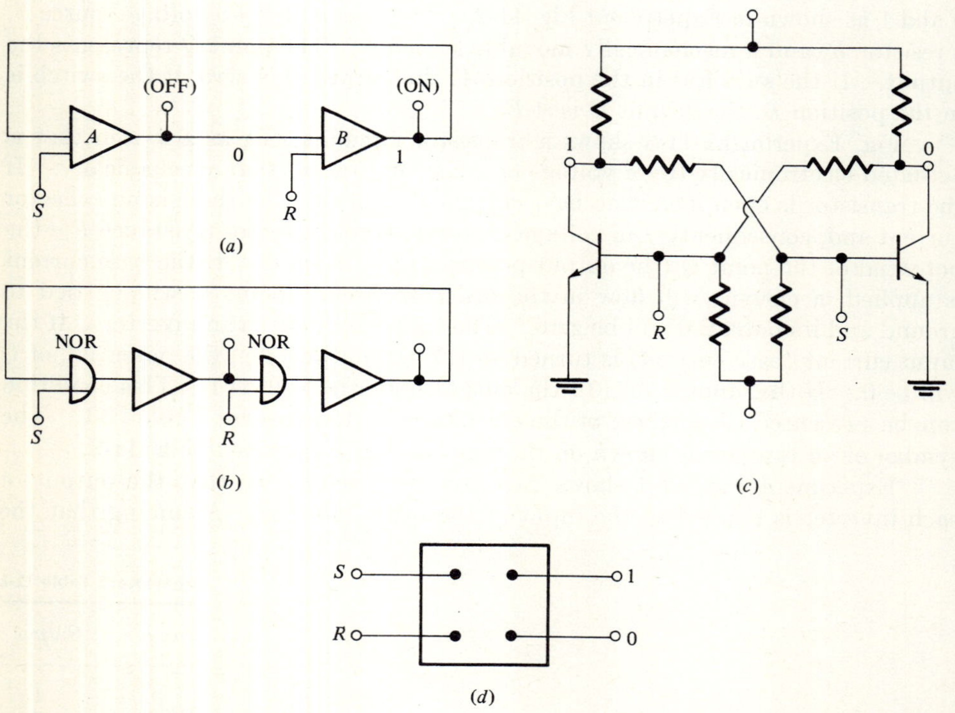

Experiment Fig. 11-5

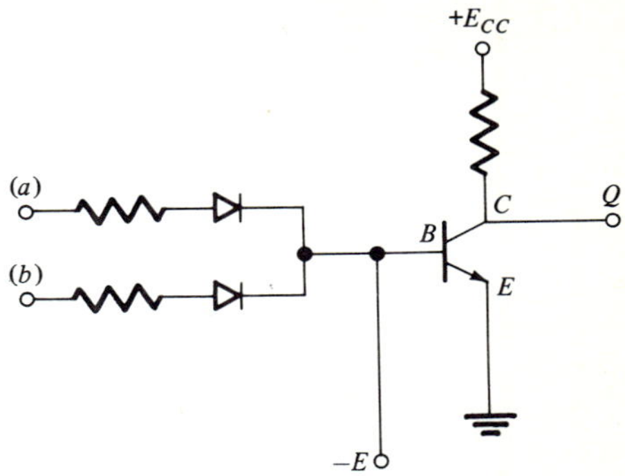

Experiment Fig. 11-6

controllable element. One of the additional inputs is called S (set), the other R (reset). The two inputs to one inverter must be connected so that the signals do not influence each other. This can be done with a resistor and a diode in each lead as shown in Experiment Fig. 11-6. If either terminal *a or b* is on, or if both are on, a base-emitter current will flow and the output Q will be 0. Only if both terminals *a* and *b* are off, will there be no base-emitter current and, consequently, no collector current, and the output from the transistor inverter will be 1. The truth table for the two-input inverter is as shown in Experiment Table 11-2.

The two-input system acts as a gate with a NOT-OR or NOR logic. Since the system has two stable modes of operation, it is called a *bistable multivibrator* or *bistable flip-flop circuit.* The circuit of Experiment Fig. 11-5*a* is redrawn in Experiment Fig. 11-5*b* by incorporating the NOR logic, and in Experiment Fig. 11-5*c* in transistor form. Experiment Fig. 11-5*d* shows the common symbol of a bistable flip-flop circuit.

By slightly modifying this flip-flop design, a complementing flip-flop can be constructed. This is a flip-flop which will change its state every time it is fed a switching trigger pulse. Experiment Fig. 11-7*a* shows the schematic diagram of a complementing flip-flop. Experiment Fig. 11-7*b* shows the same flip-flop in transistor form.

Because the flip-flop has only two discrete states, it is useful in storing information. It will remain in a single state, remembering its value until told to change by a triggering pulse. Cascading many of these flip-flop memory bits

Experiment Table 11-2

Input a	*Input b*	*Output*
1	0	0
0	1	0
1	1	0
0	0	1

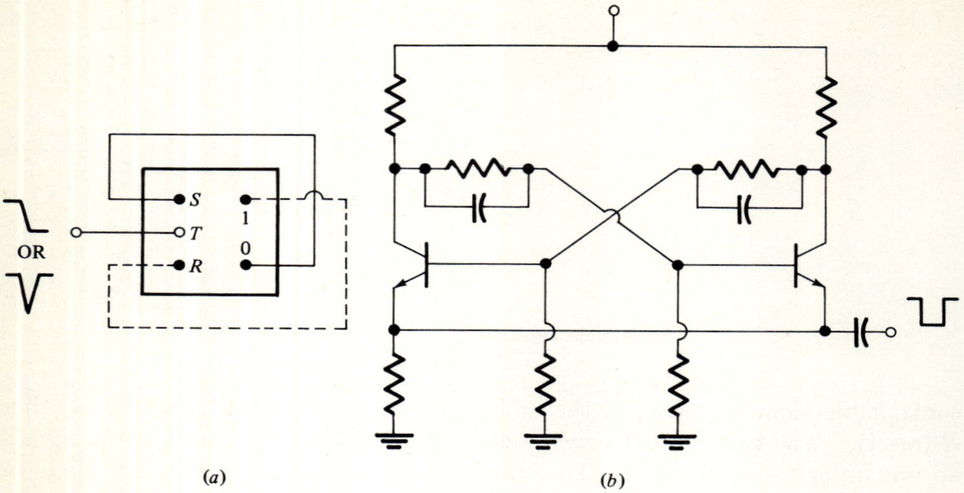

Experiment Fig. 11-7

allows more information to be stored, e.g., 1 bit has two possible states, 2 bits have 2^2 or 4 possible states, and n bits have 2^n possible states where each possible state can be considered a form of information.

Counter A binary counter is shown in Experiment Fig. 11-8. It consists of a series of complementing flip-flops whose 1 output terminals feed the trigger input of the succeeding bit. Each downward transition of one of the bits triggers the next flip-flop which complements. Experiment Table 11-3 shows the information content of the 4-bit counter as a series of downward transitions are applied to the first input.

Monostable circuits A one-shot or monostable multivibrator is a flip-flop which is normally off but which turns on for a predetermined length of time with each trigger pulse. Experiment Fig. 11-9 shows a circuit diagram of a one-shot.

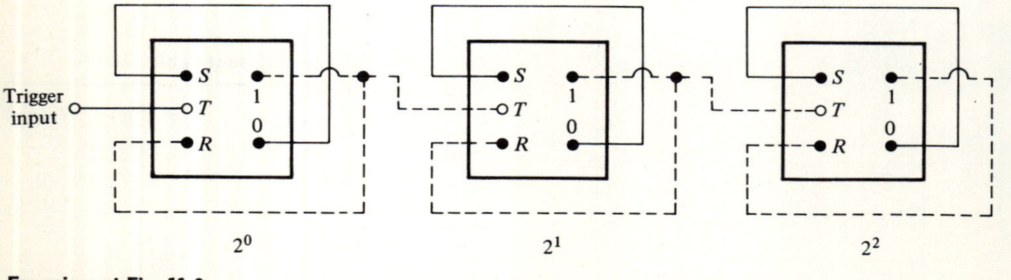

Experiment Fig. 11-8

Experiment Table 11-3

Number of pulses	2^2	2^1	2^0
0	0	0	0
1	0	0	1
2	0	1	0
3	0	1	1
4	1	0	0
5	1	0	1
etc.			

Transistor Q_1 is normally open and Q_2 is normally shorted. The collector of Q_1 is high. R_3 feeds the base of Q_2, keeping Q_2 shorted. When a positive pulse is fed into the base of Q_1, it turns the transistor on, making the capacitor potential E_b negative by the amount $E_{cc} - E_1$. This turns off Q_2 and sets the output high. Q_2 remains open and the capacitor charges up with time constant R_tC until $E_b > E_1$ which turns Q_2 on returning the output to zero. The result is a single positive pulse at the output. The pulse duration is determined by the time constant of the charging capacitor. By changing the time constant R_tC, one can change the pulse width of the one-shot multivibrator. The one-shot is useful in digital circuitry for shaping pulses and for delaying pulses in timing applications.

Astable circuits An astable multivibrator is a free-running oscillator which oscillates between its two stable states. A schematic diagram of one is shown in Experiment Fig. 11-10. It is very similar to the monostable multivibrator in Experiment Fig. 11-9 but has capacitive cross-coupling to both transistors. The charging time constant R_tC determines the frequency of oscillation.

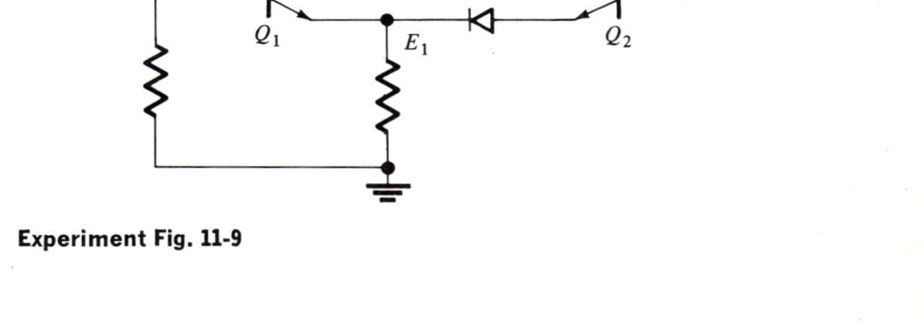

Experiment Fig. 11-9

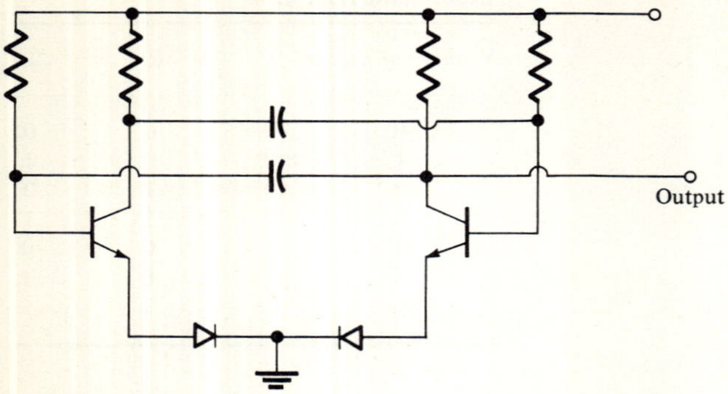

Experiment Fig. 11-10

PROBLEM

Investigate the following classes of digital circuits:

1. Bistable circuits
   (*a*) Flip-flops, shift registers, and binary coding
   (*b*) Complementing flip-flops and binary counters
2. Monostable circuits
   One-shot multivibrator
3. Astable circuits
   Free-running multivibrator

Available equipment

Science-Electronics boards
Special equipment
Power supply: −15, 0, and +15 V
Vacuum-tube voltmeter
Other equipment on request

EXPERIMENT 12: DIGITAL LOGIC AND CIRCUIT ELEMENTS

BACKGROUND INFORMATION

The basic units in digital operation, data processing, or computers are electrically controlled switches or *gates*. Each input to a gate can have one of two discrete levels, either 0 or 1 (such as 0 V or +10 V). The output likewise appears in the form of one of two discrete levels, either 0 or 1, off or on, true or false. The mechanism inside the gate, which causes the response, is of secondary importance. In general, it is a transistor switching circuit, but it may also be a relay or any other electrically actuated switch. The following basic units are most frequently used.

Experiment Fig. 12-1 shows the symbol of an AND gate for two input sig-

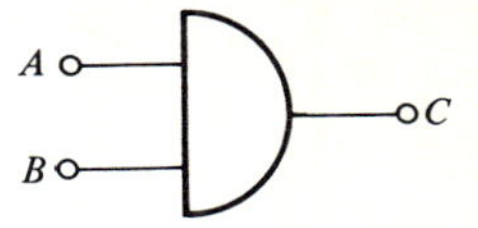

Experiment Fig. 12-1

Experiment Table 12-1

A	B	$C = A \cdot B$
0	0	0
0	1	0
1	0	0
1	1	1

nals A and B. The AND gate provides an output only when *all inputs are present* (are 1). The output C in terms of the input signals A and B is expressed by the symbolic logic equation

$$C = A \cdot B$$

(read: C equals A *and* B). Experiment Table 12-1 shows the output C for the different conditions of the inputs (truth table), if only two inputs A and B are provided.

An output of 1 occurs only if both inputs A and B are 1.

The OR gate (Experiment Fig. 12-2, for two inputs). The OR gate provides an output if either A or B (or A and B) are 1 and an output 0 appears only if all inputs are zero. This condition is expressed by the symbolic equation

$$C = A + B$$

(read: C equals A or B). A truth table of the OR gate with two inputs is shown in Experiment Table 12-2.

The inverter (Experiment Fig. 12-3), as its name implies, inverts the states

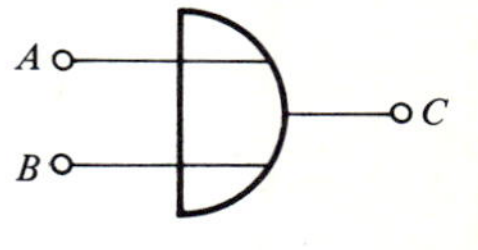

Experiment Fig. 12-2

Experiment Table 12-2

A	B	$C = A + B$
0	0	0
0	1	1
1	0	1
1	1	1

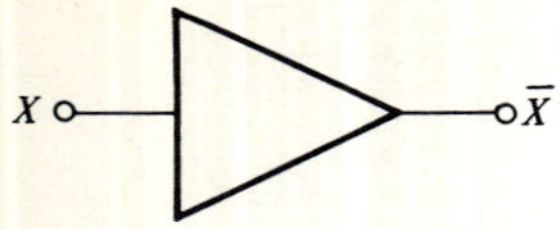

Experiment Fig. 12-3

Experiment Table 12-3

Input	*Output*
0	1
1	0

of a digital signal. If the digital input signal is 0, the output signal is 1. If the input is 1, the output is 0. If the input signal is X, the output is expressed symbolically by the sign $\bar{X}$ (in words: NOT X). The truth table for an inverter is shown in Experiment Table 12-3.

An example showing the combination of an AND gate and an OR gate is shown in Experiment Fig. 12-4. The symbolic equation for this system is

$$(A \cdot B) + C = D$$

The output D is 1 only if both A and B are 1 or if C is 1.

The NAND gate shown in Experiment Fig. 12-5 is an AND gate followed by an inverter. It is useful for solving equations of the type

$$\bar{C} = A \cdot B$$

The NOR gate is an OR gate followed by an inverter and is shown in Experiment Fig. 12-6. It solves switching problems of the type

$$\bar{C} = A + B$$

In order to design a logic circuit to describe a logic equation, one must first express the equation in its simplest form. A listing of some of the rules manipulating logic expressions follows:

1. $X + 1 = 1$ — whether X is 1 or 0, the output is 1
2. $X + 0 = X$ — when $X = 1$, the output is 1; when $X = 0$, the output is 0

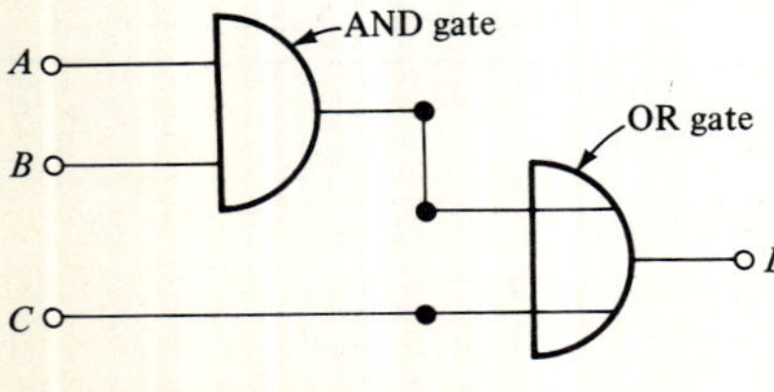

Experiment Fig. 12-4

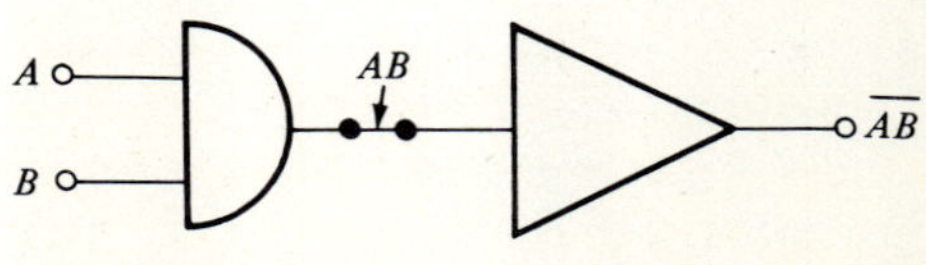

Experiment Fig. 12-5

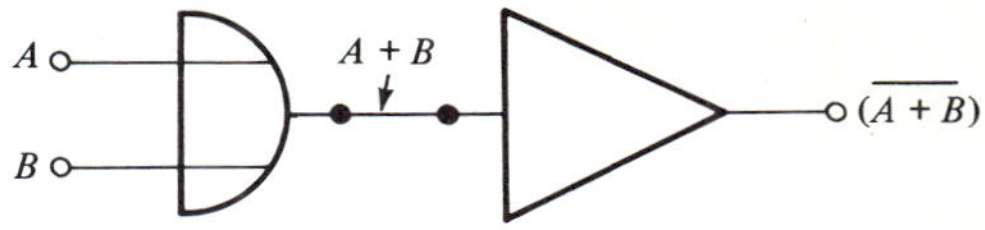

Experiment Fig. 12-6

3. $X \cdot 0 = 0$	whether $X = 1$ or 0, the output is 0
4. $X \cdot 1 = X$	when $X = 1$, the output is 1; when $X = 0$, the output is 0
5. $X + X = X$	when $X = 1$, the output is 1; when $X = 0$, the output is 0
6. $X \cdot X = X$	when $X = 1$, the output is 1; when $X = 0$, the output is 0
7. $X \cdot \bar{X} = 0$	when $X = 1$, $\bar{X} = 0$, and the output is 0; when $X = 0$, $\bar{X} = 1$, and the output is 0
8. $X + \bar{X} = 1$	when $X = 1$, $\bar{X} = 0$, and the output is 1; when $X = 0$, $\bar{X} = 1$, and the output is 1
9. $X + Y = Y + X$; $X \cdot Y = Y \cdot X$	Commutative law
10. $(X + Y) + Z = X + (Y + Z)$; $(X \cdot Y) \cdot Z = X \cdot (Y \cdot Z)$	Associative law
11. $XY + XZ = X \cdot (Y + Z)$	Distributive law

Using these rules the following expression can be greatly reduced:

1. $W = \bar{X}(YX + Z)$
2. $W = \bar{X}YX + \bar{X}Z$ distributive law
3. $W = 0 + \bar{X}Z$ since $X \cdot X = 0$
4. $W = \bar{X}Z$

De Morgan's theorem, a very useful tool for manipulating logic equations, is stated as follows:

12. $\overline{(X + Y)} = \bar{X} \cdot \bar{Y}$
13. $\overline{(X \cdot Y)} = \bar{X} + \bar{Y}$

This theorem is of considerable importance when dealing with the more common computer building blocks such as NAND gates and NOR gates.

PROBLEMS AND RECOMMENDED PROCEDURE

12-1. Investigate the AND, OR, and inverter gates and verify the truth tables using the Digiac 3010. The manual switches on the Digiac 3010 can serve as inputs to the logic elements. When the switch is up, the 1 terminal is true or one and the 0 terminal is false or zero. When it is down, the reverse is true. With the provided connecting cords, wire several switches to the logic component (AND, OR, and inverter) inputs, and use the indicator lights to observe the outputs. The indicator lamp will light when it is fed a 1 and will remain unlit when fed a 0. For example,

X	Y	$X\ Y$	Lamp
1	0	0	Unlit
0	1	0	Unlit
0	0	0	Unlit
1	1	1	Lit

12-2. Simplify and implement the following logic functions. Check the truth tables for the original and simplified expression in each case.

$$W = XYZ + X(YZ)$$
$$V = (XY + XZ)XY$$
$$A = (BC + CD)BD$$
$$V = (WX + YZ)(X + Y)$$
$$E = (AC + DC)C$$

12-3. Verify De Morgan's theorem. One way to do this is by implementing the logic equations

$$\bar{X} + Y = \bar{X} \cdot \bar{Y}$$

and

$$X \cdot Y = \bar{X} + \bar{Y}$$

When the truth tables are the same for each side of these equations, then each side is equivalent to the other.

Recommended reading

Louis Nashelsky, Digital Logic and Computer Mathematics. Laboratory Manual. Publ. Digital Electronics, Inc. (1965).

Available equipment

Digital 3010DD Advanced Digital Logic Trainer
Other equipment on request

Index